5th International Conference on Earthquake Engineering and Disaster Mitigation (ICEEDM-2022)

IOP Conference Series: Earth and Environmental Science
Volume 1244

Yogyakarta, Indonesia
27 - 28 September 2022

ISBN: 978-1-7138-7979-4
ISSN: 1755-1307

Printed from e-media with permission by:

Curran Associates, Inc.
57 Morehouse Lane
Red Hook, NY 12571

Some format issues inherent in the e-media version may also appear in this print version.

Institute of Physics
Dirac House, Temple Back
Bristol BS1 6BE UK

Phone: 44 1 17 929 7481
Fax: 44 1 17 920 0979

techtracking@iop.org

Additional copies of this publication are available from:

Curran Associates, Inc.
57 Morehouse Lane
Red Hook, NY 12571 USA
Phone: 845-758-0400
Fax: 845-758-2633
Email: curran@proceedings.com
Web: www.proceedings.com

TABLE OF CONTENTS

The 5th International Conference on Earthquake
Engineering and Disaster Mitigation
September 28-30, 2022
Yogyakarta – Indonesia

Abstract

Preface

The 5th International Conference on Earthquake Engineering and Disaster Mitigation, ICEEDM 2022, was held on 28th-29th September 2022. The 5th ICEEDM in Gadjah Mada University was profoundly enjoying the collaboration with the joint-organizing committee with the Indonesian Earthquake Engineering Association (AARGI), International Association for Earthquake Engineering (IAEE), Japan Association for Earthquake Engineering (JAEE), Bandung Institute of Technology, Indonesian Islamic University, and Atmajaya University. The conference was supported by a scientific committee from four countries and twelve keynote speakers from five countries, i.e., Canada, Indonesia, Japan, Taiwan, and the USA.

The 5th ICEEDM 2022 was held to provide an opportunity to share experiences and disseminate the latest research and development results related to earthquake engineering, which will be very useful for increasing knowledge and disaster mitigation policies. It is hoped that this conference will benefit all parties to minimize the risk of earthquake hazards in Indonesia and worldwide.

The 5th ICEEDM 2022 proceeding contains papers written by academics, professionals, researchers, and students from Indonesia and Japan. Various research works related to the conference theme were presented during the conference. A total of 41 full papers accepted upon review for the proceeding publication are categorized into the following topics:

A. Structural Engineering, including; Structural Earthquake Engineering, Structural Dynamics, Performance-Based Design and Evaluation, Structural Assessment and Retrofit, Seismic Devices and Applications, Non-engineered Buildings, Experimental & Numerical Modelling

B. Geotechnical Engineering, including; Geotechnical Earthquake Engineering, Soil-Structure Interaction

C. Seismological Engineering, including; Seismic Micro-zonation, Case Histories in Recent Earthquakes,

D. Disaster Mitigation, including; Seismic Hazard and Tsunami Disaster Mitigation, Early Warning System, Post Disaster Recovery and Reconstruction, Community-Based Disaster Risk Reduction & Management

Peer Review Statement

All papers published in this volume have been reviewed through processes administered by the Editors. Reviews were conducted by expert referees to the professional and scientific standards expected of a proceedings journal published by IOP Publishing.

- **Type of peer review:** Double Anonymous
- **Conference submission management system:** Morressier
- **Number of submissions received:** 67
- **Number of submissions sent for review:** 58
- **Number of submissions accepted:** 41
- **Acceptance Rate (Submissions Accepted / Submissions Received × 100): 61.2**
- **Average number of reviews per paper:** 2.21
- **Total number of reviewers involved:** 29
- **Contact person for queries:**
  Name: Bambang Suhendro
  Email: bsuhendro@ugm.ac.id
  Affiliation: Department of Civil and Environmental Engineering, Engineering Faculty, Gadjah Mada University

Strength Enhancement and Deformability of ConfinedHigh-Strength Steel Fiber Concrete After Burning

Purwanto[1], Antonius[2], Prabowo Setiyawan[3] and Aris Krisdiyanto[4]

[1] Associated Professor in Civil Engineering, Department of Civil Engineering, Universitas Semarang, Semarang,Indonesia, purwanto_sipilusm@yahoo.com
[2] Professor in Civil Engineering, Department of Civil Engineering, Universitas Islam Sultan Agung(UNISSULA), Semarang, Indonesia, antonius@unissula.ac.id
[3] Associated Professor in Civil Engineering, Department of Civil Engineering, Universitas Islam Sultan Agung (UNISSULA), Semarang, Indonesia, prabowo@unissula.ac.id
[4] Associated Professor in Civil Engineering, Department of Civil Engineering, Universitas 17 Agustus 1945 (UNTAG), Semarang, Indonesia, ariskrisdiyanto123@gmail.com

Abstract. This study investigates post-burning steel fibrous concrete that aims to identify the mechanical quantities affecting the behavior of confined steel fibrous concrete after burning. A total of 72 specimens of unconfined concrete and confined concrete were made by reviewing design parameters, namely concrete compressive strength (f_c'), fiber fraction and duration of combustion (1 hour, 2 hours and 3 hours). All specimens were monotonically compressed with a deformation control testing system. The parameters discussed from the experimental results represent the strength of the confined concrete ($fccf'$), the peak strain of the confined concrete (ε_{ccf}'), and the strain of the confined concrete after the peak stress is exceeded and decreases by 50% (ε_{cc50f}). The experimental results showed that for concrete with higher volume fractions of fiber (0%, 1% and 2%), the relative strength enhancement of confined concrete (*K* value) indicates an increase in the value of *K* although it is not significant. On the other hand, in general,the values of ε_{ccf}' and ε_{cc50f} tend to increase significantly at post peak response, if the confined steel fiber concrete specimen is burned for a longer time and volume fraction of fiber were higher. The increase in the strain of confined concrete is one indication that the specimen is moreductile with increasing burning time.

Keywords: steel fiber, confinement, duration of burning, strength enhancement, strain

1. Introduction

It is universally recognized that high-strength concrete is more brittle than normal-strength concrete [1,2]. These properties cause the ductility of high-strength concrete to be lower than the ductility of normal-strength concrete. However, from several conducted studies, the combination of fiber concrete with adequate confinement called confined high-strength fiber concrete will cause the concrete to be superior especially in increasing its strength and ductility [3-4]. Nowadays several analytical models of confinement has been proposed [5-7], as well as its behavior under shear [8].

Although reinforced concrete structures are principally designed to be fire resistantly in post-burned conditions it will affect the elements of the concrete structure. Because of the high temperature of the fire, strength degradation will occur [9,10]. It can also occur in passively confined fiber concrete structural components [11]. For the present the extensive research on the behavior of post-burned steel fiber, concrete is however significantly limited. The conducted research is still dominated by normal concrete (without fiber) including with its confinement behavior at various temperatures [12-19]. Therefore, it isstill absolutely necessary to properly conduct developmental research in post-burned steel fiber concrete.

The intended purpose of this study is to evaluate the contribution of steel fibers to the strength and ductility of post-burned confined steel fiber concrete. It includes determining the strength degradation behavior of unconfined and confined high-strength steel fiber concrete on the burning time with variations in the volume of fiber fraction. In addition, observations were equally performed on the deformation of confined steel fiber concrete, namely the post-peak behavior on the stress-strain curve.

2. Experimental program

A total of 36 specimens was produced with a high-strength concrete design that has a water-cement ratio (*w/c*) of 0.38 and 0.28. Mix design procedure based on Indonesian national standard (SNI-7656-2012) [20]. The design of the concrete mix is shown in Table 1. To accurately maintain the workability of theconcrete, Viscocrete is typically added with a dose of 0.5% of the cement weight. Steel fibers are purchased in the market and in the specimen the volume fraction (V_f) of the fiber varies 0%, 1% and 2%. The ratio of fiber length to diameter (*l/d*) is accurately calculated between 40-50 values.

Table 1 : Concrete mix design

Materials		w/c=0.38	$w/c=0.28$
Cement (kg /m^3)		420	485
Fly Ash (kg/m^3)		72	828
Water (lt/m^3)		160	136
Sand (kg/m^3)		464.6	622
Coarse aggregate (kg/m^3)		1045	1086
Viscocrete (kg/m^3)		7.38	10.7
Steel fiber (kg/m^3)	V_f=0%	-	-
	V_f=1%	24	24.6
	V_f=2%	48	49

Confined concrete specimens are cylindrical in shape with a diameter of 100 mm and a height of 200 mm, with 4Ø8 main reinforcement. Hoop reinforcement as a 5.5 diameter with variation of spacing (*s*) is 50 mm and 100 mm (Figure 1). Burning of specimens varied with burning time of 1 hour, 2 hours and

3 hours. The process of burning the test specimens was carried out at the Crematorium of Semarang City (Figure 2 & 3). The specimen burning was according to ASTM E 119-00a [21].

Figure 1 : Hoop reinforcement

Figure 2 : Preparing of specimens burning

Figure 3: Process of burning

All specimens were tested with a monotonic uniaxial compression load using a Universal Testing Machine (UTM) with a capacity of 2000 kN. The performing maximum load (P_{max}) was readable from the Data Loger, where the P_{max} value was correlated with the strain. The load beared by the concrete (P_{conc}) was obtained from the P_{max} value reduced by the load received by the longitudinal reinforcement based on the strain data. The test results obtained consist of several design parameters as shown in Figure 4 namely f_{ccf}' (peak stress of confined steel fiber concrete), ε_{ccf}' (peak strain of confined steel fibrous concrete), ε_{ccf50} (strain when the stress decreased by 50% in the post peak), and the value of $K=f_{ccf}'/0.85f_{cf}'$ (strength enhancement of confined steel fiber concrete). In addition, the curve of the relationship between the value of K and the strain of steel fiber concrete was also shown.

3. Experimental Results and Discussion

Tables 2, 3 and 4 show experimental results for specimens burned for 1 hour, 2 hours and 3 hours, respectively. In the table, $\boldsymbol{f_c'}$ is the compressive strength of cylindrical concrete with a diameter of 150 mm and a height of 300 mm in 28 days without fiber ($\boldsymbol{V_f=0}$) and $\boldsymbol{f_{cf}'}$ for a cylinder containing fiber ($\boldsymbol{V_f \neq 0}$). Table 1 shows that, specimens with volume fractions of 0%, 1% and 2% for 100 mm hoop spacing shows K values that is indifferent greatly from each other (difference below 5%). It indicates the fiber and hoop reinforcement serve no optimal function because the hoop spacing used is the same as the diameter of the specimen. Compared with the specimens with 100 mm spacing above, the 50 mm hoop spaced specimens showed a more excessive increase in *K* value when the V_f value increased. In most cases, there is a behavioral tendency for the values of ε_{ccf}' and ε_{ccf50} to increase when a higher volume of fiber fraction is used.

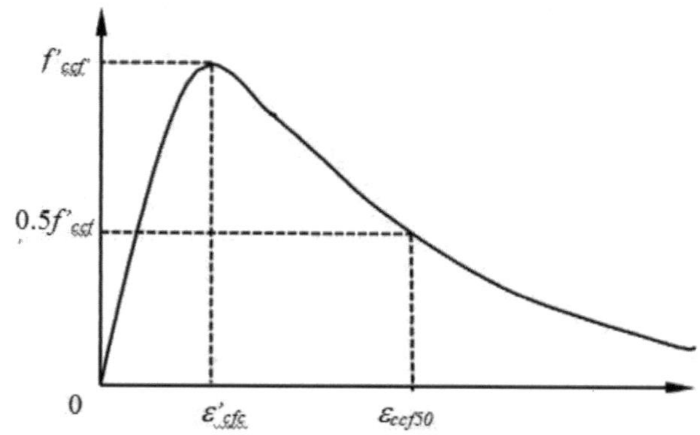

Figure 4 : Variable design of confined concrete [22]

Based on Table 2 where the specimen was burned for 2 hours, the highest *K* value occurred in the F2C15 specimen (V_f=2%, f_c'=50.6 MPa, *s*=50 mm) and the lowest occurred in the F2C7 specimen (V_f=0%, f_c'=70.5 MPa, *s*=100 mm). These results indicate the confined steel fiber concrete will be optimal at the *K* value if the lower the compressive strength of the concrete and the tighter the hoop spacing the higher the volume of the fiber fraction will be.

The experimental results of confined steel fiber concrete specimens that were burned for 3 hours are presented in Table 3. Similar to the specimens that were burned for 1 hour and 2 hours above, a 3 hour-burned specimens also resulted in an increase in the *K* value if the compressive strength of the specimen decreased. The lowest *K* value occurred in the F1C4 a lint-free specimen. Another result of the specimens burned for 3 hours represents the increase in the values of ε_{ccf}' and ε_{ccf50} along with the increase in fiber volume fraction.

Table 2 : Experimental results of specimens burned for 1 hour

No	Code	f_c'	f_{cf}'	V_f	Hoop φ-spacing	f_{ccf}'	ε_{ccf}'	:cf50	$K = f_{ccf}'/(0.85 f_{cf}')$
1	F1C4		49.4	0		38.75	0.0029	0.0051	0.92
2	F1C5	50.6	52.6	1	5.5-100	40.09	0.0032	0.0047	0.95
3	F1C6		52.3	2		45.14	0.0035	0.0059	1.02
4	F1C7		62.1	0		54.03	0.003	0.0044	1.02
5	F1C8	70.5	69.6	1	5.5-100	61.9	0.0035	0.0054	1.05
6	F1C9		70.7	2		61.3	0.0037	0.0069	1.05
7	F1C13		49.4	0		39.45	0.0052	0.0083	1.13
8	F1C14	50.6	52.6	1	5.5-50	47.86	0.0063	0.0093	1.21
9	F1C15		52.3	2		48.92	0.0064	0.0108	1.23
10	F1C16		62.1	0		59.12	0.0051	0.0081	1.12
11	F1C17	70.5	69.6	1	5.5-50	61.28	0.0062	0.0102	1.19
12	F1C18		70.7	2		61.87	0.0077	0.0117	1.22

Table 3 : Experimental results of specimens burned for 2 hour

No	Code	f_c'	f_{cf}'	V_f	Hoop φ-spacing	f_{ccf}'	ε_{ccf}'	ε_{ccf50}	K= $f_{ccf}'/(0.85f_{cf}')$
13	F2C4		36.7	0		29.55	0.003	0.0047	0.95
14	F2C5	50.6	37.2	1	5.5-100	32.19	0.0033	0.0075	1.02
15	F2C6		40	2		32.38	0.0037	0.0072	1.02
16	F2C7		55.4	0		42.7	0.0031	0.005	0.91
17	F2C8	70.5	58.4	1	5.5-100	46.69	0.0036	0.009	0.94
18	F2C9		62.1	2		49.41	0.0037	0.0076	0.94
19	F2C13		36.7	0		36.64	0.0045	0.008	1.17
20	F2C14	50.6	37.2	1	5.5-50	37.68	0.0055	0.0082	1.19
21	F2C15		40	2		40.81	0.0068	0.0155	1.20
22	F2C16		55.4	0		45.48	0.0062	0.0075	1.13
23	F2C17	70.5	58.4	1	5.5-50	47.7	0.0076	0.0104	1.14
24	F2C18		62.1	2		49.35	0.0078	0.0117	1.17

Table 4 : Experimental results of specimens burned for 3 hour

No	Code	f_c'	f_{cf}'	V_f	Hoop φ-spacing	f_{ccf}'	ε_{ccf}'	ε_{ccf50}	K= $f_{ccf}'/(0.85f_{cf}')$
25	F3C4		18.3	0		20.17	0.0051	0.0065	0.99
26	F3C5	50.6	25	1	5.5-100	21.48	0.0054	0.0076	1.01
27	F3C6		26.6	2		28.53	0.006	0.0091	1.09
28	F3C7		33	0		32.17	0.0053	0.0069	1.02
29	F3C8	70.5	33.6	1	5.5-100	30.29	0.0056	0.0089	1.06
30	F3C9		39.6	2		31.14	0.0062	0.0140	1.09
31	F3C13		18.3	0		24.92	0.005	0.0088	1.21
32	F3C14	50.6	25	1		25.12	0.0074	0.0116	1.24
33	F3C15		26.6	2	5.5-50	29.25	0.008	0.0117	1.29
34	F3C16		33	0		31.6	0.0068	0.0089	1.13
35	F3C17	70.5	33.6	1	5.5-50	32.61	0.0081	0.0112	1.14
36	F3C18		39.6	2		38.88	0.0099	0.014	1.16

3.1. Stress-strain behavior of specimens confined by hoops, spacing 100 mm

The stress-strain behavior of confined fiber concrete, where the magnitude of the stress was expressed in the value of the increase in the strength of the confined concrete (K), is sufficiently shown in the following figure. In line with the experimental results shown in Tables 2, 3 and 4 above, the value of $K$and axial deformation of confined concrete tends to increase when a higher volume of fiber fraction is used. It generally applies where the specimen is hooped with 100 mm spacing. At the pre-peak behavior, all specimens indicated behavior that almost coincided with each other. However at post peak, if the volume of fiber fraction is higher, the curve tends to be steeper and longer (Figures 5, 6, 7, 8, 9, 10). It happened to the specimens that were burned for 1 hour, 2 hours and 3 hours. This behavior shows moreductile of the confined concrete with the higher volume of fiber fraction.

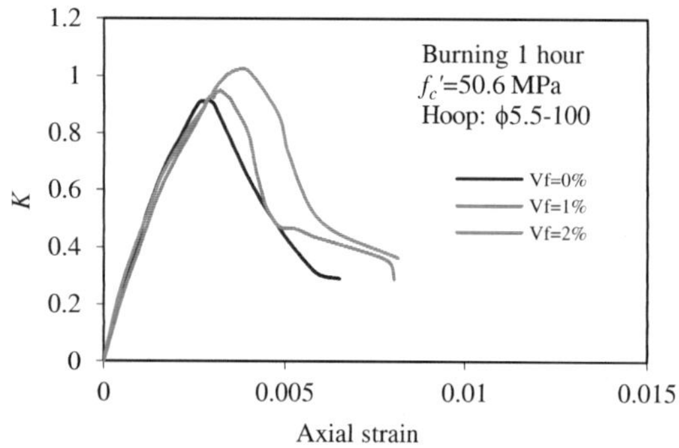

Figure 5 : Stress-strain curves of confined concrete, burning 1 hour, f'_c= 50.6 MPa

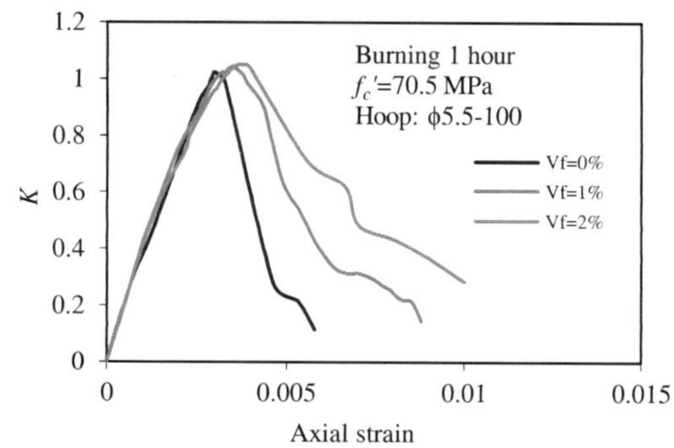

Figure 6 : Stress-strain curves of confined concrete, burning 1 hour, f'_c= 70.5 MPa

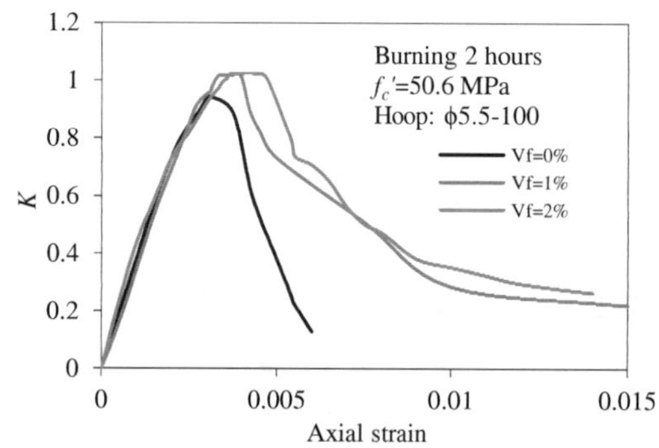

Figure 7 : Stress-strain curves of confined concrete, burning 2 hours, f'_c= 50.6 MPa

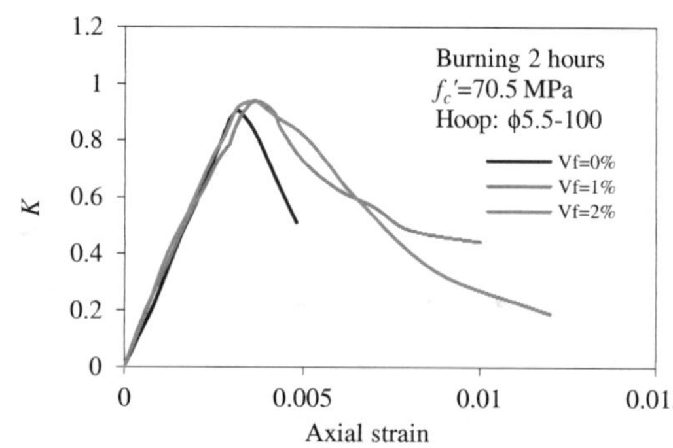

Figure 8 : Stress-strain curves of confined concrete, burning 2 hours, f'_c= 70.5 MPa

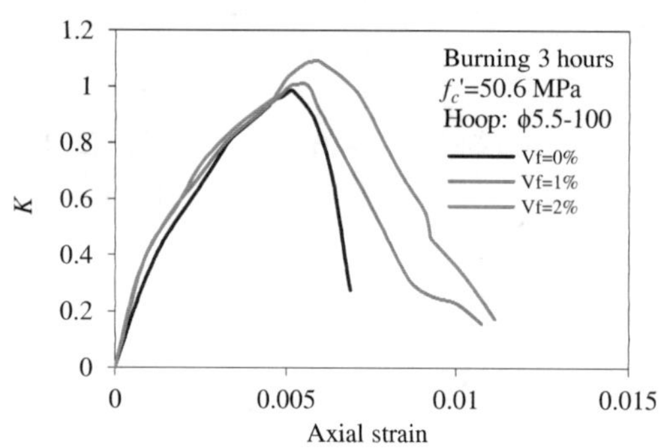

Figure 9 : Stress-strain curves of confined concrete, burning 3 hours, f'_c= 50.6 MPa

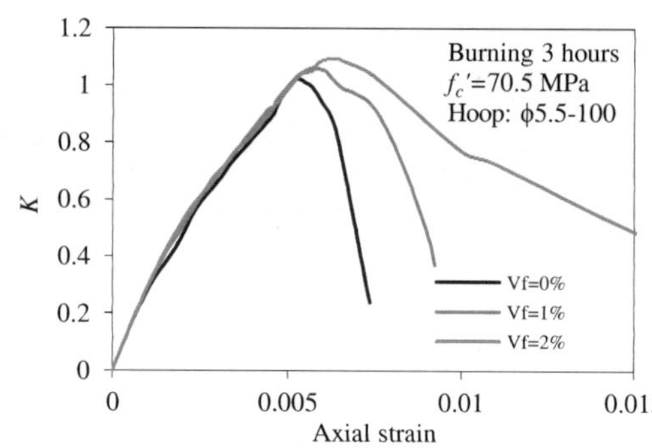

Figure 10 : Stress-strain curves of confined concrete, burning 3 hours, f'_c= 70.5 MPa

3.2. Stress-strain behavior of specimens confined by hoops, spacing 50 mm

Furthermore, for specimens with tighter hoop reinforcement (s=50 mm) shown in Figures 11, 12, 13, 14, 15 and 16. The phenomenon is almost the same as for specimens with a hoop spacing of 100 mm above, which tends to increase in value of K along with the increase in the volume of fiber fraction although the increase was not significant. The post-peak behavior indicates the fiber plays an important role in increasing the deformability of the confined concrete. This is indicated by the post-peak curve which tends to be gentler if the volume of fiber fraction in the specimen is high. This behavior generally occurs in specimens burned for 1 hour, 2 hours or 3 hours.

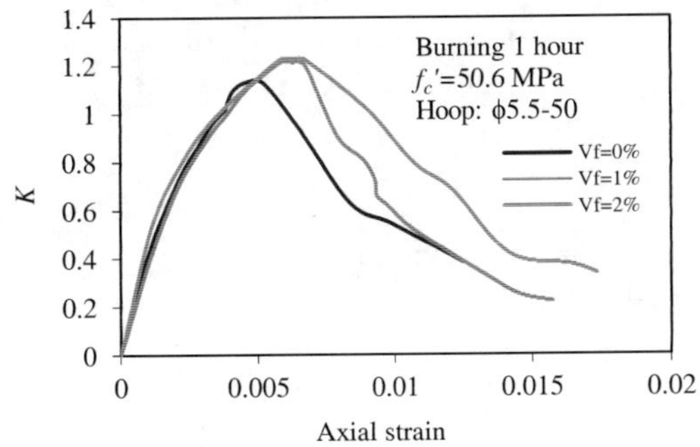

Figure 11 : Stress-strain curves of confined concrete, burning 1 hour, f'_c= 50.6 MPa

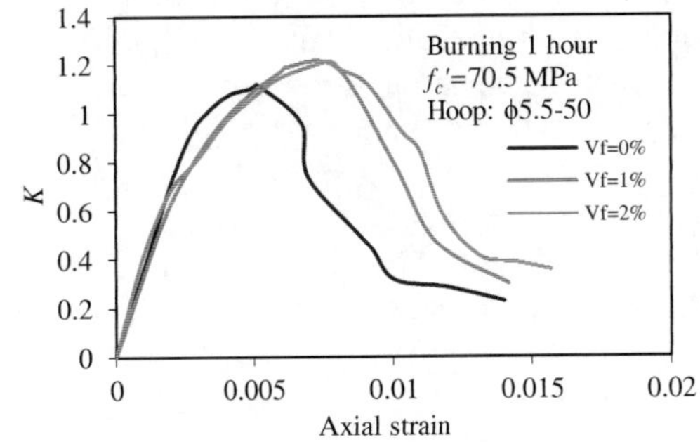

Figure 12 : Stress-strain curves of confined concrete, burning 1 hour, f'_c= 70.5 MPa

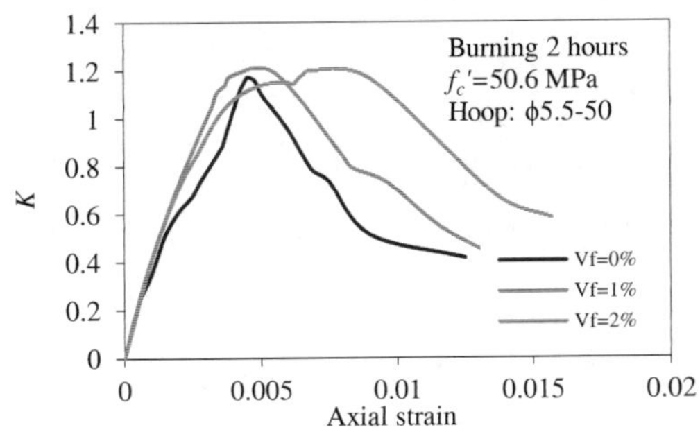

Figure 13 : Stress-strain curves of confined concrete, burning 2 hours, f'_c= 50.6 MPa

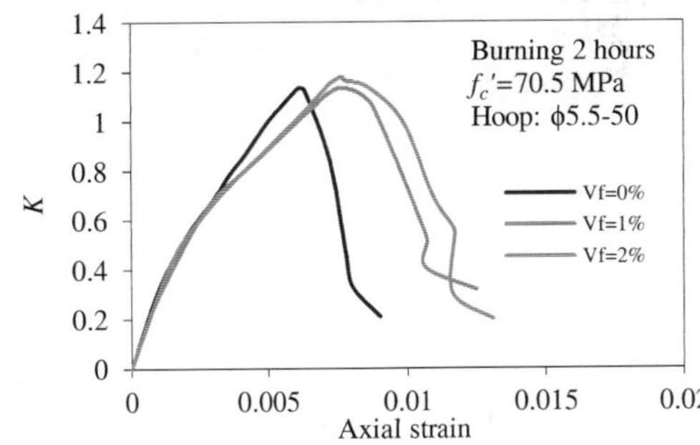

Figure 14: Stress-strain curves of confined concrete, burning 2 hours, f'_c= 70.5 MPa

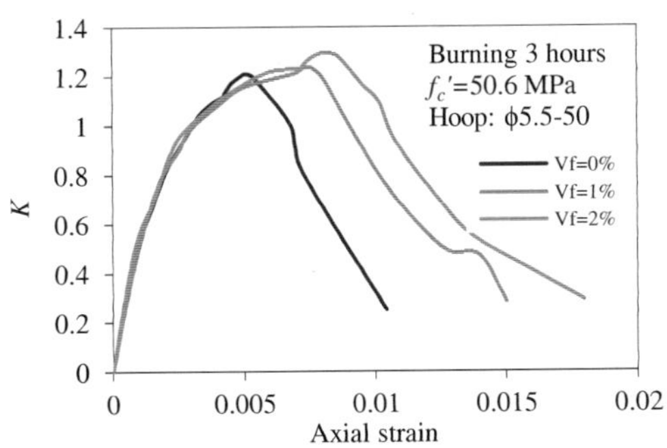

Figure 15 : Stress-strain curves of confined concrete, burning 3 hours, f'_c= 50.6 MPa

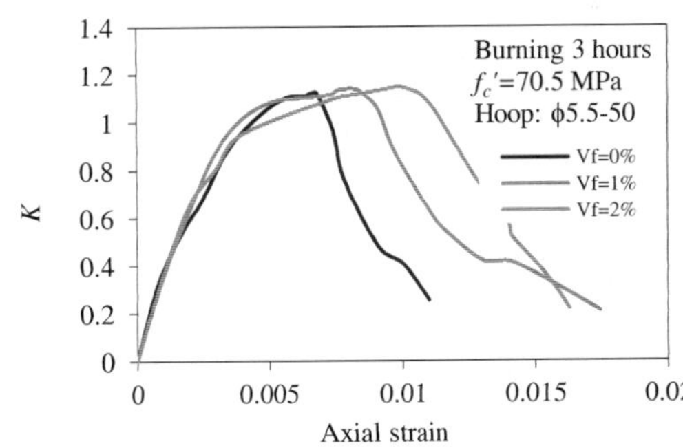

Figure 16 : Stress-strain curves of confined concrete, burning 3 hours, f'_c= 70.5 MPa

4. Conclusion

Based on the experimental results and the discussion described above, steel fiber occupies no significant role in increasing the value of *K* if high-strength fiber concrete is confined with hoops spaced equal to the diameter of the specimen. Steel fibers can play a role in increasing the value of *K* if the specimen isinstalled with a tighter hoop spacing (50 mm). However in most cases it can be stated that for each burning time, the ratio between specimens with the same spacing, the increase in the volume of the steel fiber fraction tends to increase the *K* value although the increase is insignificant. The use of the higher volume of steel fiber fraction and tighter hoop spacing will have a significant role in terms of its ability to increase the deformability of confined high-strength steel fiber concrete, which in this case is expressed in terms of ε_{ccf}' and ε_{ccf50}. Based on the stress-strain behavior of confined concrete at various specimens burning conditions, the deformability of confined high-strength steel fiber concrete tends to be lower when the concrete strength is higher.

Acknowledgements
This research funded by the Undip Alumni Foundation and Semarang University. The supports received for this research successfully is gratefully acknowledged

References

[1] Antonius, Imran I and Setiyawan P 2017 *Procedia Engineering* **171** 121
[2] Purwanto, Antonius and Sctiyawan P 2021 *International J. of GEOMATE* **21** 61
[3] Pantazopoulou S J, and Zanganeh M 2001 *J. of Materials in Civil Eng.* **13**, 340
[4] Han A L, Antonius and Okiyarta A W 2015 Procedia Engineering **125** 1030
[5] Paultre P, Eid R, Langlois Y and Lévesque Y 2010 *J. of Structural Eng.* **136** 225
[6] Antonius 2015 *International J. of Technology* **6** 217
[7] Cholida N F F, Antonius and Enggartiasto L 2022 *J. of the Civil Engineering Forum* **8** 309
[8] Hanif F and Kanakubo T 2017 *J. of the Civil Engineering Forum* **3** 113
[9] Sharma U K, Zaidi K A and Bhandari N M 2017 *Magazine of Concrete Research* **64** 631
[10] Antonius, Purwanto and Harprastanti P 2019 *International J. of Technology* **10** 428
[11] Zaidi K A, Sharma, U K and Bhandari N M 2012 *Fire Safety Journal* **48** 58
[12] Cheng F-P, Kodur V K R and Wang T-C 2004 *J. of Materials in Civil Engineering* **6** 84
[13] Li L and Purkiss J A 2005 *Fire Safety Journal* **40** 669
[14] Youssef M A and Moftah M 2007 *J. of Engineering Structures* **29** 2618
[15] Kodur V K R, Dwaikat M M S and Dwaikat M B 2008 *ACI Materials J.* **105** 517
[16] Aslani F and Bastami M 2011 *ACI Materials J.* **108** 355
[17] Chen H-J, Yu Y-L and Tang C-W 2020 *Materials MDPI* **13** 3
[18] Kodur V 2014 *ISRN Civil Engineering* 468510
[19] Antonius, Karlinasari R, Purwanto and Widhianto A 2020 *Advances in Concrete Construction* **10** 105
[20] Indonesian National Standard SNI 7656:2012 *Procedures of Mixed Selection for Normal Concrete, Heavy Concrete and Mass Concrete* (in Indonesian)
[21] ASTM E 119-00a *Standard Test Methods for Fire Tests of Building Construction and Materials*
[22] Purwanto 2021 *Behavior of Unconfined and Confined Steel Fiber Concrete at Post Burning Under Uniaxial Loadings* Dissertation, Civil Engineering Doctoral Program, Universitas Islam Sultan Agung (in Indonesian)

Numerical Study on Slender Circular Reinforced Concrete Column Clamped with Steel Sheet

Sandy I. Yansiku[1], Bambang Piscesa[2], Priyo Suprobo[3]

[1] Student in Civil Engineering, Department of Civil Engineering, Sepuluh Nopember Institute of Technology, Surabaya, Indonesia.
[2] Doctor in Structural Engineering, Civil Planning and Geo Engineering Faculty, Sepuluh Nopember Institute of Technology, Surabaya, Indonesia.
[3] Professor in Structural Engineering, Civil Planning and Geo Engineering Faculty, Sepuluh Nopember Institute of Technology, Surabaya, Indonesia.

E-mail: sandyyansiku.19031@mhs.its.ac.id

Abstract. Various methods to enhance the performance of circular reinforced concrete column have been widely investigated. However, most researches used internal strengthening method such as intensifying steel reinforcement. Others utilized carbon fiber reinforced polymer (CFRP), glass fiber and aramid to retrofit either reinforcement or concrete surface. Few studies incorporated steel to cover concrete surface but neglecting clamp pressure between steel and concrete surface. This study proposes a steel sheet strap strengthening method by initial clamping stress on steel sheets as an external confining technique through experimental works and in-depth numerical analysis. The experimental investigation includes nine steel sheet strengthened reinforced concrete specimens of 20 MPa subjected to axial concentric and axial eccentric. Retrofitting devices will be fabricated from BJ37 steel. Steel sheet straps in each specimen will be subjected to varying initial clamp pressure by fastening bolts. Applied load, lateral displacement and vertical and horizontal deformation will be measured by the instrumentation. Experimental works will observe failure mode, force displacement, stress and ductility. Section analysis based on experimental results will define compatibility equations for numerical stage. Strengthening model from experimental stage will also be analyzed in numerical study with in-house software 3DNLFEA. F-D curve and moment-curvature loading behavior and ductility will be obtained from numerical analysis. It is believed that the proposed strengthening mechanism with its geometrical and stress variations could enhance the section, member and structural ductility of circular reinforced column against vertical and lateral load. Furthermore, the developed analytical model could well predict the behavior of the proposed clamp mechanism for practical applications.

1. Introduction

Strengthening techniques in reinforced concrete had been massively introduced such as providing more reinforcements and encasing reinforcing bar with carbon fiber reinforced polymer to enhance load carrying capacity and ductility of circular column and its connection. When the connection is adequate, plastic hinge formation could occur near to the middle height of column by which reduces the displacement ductility. As axial and lateral load are imposed to the column simultaneously, the lack of displacement ductility possibly causes column failure and collapse of upper structure. Hence, critical zones generating high curvature requires sufficient ductility to allow large inelastic deformation without notable reduction in strength during service load.

Spiral reinforcing in reinforced concrete column with circular section enhances curvature and energy dissipation capacity [1][2]. Insufficient spiral reinforcement decreases the captivity of concrete core and yields crushing of concrete against lateral load. Excel deformation capacity and ductility refinement under lateral cyclic reversed loading have been observed in a fully retrofitted circular column with steel layer [3]. Reinforcing bar and spiral stirrups was possibly be confined by carbon fiber reinforced polymer (CFRP) [4] despite that these techniques may induce tube buckling and costly. Taking into consideration abovementioned issues, it is eminent to put forward retrofitting technique of circular RC column to boost load carrying capacity and to deliver ample ductility by simple manufacturing process and low cost. This research suggests an external retrofitting technique of circular reinforced concrete column by segmented clamping instruments around the surface of column utilizing thin steel plate. It is apparently that the collaboration of reinforcement, steel clamps and bolts generating initial confining pressure yields to better retrofitting methods. A numbers of finite element studies using the 3D-NLFEA will be conducted to gain better configuration for optimum ductility of the proposed model.

2. Literature review

Circular reinforced concrete column under axial and lateral load usually generates plastic hinges leading to concrete cover spalling failure [5], [6] near to the connection at column ends. Vertical member of structure subjected to laterally cyclic motion could fail in various manners. For instance, crushing of the inner wall surface of hollow reinforced column is caused by inadequate ductility [7] due to transverse reinforcement failure in confining concrete core [8], [9]. The summary results of three studies can be seen in Table 1.

Table 1. Ductility of circular RC column

Researcher	D	t/D	ρ_s	$\rho_{s\text{-act}}$	Ductility level	Failure type
	mm	test	test	test	test	
Hoshikuma and Priestley (2000)	1524	0.09	0.48%	6.35 mm dia. @ 35 mm (0.22%)	3.3	Brittle failure
	1524	0.09	1.06%		1.8	Brittle failure
Ranzo and Priestley (2001)	1524	0.1	0.49%	6.35 mm dia. @ 70 mm (0.12%)	6	Flexural failure
	1524	0.09	0.80%		3	Brittle flexural/shear failure
	1524	0.09	0.80%		2	Shear failure
Zahn et al. (1990)	400	0.24	2.56%	10-12 mm dia. @ 75-90 mm (1.13%-1.36%)	2	Brittle failure
	400	0.14	2.56%		3.2	Limit ductile failure
	400	0.19	2.56%		5.8	Ductile failure

2.1. *Internal strengthening*

The amount of transverse reinforcement of circular confined concrete reflects the curvature and considerably influences the confinement of core concrete thereby energy dissipation by means of transverse reinforcement volumetric ratio [10], [11]. Nevertheless, excessive transverse reinforcement will suspend optimum curvature capacity until the crack and crushing of concrete.

2.2. *External strengthening*

External retrofitting on reinforced concrete was studied by [12]–[17]. High strength rectangular reinforced concrete confined with single layer of CFRP shows higher load carrying capacity and displacement ductility compared to the conventional reinforced concrete column. Steel sheets also capable to rise the capacity of core, concrete ductility and prevent the cover spalling [18]. Five steel tubular confinements for normal and high strength concrete column by varying tube height between 100 mm and 120 mm and compressive strength in Figure 4(b) was observed without longitudinal and transverse reinforcement [1]. The study found brittle shear failure in all specimen under monotonic loading and rapid failure under cyclic loading in specimen under optimum compressive strength.

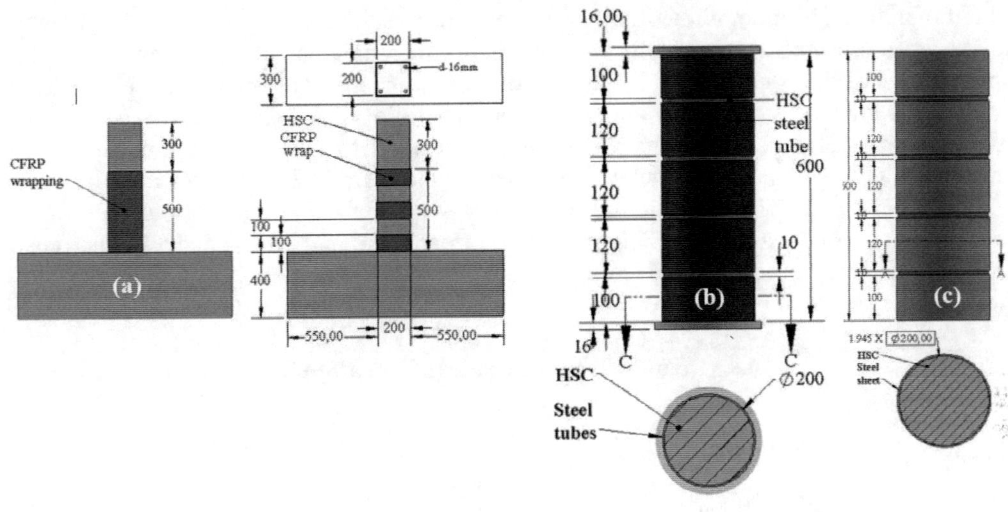

Figure 1. Segmented confinement model [18][1][19]

The behavior of steel tube confined concrete (STCC) was studied using finite element package [19], which incorporated plasticity principle [20]. This study included zero thickness element technique to determine the interface of contact surfaces as shown in Figure 4(c). The author defined hoop stress, axial stress, confining pressure and shear friction stress that influenced the behavior of the confined concrete column. Comparing fully and consecutive segmented confinement shown in Figure 4(a), the ultimate lateral load capacity and cumulative hysteretic energy dissipated by the segmented confinement was higher than that by full jacketing method despite the number of applied CFRP layer [18], [21]. Nevertheless, the displacement ductility of segmented confinement was less than full confinement. It is apparent that axial compressive load, confinement height and thickness and gap width determine this unfavorable issue. Similarly, two layers of CFRP confining rectangular reinforced concrete subjected to lateral seismic loading showed the extension of drift ratio and better ductility.

2.3. *Confined Concrete Constitutive Models*

Confined concrete constitutive model has been developed extensively [10], [12], [16], [22]–[24]. A recent model [25] can predict on how the conventional and high-performance concrete behave at any level of confinement by the variation of concrete height and aspect ratio. The concrete models confined by steel tube have been numerically studied using finite element application such as Three-Dimensional Nonlinear Finite Element Analysis (3D-NLFEA) [20]. The application utilizes plasticity fracture method encompassing compressive and tensile behaviors with failure surface provision [26]. The model could recognize passive or active confinement and cover spalling behavior to control active failure surfaces [27]. The capability of plasticity fracture method and its series of development were claimed to be superior to other concrete constitutive models in certain extends since the mechanical changes and their impact on stress-strain curve were explicitly captured and visually depicted in 3D model.

3. Methods

3.1. Experimental study

This stage investigates the axial capacity of slender circular reinforced column strengthened with steel clamp under axial concentric and eccentric loading and to assure that the strengthening mechanism possibly increases the load carrying capacity. The concrete specimen size is 100 mm in diameter and the height of 600 mm. The ratio of longitudinal reinforcement at 1% is embedded into 20 MPa concrete. Longitudinal bars are embedded without transverse reinforcement in order to investigate the pure confinement effect of thin steel sheet on axial compression capacity. The thickness and height of sheet is unvaried. Properties of steel clamp and bolts material follows SNI 2052-2017 and ASTM A325 respectively. Thin steel sheets, with its variation of thickness to height ratio and gap, are formed according to Table 2. The main parameters investigated under single slenderness ratio based on ACI 318-19 include load eccentricity and initial clamp stresses. At the top and bottom face of concrete specimen, end plates are welded to the longitudinal reinforcement and knife-edge articulation steel are positioned to apply eccentrically axial compression. The top and bottom end of the specimen are embedded into the sleeve of plate to prevent premature unintended displacement of the specimen.

Table 2. Specimen configuration

No.	Notation	λ	Bar	$h_s t_s$	g_s	f_l'	e
			mm^2		mm	MPa	mm
1	E00000000	24	$\rho = 1\%$ cover 20 mm	0.00	-	0.000	0.00
2	E10000000						e_1
3	E20000000						e_2
4	E0T1G1P1			50	10	$f_l'_1$	0.00
5	E1T1G1P1						e_1
6	E2T1G1P1						e_2
7	E0T1G1P2					$f_l'_2$	0.00
8	E1T1G1P2						e_1
9	E2T1G1P2						e_2

All specimens are tested under concentric and eccentric axial load of 157 kN. The first series of specimen undergoes increasingly monotonic axial load of with without eccentricity until failure. The second series apply eccentric axial compression loads until the failure of specimen. For higher accuracy of load application reading, a load transducer (load cell) is installed at the top of the articulation unit. Three variables: applied load, lateral displacement at middle height and vertical and horizontal deformation are measured by the instrumentation as illustrated in Figure 2. Three LVDTs quantifies vertical displacement and horizontal displacement at middle height of the specimen in tension area parallel to the direction of bending. Two LVDTs are installed at the middle segment of the specimen for better accuracy in obtaining moment-curvature measurement data. Strain gauges are set to measure strain of concrete and steel sheets respectively.

3.2. Finite element model of column

Eight node hexahedral elements are used for concrete element using Bbar technology with selective integration that disassociates shape functions to prevent volumetric locking. Specimen model is meshed such that the global size of elements is 2 mm. Concrete model utilized in this research follows multisurface plasticity fracture model. Steel yielding point can be defined using Von Mises theory. Hence, truss elements of reinforcing bars embedded in the concrete are assumed as isotropic and elastic-perfectly plastic material that follows Mises criterion [28].

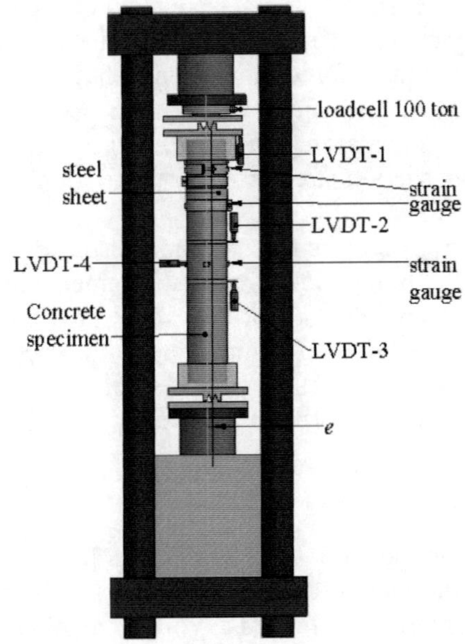

Figure 2. Illustrative drawing of testing instrument

Bottom end of column specimen is treated as pin support where three DOF u_x, u_y and u_z are restrained at cross-sectional nodes as shown in Figure 3. The DOF at top end is restrained in z direction at all nodes within the steel band. Two loading stage will be applied in the assessment of the column specimens. A constant downward axial load corresponding to the axial load ratio of $0.08f'_cA_g$ is applied at top end and followed by cyclic lateral displacement applied at top end until reaching drift ratio of 0.035. The displacement u_x and u_y at top end due to loading applications is captured to obtain stress-strain or load-displacement data.

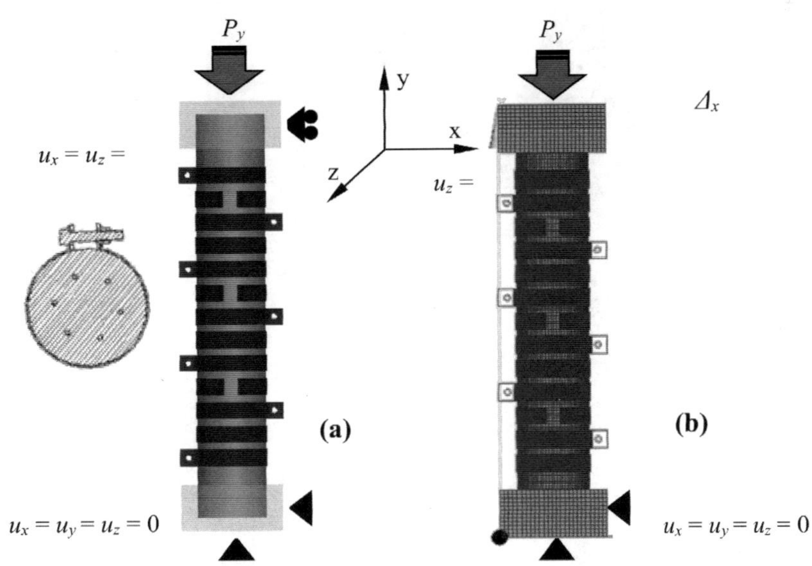

Figure 3. Boundary condition and loading plan of simply supported circular column

The following nomenclature is used to specify surface contact between element within the strengthened column system: CC = cylinder concrete, LR = longitudinal reinforcement, SS = steel strap, BT = bolt, NT = nut. Contact between different material within the system includes perfectly-bond contact types: CC-LR, SS-BT, SS-NT, BT-NT and a specific contact types: CC-SS. Specific contact type between concrete and steel clamping devices are treated with zero-thickness element [29] and the Mohr-Coulomb plasticity criterion as well as shear friction stress τ_f, internal friction angle ϕ_f and bond tensile stress σ_t within both interfaces.

Nine model will be generated to determine the optimum result for steel sheet confined RC column. The finite element analysis in this section yields force-displacement curve and flexural toughness value I_{10}, bending moment along column height, moment-curvature diagram, the maximum stress at column ends, force-displacement curve at the middle height of steel. Interaction diagrams from experimental results and FEA are used to develop analytical model by fiber-based method and formulate flexural toughness I_{10}'. Analytical model is used to predict effective confinement and generate F-D curve and M-φ diagrams for each component.

4. Results

Preliminary study using in-house application 3D-NLFEA was performed to observe stress and lateral displacement behavior of retrofitting model. Figure 4 depicts stress contours of specimen subjected to 157 kN axial compression. As specimen is subjected to vertical axial load of 157 kN, confining pressure at the surface of strengthened model is approximately 3 MPa compared to 0.34 MPa at that of unconfined model. Furthermore, stress concentration was well distributed along the height of confined model. It is in contrast to model without steel sheet since the stress concentration tend to occur at both end and at the middle height of specimen where concrete crush normally occurs during axial compression testing. The axial pressure towards z direction in Figure 5 confirms the role of confining mechanism.

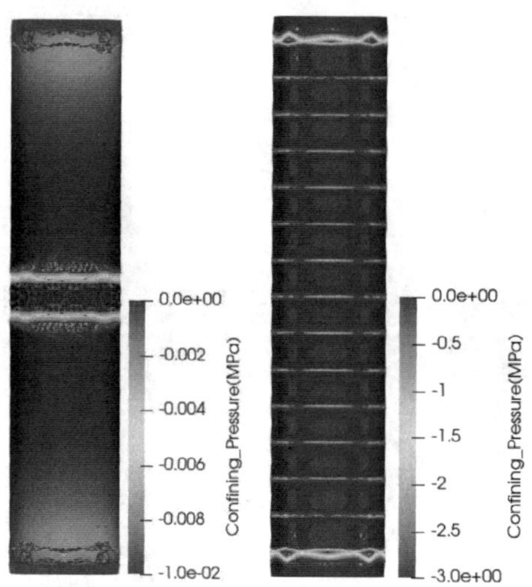

Figure 4. Resultant confining pressure of non-confinend and confined model

When visually observing along the width of cylinder model in Figure 6(a), the confining pressure on concrete surface begins with 2.5 MPa and reduces to zero within the thickness of 6 mm and entering tension zone towards the centroid before turns to compression zone at the opposite surface with similar thickness. The horizontal displacement at the midplane of mesh element located at the outer surface of unconfined concrete model is doubled than that of confined model. The displacement of both models reaching zero at the centroid as can be seen in Figure 6(b). This also explain the role of strengthening sheet in reducing lateral crushing of concrete due to vertical compression.

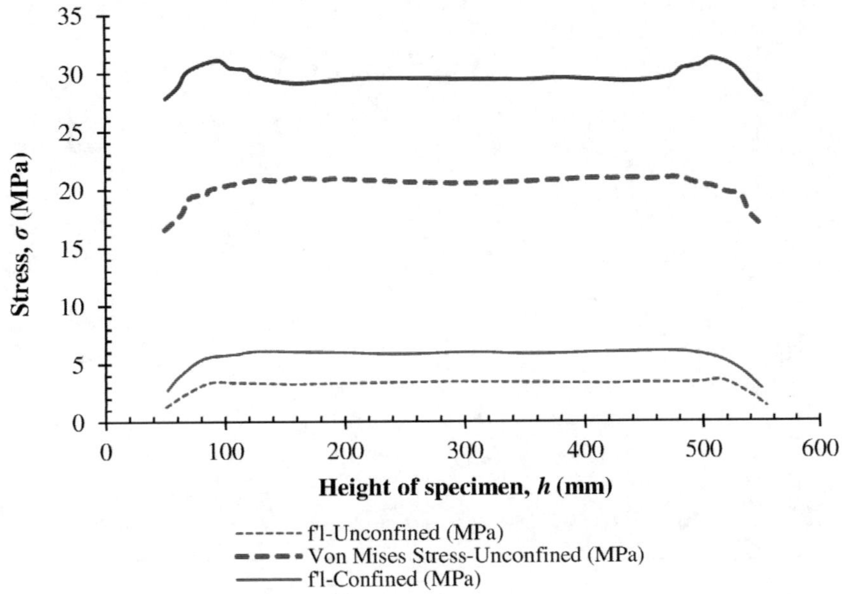

Figure 5. Axial stress along the height of model

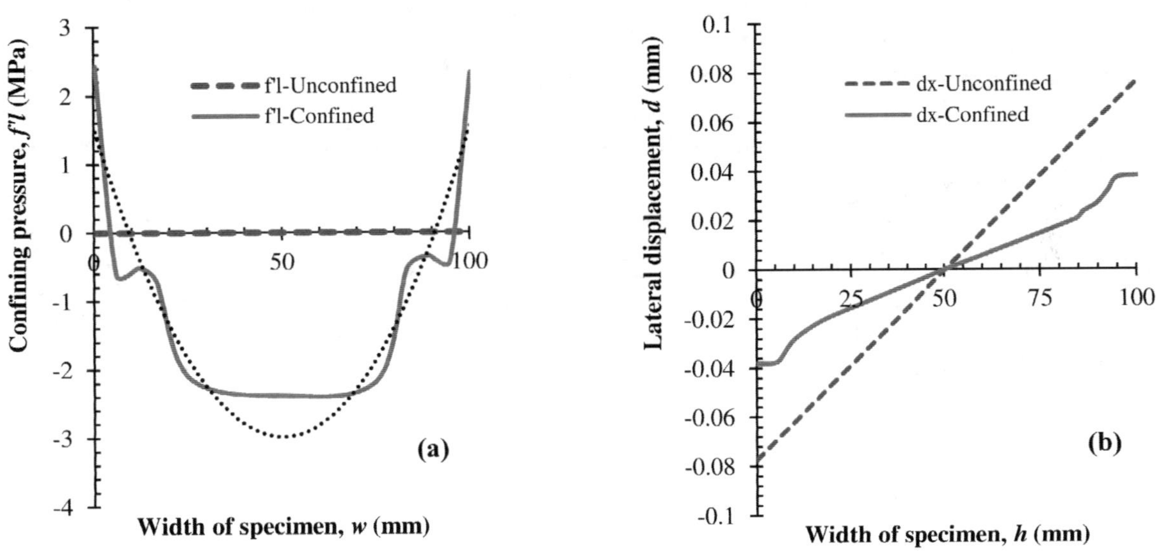

Figure 6. Horizontal stress along the height of model

Despite the lesser dense mesh size and neglecting longitudinal reinforcement in the preliminary study, Figure 7 clearly depicts that the rapid increase in axial capacity of strengthened concrete model indicates the major impact on the addition of proposed strengthening method. By the addition of only 2 mm steel sheet retrofitting apparatus on the cylinder concrete, not only the ultimate axial capacity of concrete could be tripled but also vertical displacement of concrete model can be increased. Therefore, the suggested approach is potential to be investigated for the advancement of reinforced concrete strengthening application.

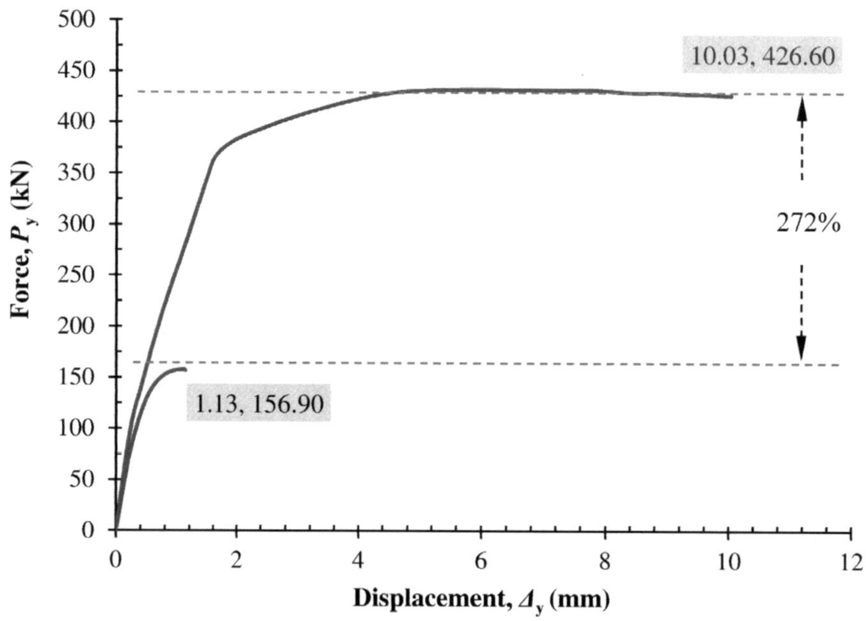

Figure 7. Force-displacement relationship

Acknowledgements
The authors acknowledge the facilities, scientific and technical support from Laboratory of Concrete, Advance Material and Computational Mechanics of Civil Engineering Department, Sepuluh Nopember Institute of Technology.

References

[1] J. Liu, Y. Teng, Y. Zhang, X. Wang, and Y. F. Chen, "Axial stress-strain behavior of high-strength concrete confined by circular thin-walled steel tubes," Constr. Build. Mater., vol. 177, pp. 366–377, 2018, doi: 10.1016/j.conbuildmat.2018.05.021.

[2] R. Park and T. Paulay, Reinforced Concrete Structure. Toronto, Canada: John Wiley & Sons Inc., 1975.

[3] C. Zhou, J. Sun, Z. Zeng, and J. Liu, "Studies on the RC beam strengthened by using the exterior-wrapping u-shaped steel plate and its application," Adv. Civ. Eng., vol. 2019, 2019, doi: 10.1155/2019/2396764.

[4] A. Hadhood, H. M. Mohamed, and B. Benmokrane, "Axial Load–Moment Interaction Diagram of Circular Concrete Columns Reinforced with CFRP Bars and Spirals: Experimental and Theoretical Investigations," J. Compos. Constr., vol. 21, no. 2, p. 04016092, 2017, doi: 10.1061/(asce)cc.1943-5614.0000748.

[5] A. Belarbi, M. Dawood, M. Bowman, and A. Mirmiran, "Synthesis of Concrete Bridge Piles Prestressed with CFRP Systems," Texas, 2017.

[6] A. Budek and G. Benzoni, "Obtaining ductile performance from precast, prestressed concrete piles," PCI J., vol. 54, no. 3, pp. 64–80, 2009, doi: 10.15554/pcij.06012009.64.80.

[7] F. A. Zahn, M. J. N. Priestley, and R. Park, "Flexural Strength and Ductility of Circular Hollow Reinforced Concrete Columns without Confinement on Inside Face," ACI Struct. J., vol. 87, no. 2, pp. 156–166, 1990, doi: 10.14359/9295.

[8] J. Hoshikuma and M. J. N. Priestley, "Flexural Behavior Of Circular Hollow Columns With A Single Layer Of Reinforcement Under Seismic Loading," California, USA, 2000.

[9] G. Ranzo and M. J. N. Priestley, "Seismic performance of circular hollow columns subjected to high shear," University of California, San Diego, La Jolla, California, 2001.

[10] J. B. Mander, M. J. N. Priestley, and R. Park, "Theoretical Stress-Strain Model for Confined Concrete," J. Struct. Eng., vol. 114, no. 8, pp. 1804–1826, Aug. 1988, doi: 10.1061/(ASCE)0733-9445(1988)114:8(1804).

[11] J. B. Mander, M. J. N. Priestley, and R. Park, "Observed Stress-Strain Behavior of Confined Concrete," J. Struct. Eng., vol. 114, no. 8, pp. 1827–1849, 1988.

[12] M. F. Ferrotto, L. Cavaleri, and F. Di Trapani, "FE modeling of Partially Steel-Jacketed (PSJ) RC columns using CDP model," Comput. Concr., vol. 22, no. 2, pp. 143–152, 2018, doi: 10.12989/cac.2018.22.2.143.

[13] G. T. Truong, J. C. Kim, and K. K. Choi, "Seismic performance of reinforced concrete columns retrofitted by various methods," Eng. Struct., vol. 134, pp. 217–235, 2017, doi: 10.1016/j.engstruct.2016.12.046.

[14] B. Olmos, J. M. Jara, G. Gómez, and G. Martínez, "Influence of steel jacket thickness on the RC bridges' seismic vulnerability," J. Traffic Transp. Eng. (English Ed., vol. 6, no. 1, pp. 15–34, 2019, doi: https://doi.org/10.1016/j.jtte.2018.09.004.

[15] A. He, J. Cai, Q. J. Chen, X. Liu, and J. Xu, "Behaviour of steel-jacket retrofitted RC columns with preload effects," Thin-Walled Struct., vol. 109, pp. 25–39, 2016, doi: 10.1016/j.tws.2016.09.013.

[16] Z. Tao, Z. Bin Wang, and Q. Yu, "Finite element modelling of concrete-filled steel stub columns under axial compression," J. Constr. Steel Res., vol. 89, pp. 121–131, 2013, doi: 10.1016/j.jcsr.2013.07.001.

[17] A. He, J. Cai, Q. J. Chen, X. Liu, H. Xue, and C. Yu, "Axial compressive behaviour of steel-jacket retrofitted RC columns with recycled aggregate concrete," Constr. Build. Mater., vol. 141, pp. 501–516, 2017, doi: 10.1016/j.conbuildmat.2017.03.013.

[18] J. Wang, J. Yang, and L. Cheng, "Experimental Study of Seismic Behavior of High-Strength RC Columns Strengthened with CFRP Subjected to Cyclic Loading," J. Struct. Eng., vol. 145, no. 2, p. 04018240, 2019, doi: 10.1061/(asce)st.1943-541x.0002251.

[19] A. H. Yudhanto, B. Piscesa, M. M. Attard, B. Suswanto, and P. Suprobo, "Finite element modeling of concrete confined with circular thin-walled steel sheet," E3S Web Conf., vol. 156, no. October, 2020, doi: 10.1051/e3sconf/202015605009.

[20] B. Piscesa, M. M. Attard, and A. K. Samani, "3D Finite element modeling of circular reinforced concrete columns confined with FRP using a plasticity based formulation," Compos. Struct., vol. 194, no. April 2017, pp. 478–493, 2018, doi: 10.1016/j.compstruct.2018.04.039.

[21] H. F. Ghatte, M. Comert, C. Demir, M. Akbaba, and A. Ilki, "Seismic Retrofit of Full-Scale Substandard Extended Rectangular RC Columns through CFRP Jacketing: Test Results and Design Recommendations," J. Compos. Constr., vol. 23, no. 1, p. 04018071, 2019, doi: 10.1061/(asce)cc.1943-5614.0000907.

[22] G. A. Chang and J. Mander, "Seismic Energy Based Fatigue Damage Analysis of Bridge Columns: Part 1 - Evaluation of Seismic Capacity," 1994.

[23] M. Saatcioglu and S. R. Razvi, "Strength and Ductility of Confined Concrete," J. Struct. Eng., vol. 118, no. 6, pp. 1590–1607, Jun. 1992, doi: 10.1061/(ASCE)0733-9445(1992)118:6(1590).

[24] Z. Diogo, B. Michele, and H. Dan, "Constitutive Model of Concrete Simultaneously Confined by FRP and Steel for Finite-Element Analysis of FRP-Confined RC Columns," J. Compos. Constr., vol. 22, no. 6, p. 4018064, Dec. 2018, doi: 10.1061/(ASCE)CC.1943-5614.0000902.

[25] A. K. Samani and M. M. Attard, "A stress–strain model for uniaxial and confined concrete under compression," Eng. Struct., vol. 41, pp. 335–349, 2012, doi: https://doi.org/10.1016/j.engstruct.2012.03.027.

[26] B. Piscesa, M. M. Attard, and A. K. Samani, "A lateral strain plasticity model for FRP confined concrete," Compos. Struct., vol. 158, pp. 160–174, 2016, doi: https://doi.org/10.1016/j.compstruct.2016.09.028.

[27] B. Piscesa, M. M. Attard, D. Prasetya, and A. K. Samani, "Modeling cover spalling behavior in high strength reinforced concrete columns using a plasticity-fracture model," Eng. Struct., vol. 196, no. April, p. 109336, 2019, doi: 10.1016/j.engstruct.2019.109336.

[28] A. B. Christianto, B. Piscesa, M. M. Attard, Faimun, and P. Aji, "3D Finite element modeling of circular reinforced concrete column confined with CFRP under different eccentric loads," IOP Conf. Ser. Mater. Sci. Eng., vol. 930, no. 1, 2020, doi: 10.1088/1757-899X/930/1/012047.

[29] R. E. Goodman, R. L. Taylor, and T. L. Brekke, "A Model for the Mechanics of Jointed Rock," J. Soil Mech. Found. Div., vol. 94, no. 3, pp. 637–659, May 1968, doi: 10.1061/JSFEAQ.0001133.

Experimental study on replaceable shear link on reinforced concrete shear wall

M Moestopo[1*], D Kusumastuti[1], and Kurniawan[2]

[1] Faculty of Civil and Environmental Engineering, Research Center for Industry, Institut Teknologi Bandung, Bandung, Indonesia
[2] Graduate Student, Faculty Civil and Environmental Engineering, Institut Teknologi Bandung, Bandung, Indonesia

*Corresponding author: mmoestopo@gmail.com

Abstract. Experimental study has been conducted on specimens of a half-scale coupled reinforced concrete shear wall with steel coupling beam and an inserted shear link, to investigate the seismic performance of the core-wall and the replaceability of the shear link as a mechanical damper to dissipate the seismic energy. The link was designed following requirements for typical shear link, while the coupling beams as well as the shear wall were designed based on the link capacity. The result showed that under the prescribed cyclic loading (AISC 341-2016), the sub-assembly performed as expected, as shown by a good hysteretic curve with the plastic strain of shear link exceeded 0.08 radian. The damage was limited on the shear link, while the steel coupling beam and steel bar in the reinforced concrete wall remained elastic. Moreover, the experimental study showed no permanent distortion on the steel coupling beam after series of shear-link replacement. This confirmed the replaceability of the shear links as an excellent seismic device in the core-wall system.

1. Introduction

The replaceable link has an economical advantage to be used as a seismic device in eccentrically braced frames, particularly in the area with frequent earthquakes. The replaceable link acts as a fuse on the structure, which develops plastic hinge and experiences damage, while other structural elements remain elastic. With the damaged elements can be easily repaired and replaced, a better structural performance can be obtained, and the structure can immediately continue its function with no or very little damage. The replaceability of steel shear link is then applied to a sub-assembly of coupled shear wall, by inserting the steel shear link at the center of its steel coupling beams. In this setup, the damage was expected to occur on the shear link only, and the coupling beams as well as other structural elements will not experience yielding. This concept provides easy access and replacement of the link

The use of steel coupling beams with replaceable steel shear link in a sub-assembly of a steel-core frame for high-rise buildings has been studied by Kusumastuti, et al [1] and Moestopo, et al [2]. The experimental work showed that the shear link yielded and dissipated energy due to cyclic loading until its maximum strength with its plastic rotation exceeded 0.08 radian, while other elements of core-frame i.e., coupling beam and steel-frame remained elastic. This showed the advantage of inserting the shear link between coupling beam in the core-frame, i.e.: (1) additional energy dissipation device and (2) replaceability of the shear link after damage. Previous works on replaceable coupling beams were conducted by Ji et al [3],[4].

This paper presents the result of further experimental work that has been conducted on specimens of a half-scale coupled reinforced concrete shear wall with steel coupling beam and shear link, to investigate the seismic performance of the structure and the replaceability of the shear links as an excellent seismic damper in the core-wall system.

2. Experimental work

This study investigates the core-wall system consists of coupling beam and shear link in a RC-wall of 15-story office building located in Jakarta, Indonesia, with building plan is shown in Figure 1. In the y-direction of the core-wall system, a steel link beam was placed in the mid-span of each coupling beam to dissipate the seismic energy by the shear yielding. The link beam was designed as a mechanical damper that was the only element to perform plastic deformation, while all other structural elements in the building were designed to remain elastic. The structural design for the office building was carried out according to Indonesian Seismic Loading Code and RC Building Codes [5], [6], and [7]. The link was designed following requirements for typical shear link, and the coupling beams as well as the shear wall were designed based on the link capacity. More explanation related to numerical modeling and design can be found in [8].

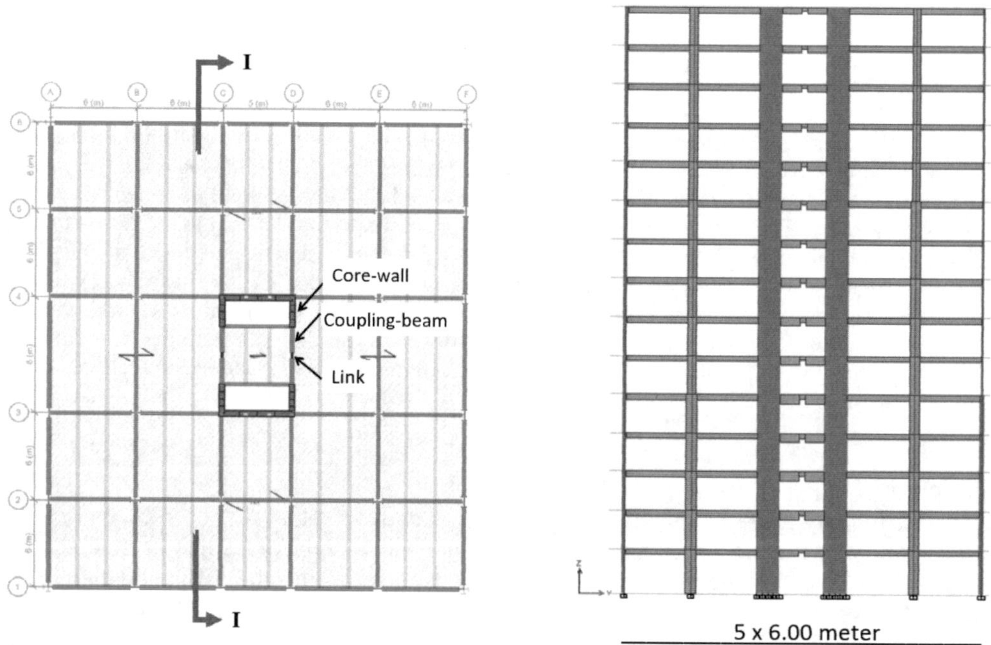

a. Plan b. View I - I

Figure 1. 15-storey RC Building

2.1. Test set-up

Experimental work was conducted on a half-scale coupled reinforced concrete shear wall with steel coupling beam and an inserted shear link representing the 15-story coupled reinforced concrete shear wall connected by coupling beam and shear link. The test set-up and the specimen are shown in Figure 2. The steel coupling beams used WF 400.200.8.13 and the shear link used WF 200.100.5,5.8, both were made from Indonesian steel BJ-41 (yield stress of 250 MPa and tensile stress of 410 MPa). The thickness of RC-wall was 400-mm with steel reinforcement of dia-13mm (horizontal) and dia-19 mm (vertical). The link beam was bolt-connected to coupling beam using high-tension bolt and was designed to be replaced after testing.

The cyclic loading follows the AISC 341-16 [9] which is the loading protocol for cyclic tests of link-beam to beam moment connections in eccentrically braced frames. The LVDT was used to monitor the plastic rotation in the shear link during the cyclic loading by measuring the relative vertical displacement between the link-ends. Several LVDTs were also used to measure the potential slip between the reaction unit and the strong floor. Strain-gages were used to measure the strain at the flanges of coupling beam and the strain at the steel reinforcing bar in the RC-wall.

Two identical specimens of sub-assembly were tested with the same loading protocol to investigate the difference in the seismic performance of the second specimen as well as the replaceability of the damaged link after the cyclic loading test of the first specimen.

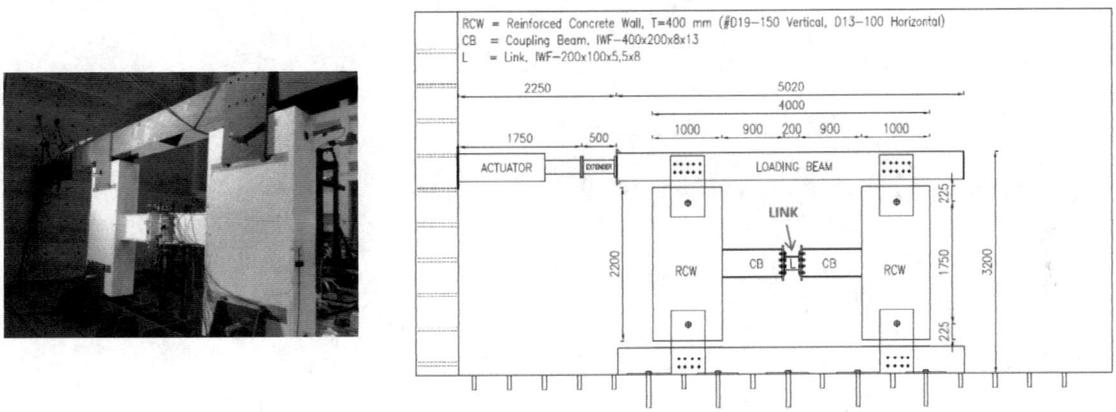

(a) Photograph (b) Half-scale sub-assembly

Figure 2. Test set-up

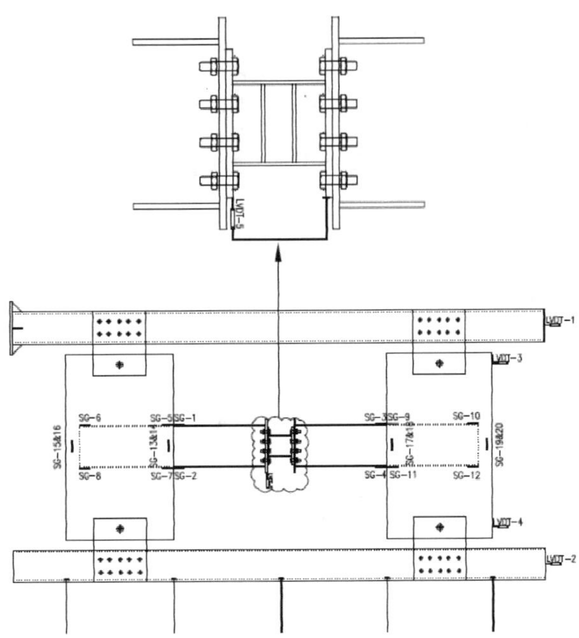

Figure 3. Location of LVDTs and Strain Gages

2.2. Material Test

Tensile tests were conducted to all steel materials used in the experiment. The results of tensile tests for shear link, coupling beam and reinforcing steel are shown in Tables 1 to 3.

Table 1. Average test result of shear link

	F_{ye} (MPa)	F_{ue} (MPa)	ε_u	R_y	R_t	F_{ye} / F_{ue}
Web	382.1	492.00	0.155	1.53	1.20	0.776
Flange	380.0	504.15	0.225	1.52	1.23	0.754

Table 2. Average test result of coupling beam

	F_{ye} (MPa)	F_{ue} (MPa)	ε_u	R_y	R_t	F_{ye} / F_{ue}
Web	435.7	545.6	0.14	1.74	1.33	0.80
Flange	388.8	518.3	0.19	1.55	1.26	0.76

Table 3. Average test result of reinforcing steel

Specimen	Position	F_{ye} (MPa)	F_{ue} (MPa)	ε_u	R_y	R_t	F_{ye} / F_{ue}
D-13	Horizontal	429.4	582.5	0.06	1.10	1.06	0.737
D-19	Vertical	451.5	611.6	0.11	1.15	1.11	0.738

3. Test Result

3.1. Seismic Performance

The link rotation for each specimen during the cyclic loading was determined using the relative displacement of link-ends that were recorded by the LVDT located at one end of the link. The hysteretic loop between link shear vs total link rotation for two specimen of sub-assembly is shown in Figure 4.

Figure 4 also shows the link rotation (γ_y) at yield shear (V_y) and the link rotation (γ_u) at maximum shear (V_u). The overstrength (Ω) and ductility (μ) of each specimen were determined as the ratio of V_u to V_y and ratio of γ_u to γ_y, respectively. Specimen-2 shows slightly higher overstrength than Specimen-1, but slightly lower in ductility.

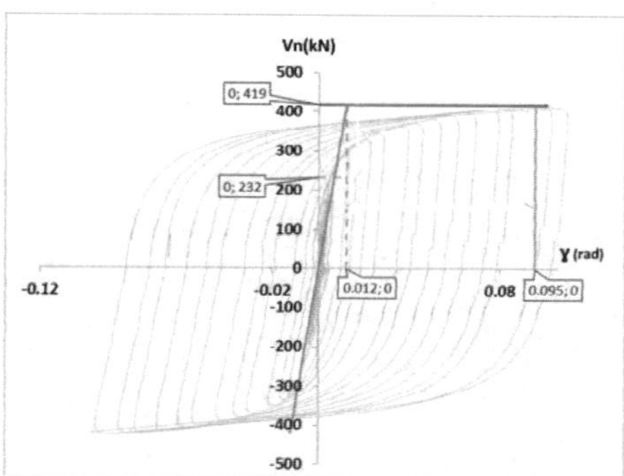

(a) Link-1

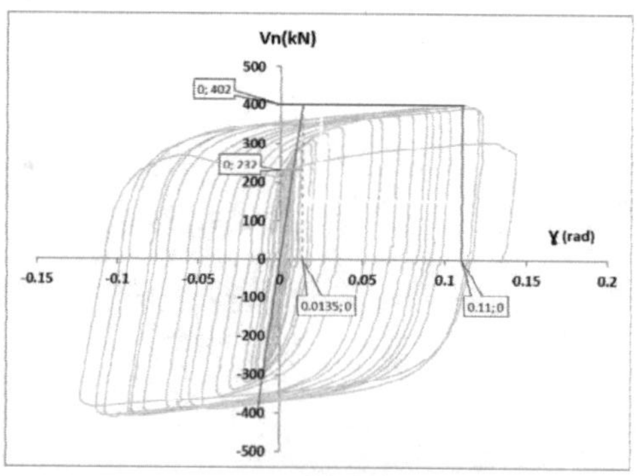

(b) Link-2

Figure 4. Shear (Vn) vs Total link rotation (γ)

Table 4. Overstrength (Ω), ductility (μ), and plastic link rotation (γ_p)

Specimen	V_y (kN)	V_u (kN)	Ω	γ_y (rad)	γ_u (rad)	μ	$\gamma_{tot\text{-}max}$ (rad)	γ_p (rad)
Link-1	232	419	1.81	0.0120	0.095	7.9	0.104	0.096
Link-2	232	402	1.73	0.0135	0.110	8.1	0.124	0.116

Figure 5 shows the plastic link rotation (γ_p) for both specimens during the cyclic loading, that was obtained by subtracting the elastic part of the total link rotation at each recorded displacement. Both specimens showed good seismic performance as the plastic link rotation (γ_p) exceeded 0.08 radian.

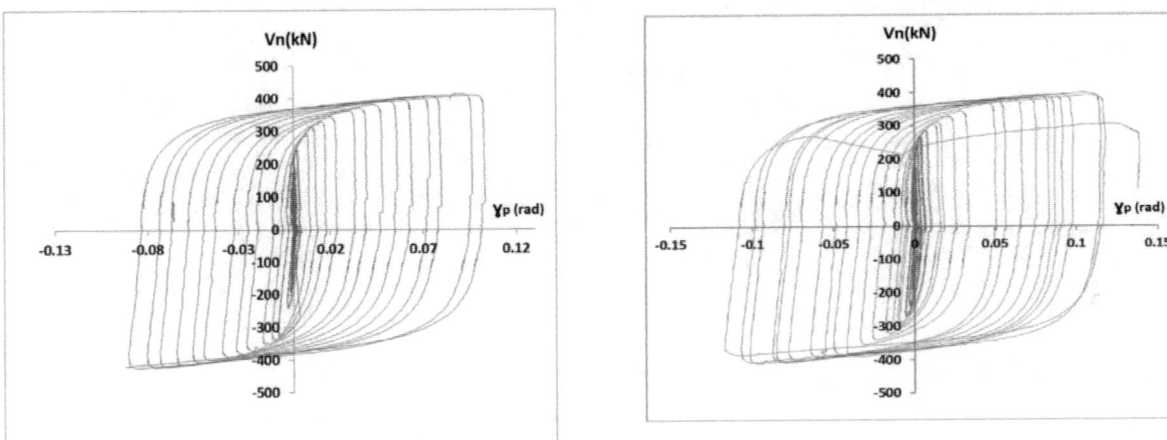

(a) Link-1 (b) Link-2

Figure 5. Shear (Vn) vs Plastic link rotation (γ_p)

Figure 6 shows that in both specimens, the damage was very visibly seen at the web of the shear link, as expected. Meanwhile, no damage was detected on either connected ends of the shear link.

(a) At $\gamma_p = 0.116$ (b) After web-fracture

Figure 6. Damaged Link-2

Figures 7 and 8 show the strain measurement data (ε) at the top and bottom-flange of coupling beams (CB) and at the reinforcing bar (RB) in RC-wall during the cyclic loading with actuator force (F). All figures show almost straight lines, indicating all steel materials were still elastic. This confirmed that only link yielded to dissipate the energy, while all other elements remained elastic.

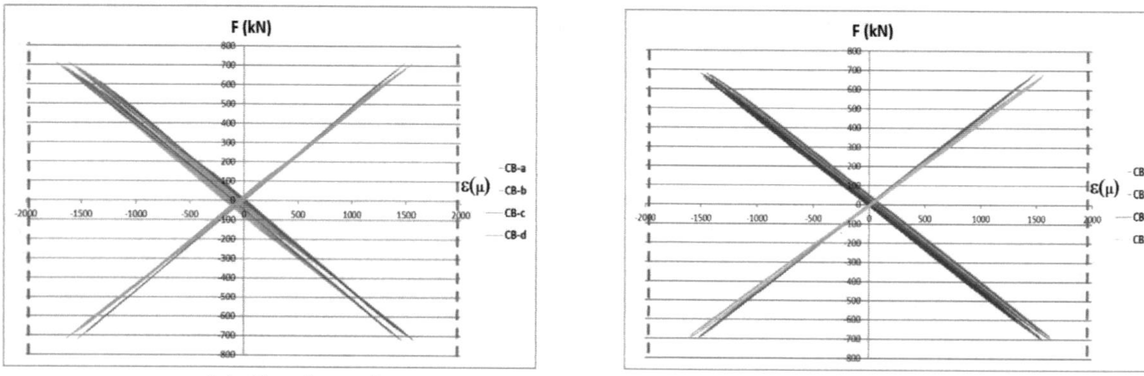

(a) Specimen-1 (b) Specimen-2

Figure 7. Strain at coupling beam

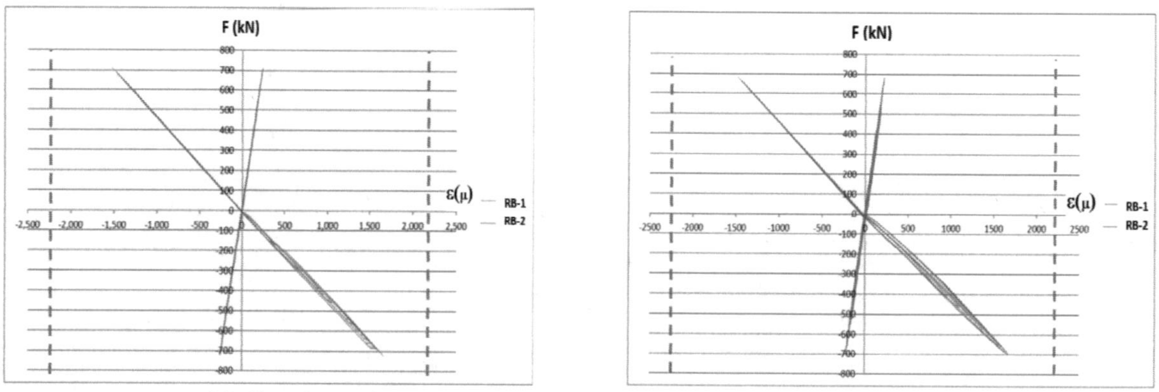

(a) Specimen-1 (b) Specimen-2

Figure 8. Strain at reinforcing bar in RC-wall

Cracks in the RC-wall were indicated when the link experienced large rotation during the cyclic loading. However, the crack path disappeared when the RC-wall was loaded in the opposite direction, and no damage was indicated in the RC-wall. Moreover, it verified the good performance of the link attached to the commonly used coupling beam system in RC-wall.

3.2. Replaceability of link

The set-up of sub-assembly of RC-wall with coupling beam used in this experimental work was designed to remain elastic during the cyclic loading until the failure of the shear link attached to the coupling beam. In this experimental work, link specimens (including the two specimens) have been loaded and removed, and then was replaced by other identical links.

Visual inspection after the completion of each testing showed the followings in both end plates of coupling beam and link: (1) no indication of damage on the connecting-surfaces as well as around the perimeter surface of bolt holes, (2) no indication of damage in all bolts, (3) no significant slip occurred on the connected surface, and (4) no permanent distortion observed on coupling beam. Meanwhile, inspection on the RC-wall indicated: (1) a regular but not severe crack pattern, (2) no slip between reinforcing bar and coupling beam to the RC-concrete. Moreover, there was no special technique required during the replacement of link, i.e., removal and installation of links, as shown in Figure 9.

Figure 9. Easy replacement of links

The experimental work verified the replaceability of the shear link in the core system of coupled RC-wall and coupling beam. Further improvement for the core system could be performed such as inserting additional diagonal web stiffeners in the link, to increase the stiffness and the shear strength as well as the energy dissipation of the link [10].

4. Conclusion

The result of the experimental study on a half-scale coupled RC-shear wall with steel coupling beam and an inserted shear link, showed that under the prescribed cyclic loading [7]:

a. The sub-assembly performed as expected, as shown by a good hysteretic curve with the plastic strain of shear link exceeds 0.08 radian.
b. The damage was limited to the shear link, while the steel coupling beam and steel bar in the RC-shear wall remained elastic.
c. The experimental study showed no permanent distortion on the steel coupling beam after series of shear-link replacements. This confirmed the replaceability of the shear links as an excellent seismic device in the core-wall system.

Acknowledgement

This research has been supported by the Applied Research Program - Ministry of Research and Higher Education Republic of Indonesia, for which the authors would like to thank for the supporting fund.

References

[1] Kusumastuti D, et al 2017 Analytical Investigation of the Applicability of Seismic Devices for Core Frames System in High-Rise Buildings *Proc.3rd Int. Conf. on Civ.Eng.Res* Surabaya Indonesia
[2] Moestopo M, et al 2018 Experimental Study on the Seismic Behavior of Replaceable Shear Links Connected to Coupling Beam *Intl J. Adv. Sci. Eng. Inf. Tech.* **8** 532-539
[3] Ji, et al 2017 Seismic Performance assessment of a Hybrid Coupled-wall System with Replaceable Steel Coupling Beam vs Traditional RC Coupling Beams, *Eq. Eng.& Str. Dyn.* **46** 517-535
[4] Ji, et al 2017 Cyclic behavior of replaceable Steel Coupling Beams, J. Str.Eng. **143** 04016169
[5] Badan Standardisasi Nasional 2019 Seismic Design Loading For Buildings and Other Structures *SNI 1726:2019 (In Indonesian)* BSN Jakarta Indonesia.
[6] Badan Standardisasi Nasional 2013 Minimum Design Loads For Buildings and Other Structures *SNI 1727:2013 (In Indonesian)* BSN Jakarta Indonesia
[7] Badan Standardisasi Nasional 2019 Specification for Structural Concrete Buildings *SNI 2847:2019 (In Indonesian)* BSN Jakarta Indonesia
[8] Prasetyo JD 2021 Numerical Study on Application of Replaceable Link in 15-story RC Building with Coupled-Wall (In Indonesian) *Master Thesis* Bandung Institute of Technology Indonesia

[9] American Institute of Steel Construction 2016 Seismic Provisions for Structural Steel Buildings *AISC 341-16* AISC Illinois USA
[10] Yurisman, et al 2010 Behavior of Shear Link of WF Section with Diagonal Web Stiffener of Eccentrically Braed Frame (EBF) of Steel Structure, ITB J.Eng.Sci. Vol.**42** 103-128

IOP Conf. Series: Earth and Environmental Science 1244 (2023) 012004 doi:10.1088/1755-1315/1244/1/012004

Application of Modal Analysis of Earthquake Force and Natural Frequency Tanjung Baru Bridge

A Mardhiyah[1], J Tarigan[1], E P Bangun[1]

[1] Department of Civil Engineering, Engineering Faculty, Universitas Sumatera Utara, Medan, Indonesia

ainilmardhiyah.1109@gmail.com

Abstract. The dynamic response of a structural system can be described by modal analysis. In the present study, the modal analysis that adopts an Undamped Degree of Freedom (Undamped-DOF) System is implemented to describe the modal descriptors such as the natural frequency and mode shape of a bridge. Furthermore, the earthquake load acting on the bridge is also evaluated and hence the possibility of a resonance associated with this load is investigated. For this objective, the earthquake data collected during the past ten years in Indonesia is explored. This study considers the Tanjung Baru Bridge, a prestressed concrete bridge situated on the Medan-Kualanamu-Tebing Tinggi highway, as the case study. The modal analysis and the evaluation of the earthquake load confirm that the possibility of resonance for the Tanjung Baru Bridge is extremely low. The study also proves that this bridge has a sufficient capacity to withstand earthquake loads.

1. Introduction

An earthquake is a phenomenon that occurs when seismic energy is released suddenly at a specific location. Shocks that do not occur continuously are one sign of an earthquake [1]–[3]. Because Indonesia is located between two large oceans and a tectonic plate (ring of fire), major earthquakes occur frequently with significant magnitude. Earthquakes could destruct a bridge with damage ranging from minor to severe [4]. As a result, the design of earthquake-resistant bridges is critical.

Proper analysis is required to avoid bridge damage from the earthquake. One type of analysis that can be used is modal analysis. This analysis is considered the most effective method for evaluating structural performance and designing various types of structures [5]. Modal descriptors such as natural frequency, mode shape, damping ratio, and earthquake loads can be used to describe the dynamic response of a structural system using modal analysis. The dynamic behavior of a bridge due to earthquake loads can be formulated in a mathematical model based on its modal descriptors [6].

The application of modal analysis to a bridge enables a more accurate assessment of the structure's structural integrity and monitoring of the structural suitability of the bridge. Natural frequency, mode shape, and earthquake loads are all required parameters for bridge design, competence, and condition evaluation. Given these parameters, particularly the earthquake loads and the natural frequency, the capability, suitability, and strength of a bridge can be determined [7]–[12]. These parameters have a high impact on the bridge's ability to withstand lateral loads, which can be driven by earthquake loads.

Natural frequency and mode shape are dynamic characteristics that are critical in bridges [13]. As a result, the natural frequency and mode shape must be examined. These two predicted dynamic characteristics will be used in a modal analysis to determine the magnitude of the earthquake load that acts on a bridge. To the authors' knowledge, existing literature lack of ttudies concerning the implementation of a modal analysis for evaluating the earthquake load on a bridge. In earlier studies, the modal analysis was explored for evaluating the effects of the earthquake loads on bridge. For instance, research conducted by Çubuk et al [14], Murdiansyah et al [15], and Zhang et al [16]. To investigate the earthquake load, earlier researchers in this field predominantly applied spectral response methods and employed analysis of time-histories as the study carried out by Kamal and Gunadi [17], Indulkas et al [18], and Hakim et al [19]. Therefore, the present study contributes to the existing literature by providing an improved understanding of how to apply a modal analysis for the analysis of earthquake load on a bridge.

2. Research methodology

Tanjung Baru Bridge is considered herein as the research subject. It is a toll bridge situated on the Medan-Kualanamu-Tebing Tinggi highway. This bridge spans an intersection of the main road, with its coordinate located at STA 34+127,351. Tanjung Baru Bridge is a prestressed concrete bridge with a Prestressed Concrete I (PCI) Girder cross-section and a total length of 62.1 m. Figure 1 shows that this bridge consists of three spans that measure 15.5 m, 31.1 m, and 15.5 m for each span, and thus has two pillars as the supporting structures. The width of the Tanjung Baru Bridge is 12.70 m. Figures 2 and 3 present the cross sections of the spans with 15.5-meter and 31.1-meter lengths, respectively. ClClose-upiews of the PCI Girder constructed at these two spans show that the height of the girder is 900 cm for the 15-meter long span (see Figure 4) and 1700 cm for the 31.1-meter long span (see Figure 5).

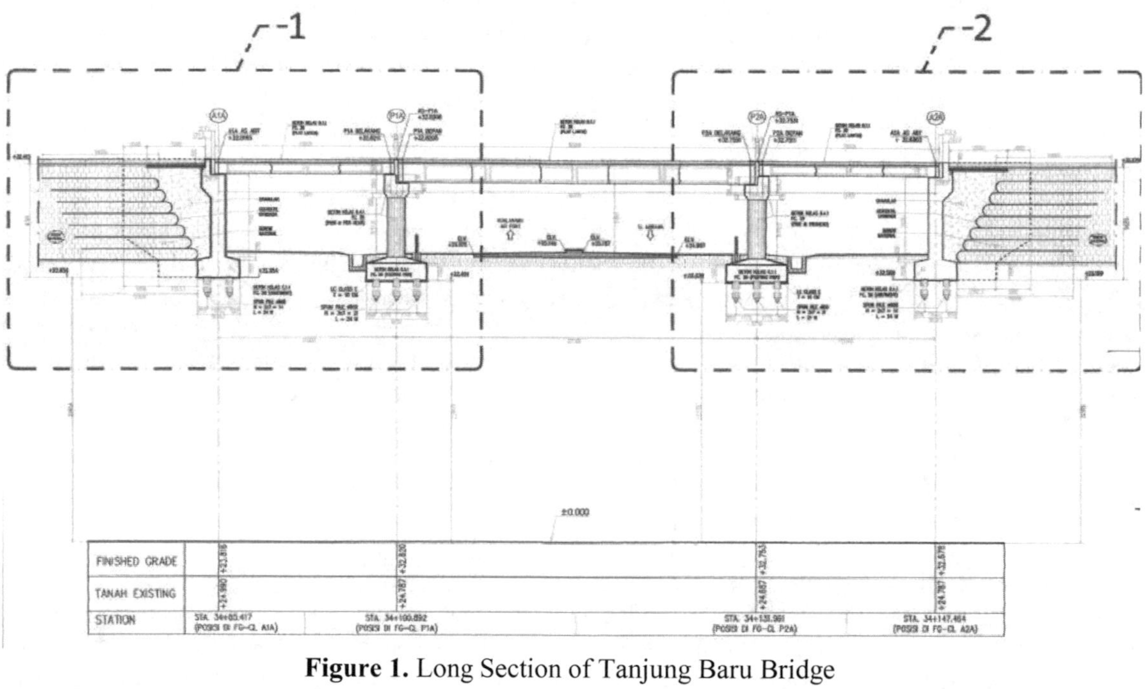

Figure 1. Long Section of Tanjung Baru Bridge

The present modal analysis implements an Undamped Multi Degree of Freedom (Undamped-MDOF) System for predicting the natural frequency and mode shape. The first three natural frequencies and mode shapes estimated using this system are expressed as Equations (1) and (2) [20], respectively. In the present study, thdal analysis is also employed to calculate earthquake loads. The formulation for this calculation is given by Equation (3) [21].

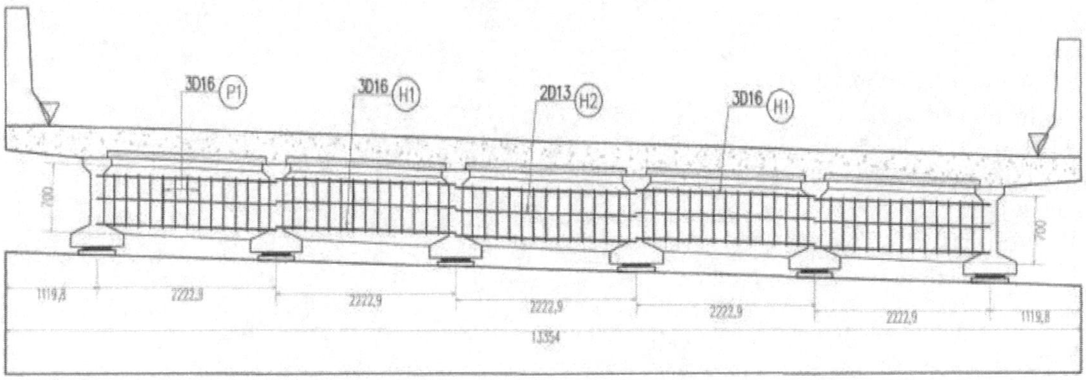

Figure 2. Cross section of Tanjung Baru Bridge (15.5 m)

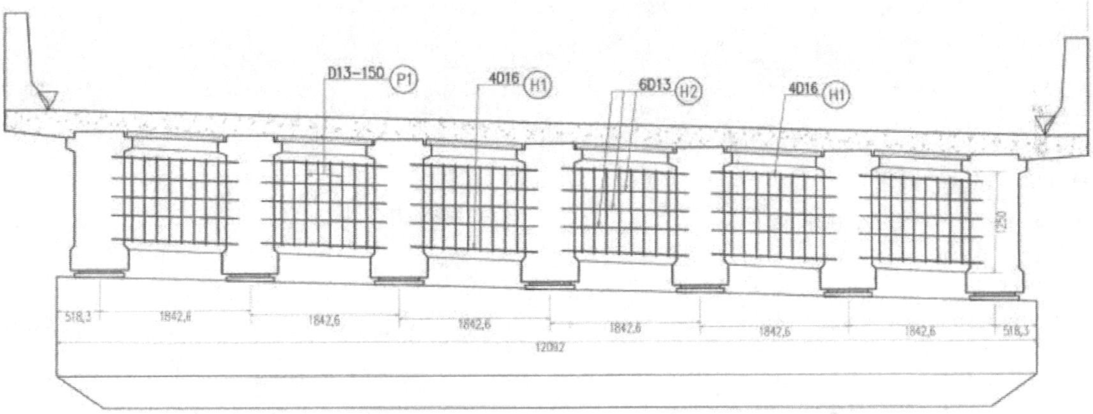

Figure 3. Cross section of Tanjung Baru Bridge (31.1 m)

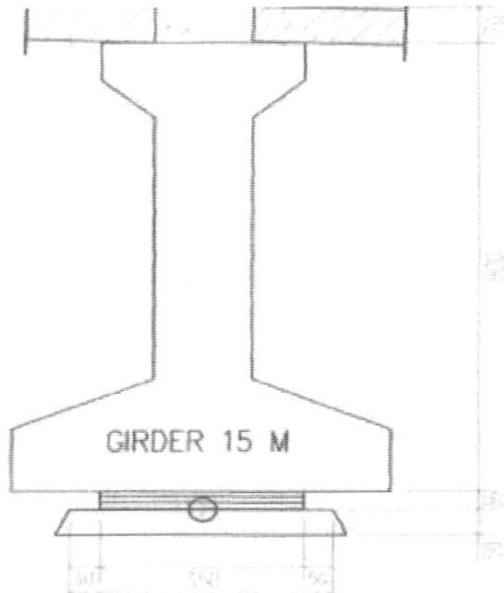

Figure 4. PCI Girder (15.5 m)

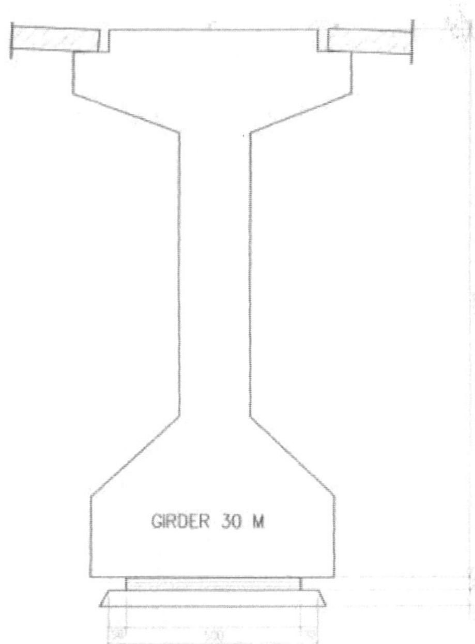

Figure 5. PCI Girder (31.1)

IOP Conf. Series: Earth and Environmental Science 1244 (2023) 012004 doi:10.1088/1755-1315/1244/1/012004

$$f_n = n^2\pi^2\sqrt{\frac{EL}{\widetilde{m}L^4}} \tag{1}$$

Where:
f_n is n[th] nathe tural frequency (Hz)
n is mode ithe ndex defined by 1, 2, 3 . . . and so forth
π is the ratio of a circle's circumference to its diameter (3.14 12 or 22/7)
E is module s of elasticity (kN/m^2)
L is length of bridge span (m)
$\widetilde{m}$ is uniform mass (kg/m)

$$\phi_n(x) = \sin\frac{n\pi}{L}x \tag{2}$$

Where:
$\phi_n(x)$ is mode shape
x is coordinate location relative to a support (m)

$$\mathrm{EL}_n = \frac{\left\lfloor \sum_{j=1}^{N} m_j\phi_{jn}\right\rfloor^2}{\sum_{j=1}^{N} m_j{\phi^2}_{jn}} \tag{3}$$

Where:
EL_n is n[th] earthquake load (kN)
m_j is mass in j on MDOF
ϕ_{jn} is nth mode shape in j
j is review point

An Undamped-MDOF System was chosen over a Damped-MDOF System because the former system considers no loss of load while the latter takes it into account. The damping element in the Damped-MDOF System is responsible for this. The lost load can occur because of the movement of molecules in the soil material or the release of energy associated with a joint friction, where the damping element functions as a shear and load spreader property of the structure [22].

The modal analysis involves the development of a mathematical model to predict the dynamic behavior of a bridge [6]. Furthermore, the modal analysis serves as an effective method for determining the modal properties of a structure. This is due to the fact that the modal analysis only requires structural measurements in normal operational conditions to represent the real dynamic response [23]. This method uses the vibration time and mode shape calculated from the natural frequency value. It proves that in the modal analysis, the natural frequency is closely related to earthquake loads.

Having quantified the earthquake load value, the bridge's ability to shear is evaluated. This analysis is required to demonstrate whether the bridge design can withstand shear forces caused by earthquakes. This analysis employs the following equation.

$$\mathrm{HE}_j = \sqrt{{EL_{j1}}^2 + {EL_{j2}}^2 + \ldots + {EL_{jn}}^2} \tag{4}$$

Where :
HE_j is the total earthquake load at point j (kN)
El_{j1} is the earthquake load at point j in the 1st mode (kN)
El_{j2} is the earthquake load at point j in the 2nd mode (kN)
El_{j3} is the earthquake load at point j in the 3rd mode (kN)

The earthquake load at point j in the 1st mode, El_{j1}, 2nd mode, El_{j2}, and 3rd mode, El_{j3}, is obtained from equation (3).

This study also investigates the possibility of resonance caused by earthquakes in the last ten years Based on the earthquake data collected from the Meteorology, Climatology, and Geophysics Agency

IOP Conf. Series: Earth and Environmental Science 1244 (2023) 012004 doi:10.1088/1755-1315/1244/1/012004

for the past ten years in Indonesia, the highest peak ground acceleration (PGA) for bridges is used to analyze the possibility of resonances. Furthermore, Equation (5) is employed to examine it. The bridge will experience a resonance if the resonance ratio, η, falls between 0.75 and 1.25 [21]. Once it is obtained, the bridge's ability to withstand earthquake loads will be more proven.

$$\eta = \frac{f_E}{f} \tag{5}$$

Where:
η is the resonance ratio
f_E is earthquake frequency (Hz)
f is the natural frequency in each mode (Hz)

The earthquake frequency, f_E, is calculated by analyzing the PGA, which starts by calculating the distance for the attenuation function, r, using equation (6) [24], [25]. The obtained attenuation distance value, r, is then entered into equation (7) to analyze the PGA. After obtaining the PGA value, it is reduced to the angular velocity, ω, using equation (8) [26] and then converted back to the earthquake frequency, fE, using equation (9).

$$r^2 = \left(\sqrt{(L_l - L_g)^2 + (B_l - B_g)^2} \times 1^0\right)^2 + 8^2 \tag{6}$$

Where:
r is the distance for the attenuation function (km)
L_l is the latitude of the research location (^{0})
L_g is the latitude of the earthquake source location (^{0})
B_{lthe} is longitude of the research location (^{0})
B_g is the longitude of the earthquake source location (^{0})

$$PGA = 10^{(0{,}71+0{,}23(\dot{M}-6)-\log(r)-0{,}0027r)} \tag{7}$$

Where:
PGA is peak ground acceleration (gall atau 0,01 m/det^2)
$\dot{M}$ is moment magnitude of the earthquake (M)

$$\omega^2 = \frac{PGA}{r} \tag{8}$$

Where:
ω is the angular velocity (rad/sec)

$$f_E = \frac{\omega}{2\pi} \tag{9}$$

3. Result and discussion

3.1. Natural frequency

The natural frequency value, f, of the Tanjung Baru Bridge is obtained using the Undamped-MDOF System. Having implemented Equation (1), the values of natural frequencies for the first three modes, f_1, f_2 and f_3, are presented in Table 1.

Table 1. Natural Frequency Values

Length (m)	Natural Frequency, *f*, [Hz]		
	f_1	f_2	f_3
15.5	2616.77	10467.10	23550.97
31.1	1234.97	4939.87	11114.70

3.2. Mode shape

The corresponding mode shapes for the first three modes, ϕ_1, ϕ_2 and ϕ_3 , are estimated from Equation (2). The values of ϕ_1, ϕ_2 and ϕ_3 are each predicted at four equally spaced points located between the two ends of a span. Table 2 presents these values for the 15.5-m long and 31.1-m long spans, while Figure 6 depicts the associated mode shapes for each of these two spans. In Figure 6, the mode shape value has been scaled to make the mode shape visible.

Table 2. Mode Shape Values

Length (m)	Mode Shape		
	ϕ_1	ϕ_2	ϕ_3
15.5	0.575	0.941	0.965
	0.941	0.639	-0.506
	0.965	-0.506	-0.699
	0.639	-0.983	0.873
31.1	0.588	0.951	0.951
	0.951	0.588	-0.588
	0.951	-0.588	-0.588
	0.588	-0.951	0.951

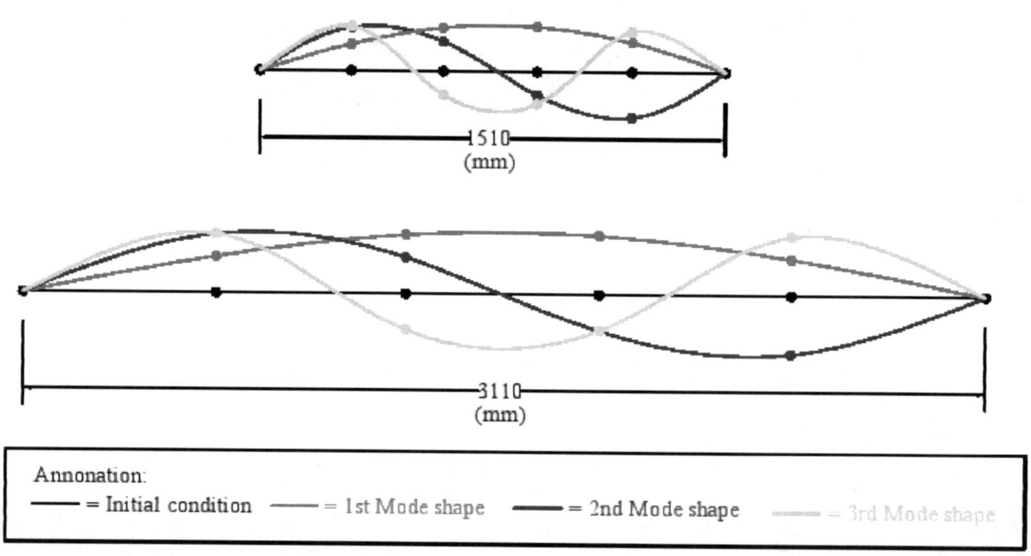

Figure 6. Mode shape of the Tanjung Baru Bridge

3.3. Earthquake load

The earthquake load values are estimated using the modal analysis at each mode shape. In the present analysis, the direction of the earthquake load analyzed is the vertical direction. This is based on the consideration that the result of the mode shape has a vertical direction. The earthquake loads for the first three modes, EL_1, EL_2, and EL_3, in each case of span length are shown in Table 3.

Table 3. Earthquake Load Values

Length (m)	Earthquake Load (kN)		
	EL_1	EL_2	EL_3
15.5	1301.72	64.51	852.60
	2130.76	43.83	-447.46
	2186.09	-34.74	-617.77
	1447.60	-67.43	771.67
31.1	1038.60	0	2690.50
	1680.49	0	-1662.82
	1680.49	0	-1662.82
	1038.60	0	2690.50

3.4. Bridge capability

Following on, the bridge capability was determined based on the total earthquake load vålue that has been evaluated using Equation (4). Table 4 provides the total earthquake load at the first review point, HE_1, the second, HE_2, the third, HE_3, and the fourth review point, HE_4.

Table 4. Bridge Capability Values

Length (m)	Shear Strength (kN)	Total Earthquake Load (kN)			
		HE_1	HE_2	HE_3	HE_4
15.5	890.611	1557.42	2131.11	2186.22	1641.82
31.1	890.611	2884.01	1679.99	1679.99	2884.01

3.5. Resonance

The possibility of resonance arising due to the earthquake load is also evaluated in the present study. This analysis employs the data from the most significant earthquake that occurred closest to the bridge location. The data used herein corresponds to the earthquake that took place on January 16, 2017. Given the earthquake frequency, f_E , based on this data, the possibility of resonance, η , associated with the natural frequencies for the first three modes is evaluated. Table 5 provides the η values for these three modes. Evidently, all the estimated values lie far outside the range of $0.75< \eta <1.25$.

Table 5. Resonance Values

Length (m)	Natural Frequency, f_1, f_2, f_3, [Hz]	Earthquake Frequency, f_E, [Hz]	Resonance, η,
15.5	2616.77	0.4	0.0002
	10467.10		0.00004
	23550.97		0.00002
31.1	1234.97	0.4	0.0003
	4939.87		0.00008
	11114.70		0.00004

4. Conclusion

Based on the data outlined in Table 1, it can be seen that the natural frequency value of a bridge increases with the number of modes. In addition, the study shows that a longer span has a smaller natural frequency value than that of a short span. Evidence of this is indicated by the natural frequency of the 31.1-m long span being estimated to be 1234.97 Hz while that of the 15.5-m long span being 2616.77 Hz. Both correspond to the frequencies for the first mode. Importantly, this effect of span length applies for the second and the third modes.

Based on the data presented on Table 2, a mode shape of the 15.1-m long span has a higher value than that of a 31.1-m long span. The mode shape of the 15.1-m long span has values of 0.575, 0.941, 0.965, and 0.639 in the first mode and 0.941, 0.639, -0.506, and -0.983 for the second mode. At the same time, the third mode shape has values of 0.965, -0.506, -0.699, and -0.873. For the 31.1-m long span, the first mode shape has values of 0.588, 0.951, 0.951, and 0.588 while the second mode shape has values of 0.951, 0.588, -0.588, and -0.951. At the same time, the third mode shape has values of 0.951, -0.588, -0.588, and 0.951. According to these results, the first and second mode shapes have the most significant deviation. Simultaneously, the third mode shape has the lowest mode shape. The two spans also show that the higher the level of the mode shape under consideration, the mode shape will be smaller.

Based on the data outlined in Table 3, the earthquake load acting on the 15.5-m long span is the highest in the first mode. In contrast, the earthquake load in the case of the 31.1-m long span is the highest in the third mode. Whereas in the second mode, spans of 15.5 m and 31.1 m have the smallest values compared to the others. The earthquake loads in the first mode are 1301.72 kN, 2130.76 kN, 2186.09 kN, and 1447.60 kN in the case of 15.5-m long span, and 1038.60 kN, 1680.49 kN, 1680.49 kN, and 1038.60 kN in the case of 31.1-m long span. The four load values for each case are applied at the four equally spaced points. In the second mode, the earthquake loads at the four points are 64.51 kN, 43.83 kN, 34.74 kN, and 67.43 kN for the 15.5-m long span, yet the loads are 0 kN for the 31.1-m long span. In the third mode, the earthquake load is 852.60 kN, 447.46 kN, 617.77 kN, and 771.67 kN for the 15.5-m long span, and 2690.50 kN, 1662.82 kN, 1662.82 kN, and 2690.50 kN for the 31.1-m long span.

The present study also shows that the values of shear strength for the two cases of span length are equal. With these values of shear strength, the total earthquake load at each point along the 15.5-m and 31.1-m long spans is greater than the bridge's shear strength value. It proves that the Tanjung Baru Bridge has a sufficient capacity to withstand earthquake loads. The total earthquake load on the 15.5-m long span is 1557.42 kN, 2131.11 kN, 2186.22 kN, and 1641382 kN, respectively. And the 31.1 m span has a total earthquake load of 2884.01 kN, 1679.99 kN, 1679.99 kN, and 2884.01 kN.

The earthquake on January 16, 2017, had the highest magnitude for the Tanjung Baru Bridge in the previous ten years. Table 5 shows that the possibility of resonance is extremely low. This understanding arises because the resonance value lies far outside the range where a resonance is very likely to occur. The resonance value is significantly much less than 0.75 for both cases of span length. The possibility of a resonance occurring on the 15.5-m long span is 0.0002, 0.00004 and 0.00002 for the first, the second and the third natural frequency, respectively. Meanwhile, the probability of a resonance occurring on the 31.1-m long span is 0.0003 at the first natural frequency, 0.00008 at the second natural frequency, and 0.00004 at the third natural frequency.

References

[1] E. R. Hartuti, *Buku Pintar Gempa*. Yogayakarta, Indonesia: DNA Press, 2009.
[2] Suharjanto, *Rekayasa Gempa (Dilengkapi dengan Analisis Beban Gempa Sesuai SNI-03-1726-2002)*. Yogyakarta: Kepel Press, 2013.
[3] A. B. Siswanto and M. A. Salim, *Rekayasa Gempa*. Yogyakarta: K-Media, 2018.
[4] H. P. Adi, S. I. Wahyudi, and E. Santoso, "Studi Tentang Kerusakan Ifrastruktur Keairan Akibat Gempa Tektonik di Kabupaten Klaten," *J. Tek. Sipil Perenc.*, vol. 11, no. 2, pp. 161–168, 2009.

[5] S. Qi, "Modal Test Analysis of Prestressed Concrete Simply Supported Beam Bridge," in *2018 International Conference on Robots & Intelligent System (ICRIS)*, 2018, pp. 86–89. doi: 10.1109/ICRIS.2018.00030.
[6] J. He and Z.-F. Fu, *Modal Analysis*. Oxford: Butterworth-Heinemann, 2001.
[7] S. Maulidiya and Rusli, "Penentuan Frekuensi Natural Dan Arah Pergerakan Gelombang (Studi Kasus: Jembatan Soekarno Hatta Kota Malang)," *J. MIPA UNSRAT ONLINE*, vol. 6, no. 1, pp. 1–7, 2017.
[8] R. Akbari, S. Maadani, and S. Maalek, "On the fundamental natural frequency of bridge decks: Review and applications," in *Proceedings of the Institution of Civil Engineers: Structures and Buildings*, Dec. 2018, vol. 171, no. 12, pp. 931–945. doi: 10.1680/jstbu.17.00102.
[9] L. Cao, J. Liu, and Y. Frank Chen, "Theoretical and Numerical Study on the Natural Frequencies of Bridges With Corrugated Steel Webs," *Structures*, vol. 15, pp. 224–231, Aug. 2018, doi: 10.1016/j.istruc.2018.07.005.
[10] W. Sutrisno, L. Chandra, and A. Deonanda, "Perbandingan Frekuensi Alami Jembatan Karangsemut Menggunakan Accelerometer Dan Sap2000," *J. Rekayasa dan Inov. Tek. Sipil*, vol. 6, no. 2, pp. 13–18, 2021.
[11] H. T. Santoso, L. F. Hidayatiningrum, A. B. Utomo, J. Hartono, and Masrianto, "Analisa Korelasi Antara Frekuensi dengan Bentang Jembatan Berdasarkan Uji Dinamik (Correlation Analysis Between Frequency and Bridge Span Based on Dynamic Test)," *J. Jalan-Jembatan*, vol. 38, no. 1, pp. 60–72, 2021.
[12] D. Prawestri, W. Sutrisno, and A. Priyanto, "Perbandingan Analisis Frekuensi Alami Jembatan Gantung Dengan Menggunakan Aplikasi Accelerometer Meter Dan Software (Studi Kasus Jembatan Gantung Kemiri Buluharjo Karangmojo)," *J. Rekayasa dan Inov. Tek. Sipil*, vol. 6, no. 1, pp. 54–56, 2021.
[13] J. Cui and A. Che, "Calculation Method of Natural Characteristics of Beam Element Considering Initial State," in *4th International Conference on Civil, Architecture and Environment Research*, 2021, p. 012140. doi: 10.1088/1755-1315/676/1/012140.
[14] E. G. Çubuk, E. Sayin, and A. Özmen, "Dynamic Analysis of Historical Masonry Arch Bridges under Different Earthquakes: The Case of Murat Bey Bridge," *Turkish J. Sci. Technol.*, vol. 17, no. 2, pp. 461–473, 2022, doi: 10.55525/tjst.1105998.
[15] L. Murdiansyah, R. Permata, and D. Essen, "Modal Pushover Analysis on Reinforced Concrete Arch Bridge to Estimate Seismic Responses," in *4th ICEEDM 2019*, 2020, p. 03005. doi: 10.1051/e3sconf/202015603005.
[16] S. Zhang, J. Li, J. Li, and J. Zhang, "Dynamic Response Analysis of The Integrated Station-Bridge Elevated Station under The Pile-Soil Interaction," in *2020 2nd International Conference on Civil Architecture and Energy Science (CAES 2020*, 2020, p. 04081. doi: 10.1051/e3sconf/202016504081.
[17] R. F. Kamal and R. Gunadi, "Analysis of Bridge Structure with Non-Linear Time History Method," in *The 1st Annual Technology, Applied Science and Engineering Conference*, 2020, p. 012012. doi: 10.1088/1757-899X/732/1/012012.
[18] N. Indulkar, P. A. N. Humnabad, and P. D. N. V. Khadake, "Comparative Study of Seismic analysis of Bridge Substructure in different Seismic Zones as per IRC Guidelines," *Ijraset J. Res. Appl. Sci. Eng. Technol.*, vol. 10, no. V, pp. 3551–3557, 2022, doi: 10.22214/ijraset.2022.43016.
[19] N. Hakim, E. K. Kertapati, Suhara, and E. V. M., "Desain Tahan Gempa Jembatan Ngarai Sianok di Bukittinggi, Sumatera Barat - Indonesia," *J. Tek. Sipil*, vol. 5, no. 1, pp. 277–290, 2019, doi: 10.26760/rekaracana.v5i1.19.
[20] R. W. Clough and J. Penzien, *Dynamics of Structures*, 3rd ed. Berkeley: Computers & Structure Inc., 1995.
[21] A. K. Chopra, *Dynamic of Structures Theory and Applications to Earthquake Engineering*. Britania Raya: Pretince Hall, 2012.

[22] M. Paz, *Dinamika Struktur*, 2nd ed. Jakarta, Indonesia: Erlangga, 1996.
[23] H.-P. Chen and Y.-Q. Ni, "Modal Analysis of Civil Engineering Structures," in *Structural Health Monitoring of Large Civil Engineering Structures*, New Jersey: Wiley-Blackwell, 2018, pp. 91–120.
[24] S. L. Kramer, *Geotechnical Earthquake Engineering*. New Jersey: Prentice-Hall Inc, 1996.
[25] R. Villaverde, *Fundamental Concepts of Earthquake Engineering*. Amerika Serikat: CRC Press, 2009.
[26] Y. Surya, *Getaran dan Gelombang-Persiapan Olimpiade Fisika*. Banten: KANDEL, 2009.

Acknowledgments
I wish to express my deep gratitude to PT. Adhi Karya has helped me by providing the research data that I used. I also want to thank Mrs. Ernie Shinta Yosephine Sitanggang as my previous mentor. Finally, I am indebted to my parents, my sister, my role model Rahmi Ramadhani and my whole family for their continuous support and encouragement in my pursuit.

Hybrid viscoelastic damper using natural rubber compound, waste of rubber tire crumb with various compacted sands (iron sand, natural sand, and silica sand) to increase damping property for non-engineered buildings

Edwin[1], **Henricus Priyosulistyo**[2*]

[1)]Student of the Department of Civil and Environmental Engineering, Universitas Gadjah Mada (UGM), Indonesia
[2)]Professor of the Department of Civil and Environmental Engineering, Universitas Gadjah Mada (UGM), Indonesia

*Corresponding author: priyo_ugm@ugm.ac.id

Abstract. To reduce lateral deformation due to earthquake shake has been carried out by strengthening techniques. Such techniques require additional structural elements such as bracing or shear wall, if not applying additional new materials on beams or columns such as CFRP. This study focuses on the formation of the hybrid rubber dampers from materials such as natural rubber compound (NRC), waste of rubber tire crumb (WRTC), compacted natural sand, iron sand, and silica sand. The main goal of this research is to investigate the dynamic properties of hybrid materials of compacted various sands encapsuled with the NRC and WRTC compound. Nonetheless, the mechanical properties of rubber, such as tensile strength and modulus of elasticity, were also tested. In this research, the specimens were in the form of a square block of 100 x 100 mm^2 base and 25 mm thickness and consisted of two types of specimens for reference i.e., fully natural rubber compound (NRC) and blended NRC with 25% by volume of the WRTC, namely MRTC. Furthermore, other three types of hybrid viscoelastic dampers (HVD) were also provided i.e., 1) compacted natural sand, 2) compacted iron sand, and 3) compacted silica sand, which were all wrapped with WRTC in a proportion of 3:4 by volume. The results show that the addition of the MRTC reduces the tensile strength and elastic modulus in comparison to the NRC but increases the damping properties. Moreover, the elaboration of compacted natural sand, iron sand, and silica sand in the HVD also shows an increase in the damping properties up to 200%.

1. Introduction

An earthquake is a natural disaster that cannot be avoided. Indonesia is located between three major tectonic plates, namely Eurasian Plate, Indo-Australian Plate, and Pacific Plate. Based on the BMKG in 2021, it was recorded that so far there were 10,570 tectonic earthquakes. Also, according to CNN, in 2021 there were earthquakes with a magnitude above 5 Richter Scale that occurred as many as 244 times, and twenty-three of them were destructive. A damper can be used to minimize damage caused by the earthquake. A damper is a tool to absorb vibrations that occur in a building [1]. Strengthening building structures using a damper system has proven to be an effective strategy for reducing vibrations caused by earthquakes [2]. It was stated in reference [3] that there are three types of dampers, namely base isolation damper, mass damper, and wall/brace type damper. A viscoelastic damper is one type of damper which comes in the form of a wall/brace damper.

On the other hand, there are also environmental problems related to waste of used tires as the number of motorized vehicles continues to grow over times. Referring to Badan Pusat Statistik (BPS), in 2019, the total number of motorized vehicles in Indonesia was 133,617,012 while in 2020 there were 136,137,451. This means that there was an increase of 1.9% in the number of motorized vehicles.

There are several studies related to the use of WRTC that have been done before. Previous research stated that WRTC had a modulus of elasticity (E) of between 1.8 and 2.6 MPa and a shear modulus (G) of between 0.9 and 1.7 MPa [4]. A material that had a low modulus of elasticity and a low shear modulus indicated a material that had high damping ability [5]. Moreover, the addition of the WRTC caused a decrease in the mechanical properties of the rubber, but it increased the damping ratio of the rubber [6-7]. Meanwhile, based on [8], the addition of the WRTC increased the hardness of the rubber but decreased the tensile strength of the rubber. This was consistent with the findings of a study [9], which found that the hardness of rubber was inversely proportional to its tensile strength. The addition of natural sand to the rubber increased the damping through friction between its grains [10]. Not only natural sand, blended rubber compounds with silica sand to make composite rubber also showed an increase in the damping ratio of rubber [6]. Another type of sand, iron sand, was also able to increase the damping ratio of rubber, where iron sand was used as a filler in the process of making rubber compound [11].

The WRTC can be used as a compound to make viscoelastic dampers to reduce manufacturing costs, but it is also intended to solve the environmental problem. Therefore, this study has utilized the waste of vehicle tires in the form of waste of rubber tire crumb (WRTC). In addition, this study also wanted to find out the effect of adding compacted natural sand, iron sand, or silica sand to increase the damping ability of rubber. In this study, the composition and process of creating rubber specimens to be formed as composite rubber differ from previous research. The other difference is that the process of making the test specimen is expected to optimize the friction that occurs due to the compacted sand in the test specimen and increase the damping ratio of composite rubber.

2. Materials and experimental procedure

2.1. Materials

Five materials were used in this study. A natural rubber compound (NRC) was the first material used in this study. The main component of the compound was natural rubber in the form of ribbed smoked sheets (RSS) with a mixture of synthetic rubber, namely styrene-butadiene rubber (SBR), as shown in Figure 1(a), which was obtained from the workshop of the Polytechnic ATK Yogyakarta. The second material was the waste of rubber tire crumb (WRTC), as shown in Figure 1(b). The WRTC was obtained from the outside part of used vehicle tires. There are three methods of processing used tires, namely cryogenic, water, and ambient processes [12]. This rubber tire crumb was obtained by an ambient process. Through this process, the WRTC was obtained with a grain size of 0.15 mm (#100).

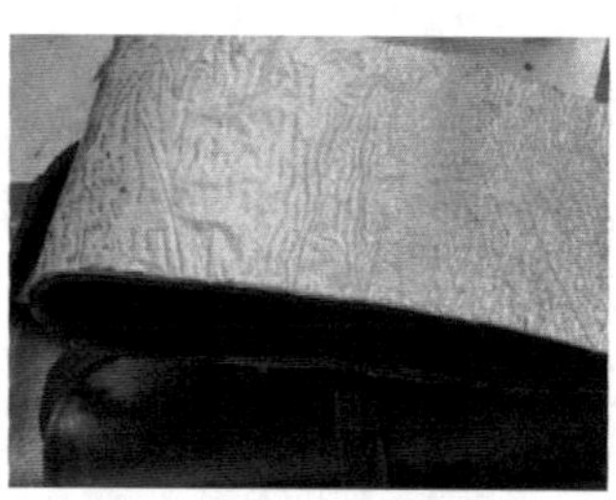
(a)

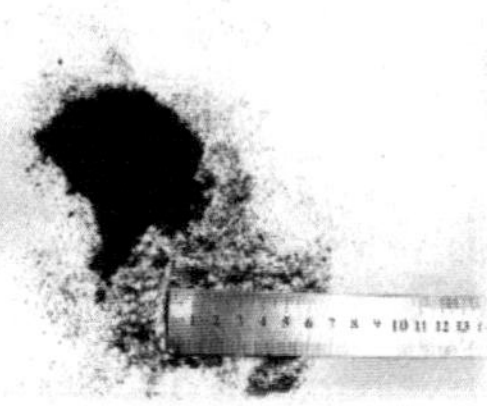
(b)

Figure 1. (a) Rubber compound, (b) waste of rubber tire crumb (WRTC).

The third material was natural sand, as shown in Figure 2(a). This natural sand was from the Krasak River, which flows from Mount Merapi in Central Java. Prior to the casting of the specimens, the aggregate gradation test was carried out first according to SNI 03-1968-1990. The results were obtained as shown in Table 1, with notations of WR, CR, and CP for weight retained, cumulative weight retained, and cumulative weight passing, respectively. The natural sand had a fineness modulus of 3.034.

Table 1. Fineness modulus of natural sand.

Sieve	**WR (g)**	**WR (%)**	**CR (%)**	**CP (%)**
3/8 inch (9.5 mm)	0	0.000	0.000	100.000
No. 4 (4.75 mm)	109.700	7.936	7.936	92.064
No. 8 (2.36 mm)	204.000	14.758	22.694	77.306
No. 16 (1.18 mm)	200.700	14.519	37.213	62.787
No. 30 (0.60 mm)	363.200	26.275	63.488	36.512
No. 50 (0.30 mm)	211.900	15.330	78.818	21.182
No. 100 (0.15 mm)	200.000	14.469	93.287	6.713

The fourth material was iron sand, as shown in Figure 2(b). This iron sand was obtained from West Java. Before being used, the aggregate gradation test was carried out, and the results showed that iron sand had a uniform grain size, as the test results showed that all iron sand grain passed through the smallest sieve, which was a sieve with a size of 0.15 mm (#100).

As shown in Figure 2(c), the fifth material was silica sand. This silica sand was white in color and was obtained from a local shop in Yogyakarta. Before being used, this material was also tested for aggregate gradation using the SNI 03-1968-1990 standard, and the results were obtained as shown in Table 2. The results show that the fineness modulus of silica sand was 2.360.

Table 2. Fineness modulus of silica sand.

Sieve	WR (g)	WR (%)	CR (%)	CP (%)
3/8 inch (9.5 mm)	0	0.000	0.000	100.000
No. 4 (4.75 mm)	0.000	0.000	0.000	100.000
No. 8 (2.36 mm)	0.000	0.000	0.000	100.000
No. 16 (1.18 mm)	10.200	0.805	0.805	99.195
No. 30 (0.60 mm)	724.100	57.133	57.938	42.062
No. 50 (0.30 mm)	298.100	23.521	81.458	18.542
No. 100 (0.15 mm)	181.900	14.352	95.810	4.190

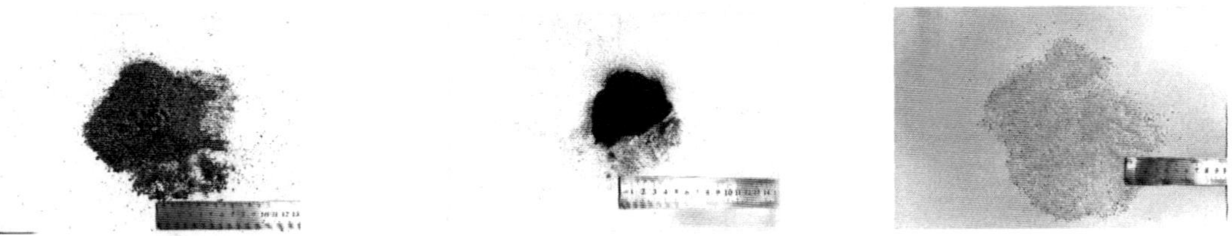

(a) (b) (c)

Figure 2. (a) Natural sand, (b) iron sand, (c) silica sand.

2.2. Mixture proportions

In this study, there were five mixture variations of materials used to make test specimens in the form of a square block of 100 x 100 mm^2 base and 25 mm thickness, as seen in Table 3. Four rubber test specimens were made for every mixture variation. The composition of the mixture used was based on the ratio of mass (in grams) among the natural rubber compound, the WRTC and the volume (in mm3) for the compacted natural sand, iron sand, or silica sand.

Table 3. Mixture component for 100 mm x 100 mm x 25 mm test specimens.

Specimens Code	Rubber Compound (g)	WRTC (g)	Total of Rubber Compound and WRTC (mm^3)	Natural Sand (mm^3)	Iron Sand (mm^3)	Silica Sand (mm^3)
R1	280	0	250,000	0	0	0
R2	210	70	250,000	0	0	0
R3	165	55	186,496	63,504	0	0
R4	165	55	186,496	0	63,504	0
R5	165	55	186,496	0	0	63,504

2.3. Mixing and curing procedure

The test specimens were created at the workshop of the Polytechnic ATK, Yogyakarta. First, a rheology test was carried out on the rubber compound being used. Next, the rubber compound was mixed with the WRTC using a two mill roll machine, as shown in Figure 3(a). Following that, the vulcanized rubber compound was inserted into the mold for casting to form a bowl. The natural sand, iron sand, or silica sand were then poured into the bowl as much as a third of the specimen's height, as shown in Figure 3(b) (the sand was wrapped with the rubber compound in a proportion of 3:4 by volume). The next step was to carry out the vulcanization process of the rubber compound using a press machine, as shown in Figure 3(c), with a temperature of 130oC and an optimum time (tc90) of 17 minutes 30 seconds.

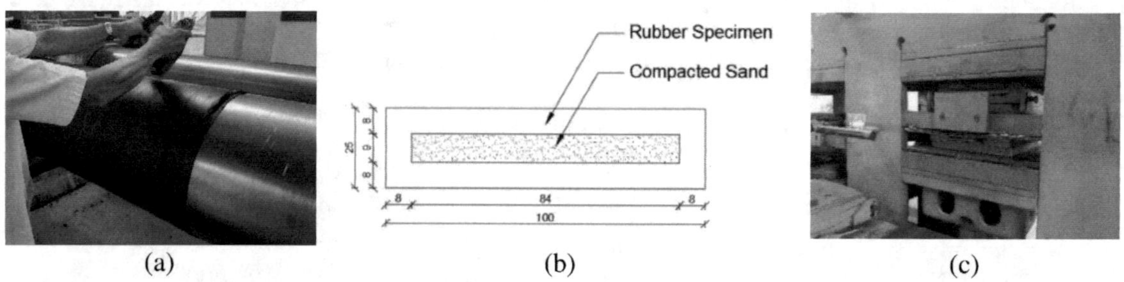

Figure 3. (a) Two roll mills, (b) test specimen with compacted sand, (c) molding press machine.

2.4. Test specimens

In this study, the hardness test (shore A) was used for the test of specimens having a size of 100 mm x 25 mm x 10 mm. The tensile strength test was performed on dumbbell-shaped specimens, as shown in Figure 4(a). Damping ratio test was carried out on two sets of rubber specimens in the form of a square block of 100 x 100 mm^2 base and 25 mm thickness, where the two sets of rubber were glued to three pieces of steel plate as shown in Figure 4(b).

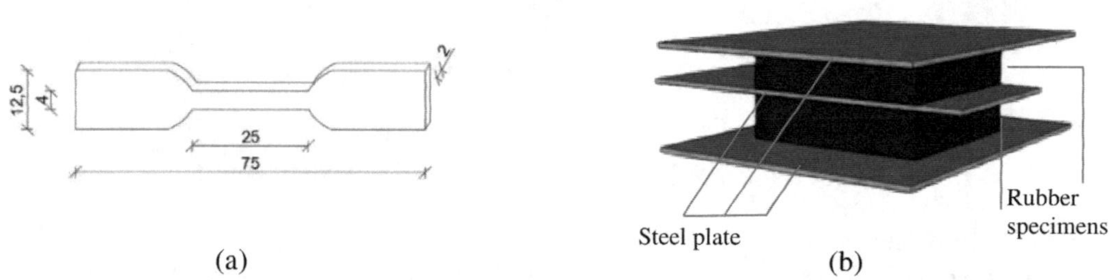

Figure 4. (a) Specimen for tensile strength test, (b) specimen for damping test.

IOP Conf. Series: Earth and Environmental Science 1244 (2023) 012005 doi:10.1088/1755-1315/1244/1/012005

3. Testing procedure

3.1. Hardness (shore A)

There are two types of rubber hardness tests: the Durometer test and the International Rubber Hardness test [13]. In this study, the Durometer test was used to measure rubber hardness parameters. There are several types of Durometer tests, namely A, B, C, D, DO, O, and OO scales. In this test, a durometer with a scale of A was used, as shown in Figure 5(a). The ISO 7619-1:2010 standard was used for the hardness test.

3.2. Tensile test

The tensile strength test followed the ISO 37:2011 standard. This test was carried out at the same time as determining the modulus of elasticity of the compound used, namely the fully natural rubber compound (NRC) and the blended NRC with 25% volume of the WRTC. The test was carried out at the Physical Laboratory of the Central Rubber, Leather, and Plastics Centre located in the Special Region of Yogyakarta using tools as shown in Figure 5(b). Equations 1 and 2 show the equations used to calculate the stress and modulus of elasticity of rubber according to [14].

$$\sigma = \frac{P}{A_0} \tag{1}$$

$$E = \frac{\sigma}{\varepsilon} \tag{2}$$

where, σ: the tensile stress (MPa), P: the maximum load (N), A_0: the tensile area (mm^2), E: the modulus of elasticity (MPa), ε: the tensile strain.

3.3. Damping test

Damping properties test was carried out on the hybrid viscoelastic damper specimens in five variations of the rubber mixture, namely R1 to R5, with two specimens for each variation. The damping properties test was carried out at the Structure Laboratory of the Department of Civil and Environmental Engineering, Faculty of Engineering, Gadjah Mada University. The test was carried out by running a sinusoidal signal on a shaking table with a frequency of 0.5 Hz with various lateral deformations i.e., 4 mm, 6 mm, 8 mm, 10 mm, and 15 mm. The specimen was positioned as shown in Figure 5(c) prior to the damping test. The damping property in this study was calculated based on the hysteresis loop method using equation 3.

$$D = \frac{1}{4\pi}\frac{W_D}{W_S} \tag{3}$$

where, D: the damping ratio, W_D: the area of the hysteresis loop, W_S: the maximum strain energy stored.

The value of energy dissipation (the area inside the hysteresis loop) is proportional to the velocity value [15]. However, since the test was carried out at the same frequency of 0.5 Hz, this study may use the load vs. lateral deformation relationship in calculating the amount of energy dissipation.

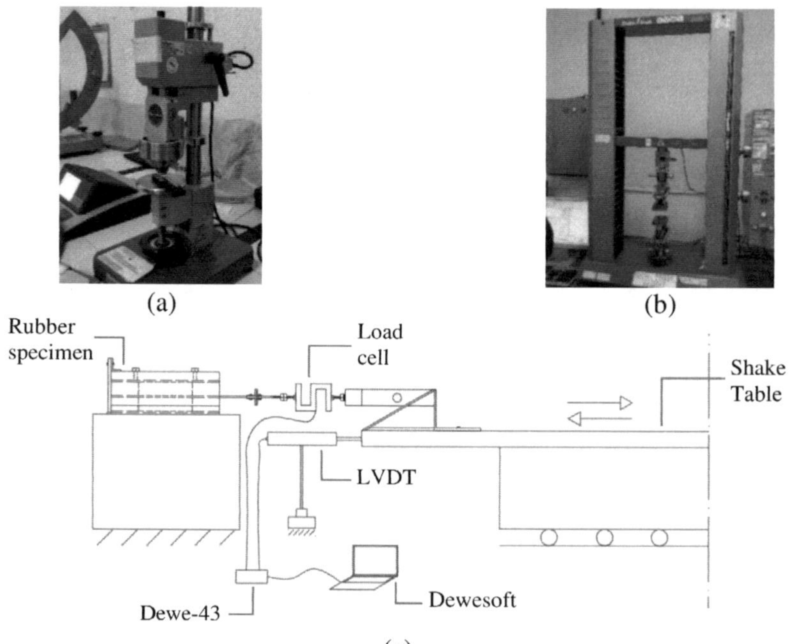

Figure 5. (a) Durometer test machine, (b) tensile strength test machine, (c) damping test configuration.

4. Results and discussion

4.1. Hardness (shore A)

The hardness test was carried out on two test specimens i.e., fully natural rubber compound (NRC), coded R1, and blended NRC with 25% by volume of the WRTC, namely the MRTC (R2). The results shown in Figure 6 prove that the addition of the WRTC results in an increase in rubber hardness. The hardness test results were 49.18% for natural rubber compound and 54.62% for a mixture of the rubber compound with 25% WRTC. There is an increase in hardness of 11.06% in the MRTC (R2) specimen. These results are consistent with previous research that found that adding waste of rubber tire crumb as an inactive filler increased the hardness of rubber. An inactive filler will cause an increase in the hardness of the rubber [8].

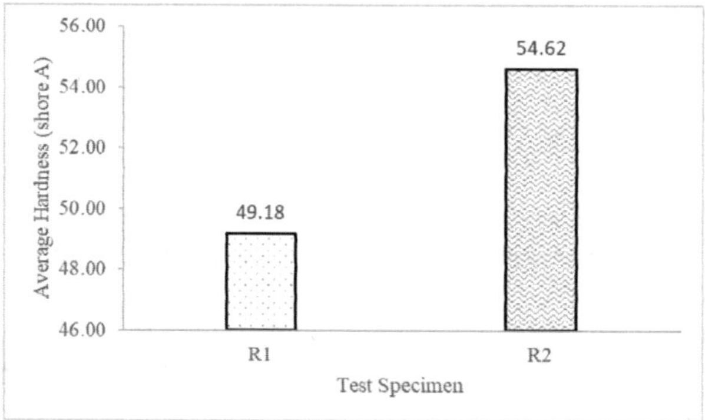

Figure 6. Hardness test results (R1: NRC, R2: MRTC).

4.2. Tensile properties

The tensile strength test in this study, like the hardness test, was performed on two test specimens, namely, fully natural rubber compound (NRC), coded R1, and blended NRC with 25% by volume of the WRTC, namely the MRTC (R2). The addition of WRTC caused a decrease in the tensile strength of the rubber, as shown in Figure 7(a). The tensile strength for the NRC (R1) is 11.02 MPa, and the tensile strength for the MRTC (R2) is 8.93 MPa. The addition of the WRTC in the rubber compound reduces the tensile strength by 18.99%. The addition of the WRTC also caused a decrease in the modulus of elasticity of rubber, as seen in Figure 7(b). The modulus of elasticity for the NRC is 2.81 MPa, while in the MRTC it is 2.67 MPa. The addition of the WRTC results in a decrease of 4.99% in the modulus of elasticity of rubber. This is also because the WRTC used in this study was an inactive filler. An inactive filler will cause an increase in the hardness of the rubber, but it will decrease the tensile strength of the rubber [8]. This finding is consistent with previous research that found tensile strength to be inversely proportional to rubber hardness [9].

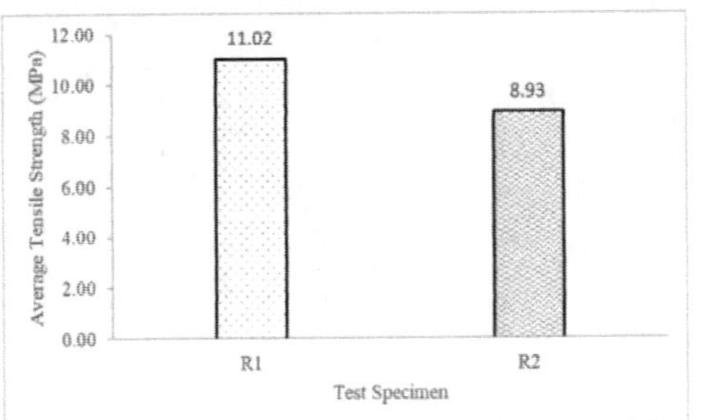

(a)

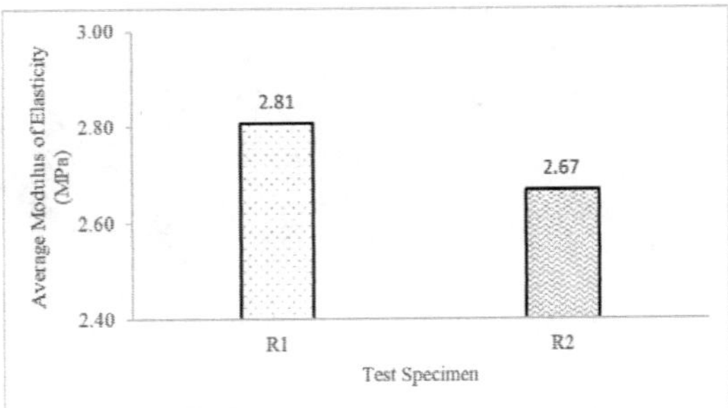

(b)

Figure 7. (a) Tensile strength test results, (b) modulus of elasticity test results (R1: NRC, R2: MRTC).

4.3. Damping properties

From the damping test, the area of the hysteresis loop is getting larger as the deformation increases during the test. This can be seen in the Figure 8(a) for the NRC specimen (R1) and Figure 8(b) for the MRTC specimen (R2). The area of the hysteresis loop for the NRC (R1) specimen is similar to the MRTC specimen. It demonstrates that the addition of WRTC has no significant effect on the ability of rubber to dissipate energy. The area of the hysteresis loop is larger in the MRTC specimen with the addition of natural sand (R3) in Figure 9(a), the MRTC specimen with the addition of iron sand (R4) in Figure 9(b), and the MRTC specimen with the addition of silica sand (R5) in Figure 9(c). This shows that the addition of sand has an effect on dissipating energy in the test specimen.

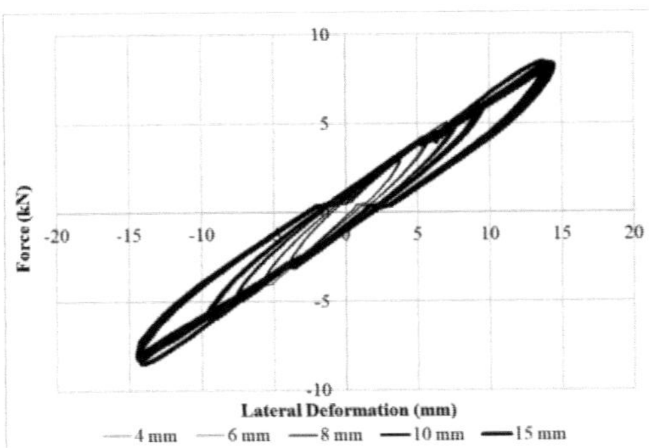

(a)

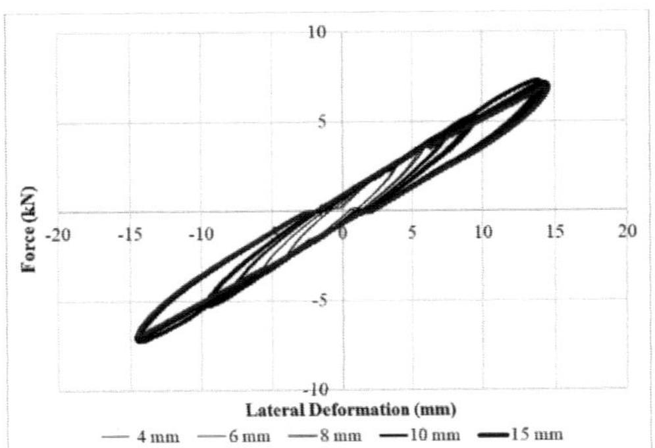

(b)

Figure 8. (a) Hysteresis loop for R1 specimen, (b) hysteresis loop for R2 specimen (R1: NRC, R2: MRTC).

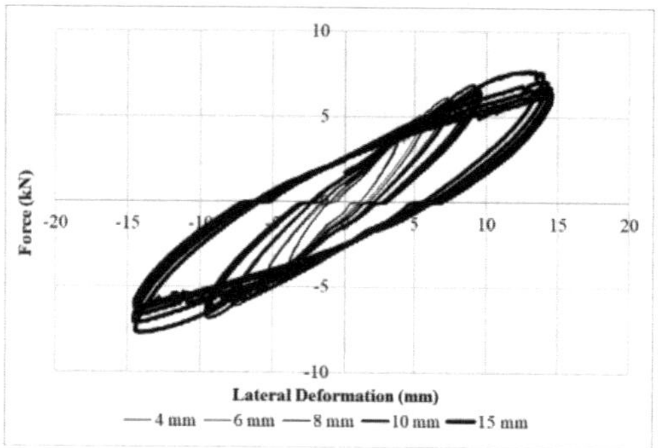

(a)

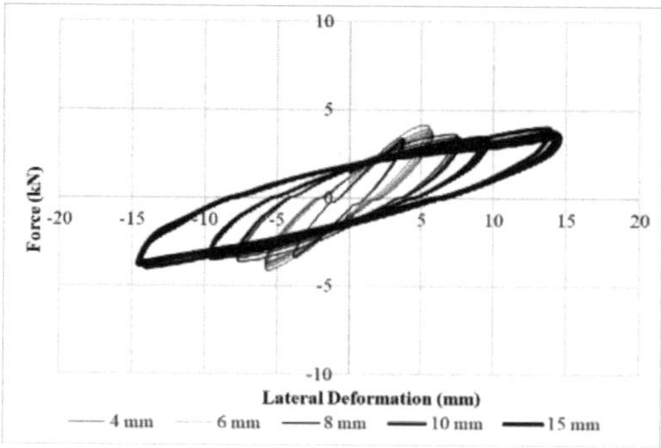

(b)

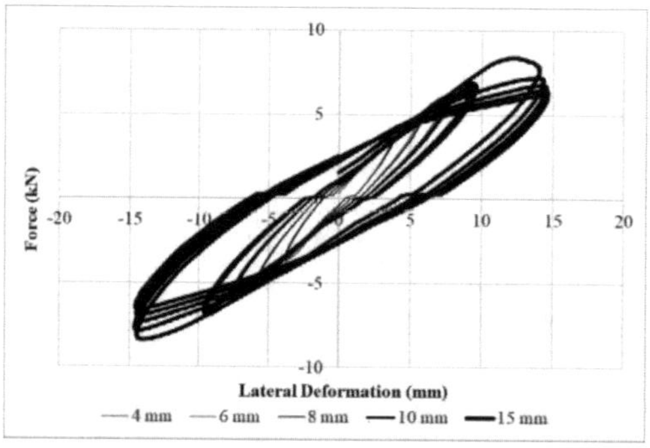

(c)

Figure 9. (a) Hysteresis loop for R3 specimen, (b) hysteresis loop for R4 specimen, (c) hysteresis loop for R5 specimen (R3: MRTC with natural sand, R4: MRTC with iron sand, R5: MRTC with silica sand).

After calculating the damping ratio, the results show that there is a decrease in the damping ratio as the lateral deformation increases from 4 mm to 15 mm, as shown in Figure 10 for the NRC (R1) specimen and the MRTC (R2) specimen. The highest damping ratio is indicated at a lateral deformation of 4 mm and gradually decreases at a lateral deformation of 15 mm. This is due to the bonding between the rubber and the steel plates being damaged (peeled off at the edge) as the applied load rises.

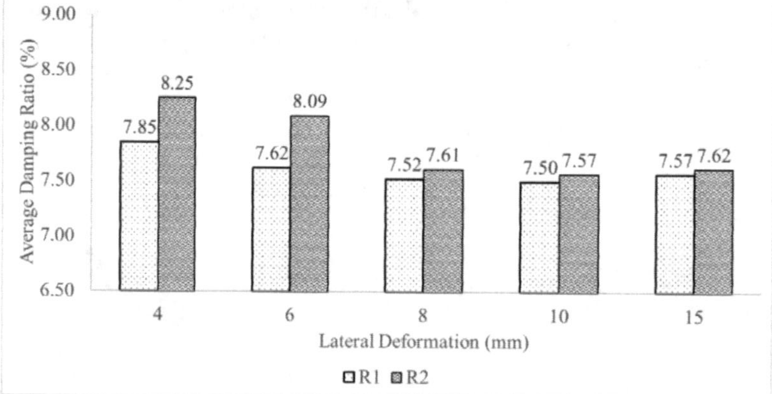

Figure 10. Damping properties of the NRC specimens (R1) and MRTC specimens (R2).

From the test results, it could be seen that there is a significant increase in the damping ratio of the specimens with a mixture of the MRTC with compacted natural sand (R3), iron sand (R4), or silica sand (R5), as shown in Figure 11. It shows that the addition of sand in the test specimen influences increasing the damping ratio of rubber. At the lateral deformation of 15 mm, the average damping ratio from the two specimens with a mixture of the MRTC with compacted natural sand (R3), iron sand (R4), or silica sand (R5) is respectively 16.92%, 23.23%, and 18.29%, while for the specimens with natural rubber compound (R1), it is 7.57%. This shows that the damping ratio increases with the addition of compacted natural sand, iron sand, or silica sand: 123.37%, 206.73%, or 141.52%, respectively.

As shown in Figure 11, the test specimens with the largest to smallest damping ratio were the R4, R3, and R5 specimens. This is caused by the difference in fineness modulus for natural sand, iron sand, and silica sand, where iron sand has the smallest fineness modulus, followed by silica sand and natural sand. The smaller the fineness modulus of sand, the greater the friction that happens because the frictional surface area between the sand grains is bigger.

The optimum damping ratio for specimens made with the MRTC and the compacted iron sand (R4) is obtained at a lateral deformation of 10 mm, which was 23.48%, but began to show a decrease at a lateral deformation of 15 mm. Meanwhile, on the test specimens of the MRTC with the addition of the compacted natural sand (R3) and silica sand (R5), it shows that the damping ratio is still increasing until the lateral deformation approaches 15 mm, as shown in Figure 11. The decrease of damping ratio in the R4 specimen was caused by the adhesive between the rubber test object and the iron plate starting to come off.

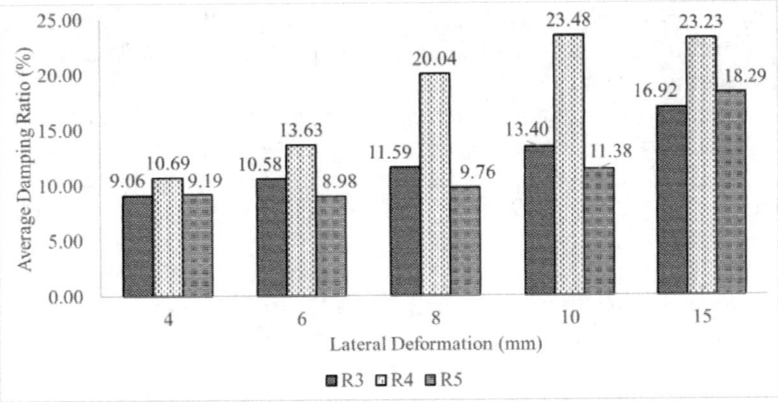

Figure 11. Damping properties of MRTC specimens with compacted natural sand (R3), compacted iron sand (R4), or compacted silica sand (R5).

5. Conclusion

1. The addition of 25% WRTC affects the mechanical properties of the rubber in the form of a decrease in the tensile strength and modulus of elasticity of the rubber, but an increase in the rubber hardness and damping ratio, although it is not significant.
2. Since the damping properties of the NRC and MRTC are not significant, but the use of WRTC will reduce waste tires and solve the environmental problem.
3. The addition of sand to form a hybrid viscoelastic damper (HVD) can increase the damping ratio of rubber. The bigger the lateral deformation at the time of testing, the greater the value of damping ratio especially for the specimen of MRTC with addition of sand.
4. At the lateral deformation of 15 mm, the addition of natural sand, iron sand, or silica sand increased the damping ratio by 123.37%, 206.73%, and 141.52%, respectively, in comparison to the NRC specimen.

Acknowledgments
We would like to thank the Gadjah Mada University Research Directorate for supporting this research through the Final Project Recognition (RTA) program. Thank you is also addressed to the workshop of the Polytechnic ATK Yogyakarta for preparing the test specimens, as well as the Structure Laboratory of Gadjah Mada University and the Physical Laboratory of the Rubber, Leather, and Plastics Centre Special Region of Yogyakarta in the process of testing the specimens.

References

[1] Ezzaki N, Stoica D, Rece L, Legendi A 2019 The use of damper systems in buildings with reinforced concrete structures *Proc. Int. Conf. on Innovations in Science and Education* vol 7 (Prague/Central Bohemia University) p 912

[2] Whittle J K, Williams M S, Karavasilis T L, Blakeborough A 2011 A comparison of viscous damper placement methods for improving seismic building design *J. Earthq. Eng.* **16** 540

[3] Heysami A 2015 Types of dampers and their seismic performance during an earthquake *Curr. World Environment* **10** 1002

[4] Faizah R 2021 Perilaku dinamik dinding pasangan bata in-plane menggunakan spesi mortar serutan karet ban bekas *Dissertation* Gadjah Mada University, Yogyakarta

[5] Siddique R, Naik T R 2004 Properties of concrete containing scrap-tire rubber-an overview *Waste Management* **24** 563

[6] Putra E G, Priyosulistyo H, Darmayadi D 2021 Mechanical and damping properties of composite rubber with waste crumb rubber tire and silica sand *J. Phys: Conf. Ser.* **1858** 012071

[7] Tirtoputro S, Priyosulistyo H, Faizah R 2021 Mechanical and dynamic properties of composite rubber (rss and sbr) with waste rubber crumb and copper fiber *J. Phys: Conf. Ser.* **1858** 012072

[8] Suharto A, Sedyono J 2021 Sifat mekanis pada kompon karet alam variasi campuran serbuk ban bekas untuk aplikasi bantalan mesin *Creative Research in Engineering* **1** 24

[9] Nasruddin, Susanto T 2020 Study of the mechanical properties of natural rubber composites with synthetic rubber using used cooking oil as a softener *Indones. J. Chem.* **205** 967

[10] Christianto D, Lase Y, Yeospitta 2013 Pengaruh pasir terhadap peningkatan rasio redaman pada perangkat kontrol pasif *Proc. Konferensi Nasional Teknik Sipil* vol 1 (Surakarta/Universitas Sebelas Maret) p S-315

[11] Khimi R 2015 Development of elastomeric composites from iron sand and natural rubber for vibration damping *PhD Thesis* University of Waikato, Hamilton

[12] Forest, M 2014 *Recycling and Re-use of Waste Rubber*, ed E Garmson and H Chavaroche (Shropshire: Smithers Rapra Technology Ltd) p 139

[13] Gent A N 2012 *Engineering with Rubber: How to Design Rubber Components, 3rd Edition* (Munich: Hanser Publisher) p 375

[14] Stanton J F, Roeder C W 1982 *Elastomeric Bearing Design, Construction, and Materials* (Seattle: Transportation Research Board, National Academy of Sciences)

[15] Priyosulistyo H 2022 *Analisis Dinamika Struktur dan Aplikasinya*, ed H Pras (Yogyakarta: UGM Press) p 24

The influence of reverse loading sequence on the behaviour of beam-column joints

Purwanto[1], AyLie Han[1] and B H W Hadikusomo[2]

[1]Department of Civil Engineering, Universitas Diponegoro, Semarang, 50275, Indonesia
[2]Department of Civil Engineering, Asian Institute of Technology, Pathum Thani, 12120, Thailand

Corresponding author's e-mail: purwatrend@lecturer.undip.ac.id

Abstract. The formation of plastic joints in the reinforced beam-to-column joint is influenced by a number of factors. The behaviour and location of this joint is of major importance since it defines the overall performance of the rigid framework when subjected to loading. Mandated by the majority of codes and standards, the strong-column weak-beam philosophy is used and recognized by the formation of this joint in the beam. This study looked into the effect of reverse loading to the behaviour of plastic joints as compared to a monotonic loading protocol. The study analysed the load-deformation behaviour and ultimate load carrying capacity while additional information is collected through visual observation on the failure and cracking pattern. The study concluded that the reverse loading negatively influenced the load carrying capacity, but negligibly affected the overall behaviour of the joint.

1. Introduction

The reinforced concrete frame structure is a system widely used for multi-story buildings yet its ductility is lower than steel frame structures [1]. One of the most important components in the moment-resisting frame structure is the beam-column joint. The beam-to-column joints which not strong enough may cause the collapse of the entire structural system [2]. Collapse at beam-column joints becomes the main focus, so the design for beam-column joints needs to be carefully designed to meet the capacity and ductility requirements [3].

Earthquake is one of the major causes of structural damage or even lead to structural failure. Indonesia is an earthquake-prone area so earthquake regulations must be updated due to the development of the earthquake phenomenon. The change in the earthquake zone map is clear evidence that adjustments to the regulations of earthquake loading on building structures need to be made. One of the requirements of earthquake-resistant structure is ductility because the design criteria for earthquake-resistant building structures require that building structures must be designed to be able to withstand earthquake load with a return period of 2500 years, this is required by the Indonesian earthquake standard, namely SNI 1726-2019 [4, 5]. One of the loads that represent the earthquake force is a load with two opposite directions.

Many studies have been discussed about beam-column joint behaviour. Thurston [6] stated that the most vulnerable element in the concrete system structure is the beam-column joint, especially when an earthquake occurs. Beam-column joint needs to be designed to meet the strength and ductility requirements so that the beam-column joint will not fail. Adnan *et al.* [7] did an assessment of the performance of beam-column joints in a building structure under earthquake load. Tang *et al.* [8] discussed modification of beam-column joint types namely square concrete-filled steel tubular (CFST) to overcome the decrease in beam-column joint capacity due to earthquake load. [9] discussed the prediction of shear strength in beam-column joints, especially in areas experiencing plastic joints.

Studies were held by [6], [7], [8], and [9] in observing the behaviour of beam-column joints will be used as the guidelines of the earthquake regulations to design a building structure. Certain elements of the building structure are allowed to undergo plastification as a means for dissipation of earthquake energy received by the structure, however, the structure must not collapse [10]. As a result, energy dissipation can occur effectively in certain elements, that are capable to undergo large deformations before they collapse. Energy dissipation occurs when the structural element undergoes a plastic joint, where the location of the plastic joint must be designed at a certain location so that the damage to the element does not result in the collapse of the overall structure.

Reinforced concrete requires to be ductile enough to meet the requirements stated by the regulations. Ductility is defined as the ability to deform before it collapses [3, 11]. To ensure the ductility requirement, the column needs to be designed stronger than the beam. The failure of the beam is expected to be a flexural failure because the flexural load will lead to large deformation of the beam and ensure the ductility requirements [5, 12]. Shear failure is not expected to occur on the beams because it will be brittle failure.

The beam-column joint in the moment-resisting-frame structure is the most critical zone because there is a combination of maximum flexural and shear stress. The basic concept of strong column weak beam is the column design must have a higher capacity than the beam because plastic joints will form on the beam in front of the column (Figure 1) so that the process of yielding and the influence of Poisson's ratio on the tensile reinforcement results in the debonding.

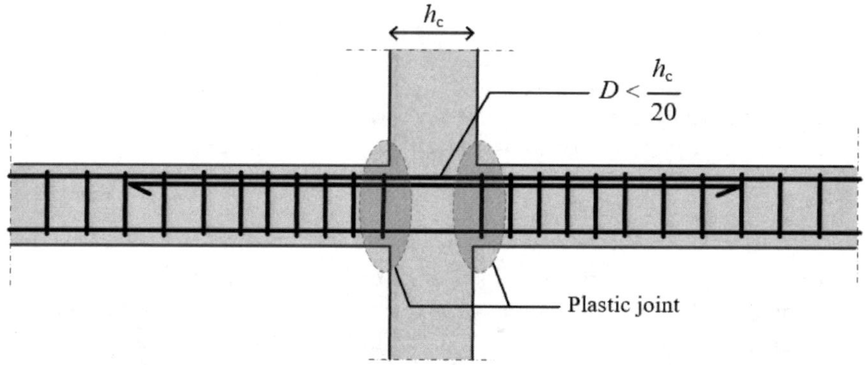

Figure 1. Plastic joint area in front of the column [13].

The beam is chosen as the location for energy dissipation in the Special Moment-Resisting Frame System (SMRF). The beam is chosen because the plastification process that occurs in the beam results in a more ductile failure than the failure in the column, this is the concept of strong column weak beam. If the earthquake load exceeds the design load, then the structure will respond with the formation of a plastic joint that occurs at the end of the beam and the lowermost column near the foundation (beam sway) seen in figure 2(a) and the behaviour of soft storey should be avoided seen in figure 2(b).

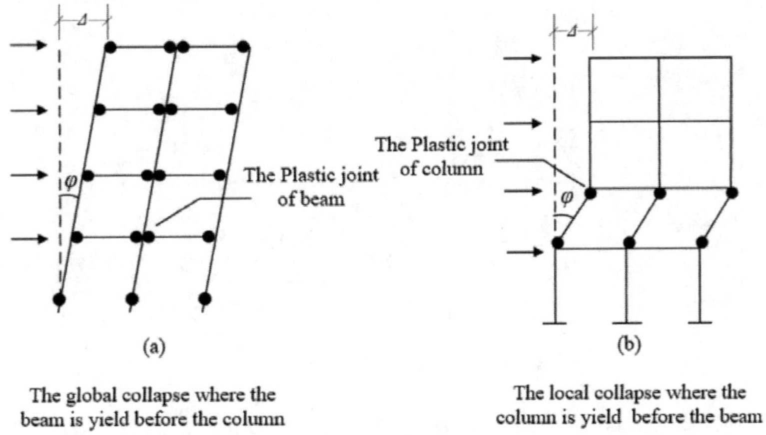

Figure 2. Reinforced concrete structure frame plastification mechanism: (a) beam sway (b) soft storey [14].

This study continues the work of [13] with a numerical approach instead of the experimental approach. Purwanto [13] used a beam that represents the beam-column joint and monotonically loaded the beam with the direction of gravity. To simulate the earthquake load, this study uses reverse loading. Reverse loading is a load model with an opposite direction of gravity. This study uses a numerical model to simulate the beam using DIANA. The experimental data from [13] is used to compare the effect of load direction. The output of this study is the comparison of the capacity, plastic joint behaviour and crack pattern of the beam due to experimental monotonic loading and reverse loading. The significance of this study is to correctly analyze the contribution of compression reinforcement when reverse loading is applied on the beam using numerical approach.

2. Methodology

2.1. Material properties

The material properties used in this study are assumed to be the same as [13]. The concrete compressive strength at the age of 28 days is 31 MPa and the modulus of elasticity is 24559 MPa. The tensile strength of concrete is 2.072 MPa. FIB 2010 is used to model the compression curve of the concrete, which is calculated using equation (1)-(3) [15].

$$\sigma = \left(\frac{k \cdot \eta - \eta^2}{1 + (k-2) \cdot \eta} \right) \cdot f'_c \tag{1}$$

$$\eta = \frac{\varepsilon_c}{\varepsilon_{c1}} \tag{2}$$

$$k = \frac{E_c}{E_{c1}} \tag{3}$$

σ is the stress of concrete at arbitrary strain (MPa), ε_c is the strain of concrete, ε_{c1} is the strain at f'_c, E_c is the elastic modulus of concrete (MPa) and E_{c1} is the secant elastic modulus at f'_c (MPa). The tensile curve is assumed to be linear until the tensile strength of concrete (f_t) is achieved and followed by a linear softening branch. The constitutive model of concrete is shown in figure 3.

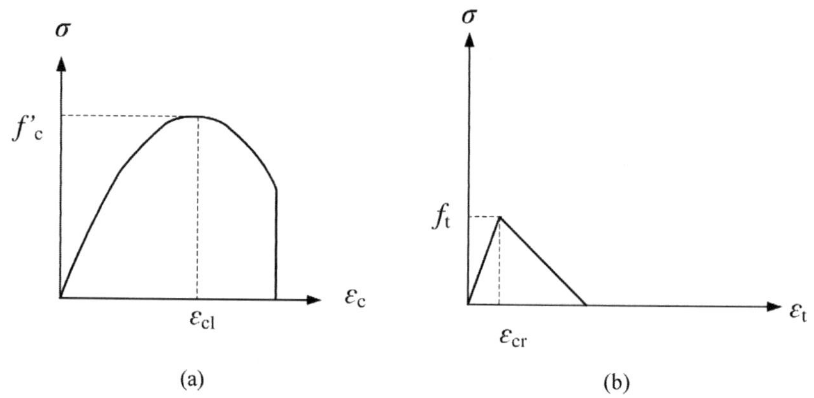

Figure 3. Constitutive model of concrete (a) compression (b) tensile.

The flexural and the shear reinforcement diameter are 13 and 8 mm, respectively. The tensile strength of the reinforcement (f_y) is 360 MPa and the ultimate strength (f_u) is 460 MPa. The modulus of elasticity of the reinforcement is 180000 MPa. Bi-linear curve is assumed for the reinforcement constitutive material to simulate the strain hardening of the steel (Figure 4).

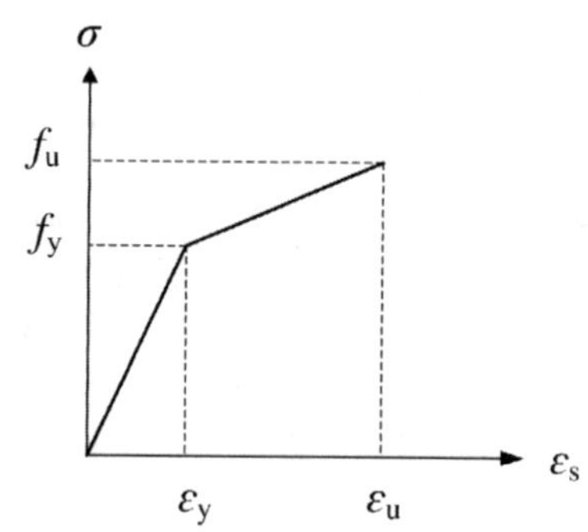

Figure 4. Constitutive model of reinforcement.

2.2. Geometry and loading setup of the specimen

The dimension of the specimen is 3800 mm in length and 3400 mm in beam span. The cross-section's dimension is 170 mm in width and 340 mm in height. The support used on this beam is a hinge and roll. The beam is designed to be in under reinforced condition. The configuration of the reinforcement used for experimental monotonic loading consist of 4 tensile and 2 compression reinforcement. As for reverse loading, it uses the opposite configuration. The dimension's detail and reinforcement configuration can be seen in figure 5.

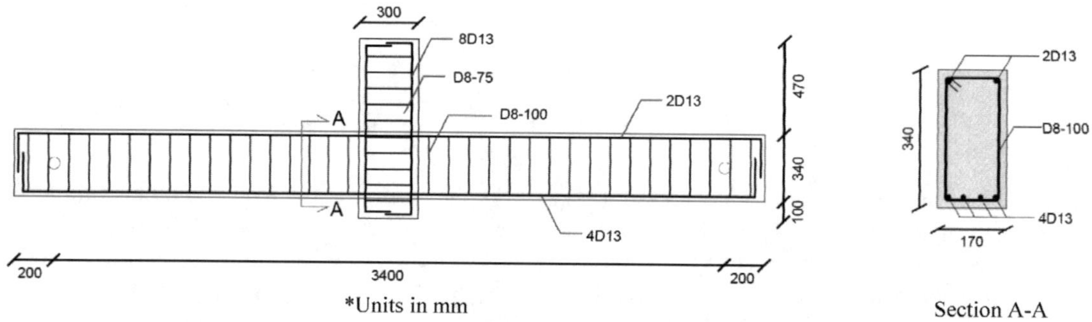

Figure 5. Specimen geometry and reinforcement configuration [13].

The internal forces diagram of the actual beam-column joint is shown in figure 6(a), however, because of the laboratory limitation, the actual geometry of the beam-column joint can not be tested in the laboratory. The specimen needs to be adjusted so that specimen can be tested at the laboratory but still retain the actual behaviour of the structure, hence [13] used the setup shown in figure 6(b) to test the specimen. In order to understand the effect of reverse loading on the specimen from [13], this study applied a load with the opposite direction of gravity as can be seen on the setup (Figure 6(c)).

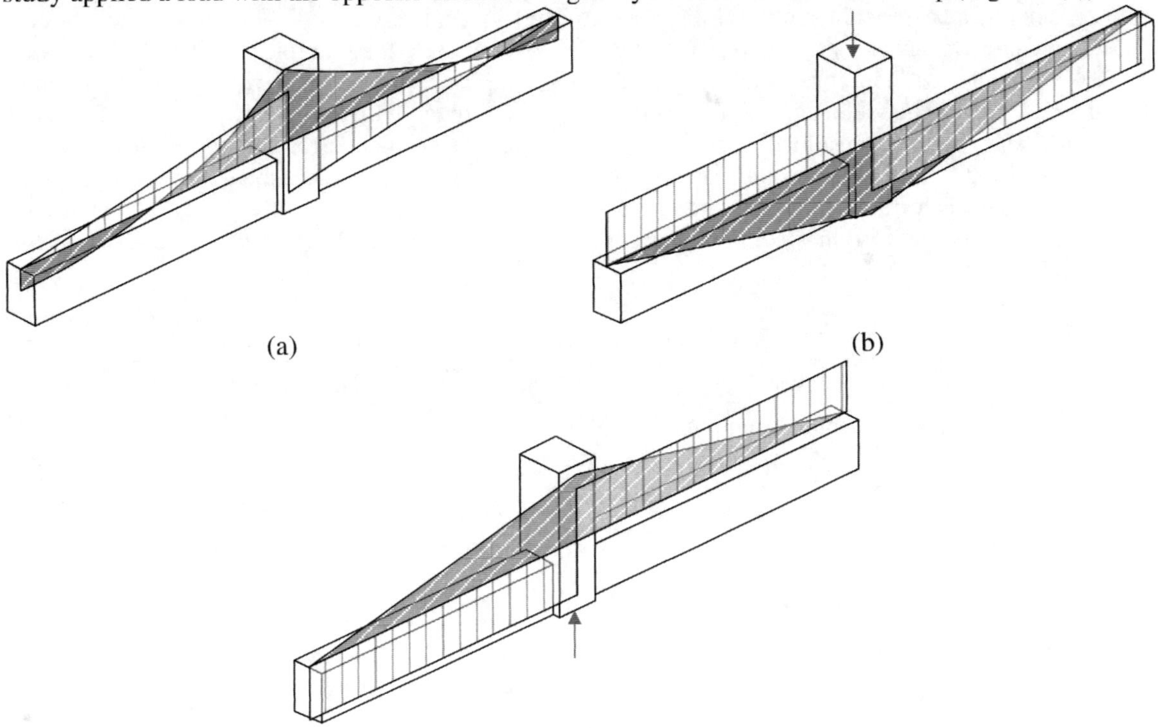

(a) (b) (c)

Figure 6. Internal forces diagram of (a) actual beam-column joint (b) experimental setup of [13] (c) reverse loading setup.

Deformation control is used as the loading type of the model in DIANA. The direction of the load is reversed from the experimental work of [13] (the opposite direction of gravity). The rate of the load is 0.2 mm/step similar to the experimental.

2.3. Method of analysis

The theoretical analysis by [3] is used to calculate the capacity of the beam. To ensure the theoretical analysis, the experimental result (4 tensile reinforcements) of the beam is used to verify the theoretical calculation. The validated theoretical calculation used to calculate the reverse loading beam (2 tensile reinforcements), the result will be compared with the numerical result to verify the model. The validated model will be used to obtain responses of the reverse loading beam. The data used for the analysis are capacity, tensile reinforcement strain and the crack pattern. Those data will be compared with the experimental results from [13] to understand the effect of reverse loading. Theoretical analysis used to calculate the yield capacity of the beam can be calculated using equation (4).

$$P = \frac{4\left[C_c.\left(d - \frac{c}{3}\right) + C_s.\left(d - d'\right)\right]}{L} \tag{4}$$

P is the capacity of the beam (kN), C_c is the resultant force of concrete in section (kN), C_s is the resultant force of compressive reinforcement section (kN), d is the effective depth of the section (mm), *d'* is the distance from extreme compression fiber to centroid of compression reinforcement (mm), *c* is neutral axis distance from compression fiber (mm) and *L* is beam span (mm).

3. Result and discussion

3.1. Monotonic and reverse loading beam capacity

The capacity of a beam is the ability of the beam to carry loads due to reverse and monotonic loading. Based on data from previous research [13] experimental monotonic loading for the yield condition was 56.724 kN while for theoretical analysis was 58.393 kN. The value of data error due to monotonic loading is 2.942%. The yield capacity of reverse loading simulated with DIANA was 31.53 kN and the theoretical analysis of reverse loading was 34.772 kN. The value of data error due to reverse loading is 10.282 %. From the results of the capacity both experimental test and model have proven that theoretical analysis results can be used due to the value of data error is relatively low. The decrease of the beam capacity due to reinforcement configuration difference is about 38.699 %. The result of the analysis can be seen in table 1.

Table 1. Beam capacity at yield condition.

No.	Type	Yield capacity (kN)	Displacement (mm)
1	Experimental monotonic	56.724	125.400
2	Theoretical monotonic	58.393	-
3	Reverse loading model	34.772	120.751
4	Theoretical reverse loading	31.530	-

The load-displacement of the experimental from [13] and reverse model can be seen at figure 7. Both of the results show similar pattern with linear behaviour until it reached yield condition. Post yield shows strain hardening on experimental [13] and reverse model simulates the strain hardening using the bi-linear curve.

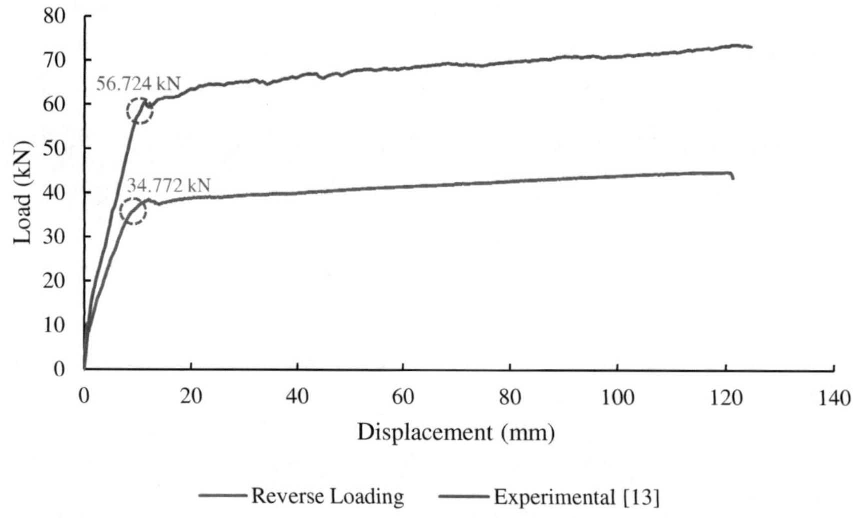

Figure 7. Load-displacement relationship of the beam.

3.2. Plastic joint behaviour

The plastic joint behaviour of the beam can be measured by the value of the strain from tensile reinforcement. The plastic joint's location can be defined from the location of tensile reinforcement which reaches the yield strain (0.002) at the earliest. The strain value for the reverse loading model of 0.00206 was obtained at the load of 34.772 kN, while the strain value for monotonic loading of 0.00201 was obtained at the load of 56.724 kN [13]. The strain distribution of the tensile reinforcement is shown in figure 8 and figure 9.

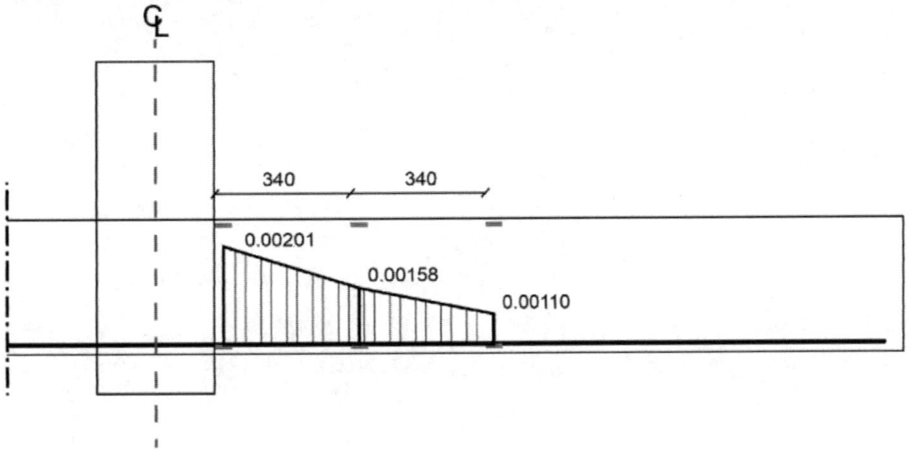

Figure 8. Strain distribution of experimental monotonic loading [13].

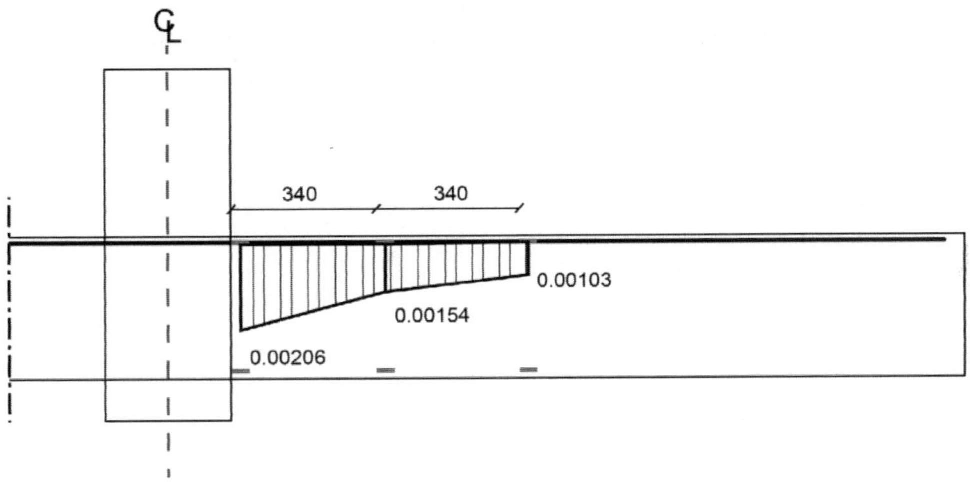

Figure 9. Strain distribution of reverse loading model.

Based on figure 8 and figure 9, the direction of the load does not affect the location of the plastic joint as can be seen by the strain value. The highest value of the strain distribution is located in front of the column in either the experimental or reverse loading model.

3.3. Crack pattern

The crack pattern can be used to understand the type of failure that occurs on the beam. The crack pattern observed from the experiment and model was taken at the yield and ultimate condition. The crack pattern is shown in figure 10 and figure 11.

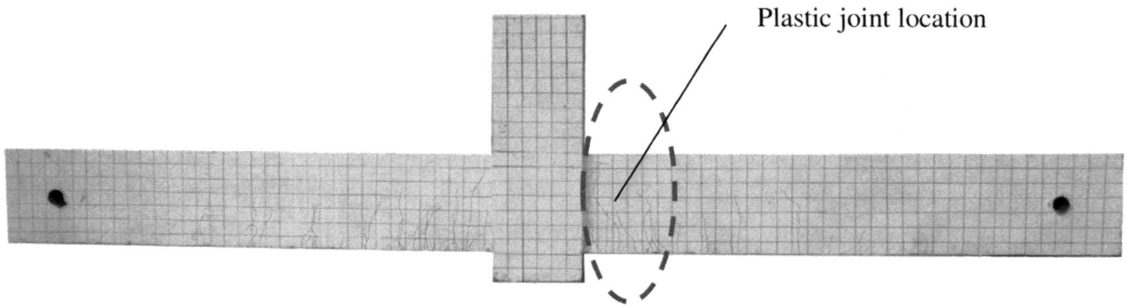

(a)

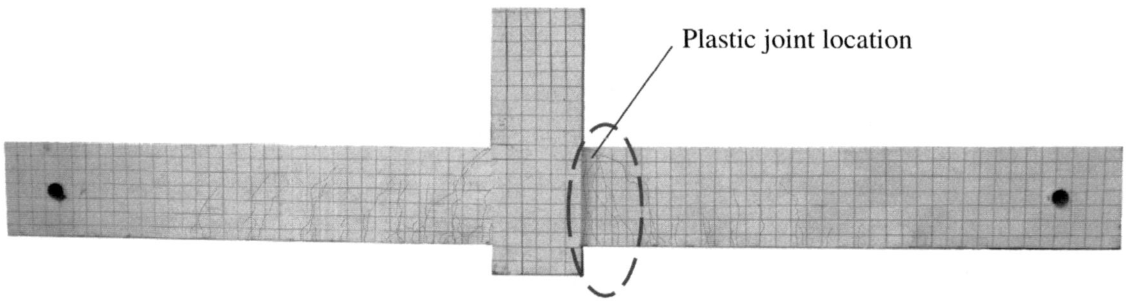

(b)

Figure 10. The crack pattern of experimental monotonic loading (a) yield (b) ultimate condition [13].

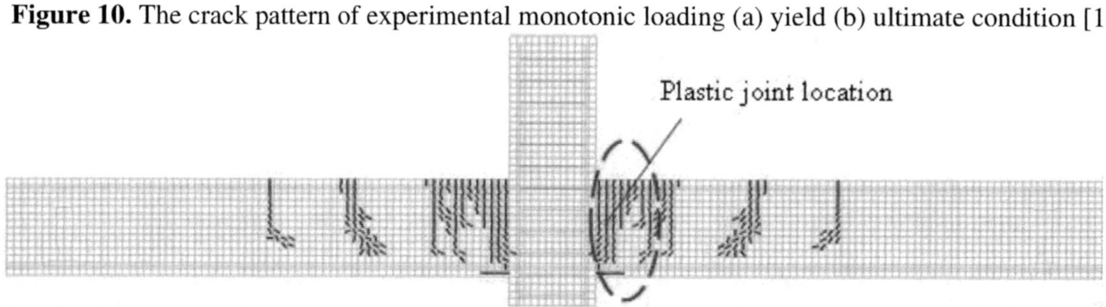

(a)

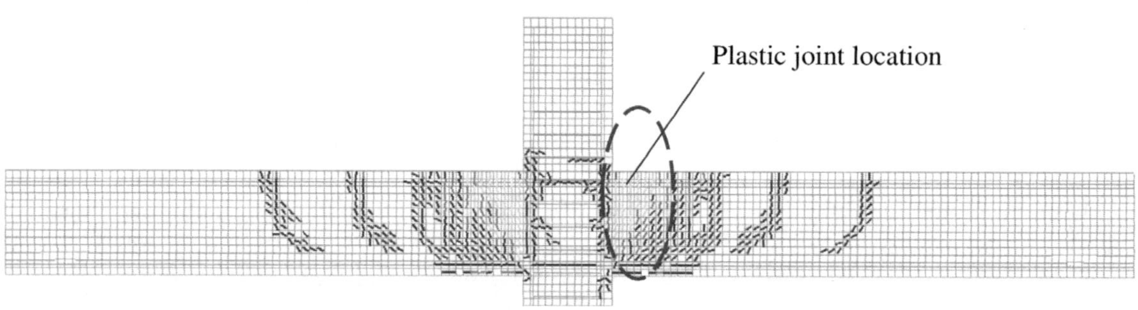

(b)

Figure 11. The crack pattern of reverse loading model (a) yield (b) ultimate condition.

Based on figure 10 and figure 11 show that the crack pattern at the yield and ultimate conditions were concentrated in front of the column. On the reverse loading model, the crack pattern does not significantly distinct from the experimental monotonic loading. The type of crack pattern near the column in reverse loading and monotonic loading is relatively perpendicular to the cross-section of the beam because the crack is dominated by flexural cracks due to the influence of the bending moment.

4. Conclusion

Based on the results of the experimental monotonic loading and reverse loading model, the conclusions were obtained as follows:

- The reverse loading model has a smaller capacity of 38.699 % than the experimental monotonic loading capacity in yield condition because the tensile reinforcements are less than the experimental monotonic.
- The location of the plastic joint at the reverse loading model and experimental monotonic loading is located at the same location and it is located in front of the column even though the direction is reversed.
- The crack pattern concentration that occurs in the reverse loading model and the experimental monotonic loading remains the same and it is located in front of the column both on reverse and experimental monotonic loading. The crack pattern is dominated by the flexural crack especially at the near column.

5. References

[1] Sudarsana I K, Susila I G A and Sastraningsih N W 2021 Perilaku seismik struktur rangka komposit RCS dengan variasi rigidity zone factor pertemuan balok – kolom *Media Komunikasi Teknik Sipil* **27** 41 – 50

[2] Yuliati N C E, Murni D S and Wisnumurni A W 2018 Comparative study of the behaviour of reinforced concrete beam-column joints concerning the monolithic and nonmonolithic connection *The 4th Int. Conf. on Rehabilitation and Maintenance in Civil Eng. (Solo Baru, Indonesia)* vol 195

[3] Park R and Paulay T 1975 *Reinforced Concrete Structures* (Canada: Wiley–Interscience)

[4] BSN 2019a *SNI 1726:2019 Tata Cara Perencanaan Ketahanan Gempa untuk Struktur Bangunan Gedung dan Non-Gedung* (Jakarta: Badan Standar Nasional)

[5] Budiono B, Dewi N T H, Kristalya M, Manik S L C and Ong E H K 2017 *Contoh Desain Bangunan Tahan Gempa dengan SRPMK dan Sistem Dinding Struktur Khusus di Jakarta* (Bandung: ITB Press)

[6] Thurston S J 1982 Cyclic Load Testing of Three Haunched Reinforced Concrete Beam-Column Assemblies *Bulletin of the New Zealand National Society for Earthquake Engineering* **15**

[7] Adnan M A, Tahir M M, Sam A R B M and Razak S M B S A 2015 Seismic Performance of Ductility Classes Medium RC Beam-column joints with Continuous Rectangular Spiral Transverse Reinforcements *Latin American Journal of Solids and Structures* **12** 787 – 807

[8] Tang X L, Cai J, Chen Q J, Liu X, and He A 2016 Seismic Behaviour of Through-Beam Connection Between Square CFST Columns and RC Beams *J. of Constructional Steel Research* **122** 151 – 66

[9] Tran M T, Bui Q B, Sentosa B, Nguyen N T T H, Duong T H and Ple O 2018 Sustainable RC Beam-column joints with Headed Bars: A Formula for Shear Strength Evaluation *Sustainability* **10** 1 – 14

[10] Arfiadi Y 2014 Pengaruh Penetapan SNI Gempa 2012 pada Desain Struktur Rangka Momen Beton Bertulang di Beberapa Kota di Indonesia *Seminar HAKI Indonesia Siaga Gempa (Medan, Indonesia)*

[11] MacGregor J G and Wight J K 2009 *Reinforced Concrete: Mechanics and Design (4th ed.)* (Prentice Hall)

[12] ACI 2014 *ACI 318-14: Building Code Requirements for Structural Concrete and Commentary* (ACI)

[13] Purwanto 2021 *Self Compacting Geopolymer Concrete (SCGC) Haunch Sebagai Perkuatan Balok Lentur di Zona Dekat Muka Kolom* (Universitas Diponegoro)

[14] Kam W Y, Akguzel U, Jury R and Pampanin S 2013 Displacement-based Seismic Assessment: Practical Considerations *2013 NZSEE Conf. (Wellington City, New Zealand)*
[15] FIB 2013 *fib Model Code for Concrete Structures 2010* (FIB)

Acknowledgments
This research was conducted with the support of the Diponegoro University, the World Class Research funding through the decreed nr. 11-20/UN7.6.1/PP/2021.

Shear Capacity of Hollow Reinforced Concrete Members without Stirrups

R Thamrin[1], SH Dewi[2*], A Asrianur[3], FP Yastari[4], Z Zaidir[1] and S Haris[1]

[1] Lecturer in Civil Engineering Department, Universitas Andalas, Padang, Indonesia
[2] Doctoral Student in Civil Engineering Department, Universitas Andalas, Padang, Indonesia
[2] Lecturer in Civil Engineering Department, Universitas Islam Riau, Pekanbaru, Indonesia
[3] Magister Student in Civil Engineering Department, Universitas Andalas, Padang, Indonesia
[4] Magister Student in Civil Engineering Department, Universitas Andalas, Padang, Indonesia

*Corresponding author: srihartatidewi@eng.uir.ac.id

Abstract. This paper presents experimental results on the shear capacity of hollow reinforced concrete members without stirrups. The area of the hollow was 2.32% of the concrete cross-sectional area. The hollow size used in this study is smaller than the maximum percentage required in the code, where the maximum hollow area was 4% of the total cross-sectional area. The longitudinal reinforcement used in the specimen varies with the three types of reinforcement ratios. The test was carried out using the four-point shear test to achieve the expected shear failure mode. The loading was given monotonically until the specimen failed. Analytical flexural capacity was also calculated using the fiber element method. The specimens failed in shear failure mode, indicated by large diagonal cracks in the observed shear span. The influence of the hollow on the shear strength was also compared with the theoretical shear capacity proposed by the code. This comparison shows that the calculated theoretical shear capacity is conservative for hollow reinforced concrete structural elements with a hollow area smaller than 4%. The results of the flexural analysis show that all the specimens did not reach the flexural capacity due to premature shear failure.

Keywords: *four-point shear test, hollow reinforced concrete member, shear capacity, shear failure, analytical flexural capacity*

1. INTRODUCTION

Circular sections are widely used in reinforced concrete structural members because they have many advantages. The circular cross-section has advantages: easy work, high moment of inertia, and identical strength characteristics in all directions in carrying lateral and vertical loads [1,2].

Generally, circular sections are used for structural elements that carry axial compression loads, such as columns or piles. In addition to carrying axial loads, the structure also carries large lateral loads such as wind, earthquake, and other lateral loads that cause the columns to carry shear loads. Therefore the adequate shear capacity of the column must be calculated carefully [3].

Several variables that affect the shear capacity of a reinforced concrete section are the compressive strength of the concrete, the cross-sectional area, and the amount of longitudinal and transversal reinforcement [4,5]. Therefore, several studies used these variables to investigate their effect on the shear capacity of reinforced concrete structural elements.

However, in construction applications, other variables affect the shear capacity of reinforced concrete elements, such as the presence of a hollow in the middle of the cross-section. These hollows are mainly used for electrical installations, plumbing, or other purposes. In addition, the cross-section of the column with a hollow adds to the efficiency of using concrete materials [6].

The previous standard (SNI 03-2847-2002 Chapter 8) requires that the hollows for drains and pipes embedded in the column do not exceed 4% of the cross-sectional area [7]. Furthermore, the newer code (SNI-2847-2013 Chapter 6) states the same [8]. However, the latest standard (SNI-2847-2019) does not regulate the maximum requirement of the hollow area [9].

Although the shear capacity of reinforced concrete structural elements with a circular cross-section has been studied by several previous researchers [10,11]. This topic is still quite interesting to study with the additional variable, the effect of hollow area on the shear capacity of reinforced concrete columns. Therefore, this study focused on the effect of hollows in the circular cross-section on the shear capacity of reinforced concrete structural elements. The contribution of concrete in carrying shear loads with a hollow cross-section was investigated by setting the specimen without shear reinforcement. The work conducted in this study was laboratory-based to determine the shear capacity of reinforced concrete structures with circular cross-sections with three reinforcement ratios and two different concrete cross-sectional areas.

2. Experimental Setup

This research has six reinforced concrete column specimens with a circular cross-section and dimensions of 2300 mm long and 250 mm diameter. The test variable consisted of the ratio of longitudinal reinforcement and cross-section type (with and without a hollow).

The diameter of the longitudinal reinforcement used is 13, 16, and 19 mm, while the hollow diameter is 1.5 inches. The pipe hollow causes a reduction in the cross-sectional area of concrete by 2.32% of the cross-sectional area of concrete. Fresh concrete with a concrete compressive strength of 40 MPa was ordered from a ready-mix concrete company. The beam part observed in this study is located in the center of the column with a length of 1000 mm. As shown in Figure 1, the zone under consideration is not provided with shear reinforcement to realize shear failure.

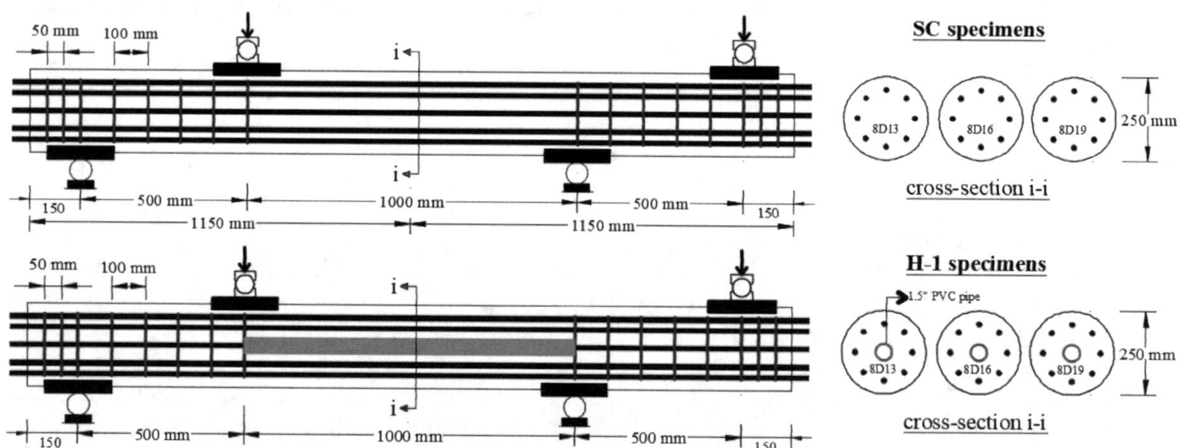

Figure 1 Specimens detail

In this paper, specimens without hollows use the SC code, and the hollow specimens are coded H. The hollow area in the H-1 specimen is below the maximum value required by the standard, which is 4%. Details of all specimens are shown in Table 1 below.

Table 1 Specimens data

Specimens data							Longitudinal reinforcement			
Specimens ID	fc' (MPa)	D_c (mm)	Ac (mm^2)	d (mm)	a (mm)	a/d	ϕ (mm)	N	f_y (MPa)	As (mm^2)
SC-01	40	250	50000	200	500	2.5	13	8	427	398
SC-02				200			16		492	603
SC-03				200			19		416	851
H-1-1			42380	200			13		427	398
H-1-2				200			16		492	603
H-1-3				200			19		416	851

The position of specimen, when the test was carried out, was placed horizontally with two concentrated loads and two supports, as shown in Figure 2. This loading position produces the desired combination of internal forces to cause shear failure. LVDTs and load cell were installed to measure deflection and load, which were connected to a data logger. Among the three different LVDT locations, the deflection data from LVDT 2 are plotted and reported in this paper.

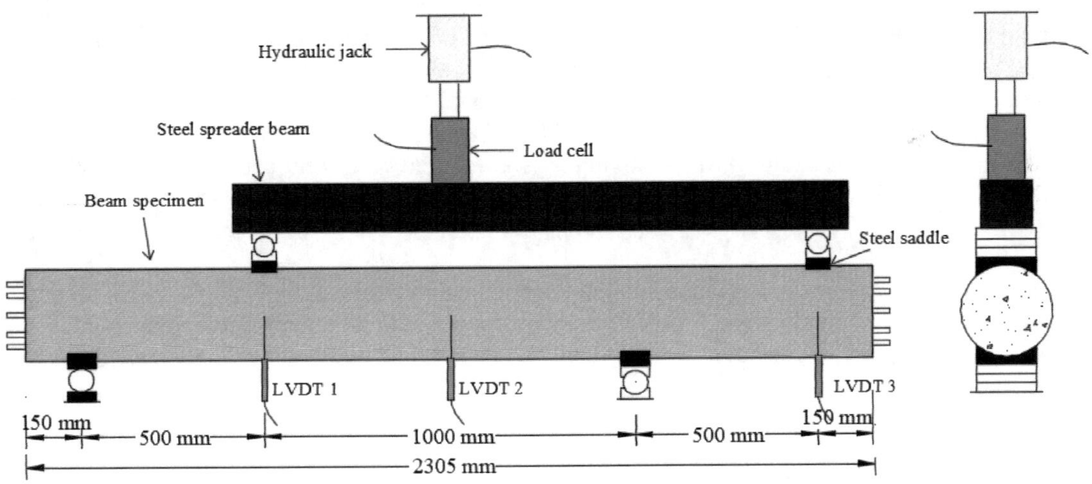

Figure 2 Schematic pictures of the tested specimen, dimension, and loading position

3. Analytical Study

The flexural analysis of reinforced concrete cross-sections performed numerically is used in analytical studies to obtain predictions of the flexural capacity of the specimens.

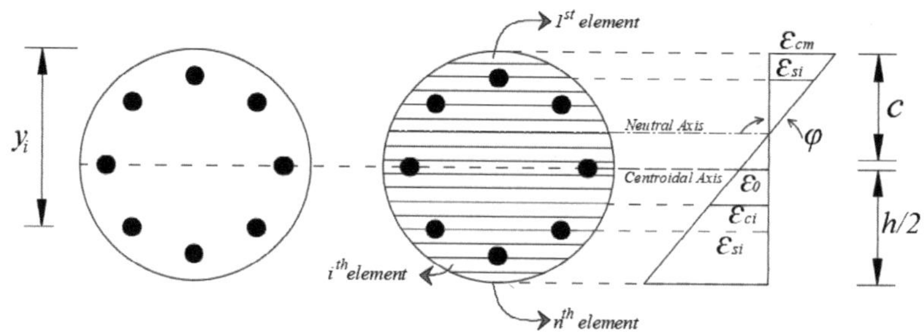

Figure 3 Solid circular cross-section with elements and strain distribution

The flexural analysis of the reinforced concrete cross-section is also assisted by a computer program developed by the author [12-14]. The layered element method divides the cross-section and reinforcement into elements, as shown in Figure 3.

4. Theoretical Shear Capacity

The nominal shear capacity of a reinforced concrete cross-section can be calculated theoretically using the equation proposed in the codes. The contribution of concrete to shear capacity is calculated based on ACI 318-19 [15], using Equation 1 below.

$$V_c = \left[0.66\lambda(\rho_w)^{\frac{1}{3}}\sqrt{f_c'} + \frac{N_u}{6A_g}\right] b_w d \quad (1)$$

Where V_c is the shear capacity of the concrete cross-section, λ is a modification factor to reflect the reduced mechanical properties of normal-weight concrete of the same compressive strength, f_c' is the specified compressive strength of concrete, ρ_w is the ratio of the total area of longitudinal reinforcement to the cross-sectional area of concrete, D is the cross-sectional diameter of the concrete, N_u is the factored axial force normal to cross-section occurring, and A_g is the gross area of the concrete section.

In the circular cross-section, b_w in Equation 1 is converted to D and d to $0.8D$. Where D is the diameter of the circular cross-section, hence, Equation 1 can be written as Equation 2:

$$V_c = \left[0.66\lambda(\rho_w)^{\frac{1}{3}}\sqrt{f_c'}\right] 0.8\, D^2 \quad (2)$$

Meanwhile, based on ACI 318-2019 for the shear capacity of the hollow circle cross-section, the code provided that the b_w value equals twice the wall thickness of the hollow circular section [15]. This study only reviews the shear capacity contributed by concrete. For this purpose, the specimens in this study did not use stirrups.

5. Test Result and Discussion

The failure mode in this study was observed visually, and the crack patterns in all specimens were almost the same, as shown in Figure 4. The failure that occurred in all specimens was a shear failure. It is shown by the direction and the wide diagonal shear crack propagating from the loading point. The shear failure occurs suddenly, followed by a wide shear crack in the observed zone. The failure location of all test specimens also appears in the same zone in the observed shear span.

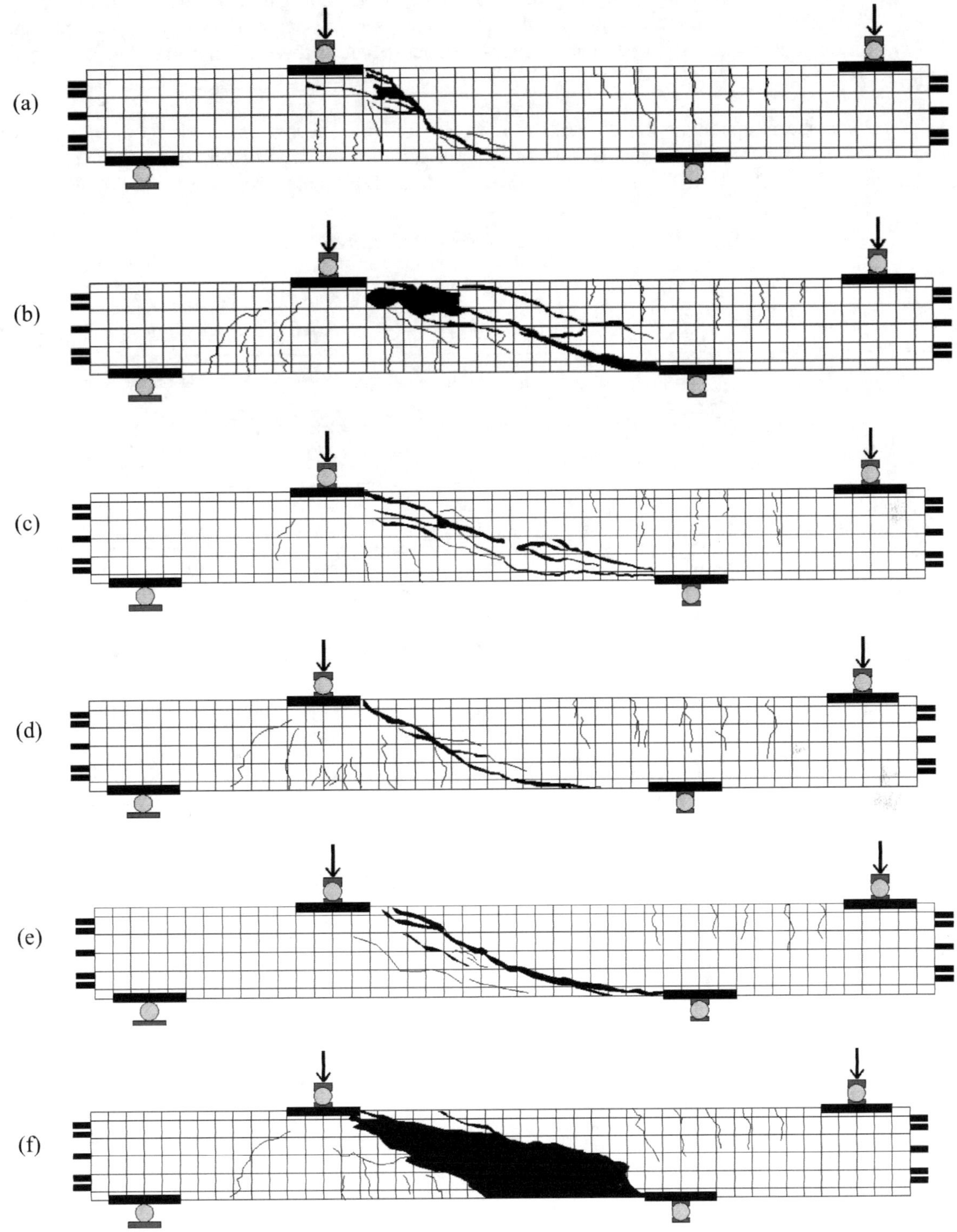

Figure 4 Crack patterns of the specimens after the test (a) SC-1, (b) SC-2, (c) SC-3, (d) H-1-1, (e) H-1-2, (f) H-1-3

From the observations, the first crack appears in the maximum bending moment zone with an average value of 23.23 kN. Then with increasing load, the flexural crack in the flexural zone extends to the shear span with a diagonal cracking mode. The test was stopped because the specimen had failed in shear mode.

Theoretical shear capacity and test results in terms of concrete shear capacity (V_c), first crack load, maximum shear load, diagonal crack angle, and identification of the type of failure of each specimen can be seen in Table 2 below. The failure mode with the letter S indicates the shear failure mode.

Table 2 Specimen theoretical capacities and experimental results

Specimens ID	Calculated Shear Capacity ACI 318M-19 V_c (kN)	Average load at first crack (kN)	Diagonal Crack Load V_{ce} (kN)	Max. Load V_{max} (kN)	The major crack angle θ°	Failure Mode
SC-01	42.2	23.23	57.83	72.3	45°	S
SC-02	48.5		67.50	87.1	22°	S
SC-03	54.3s		82.63	83.7	21°	S
H-1-1	37.8		63.13	69.2	26	S
H-1-2	43.4		60.97	82.3	24	S
H-1-3	48.7		75.93	91.3	23	S

The results in Table 2 show that the shear strength of the hollow specimen is not too different from that of the solid specimen, as shown in Figure 5. It means that the hollow with an area of 2.32% (smaller than 4%) has no significant effect on the shear strength of the specimen. In addition, the test results show that the greater the reinforcement ratio, the higher the shear capacity obtained.

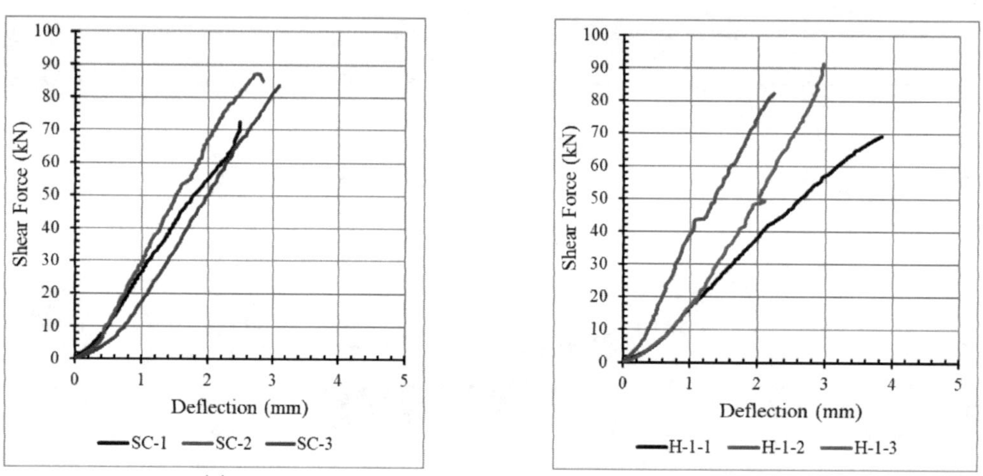

(a) (b)

Figure 5 Load-deflection curve of the tested specimens (a) Control specimens (SC), (b) Hollow specimens (H-1)

Analytically, the flexural capacity of specimens with different reinforcement ratios is very different. Figure 6 shows that the greater the reinforcement ratio, the greater the flexural capacity. However, due to premature shear failure, all specimens were proven not to reach the flexural capacity obtained analytically.

The theoretical V_c value can be seen in Figure 6 as a straight solid line. As can be seen, this theoretical shear capacity value is below the maximum shear load value obtained from the test. It proves that hollows with an area below 4% of the cross-sectional area are still safe for use in reinforced concrete structural elements.

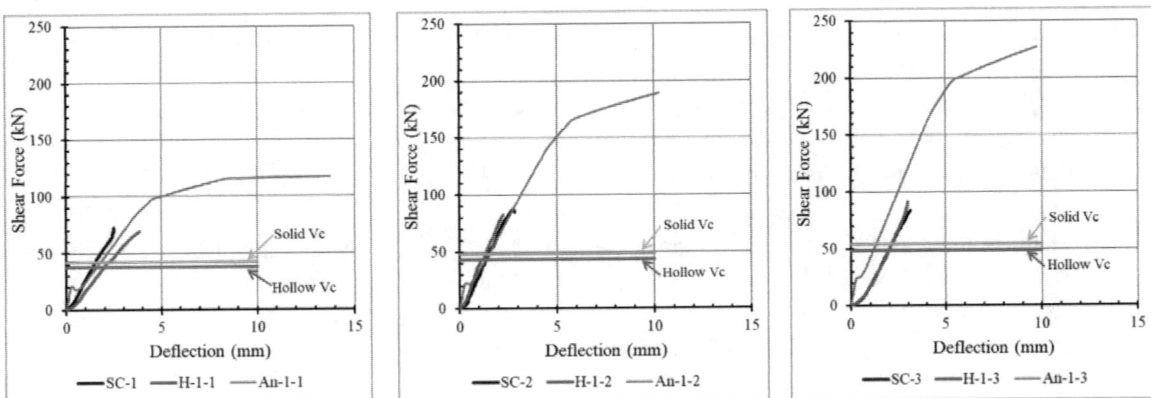

Figure 6 Load-deflection curves of specimens based on reinforcement ratio and analytical flexural capacities.

6. Conclusions

Based on the results of tests and analytical work carried out in this study, several points can be concluded as follows:

1. All specimens in this study failed in the shear mode, as indicated by a sudden wide diagonal crack.
2. Due to premature shear failure, the specimen does not reach its flexural capacity.
3. The maximum hollow area limitation provided in the code (4%) offers conservative results for the shear strength of hollow reinforced concrete structural elements.

Acknowledgment

The author would like to thank the research funds that Andalas University has provided through Penelitian Terapan Unggulan Universitas Andalas Klaster Riset-Publikasi Guru Besar with contract number T/14/UN.16.17/PP.IS-PTU-KRP1GB-Unand/2022.

References

[1] Mohamed, HM, Ali AH, and Benmokrane, B. (2016). "Behavior of circular concrete members reinforced with Carbon-FRP bars and spirals under shear," *J. Compos. Constr,* pp. 1-12.

[2] Ali AH, Mohamed, HM, and Benmokrane, B. (2016). "Shear Strength of Circular Concrete Beams Reinforced with GFRP Bars," in *Resilient Infrastructure*, London.

[3] Thamrin, R, Kurniawan, R, and Melinda, AP (2017). "Shear and flexural capacity of reinforced concrete members with circular cross-section," in *Procedia Engineering.*

[4] Thamrin, R, Tanjung, J, Aryanti, R, Nur, OF, and Devinus, A. (2016). "Shear Strength of Reinforced Concrete T-beams without Stirrups," *Journal of Engineering Science and Technology,* vol. 11, no. 4, pp. 548 – 562.

[5] Dewi, SH, Thamrin, R, Zaidir and Taufik, (2020)"Effect of stirrup type on shear capacity of reinforced concrete members with circular cross-section," in *E3S Web of Conferences; 4th ICEEDM 2019*.

[6] Liang, X, and Sritharan, S. (2018). "Effects of Confinement in Circular Hollow Concrete Columns," *J. Struct. Eng.,* pp. 1-13.

[7] SNI 03-2847-2002; Tata Cara Perhitungan Struktur Beton untuk Bangunan Gedung (Beta Version), Bandung, 2002.

[8] SNI 2847:2013-Persyaratan Beton Struktural untuk Bangunan Gedung, Jakarta: Badan Standarisasi Nasional, 2013.

[9] SNI 2847:2019-Persyaratan beton struktural untuk bangunan gedung dan penjelasan, Jakarta: Badan Standarisasi Nasional, 2019.

[10] Clarke, JL and Birjandi, FK. (1993). "The behaviour of reinforced concrete circular sections in shear," in *The Structural Engineer.*

[11] Feltham, I. (2004). "Shear in reinforced concrete piles and circular columns," in *The Structural Engineer*.

[12] Thamrin, R. (2018). " Effect of Strengthening Method and Development Length on Flexural Strength of RC Beams With Steel Plates," *Journal of Engineering Science and Technology,* vol. 13, no. 11, pp. 3781-3794.

[13] Thamrin, R. (2017). "Analytical Prediction on Flexural Response of RC Beams Strengthened with Steel Plates," in *MATEC Web of Conferences*.

[14] Zaidir, Thamrin, R and Dalmantias, E. (2017) "Evaluation of the Pre-Cracked RC Beams Repaired with Sealant Injection Method," *International Journal on Advanced Science Engineering Information Technology,* vol. 7, no. 2, pp. 380-386.

[15] Building Code Requirements for Structural Concrete (ACI 318-19), Farmington Hills: American Concrete Institute, 2019.

IOP Conf. Series: Earth and Environmental Science 1244 (2023) 012008 doi:10.1088/1755-1315/1244/1/012008

The Effect of Adding Gypsum and Lime on Lightweight Interlock Bricks with Merapi Sand as Fine Agregat

A Rakhmawati[1], A Murtopo[1*], A Sanviani[1], D S A Yuwana[1], M Amin[1] and Y Arnandha[1]

[1]Department of Civil Engineering, Universitas Tidar, Central Java, Indonesia

*corresponding author: a.m@untidar.ac.id

Abstract. Lightweight materials will reduce the earthquake forces received by the structure. Interlock lightweight brick is a new innovation in the building industry that functions as wall material. Utilization of local materials in Central Java-Yogyakarta, namely Merapi sand, needs to be developed to maximize its potential utilization as an aggregate of lightweight interlock brick. In many research, lime and gypsum as additives have been shown to improve the quality of concrete. This study aims to determine the characteristics of the addition of gypsum and lime as an added material for lightweight interlock bricks with fine aggregate of Merapi sand. The standard of test object and the test refers to SNI 8640-2018. The results showed that in dry weight all the mixed variations were categorized in the 900 kg/m3 and 1100 kg/m3 classes. All variations also complied with water absorption requirements in SNI 8640-2018, but only variations of 2.5% gypsum addition and 5% lime addition meet the compressive strength requirements. This mixture has the highest average of compressive strength value compared to other mixtures. Based on SNI 8640-2018 lightweight interlock bricks with the addition of 2.5% gypsum and 5% lime can be used as wall bricks.

Keywords. brick, gypsum, lime, lightweight interlock brick, lightweight brick

1. Introduction

The development of lightweight materials as supporting materials for earthquake-proof buildings will increase the resilience to earthquake disasters. Earthquake that occurred in several areas in Indonesia, one of which was in Yogyakarta in 2006 needs attention because it has claimed many lives and damaged infrastructure. Several studies have also shown that there are many active faults that have not been mapped with visible movements from images and outcrops [1] [2].

One of the lightweight materials that has been widely developed to support earthquake-proof buildings is brick wall which is included in the lightweight category. Lightweight category brick is a porous brick with lighter density than normal brick. Lightweight brick is a concrete brick weighing between 400 to 1400 kg/m3 [3]. There are several types of lightweight bricks on the market, one of which is Foamed Cellular Lightweight Concrete (FCLC). FCLC is a cellular lightweight concrete whose manufacturing process uses foam to create small cavities in the concrete. FCLC characteristics are influenced by foam, sand type, binder type, water-cement ratio and age [4].

Lightweight interlock brick is a new innovation in the development of lightweight construction material. This type of brick is light and is specially designed so the bricks can be interconnected with each other. The advantages of interlock bricks are easy to install, multidimensional, permanently connected, soundproof, earthquake resistant and environmentally friendly. Interlock bricks have characteristics and advantages compared to normal bricks, as they have hooks or interlocks. This interlocking system has two functions: the bricks will be interconnected so they can bind to each other (self-aligning) and have a more effective locking than normal bricks, so it will allow for a straight, precise, and stable wall arrangement [5].

As the technology keeps evolving, many new innovations are found in lightweight bricks production. One way to make lightweight bricks is by adding foam to the concrete slurry. The addition of foam can lighten the concrete because the foam will become a cavity in the concrete, but the strength will be reduced. [6].

Gypsum has many uses, as a substitute for wood in building materials, as an adhesive, drywall, hardener for building materials, and as one of the raw materials for making chalk and Portland cement [7]. In addition, some waste additives affect the physical and mechanical properties of gypsum composites [8]. In several studies, the addition of gypsum was effective in increasing the compressive strength of foamed concrete and reducing the loss of strength due to thermal [9] [10].

Adding lime to the lightweight category of concrete bricks can produce a lighter weight than normal bricks without the lime addition. The addition of lime can increase the compressive strength of concrete so that in light foam concrete the weight of the concrete can be reduced to obtain the same compressive strength. Lime was chosen as a partial replacement of cement and because lime has properties that can bind particles quite well. Moreover, the use of lime in the production of lightweight bricks can be more efficient in manufacturing costs because the price of lime is cheaper than the price of cement [11]. Concrete containing lime can increase strength and resistance to chloride compared to normal cement concrete at an advanced age. Besides the relatively low cost and low energy consumption which can decrease global warming, it can increase the initial strength and workability of concrete and the long-term chloride permeability of concrete containing limestone powder is higher than that of ordinary cement concrete [12] [13] [14].

The In this study, Merapi sand was used as a fine aggregate of lightweight interlock bricks. Mount Merapi sand is good for mixing concrete materials. Particles are formed from the pointed and angular ends of silica. The bond between Mount Merapi sand and cement is stronger because Merapi sand has an angular pattern of particles [15]. This research was conducted with the aim to know the effect of adding gypsum and lime to the compressive strength, bulk density and water absorption of lightweight interlock bricks using fine aggregate of Merapi sand.

2. Research Methods

The standard of test object and the test used in this study were based on the standards applied in Indonesia, namely SNI 8640:2018 concerning Specifications of lightweight bricks for wall structures. This research method used an experimental method for testing lightweight interlock bricks. The interlock design used was a straight double interlocking block design of 12 x 20 x 40 cm, as shown in Figure 1. The advantage of this design is a vertical and horizontal locking system between bricks in the form of a tongue and groove that extends on the right and left sides of the brick, and the top and bottom of the brick so it can withstand both vertical and horizontal forces and saves adhesives. The simple locking design can simplify the preparation process without the need for special techniques [16].

Figure 1. Lightweight Interlock Bricks.

The ingredients used in producing lightweight interlock brick are cement, Merapi sand, lime, gypsum and water. The ratio of cement to sand is 1 : 1 Water per cement ratio is 0.6. Foaming agent is used to expand the volume. The mix design and number of test objects can be seen in Table 1.

Table 1. Mix Design.

Variations	Cement	WCR	Fine Aggregate	Lime	Gypsum	Number of test objects
KK10KG2.5	1	0.6	1	10.00%	2.50%	4
KK5KG5	1	0.6	1	5.00%	5.00%	4
KK10KG5	1	0.6	1	10.00%	5.00%	4
KK5KG2.5	1	0.6	1	5.00%	2.50%	4
KK0KG0	1	0.6	1	0.00%	0.00%	4
KK5KG0	1	0.6	1	5.00%	0.00%	4
KK10KG0	1	0.6	1	10.00%	0.00%	4
KK0KG2.5	1	0.6	1	0.00%	2.50%	4
KK0KG2.5	1	0.6	1	0.00%	5.00%	4

2.1. Foaming Agent

In producing the foaming agent, it is necessary to be attentive to the structure of the foam used. The structure of the foam will affect the characteristics of the foam [17]. The foam structure is created from the assembly of surfactants that line up with each other into a rigid structure, until a foam structure is formed. A surfactant concentration specific foam is used as a developer whose main viscosity plays an important role in determining the foam stability [18]. The void in the center of the foam is filled with air to ensure the foam is lighter and has more volume. Several phenomena were identified that contribute to foam stability such as drainage and coalescence effects [19].

The foam structure used in this study is produced from foaming agent and water which is processed into foam by a foam generator. The foam produced from the generator tube must be immediately put into the mixture, about 15-30 seconds and is not recommended for foams with large bubbles. The use of foaming agent can be seen in Figure 2.

Figure 2. Foam from Foaming Agent.

2.2. Testing

Foamed concrete molds are opened after the bricks are 3 days old. This is done so that when the mold is opened, the brick is not damaged. After that, the lightweight interlock bricks are stored in a place with room temperature. Tests carried out included testing for bulk density, compressive strength and water absorption. All tests were carried out when the lightweight interlock brick was 28 days old. Each mixture was taken 4 samples. The compressive strength test was carried out by cutting the interlock bricks into cubes of 15 x 15 x 15 cm. One lightweight interlock brick test object can be cut into 3 cubes. The test was carried out on brick cubes that were 28 days old. Prior to the compression test, the brick cubes were first immersed in water for 24 hours ± 30 minutes [3]

3. Results and Discussion

3.1. Bulk Density

The bulk density test was carried out on lightweight interlock bricks which were cut into cubes of 12 cm x 12 cm x 12 cm at the age of 28 days. The specimens for each variant were 4 specimens and the total were 36 specimens. The results of the bulk density of lightweight interlock brick can be seen in Figure 3.

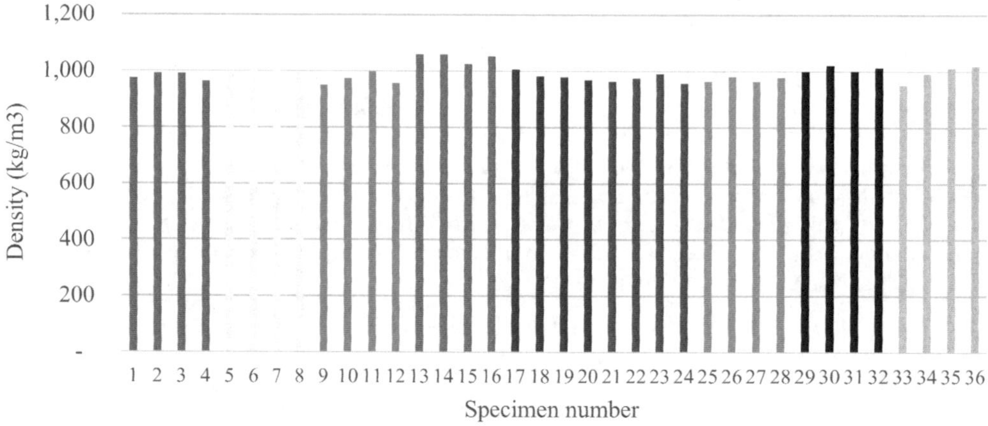

Information :
KK10KG2.5 (1 to 4)
KK5KG5 (5 to 8)
KK10KG5 (9 to 12)
KK5KG2.5 (13 to 16)
KK0KG0 (17 to 20)
KK5KG0 (20 to 24)
KK10KG0 (25 to 28)
KK0KG2.5 (29 to 32)
KK0KG5 (33 to 36)

Figure 3. Density of Lightweight Interlock Brick.

It can be seen in Figure 3 that the average bulk density in oven-dried condition of all samples was 992 kg/m3. The lowest bulk density was 948 kg/m^3 and the highest bulk density was 1060 kg/m^3. When the test object is still a mortar slurry, its weight is greater than the weight of the oven-dried lightweight interlock brick. As the mortar hardens, the water content in the mortar will evaporate and become air cavities, See in Figure 4. The air cavity is formed from the foam produced by the foam agent. It is necessary to pay attention to the addition of foam agent doses, the more foam agent added the larger the pores formed and the more interconnected they will be [20]. This causes decrease in the brick weight. When viewed from the mix design, the density of lightweight interlock bricks is shown in Figure 5.

Figure 4. Cavities in interlock bricks

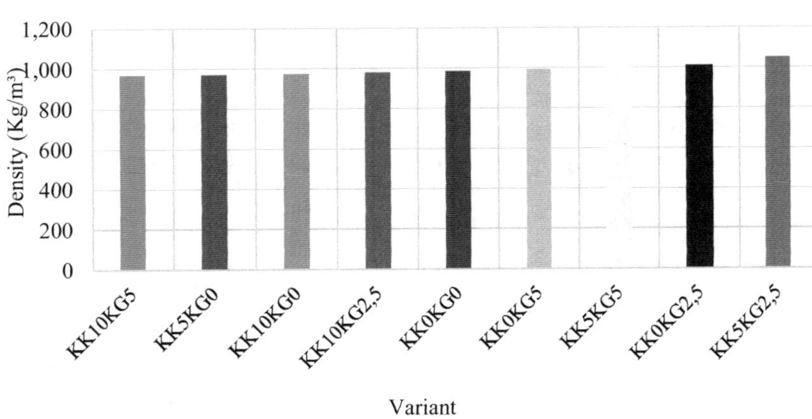

Figure 5. Average Density of Each Variant.

3.2. Compressive Strength

The compressive strength test was carried out on lightweight interlock bricks which were cut into cubes of 12 x 12 x 12 cm. Each variant consists of 4 specimens. Prior to the compression test, the specimens were first immersed in water at room temperature for ± 24 hours [3]. The results of the average compressive strength test of lightweight interlock bricks can be seen in Figure 6.

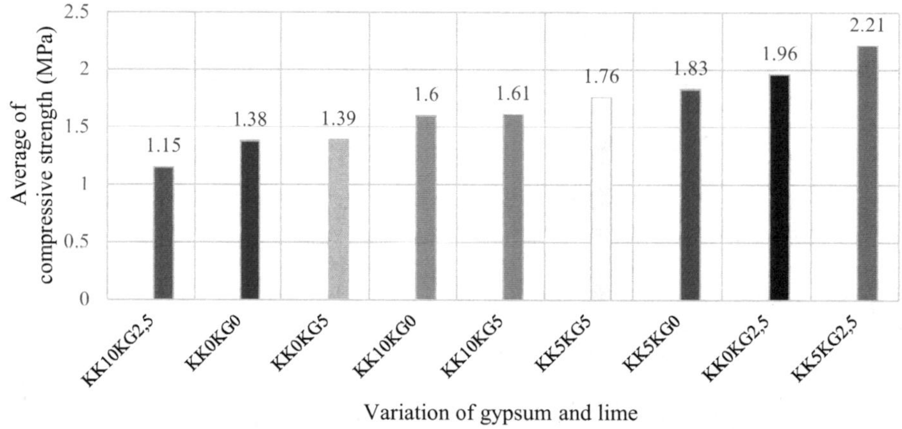

Figure 6. Average of Compressive Strength of Lightweight Interlock Brick.

It can be seen in Figure 5 that the highest compressive strength of lightweight interlock brick is in specimen KK5KG2.5 (5% lime and 2.5% gypsum). Compared to the specimens without added lime and gypsum, almost all specimens with added lime and gypsum experienced an increase in compressive strength, except for the specimen KK10KG2.5. This proves that the addition of lime and gypsum as a mixture of lightweight interlock bricks can increase its compressive strength. However, the addition of lime and gypsum should not be too high. This is because the maximum addition of lime which can increase the compressive strength is only 10%, the rest will decrease the compressive strength [12].

3.3. Water Absorption

The water absorption test was carried out on intact lightweight interlock bricks. Each variant consists of 4 specimens. The results of the average water absorption of lightweight interlock bricks for all variations is shown in Figure 7. In Figure 7, it can be seen that the addition of lime and gypsum affects the water absorption of bricks. The addition of lime and gypsum can reduce the water absorption of bricks. In specimen KK5KG2.5 the decrease of water absorption capacity without the addition of lime and gypsum (KK0KG0) is about 2%. However, the addition of lime and gypsum should not be too much because excessive addition can reincrease the water absorption [21].

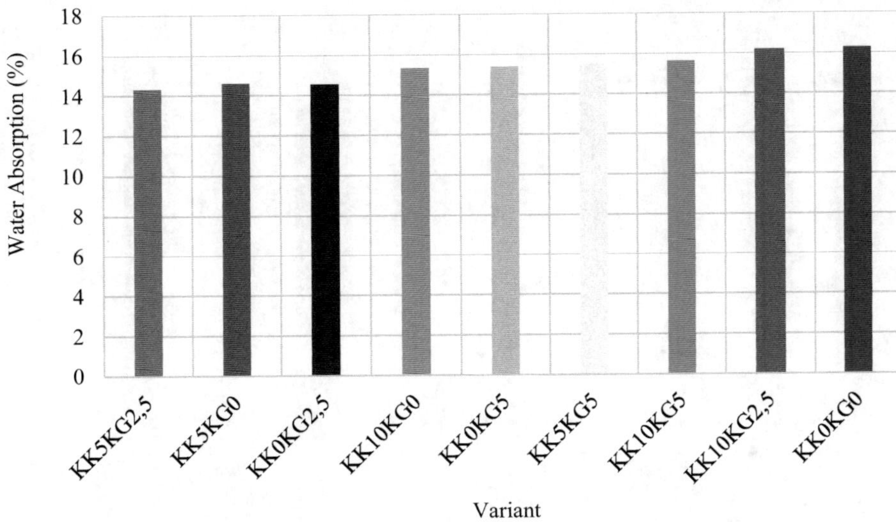

Figure 7. Water Absorption.

3.4. Lightweight Brick Based on SNI 8640-2018

SNI 8640-2018 provides several requirements related to lightweight bricks that can be used as wall pairs. These requirements are related to density, single compressive strength, average compressive strength, and water absorption [3].

The density level of bricks is classified based on the average density of lightweight bricks. The test results showed that the variations of lightweight interlock bricks include in density class category 900 and 1100 kg/m3, as shown in Table 2.

Table 2. Density Level Based on SNI 8640-2018.

Variations	Density (kg/m^3)	Density Class (kg/m^3)
KK10KG2.5	984	900
KK5KG5	1,008	1100
KK10KG5	992	900
KK5KG2.5	971	900
KK0KG0	1,049	1100
KK5KG0	1,006	1100
KK10KG0	972	900
KK0KG2.5	980	900
KK0KG2.5	968	900

The individual compressive strength of each design mix required by SNI 8640-2018 is 1.8 MPa. Mix designs that meet the requirements of SNI 8640-2018 are KK5KG2.5 and KK0KG2.5, see Figure 8.

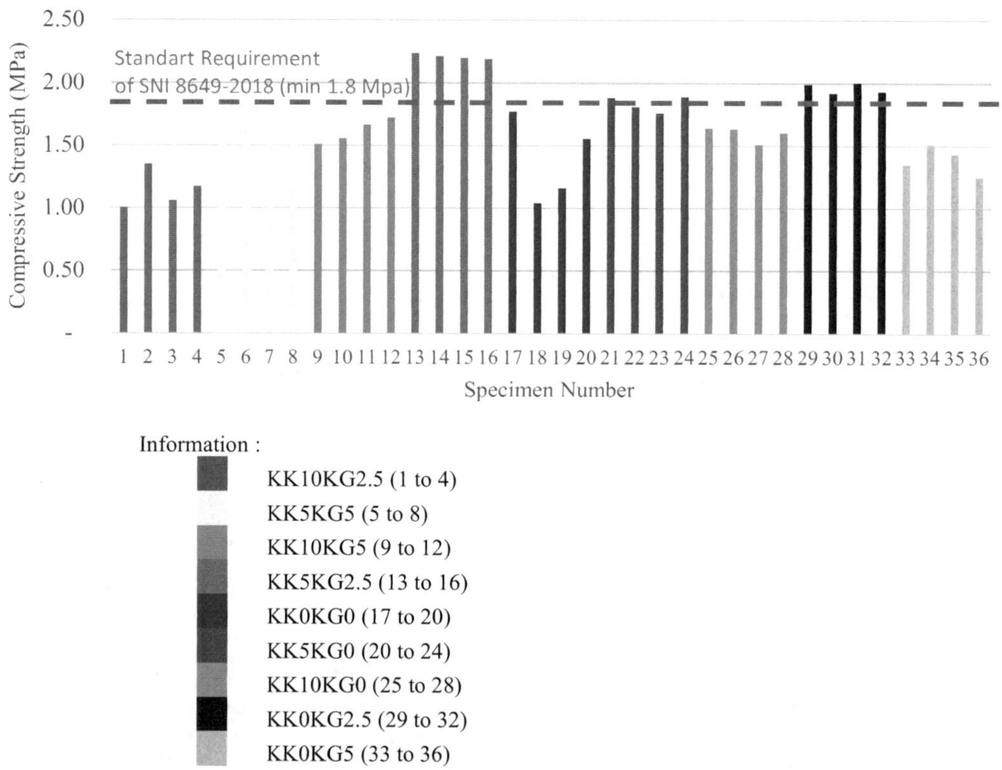

Figure 8. Individual compressive strength agains SNI 8640:2018.

The average compressive strength of each design mix required by SNI 8640-2018 is 2.0 MPa. Mix designs that meet the requirements of SNI 8640-2018 are KK5KG2.5, see Figure 9.

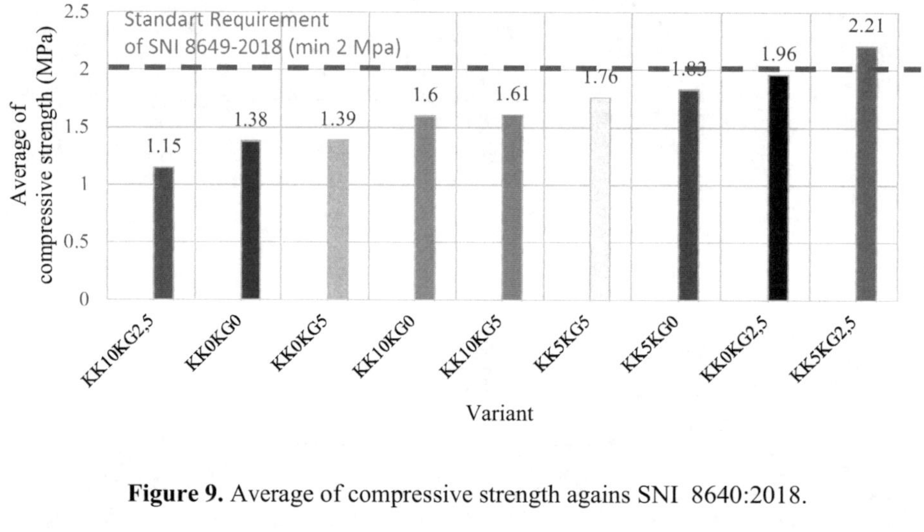

Figure 9. Average of compressive strength agains SNI 8640:2018.

The average water absorption of each mix design required by SNI 8640-2018 is a maximum of 25%. In the case of water absorption, all mix designs meet the requirements of SNI 8640-2018.

Based on the test results, it was found that the specimens which meet the requirements of SNI 8640-2018 was only KK5KG2.5, as shown in Table 3. It can be concluded that with a mix design of 1 cement : 1 fine aggregate, water per cement ratio of 0.6, the addition of lime and gypsum is recommended so that the brick meets the criteria of SNI 8640-2019 is 5% lime and 2.5% gypsum from cement. These bricks can be used as exposed or unexposed non-structural brick walls.

Table 3. Physical Requirements of Lightweight Interlock Bricks Based on SNI 8640-2018.

Physical Requirements	Unit	Condition	Variation				Information
			4A	4B	4C	4D	
Average compressive strength	MPa	Min 2 MPa	2.21				OK
Single compressive strength	MPa	Min 1.8 MPa	2.24	2.21	2.2	2.19	OK
Water absorption	%	Max 25%	14.36	14.3	14.29	14.38	OK

4. Conclusion

The lightweight interlock bricks which have been studied have density level from 948 kg/m3 to 1060 kg/m3. Based on SNI 8640-2018, lightweight interlock bricks are included in heavy duty class category of 900 kg/m3 and 1100 kg/m3. The addition of lime and gypsum to bricks can increase the compressive strength and reduce the water absorption capacity of the brick, but excessive addition works in the opposite way. Mix design of 1 cement: 1 fine aggregate, water per cement ratio 0.6, addition of lime 5% and gypsum 2.5% of cement is the recommended mix design so the brick meets the criteria of SNI 8640-2019 as exposed or unexposed non-structural brick.

5. References

[1] Diambama A D, Anggraini A, Nukman M, Luhr B-G, Suryanto W 2018 Velocity structure of the earthquake zone of the M6.3 Yogyakarta earthquake 2006 from a seismic tomography study *Geophys. J. Int.* (2019) 216, 439–452.

[2] Saputra A, Gomez C, Delikostidis I, Reza P Z, Hadmoko D S, & Sartohadi J 2020 Preliminary identification of earthquake triggered multi-hazard and risk in Pleret Sub-District (Yogyakarta, Indonesia) *Geo-spatial Information Science* Volume 24, 2021 - Issue 2.

[3] SNI 8640:2018 Spesifikasi Bata Ringan Untuk Pasangan Dinding *Badan Standarisasi Nasional* (in bahasa).

[4] Ashrafian A, Shokri F, Amiri M J T, Yaseen Z M, Balf M R 2010 Compressive strength of Foamed Cellular Lightweight Concrete simulation: New development of hybrid artificial intelligence model *Construction and Building Materials* Volume 230.

[5] Budiyani A G and Prastyatama B 2020 Evaluation and Experiment of Interlocking Brick Module Design to Obtain Varieties of Ventilation Opening Area on Wall *Jurnal RISA (Riset Arsitektur)* vol. 04 no 3, 269–287.

[6] Moniri S N H and Usman F 2019 Characteristics of Lightweight Foamed Concrete Brick Mixed with Fly Ash *SSRG International Journal of Civil Engineering* Vol. 6 Issue 3, 22-28.

[7] Permana I 2017 Pemanfaatan Limbah Gypsum Board Sebagai Pengganti Sebagian Semen Terhadap Kuat Tekan Batako *Thesis of bachelor degree, Program Studi Pendidikan Teknik Bangunan Fakultas Teknik Universitas Negeri Jakarta* (in bahasa).

[8] Khalil A A, Tawfik A, Hegazy A, and El-Shahat M F 2014 Effect of some waste additives on the physical and mechanical properties of gypsum plaster composites *Construction and Building Materials* 68:580 - 586.

[9] Cai Q, Ma B, Jiang J, Wang J, Shao Z, Hu Y, Qian B, Wang L 2021 Utilization of waste red gypsum in autoclaved aerated concrete preparation, *Construction and Building Materials*, Volume 291.

[10] Chernyisheva NV, Shatalova SV, Yu M. Drebezgova, Lesnichenko EN 2019 Thermal Insulating and Constructive Foamed Concrete on a Composite Gypsum Binder, *Materials Science Forum* Vol 974 125-130

[11] Jusi U, Maizir H, Ilham M and Saily R 2021 Pengaruh Penambahan Kapur Tohor Terhadap Sifat Mekanis Bata Ringan *Indonesian Journal of Construction Engineering and Sustainable Development* vol 4 no. 1 (in bahasa).

[12] Wang Q, Yang J, and Chen H 2017 Long-term properties of concrete containing limestone powder *Materials and Structures* vol. 50 no. 3.

[13] Zhou J, Qian S, Beltran, M G S et al. 2010 Development of engineered cementitious composites with limestone powder and blast furnace slag *Mater Struct* 43 pp 803–814.

[14] Celik K, Jackson M D, Mancio M, Meral C, Emwas A H, Mehta P K, Monteiro P J M 2013 High-volume natural volcanic pozzolan and limestone powder as partial replacements for portland cement in self-compacting and sustainable concrete *Cement and Concrete Composites* Volume 45 pp 136-147.

[15] Lasino and Cahyadi D 2016 Pemanfaatan Pasir dan Abu Merapi Untuk Pembuatan Bata Beton (Conblock) *The Lost World Historical Continuity for Sustainable Design* (in bahasa).

[16] Pérez-Peña A M et al 2009 Interlocking Stabilised Soil Blocks: Appropriate Earth Technologies in Uganda *United Nations Human Settlements Programme* UN-HABITAT Kenya.

[17] Ali M H M, Arshad M F, Ridzuan R and Nurliza J 2016 Discovery of Used Cooking Oil as Foaming Agent Admixture for Lightweight Foamed Concrete *InCIEC* 2015 (pp.621-629).

[18] Shrestha L K, Shrestha R G, Sharma S C, and Aramaki K 2008 Stabilization of nonaqueous foam with lamellar liquid crystal particles in diglycerol monolaurate/olive oil system *Journal of Colloid and Interface Science* Volume 328 Issue 1 Pages 172-179.

[19] Jalmes A S, Peugeot M L, Ferraz H, & langevin D 2005 Differences between protein and surfactant foams: Microscopic properties, stability and coarsening *Colloids and Surfaces A: Physicochemical and Engineering Aspects* Volume 263 Issues 1–3 Pages 219-225.

[20] Lee, H.-S.; Ismail, M.A.; Woo, Y.-J.; Min, T.-B.; Choi, H.-K. Fundamental Study on the Development of Structural Lightweight Concrete by Using Normal Coarse Aggregate and Foaming Agent. *Materials* 2014, 7, 4536-4554. https://doi.org/10.3390/ma7064536.

[21] Elgalhud A A, Dhir R K, and Ghataora G 2016 Limestone addition effects on concrete porosity *Cem. Concr. Compos*. vol. 72 pp. 222–234

IOP Conf. Series: Earth and Environmental Science 1244 (2023) 012009 doi:10.1088/1755-1315/1244/1/012009

Structural Performance of Buildings with Tension-Only Braced Frame Under Seismic Loading

D Kusumastuti[1*] **and F I Maura**[2]

[1]Faculty of Civil and Environmental Engineering, Institut Teknologi Bandung, Indonesia
[2]Graduate School of Civil Engineering, Faculty of Civil and Enviromental Engineering, Institut Teknologi Bandung, Indonesia

*Corresponding author: dkusumastuti@itb.ac.id

Abstract. In accordance with SNI 1726-2019, structures with tension-only braced frames (ToBF) can be used as seismic force-resisting systems for seismic design category (SDC) A to C, with a height of up to 10m. Limited study was done on this subject, hence a numerical study was conducted to better understand the structural performance. The models studied were of regular structures of SDC D, varying in height. The analyses used modal response spectrum analysis (MRSA) for design, and linear response history analysis (LRHA) to evaluate the response. Finally, pushover analyses were conducted to determine the structural performance levels. The study shows that structural design can be done for all models using MRSA, satisfying all requirements. When evaluated using LRHA, results reveal that requirements were not satisfied for structures with heights exceeding 10m. Analysis when reviewed using pushover analysis, results show that the performance levels for all buildings are Immediate Occupancy. However, for models with height exceeding 10m, first yielding occurs at a lower base shear than the design base shear, indicating that the value of R used may not be adequate. Further review reveals that the ToBF can be used as a seismic force-resisting system, with proper design and arrangements.

1. Introduction

Structures with tension-only brace frames (ToBF) are designed such that the braces only resist tensile forces and have no capacity in resisting compression forces (*Bruneau et al, 2011*) . Since all ToBF braces are assumed to only resist tensile forces, the requirement for brace slenderness is as a slender component, with $KL/r > 200$ and $KL/r < 300$. The susceptibility for buckling on braces may lead to brittle failure of structures, especially under cyclic loading. Based on AISC 341-16, ToBF structure systems are only permitted for concentrically braced frames (CBFs) used for low-rise buildings, or for buildings with seismic design category (SDC) A to C. For structures with SDC D, using ToBF for seismic resistant structures is not recommended, nor for medium and high-rise buildings.

ToBF structure was originally used as a lateral force-resisting system for wind loads in low-rise steel buildings due to its simple construction. The ToBF main advantage comes from the slenderness of the structural elements designed to only resist tension (compression force is ignored). The result of alternating between tension yielding and compression buckling of the bracing leads to deteriorating and pinched hysteretic behavior in compression cycles while maintaining full loops for tension cycles, providing the main characteristic of ToBF (*Shariati et al, 2020*). The typical hysteretic curve for ToBF braces can be seen in Figure 1.

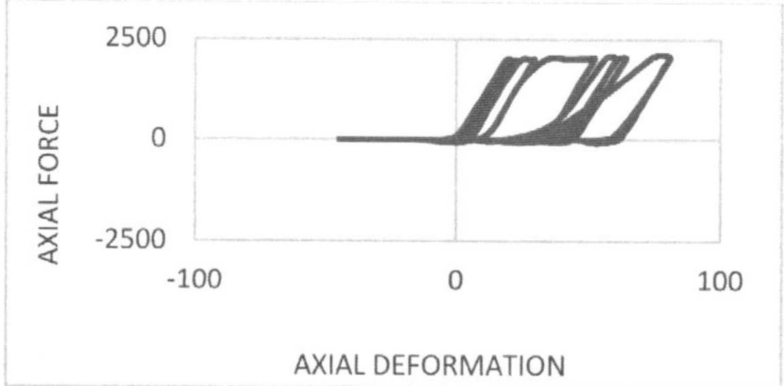

Figure 1. Hysteretic Model of Tension-only Brace Frame

The use of ToBF in seismic-resistant system is somewhat limited, as shown by its unassigned seismic parameters of R, Cd and Ωo for such system. As such, the selection of the appropriate response modification factor (R) is important in estimating the seismic load, since current seismic design practice has relied on the capability of structures to dissipate energy by inelastic behavior. The R factor proposed in the earlier seismic design codes was selected based on the observed performance of seismic resistant system during real earthquakes and engineering judgements. From previous research, the calculation focused on the R factor as the product of the strength factor as well as ductility, and it concluded the use of R = 3 and Ωo = 2 for structures with ToBF (*Shariati et al, 2020*). However, this study was conducted for buildings with relatively low stories, i.e., 1-story, 3-story, and 5-story. Hence, it is interesting to check whether the same value can be applied for buildings with higher stories.

2. Methodology

In this study, the typical structure was assumed to be based on an ordinary concentrically braced frame (OCBF) used for office buildings. Referring to SNI 1726-2019, there is a height limitation of 10 m for seismic design category D, and the R factor for such structure is R = 3.25. The models were developed with various stories and heights, i.e., 3-story (10,5m), 8-story (28m) and 15-story (52.5m). The structural height and stories were selected to evaluate the performance of structure with ToBF for low, medium, and high-rise buildings, as well as to find out whether structures with ToBF with height exceeds 10m can be designed and still satisfy the code requirements.

All structural models were 3D models. The ToBF was applied to the OCBF structure with a diagonal braced configuration and assumed to have no compressional capacity. The plan and elevation views of all structures are presented in Figure 2. The specification of the steel material to be used for beams and columns was ASTM A992 with the yield strength of 345 MPa, while braces material was of ASTM A500 Grade C with the yield strength of 317 MPa.

The applicable building loads used in modeling the structures followed SNI 1727-2020 for building loads, and SNI 1726-2019 for seismic load. Assuming that the structures were in Bandung region, the response spectra curves with moderate soil condition yielded values of Ss = 1.129 dan S1 = 0.492 to be used for analysis.

The design of structural elements was conducted by selecting elements with moderately ductile members and should satisfy the requirements of AISC 314-16 and AISC 360-16. The brace design was first carried out to find the brace strength that could resist the ultimate loads. It is also necessary to adhere to the required brace slenderness, i.e., $KL/r > 200$ and $KL/r < 300$, hence maintaining the characteristic of ToBF. Next, the design of beam and columns was carried out after the axial tensile capacity of the braces was obtained, because the design of these members utilized the capacity design concept to ensure that yielding first occurred on braces.

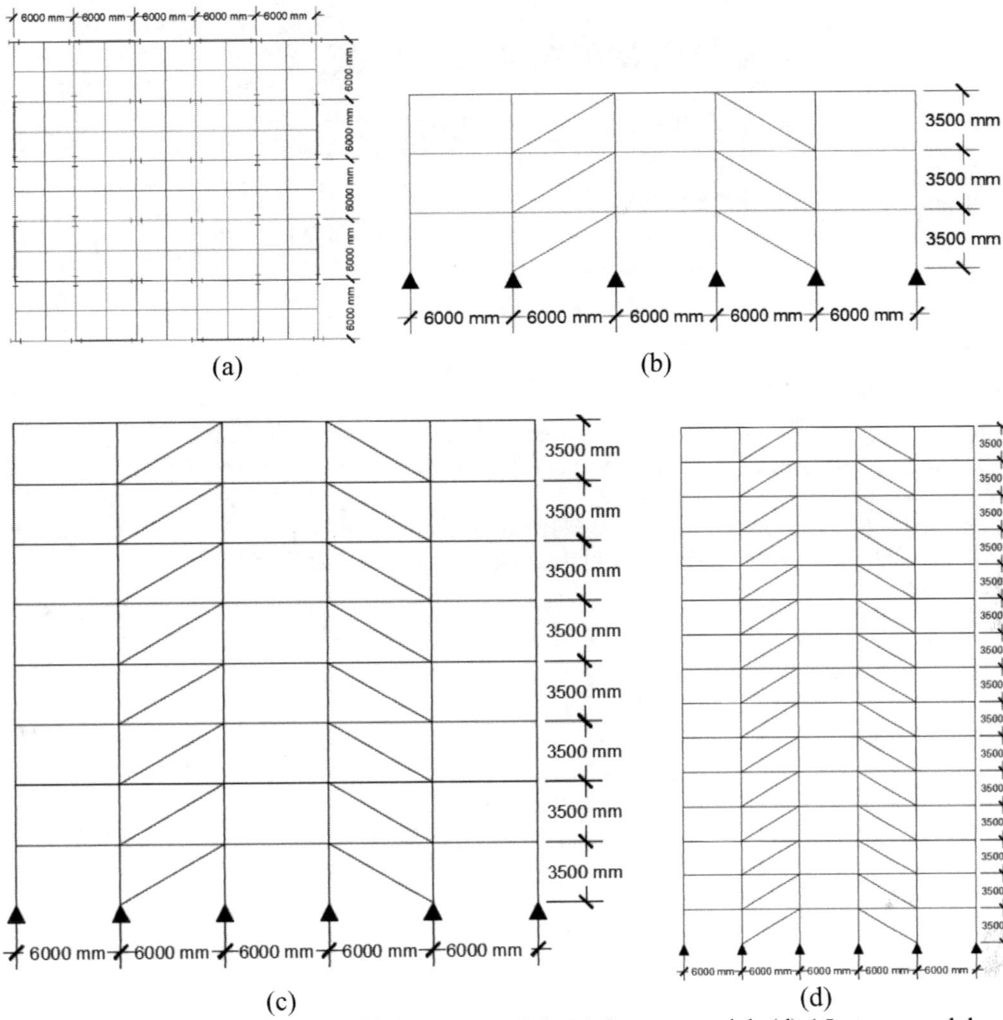

Figure 2. (a) Plan Layout; (b) 3-story model; (c) 8-story model; (d) 15-story model

3. Analytical Results

The analyses were conducted using modal response spectrum (MRSA) for designing the structures. The designs were then evaluated using linear response history analysis (LRHA) to determine the maximum axial force of the braces and the inter-story drift. Finally, pushover analyses were conducted to determine the performance level of structures. Each analysis will be presented below.

3.1. Modal Response Spectrum Analysis

MRSA was used to design the structure. The dynamic analysis was conducted for all models, including for structures with structural height considered violating code limitations. The parameters that affect the analysis were the mass of the building and the lateral stiffness of the building. According to SNI 1726-2019, the analysis should include the minimum amount of variance to achieve a combined mass participation of 100% of the mass of the structure. Then the base shear from the dynamic analysis should be checked to ensure that it is no less than 100% of the force obtained from static analysis.

Results from the analysis show that the base shear force (V_d) for the structure with 3-story V_d = 3619.6 kN, for the structure with 8-story V_d = 9785.4 kN, and for the structure with 15-story V_d = 12219.7 kN. Figure 3 shows the base shear obtained for all models.

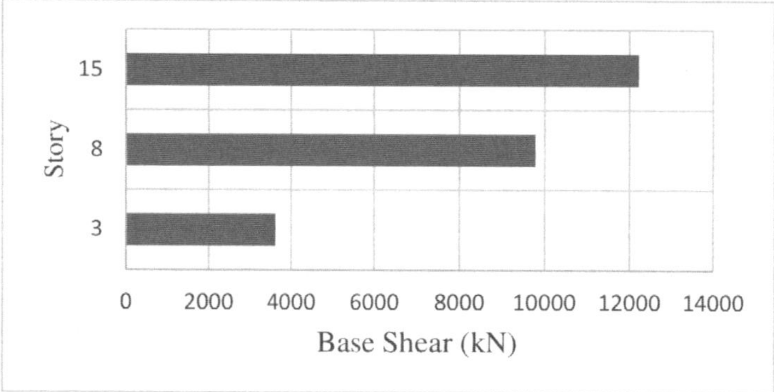

Figure 3. Comparisons of Base Shear for All Models

The analysis also reveals that the ToBF system resisted approximately 98.5% to 99% of the lateral force, while the non-seismic frames resisted only 1 to 1.5% of seismic forces, hence confirming that ToBF was the main lateral resisting system.

The code states that the inter-story drift should not exceed the inter-story drift permit level with risk category II, $\Delta_a = 0.020h_x$. Figure 4 shows that all structures meet the drift requirements.

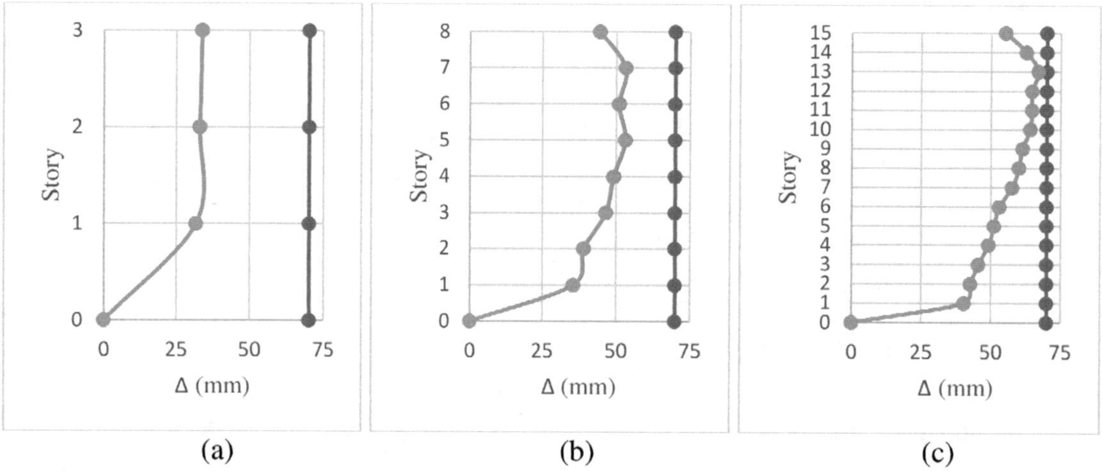

Figure 4. Inter-story Drift (a) 3-story model; (b) 8-story model; (c) 15-story model

The structural design conducted was able to obtain sectional properties for the braces, beams, and columns. All members showed capacity that satisfy the requirements of demand-capacity ratio (DCR) < 1. Hence all structural elements have adequate capacity to resist applicable building loads. Table 1 to Table 3 shows the final design for all structural members used in the models.

Table 1. Dimension and Demand-Capacity Ratio of Element for 3-story Model

Story	Brace		Beam		Column	
	Dimension	DCR	Dimension	DCR	Dimension	DCR
3	HSSX2,5X0.25	0.794	W14X48	0.426	W10X68	0.268
2	HSS5X3X0.5	0.773	W14X61	0.801	W10X68	0.454
1	HSS7X3X0.5	0.812	W14X61	0.873	W10X100	0.753

Table 2. Dimension and Demand-Capacity Ratio of Element for 8-story Model

Story	Brace		Beam		Column	
	Dimension	**DCR**	**Dimension**	**DCR**	**Dimension**	**DCR**
8	HSS7X3X0.5	0.438	W18X86	0.483	W12X170	0.209
7	HSS8X3X0.5	0.791	W18X86	0.577	W12X170	0.394
6	HSS14X3X0.5	0.728	W18X97	0.951	W14X233	0.475
5	HSS14X3X0.5	0.841	W18X97	0.797	W14X233	0.674
4	HSS18X3X0.5	0.825	W18X119	0.875	W14X283	0.757
3	HSS18X3X0.5	0.881	W18X119	0.800	W14X283	0.961
2	HSS22X3X0.5	0.830	W18X130	0.941	W14X426	0.794
1	HSS22X3X0.5	0.889	W18X130	0.872	W14X426	0.950

Table 3. Dimension and Demand-Capacity Ratio of Element for 15-story Model

Story	Brace		Beam		Column	
	Dimension	**DCR**	**Dimension**	**DCR**	**Dimension**	**DCR**
15	HSS10X3X0.5	0.275	W18X86	0.671	W14X257	0.168
14	HSS10X3X0.5	0.562	W18X86	0.686	W14X257	0.305
13	HSS10X3X0.5	0.743	W18X86	0.686	W14X257	0.453
12	HSS16X3X0.5	0.663	W18X97	1.000	W14X283	0.588
11	HSS16X3X0.5	0.706	W18X97	0.889	W14X283	0.771
10	HSS16X3X0.5	0.750	W18X97	0.890	W14X283	0.952
9	HSS18X3X0.5	0.739	W18X106	0.945	W14X426	0,764
8	HSS18X3X0.5	0.780	W18X106	0.902	W14X426	0.896
7	HSS18X3X0.5	0.817	W18X106	0.902	W14X426	0.987
6	HSS22X3X0.5	0.790	W18X119	1.000	W14X730	0.682
5	HSS22X3X0.5	0.837	W18X119	0.950	W14X730	0.772
4	HSS22X3X0.5	0.893	W18X119	0.951	W14X730	0.862
3	HSS24X3X0.5	0.914	W18X130	0.976	W14X873	0.799
2	HSS24X3X0.5	0.958	W18X130	0.941	W14X873	0.881
1	HSS24X3X0.5	1.000	W18X130	0.941	W14X873	0.962

3.2. Linear Response History Analysis
LRHA were conducted on structures that had been designed using MRSA. In LRHA, three pairs of acceleration earthquake history data were utilized. The earthquake history data were spectrally matched with the design response spectra. The earthquake history data used in this study were: El Centro-Imperial Valley (1979), Loma Prieta-USA (1989), and El Mayor-Mexico City (2010).

For LRHA, the parameters reviewed were the inter-story drift structure and the axial force of the braces. The value of the inter-story drift being considered was the largest drift values obtained from all stories. Figure 5 shows the inter-story drift for all models. As can be seen, the limit for inter-story drift is exceeded for models with 8-story and 15-story, of which both models were with building heights of more than 10m.

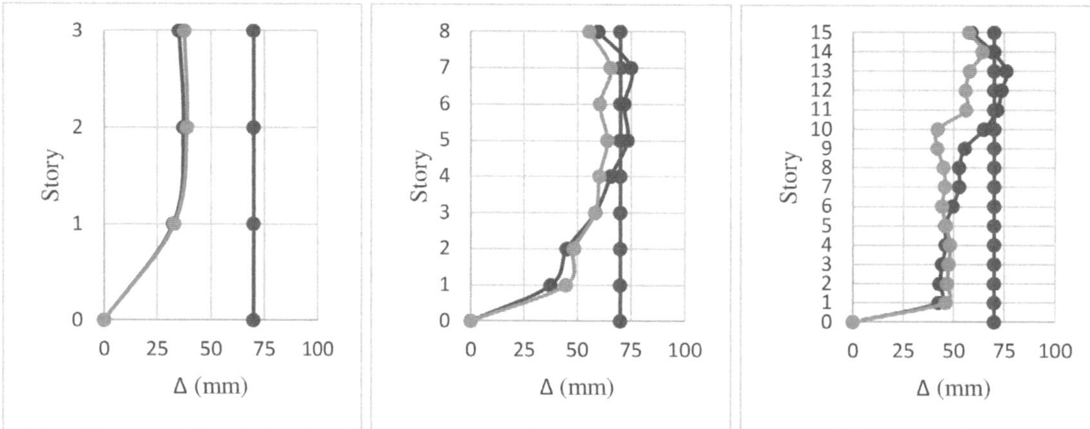

(a) (b) (c)

Figure 5. Inter-story Drift LRHA (a) 3-story model; (b) 8-story model; (c) 15-story model

From the comparison of axial forces on the ToBF elements of structures with 3-story, 8-story and 15-story, the results show that the axial forces occurred in the elements obtained using LRHA were greater than those of MRSA. This finding leads to necessities in redesigning some of the structural elements in order to meet the requirements, namely, DCR < 1.

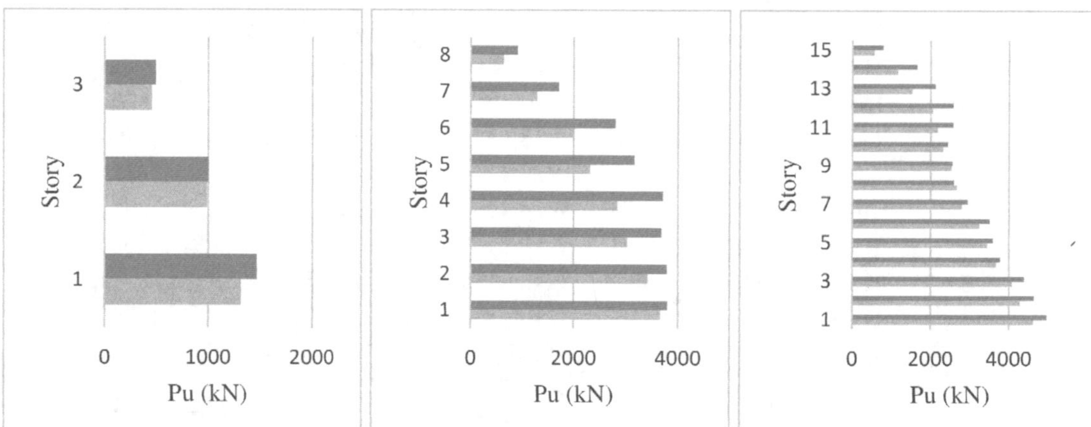

(a) (b) (c)

Figure 6. Axial Force of Braces from LRHA (a) 3-story model; (b) 8-story model; (c) 15-story model

3.3. *Performance Level of Structures with ToBF*

The performance of the structure with ToBF under seismic loading was obtained using a pushover analysis. The pushover analysis can provide an overview of the collapse mechanism of the structure, which is characterized by the formation of plastic hinges on the structural elements and can be used to determine the performance level of the structure. To determine the performance level of the structure, the performance levels for Immediate Occupancy (IO), Life Safety (LS) and Collapse Prevention (CP) criteria on the capacity curve was utilized, based on ASCE 41-13. The performance point was calculated based on the structural response due to the demand spectra, and then compared to the performance levels.

Figure 7 shows the capacity curves obtained from all models with the performance levels marked on the curves. The results of the pushover analysis show that the performance point for all models is Immediate Occupancy (IO). Therefore, it seems that all structures could be expected to perform satisfactorily under the design earthquake.

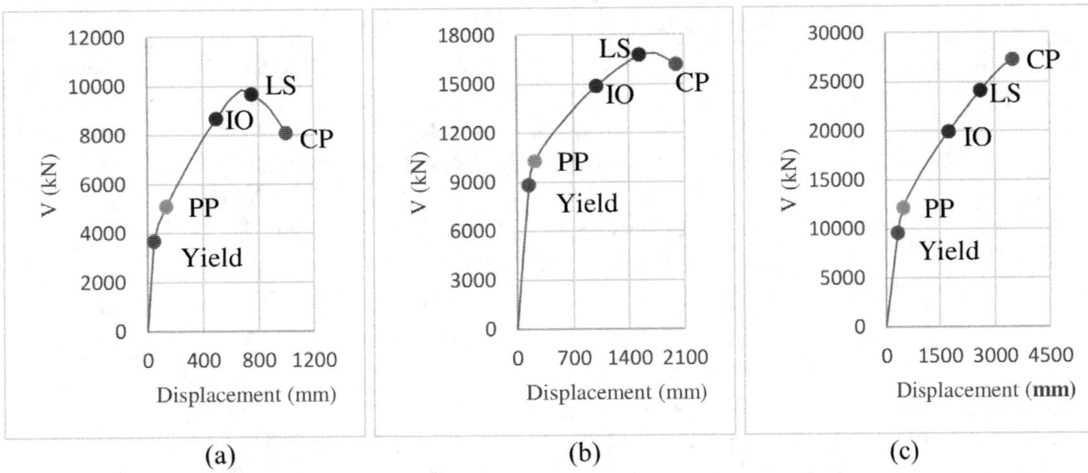

Figure 7. Capacity Curve (a) 3-story model; (b) 8-story model; (c) 15-story model

3.4. Comparison of Analytical Study

The structural design using MRSA for all three structures showed that the brace dimensions used for the 3-story structure were still available from AISC sections, while the brace dimensions for the 8-story and 15-story structures were customs and had to be designed manually. The column dimensions used in the 8-story and 15-story structural design already use the maximum column dimensions available in the AISC column-sections. The results of the analysis using MRSA show that all three structures satisfy the requirements of SNI 1726-2019. Thus, it shows that buildings with ToBF that exceed the height limit of 10m for seismic design category D can be designed with MRSA (structures were assumed to be OCBF in SNI 1726-2019), such as model structures of 8-story (28 m) and 15-story (52.5 m).

When all models were re-evaluated using LRHA, the results show that the requirements were not satisfied in terms of demand-capacity ratio and inter-story drift for structures with height exceeding 10m (8-story and 15-story). Hence, the height limitation is valid for this structure, which shows that structures with ToBF exceeding 10m do not satisfy code requirements.

The structural performance level for each model was obtained using pushover analysis. The study reveals that structures with heights exceeding the 10m height limit in SNI 1726-2019 have lower base shear at first yield when compared to the design base shear. This result indicated that such structure is weaker than it was designed, hence yielded at a lower base shear. Thus, the structures would require smaller R value to be used for design than what was utilized in this study (R=3.25). Meanwhile, structures adhered to the height limitation did not encounter similar problems. Nevertheless, the sequence of plastic hinges development for all structures was as expected, with yielding limited to the braces only. Moreover, the study reveals that the ToBF systems were successful for seismic resistant systems, with the performance level for all structures is Immediate Occupancy, even when the height limitation was omitted. Therefore, it can be concluded that the structures with ToBF were adequate to be used in seismic-prone areas with seismic design category (SDC) D and the building height exceeds the height limitation in SNI 1726-2019, provided rigorous analysis were conducted to ensure the structural behavior is satisfactory.

Table 4 provides the overall results for this study.

Table 4. Summary of Analytical Study

Analysis		**Structure 3-story**	**Structure 8-story**	**Structure 15-story**
Modal Response Spectrum Analysis	Brace Dimension	Available Section	Built-Up Section	Built-Up Section
	Column Dimension	W10X100	W10X416	W14X873
	Inter-story drift $<0.2h_x$	OK	OK	OK
	Base Shear (kN)	3619.6	9785.1	12219.7
	P-delta, $\theta < 0.1$	OK	OK	OK
	DCR of the Brace < 1	OK	OK	OK
	Percentage in Resisting Seismic Force	98.50%	99.45%	99.39%
Linear Response History Analysis	Inter-story drift $< 0.2h_x$	OK	Not OK	Not OK
	DCR of the Brace < 1	OK	Not OK	Not OK
Performance Level of Structure	$V_{yield} > V_{design}$	OK	Not OK	Not OK
	Performance Level	IO	IO	IO

Based on the above results, it is necessary to re-evaluate the R factor used for structures with ToBF system for general application. If ToBF system is to be used for seismic resistant structure with KDS D and omitting the height limitation set in the building code, the required R factor for this type of structure may need to be smaller than the R factor used in this study (R = 3,25), so that the structure can perform satisfactorily.

4. Conclusion

From the analytical study conducted on structure with ToBF using MRSA, LRHA and pushover analysis, the following conclusions can be drawn:

1. The numerical study of the ToBF structure was carried out for buildings in SDC D with various floors and heights, i.e. 3 floors (10,5 m), 8 floors (28 m) and 15 floors (52,5 m). Although SNI 1726-2019 states that ToBF system can only be used as a seismic load-bearing system for SDC A to C, with a height of up to 10 m, the study conducted that structural design can be carried out for the three ToBF structural models using MRSA and still satisfy all requirements of SNI 1726-2019.
2. The ToBF structures designed using MRSA, when reviewed with LRHA, indicated that the structure with height exceeding 10 m could not satisfied the requirements in terms of axial capacity of braces and the inter-story drift. Following redesign of structures based on results from LRHA, the revised models were able to comply with code requirements. Hence, LRHA provides better estimate for structural responses of ToBF structures, especially when code limitations were not adhered to.
3. The ToBF structures designed using MRSA when reviewed for their performances using pushover analysis according to ASCE 41-13, show that the performance levels of all buildings are Immediate Occupancy (IO) level. However, for structural models with height exceeding 10m, first yielding occurred at a lower base shear compared to the design base shear. This indicates that the R value used may need to be revised with a lower value.
4. Further evaluations regarding the inter-story drift, the performance level of the elements, and the sequence of plastic hinges development reveal that ToBF designed using LRHA can be used as a seismic force-resisting system. However, the value of R used for ToBF is recommended to be taken at a smaller value (lower than such value for OCBF or other structural system) to ensure that the structural performance is satisfactory.

References

[1] Bruneau M., Uang C M, and Sabelli R, 2011. *Ductile Design of Steel Structures 2nd Edition* (McGraw-Hill)
[2] Shariati M, Majid L, Shervin M. Ali S, Nguyen T, 2020. *Evaluation of Seismic Performance Factors for Tension-only Braced Frames*. Techno Press, 35 (4) 599-609.
[3] American Institute of Steel Construction 2016. *AISC 314-16 Seismic Provisions for Structural Steel Buildings* (AISC).
[4] American Institute of Steel Construction 2016. *AISC 360-16 Specification for Structural Steel Buildings* (AISC).
[5] Badan Standarisasi Nasional 2020. *SNI 1727-2020 Beban Desain Minimum dan Kriteria Terkait untuk Bangunan Gedung dan Struktur Lain* (Standar Nasional Indonesia)
[6] Badan Standarisasi Nasional 2019. *SNI 1726-2019 Standar Perencanaan Ketahanan Gempa untuk Struktur Bangunan Gedung dan Non Gedung* (Standar Nasional Indonesia)

Acknowledgement

This study was supported by the Faculty of Civil and Environmental Engineering, Institut Teknologi Bandung. The support is gratefully acknowledged.

IOP Conf. Series: Earth and Environmental Science 1244 (2023) 012010 doi:10.1088/1755-1315/1244/1/012010

Performance Evaluation of Structures with Buckling Restrained Braced Frame Under Earthquake Load

D Y Sari[1] and E Lim[2]

[1]Graduate School of Civil Engineering, Faculty of Civil and Environmental Engineering, Institut Teknologi Bandung, Indonesia
[2]Faculty of Civil and Environmental Engineering, Institut Teknologi Bandung, Indonesia

Email: deayunitasari8@gmail.com, erwin_petrus@yahoo.com

Abstract. Structures with buckling-restrained braced frames (BRBF) resist lateral load, i.e., earthquake, using special bracings that yield under axial tension and compression loads. Following the applicable building code SNI 1726-2019, the design for the BRBF structure has a limitation on the structure height for seismic design category D, which is 48 meters. Three structures are modelled: a 5-story building, a 10-story building, and a 20-story building, to evaluate the height limitation on the code. The initial design of the structures used modal response spectrum analysis (MRSA), then the structures were checked using linear response history analysis (LRHA). Finally, a pushover analysis was conducted to obtain the structural performance level. The study revealed that the structural design for BRBF with MRSA satisfies all requirements of SNI 1726-2019. However, the results of the evaluation of BRBF with LRHA show a structure exceeding the height limit does not meet the requirement for inter-story drift. The pushover analysis showed that the structure's performance reached the Immediate Occupancy (IO) level. Thus, all structures with BRBF in this study that were designed using MRSA have satisfactory performance under the seismic design load, even though the height limit has been exceeded.

1. Introduction

The buckling restrained braced frame system (BRBF) is an improvement of the concentrically braced frame system (CBF) to solve the buckling problem in conventional braces [1]. The buckling restrained braces have the typical tension and compression forces based on yielding, so that the buckling restrained braces have a symmetrical hysteretic character (figure 1).

A typical BRB element consist of a steel core-plate that has ductile properties, and steel tube casing filled with grout or concrete to resist the buckling of the steel core, with special connection at both ends (figure 2). The entire effective length of the steel core that receives the axial force will be restrained by the casing so that there is no slenderness in the steel core and eliminate the buckling failure mode of the core [2]. From all components of the buckling-restrained brace, the steel core of the buckling restrained brace is the main component to define the member capacity, and it will be designed in this research.

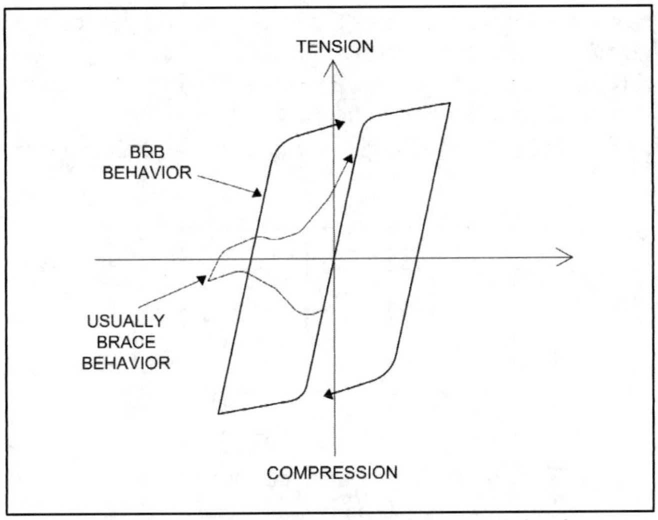

Figure 1. Buckling-Restrained Brace Behavior

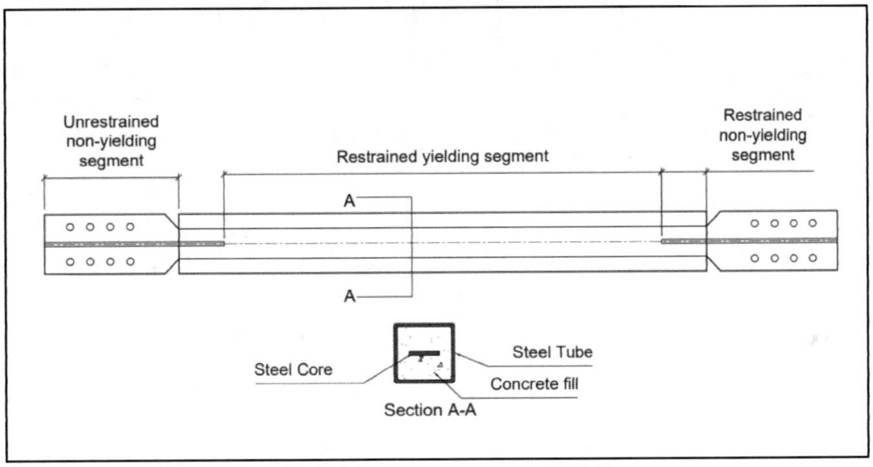

Figure 2. Components of BRB

The BRBs are the yielding element, which is expected to undergo significant inelastic deformation during a design-level earthquake, while all other elements in the system are capacity designed [3]. To design an earthquake-resistant structure with BRBF in Indonesia, it must meet the requirement of SNI 1726-2019. Based on this code, the design for the BRBF structure as a seismic force-resisting system is to be conducted with the seismic parameters of R = 8, Cd = 5, and Ωo = 2,5. Moreover, the seismic design categories D, E, and F on BRBF have structural height limitations. The limitation on the structure height for seismic design category D is 48 meters [4]. Therefore, the purpose of this research is to design a buckling-restraint brace frame system using a procedure suggested by the code, i.e., modal response spectrum analysis, then check the result using another design procedure, i.e., linear response history analysis, and evaluate the result. Next, conduct a pushover analysis to find out the failure mechanism and the structural performance of the BRBF system related to the height limitation in SNI 1726-2019.

2. Structural Models

The structure designed is an office building with a symmetrical floor plan measuring 30 x 30 meters (figure 3) with an inter-span length of 6 m and a height between floors of 3.5 meters. The building is assumed to be in the city of Bandung.

The structural model is 3D, with the BRBF as the single lateral load-resisting system. The bracing configuration used is diagonal braces. In this study, the structural design of the BRBF was varied with regards to building heights, that is, a 5-story building with a height of 17.5 m, a 10-story building with a height of 35 m, and a 20-story building with a height of 70 m (figure 4). The variation was opted to determine whether structures that exceed the height limitation can still meet the code requirements under earthquake loads.

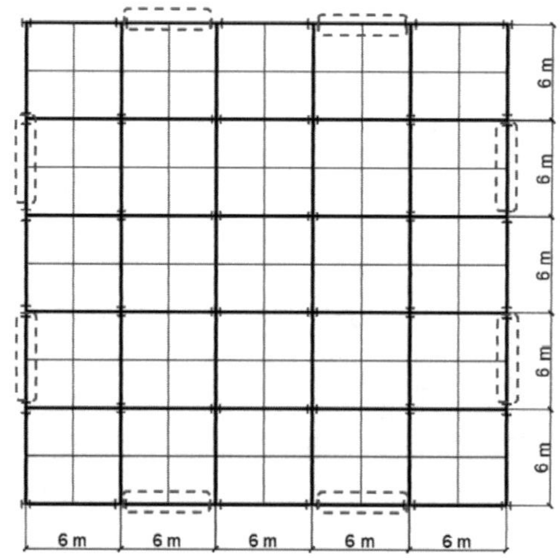

Figure 3. Floor Plan

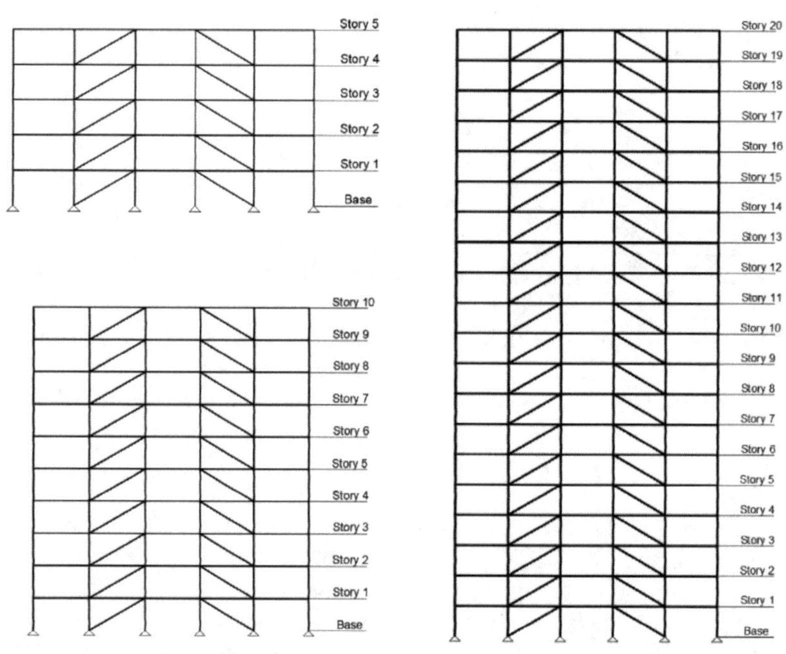

Figure 4. (a) 5-story building; (b) 10-story building; and (c) 20-story building

The material specifications for beams and columns use ASTM A992 Gr.50 with a yield stress of 345 MPa, while the buckling-resistant braced steel core material uses ASTM A36 with a yield stress of 262 MPa.

3. Structural Analysis

3.1. Modal Response Spectrum Analysis

All BRBF structural systems were designed using modal response spectrum analysis (MRSA). The analysis must include sufficient modes to obtain combined mass participation of 100% of the structure's mass, and the base shear from this analysis was scaled using the result from static analysis. The base shear force obtained after scaling the base shear is as follows: for the 5-story building, $V_d = 2343.52$ kN; for the 10-story building, $V_d = 2892.918$ kN; and for the 20-story building, $V_d = 4344.216$ kN. This study found that the BRBF structure resists about 98-99% of earthquake load, thus clearly indicating them as the lateral resisting system for the structure. The story drift was also evaluated and found that they did not exceed the story drift limit, $\Delta_a = 0{,}020h_{xx}$ (figure 5).

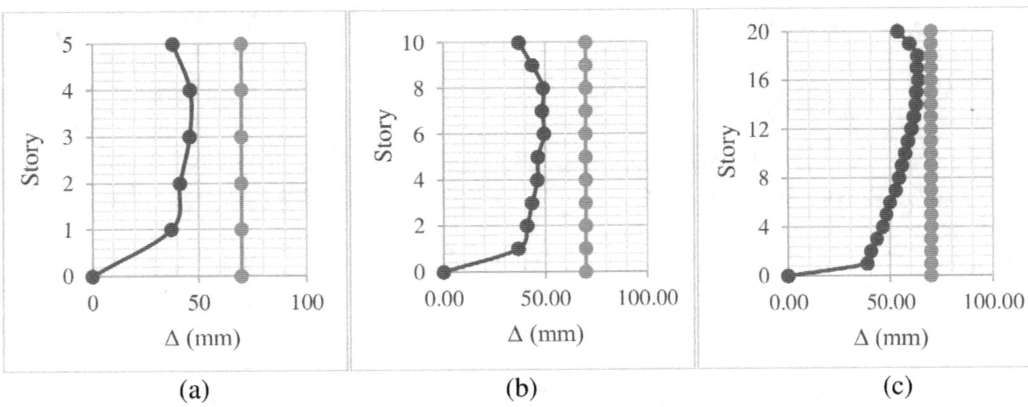

Figure 5. Interstory Drift (a) 5-story building; (b) 10-story building; dan (c) 20-story building

Table 1. Results of the Design of Structural Elements in a 5-Story Building

Story	Brace		Beam		Column	
	Dimension	Ratio	Dimension	Ratio	Dimension	Ratio
5th story	CoreBRB 1.75	0.735	W12X50	0.165	W10X68	0.144
4th story	CoreBRB 3	0.941	W12X50	0.421	W10X68	0.394
3rd story	CoreBRB 4.25	0.946	W12X50	0.526	W12X87	0.470
2nd story	CoreBRB 5.5	0.898	W12X50	0.627	W12X87	0.683
1st story	CoreBRB 6.5	0.931	W12X50	0.710	W12X87	0.895

Table 2. Results of the Design of Structural Elements in a 10-Story Building

Story	Brace		Beam		Column	
	Dimension	Ratio	Dimension	Ratio	Dimension	Ratio
10th story	CoreBRB 2	0.516	W12X58	0.155	W14X120	0.080
9th story	CoreBRB 3.5	0.668	W12X58	0.410	W14X120	0.238
8th story	CoreBRB 3.5	0.832	W12X58	0.363	W14X120	0.364
7th story	CoreBRB 4.75	0.806	W12X58	0.503	W14X120	0.513
6th story	CoreBRB 4.75	0.862	W12X58	0.458	W14X120	0.659
5th story	CoreBRB 6	0.800	W12X58	0.590	W14X176	0.563
4th story	CoreBRB 6	0.866	W12X58	0.544	W14X176	0.676
3rd story	CoreBRB 7	0.865	W12X58	0.646	W14X233	0.606
2nd story	CoreBRB 7	0.903	W12X58	0.601	W14X233	0.704
1st story	CoreBRB 8.5	0.909	W12X58	0.747	W14X233	0.805

Design structural elements with the BRBF system according to AISC 341-16 [5]. The structural design obtained cross-sectional dimensions of braces, beams, and columns for all models are given in table 1 to table 3. The demand-capacity ratio (DCR) for all elements meets the requirements of <1.

Table 3. Results of the Design of Structural Elements in a 20-Story Building

Story	Brace		Beam		Column	
	Dimension	Ratio	Dimension	Ratio	Dimension	Ratio
20th story	CoreBRB 2	0.591	W18X60	0.271	W14X193	0.055
19th story	CoreBRB 3.5	0.703	W18X60	0.488	W14X193	0.095
18th story	CoreBRB 3.5	0.813	W18X60	0.426	W14X193	0.257
17th story	CoreBRB 6	0.692	W18X60	0.781	W14X193	0.373
16th story	CoreBRB 6	0.689	W18X60	0.664	W14X193	0.487
15th story	CoreBRB 7.5	0.638	W18X76	0.577	W14X342	0.354
14th story	CoreBRB 7.5	0.635	W18X76	0.530	W14X342	0.435
13rd story	CoreBRB 8.5	0.614	W18X76	0.620	W14X342	0.522
12nd story	CoreBRB 8.5	0.622	W18X76	0.579	W14X342	0.609
11st story	CoreBRB 9.5	0.621	W18X76	0.671	W14X342	0.700
10th story	CoreBRB 9.5	0.634	W18X76	0.633	W14X426	0.638
9th story	CoreBRB 10	0.654	W18X76	0.673	W14X426	0.714
8th story	CoreBRB 10	0.691	W18X76	0.654	W14X426	0.790
7th story	CoreBRB 10	0.726	W18X76	0.647	W14X426	0.865
6th story	CoreBRB 11	0.739	W18X76	0.724	W14X426	0.943
5th story	CoreBRB 11	0.770	W18X86	0.607	W14X605	0.715
4th story	CoreBRB 11	0.811	W18X86	0.601	W14X605	0.772
3rd story	CoreBRB 12	0.822	W18X86	0.663	W14X605	0.831
2nd story	CoreBRB 12	0.846	W18X86	0.628	W14X605	0.890
1st story	CoreBRB 13	0.915	W18X86	0.696	W14X605	0.948

3.2. Linear Response History Analysis

The structure was analysed by linear response history after being designed with modal response spectra analysis to verify the design. Linear response history analysis requires three pairs of earthquake history data. The earthquake history data must be spectrally matched to the design response spectra. The earthquake records data used are:

a. El-Centro, Imperial Valley, California USA 1979
b. Chi-Chi, Taiwan 1999
c. Northridge USA 1994

The structural responses were evaluated for all three models. One of the response parameters to be checked is the inter-story drift. The story drift is reviewed based on the envelope value, and as shown in figure 6, for the 20-story model, the limit was exceeded.

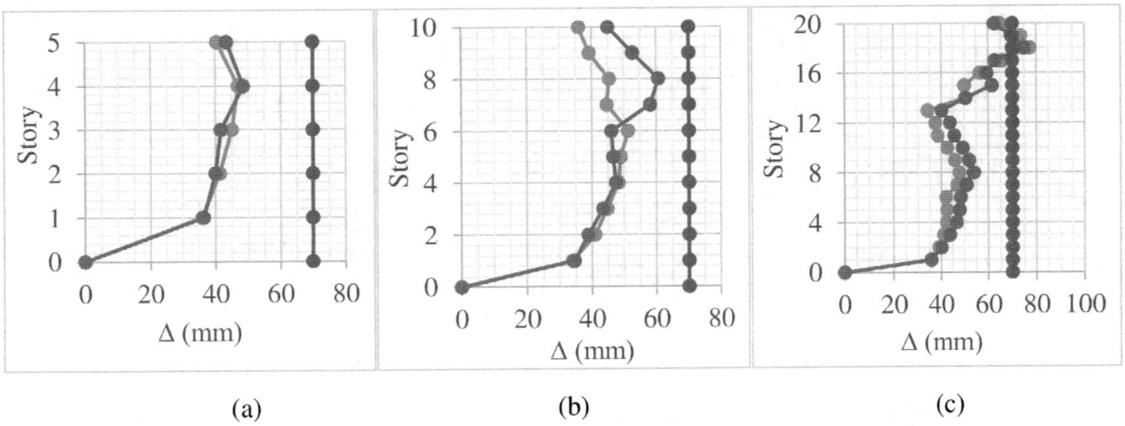

Figure 6. Inter-story drift with LTHA (a) 5-story building; (b) 10-story building; dan (c) 20-story building

The axial force of braced elements with LRHA in the three-building models is larger when compared to the axial force of braces using MRSA, and the difference is significant for taller models. Checking on the bracings' capacity, the axial forces calculated by LRHA do not meet the requirements that the DCR < 1.0. That means that the previous braces' cross-section designed with MRSA must be redesigned using LRHA.

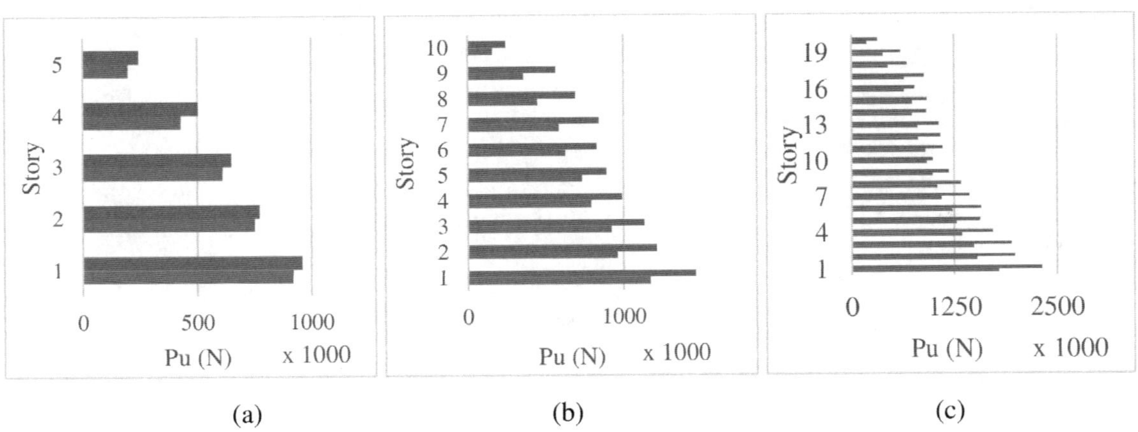

Figure 7. The axial force of braced (a) 5-story building; (b) 10-story building; dan (c) 20-story building

3.3. Structure Performance Level

The performance of the structure was evaluated using pushover analysis, utilizing the force-displacement relationship capacity curve and the demand forces. The structural performance points are obtained from this curve, which is then compared with the structural performance levels based on ASCE 41-17 [6]. The capacity curve, performance levels, and performance points for all three models are presented in figure 8. The result shows that the structural performance of the three-building models is Immediate Occupancy (IO), and none of the elements reached the Collapse Prevention (CP) level. A closer look at the distribution of the plastic hinges in the three models reveals that the braced element yields earlier than the other elements, as expected in the design.

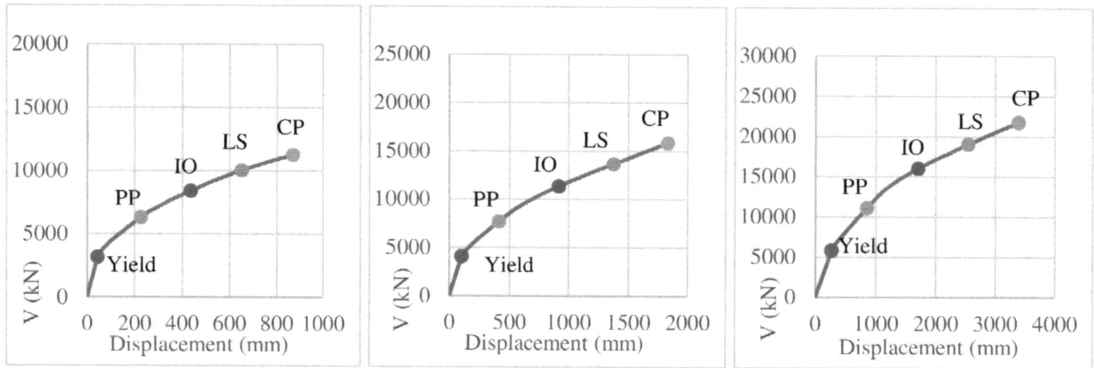

(a) (b) (c)

Figure 8. Capacity Curve (a) 5-story building; (b) 10-story building; dan (c) 20-story building

3.4. Comparison of Structural Analysis

The analytical results are presented in table 4. It can be summarized that all 5-story buildings, 10-story buildings, and 20-story buildings can be designed with MRS analysis, and the design results have met the requirements of SNI 1726-2019. However, when reviewed with LRHA, the inter-story drift exceeds the drift limit for the 20-story model. Moreover, the axial force of the braces has increased so the section must be redesigned if using LRHA.

The analytical results suggest that although all buildings show adequate performance, LRHA is the more conservative approach in designing the BRB models. It offers advantages for buildings that have a height exceeding the requirements in SNI, to ensure that the dimensions of the designed structure are appropriate and meet the requirements.

If viewed from the results of the pushover analysis, the level of structural performance in a 5-story building, a 10-story building, and a 20-story building is at the Immediate Occupancy (IO) level. It shows that BRBF designed using MRSA can be used in high-rise buildings because of its satisfactory structural performance, even when the building height exceeds the limitation.

Table 4. Summary of Structural Analysis Result

Analysis		5-Story Building	10-Story Building	20-Story Building
Model Response Spectrum Analysis	Base Shear (kN)	2343,52	2892,92	4344,22
	Percentage of resist earthquake forces	98,85 %	99%	97.65%
	P delta effect ($\theta < 0{,}1$)	OK	OK	OK
	Inter-story drift < 0,02 hs	OK	OK	OK
	Brace ratio capacity < 1	OK	OK	OK
Linear Response History Analysis	Inter-story drift < 0,02 hs	OK	OK	Not OK
	Brace ratio capacity < 1	Not OK	Not OK	Not OK
Pushover Analysis	Maximum deformation structure performance level	IO	IO	IO

4. Conclusion

Based on the results of the numerical study of buildings with buckling-resistant braced frame systems using modal response spectrum analysis (MRSA), linear response history analysis (LRHA), and pushover analysis, the conclusions are as follows:

1. The structural design for the three BRBF structural models was conducted using MRSA, and all structures meet SNI 1726-2019 requirements, including the 20-story model, which exceeds the height limitation.
2. The structures obtained from MRSA, when viewed with LRHA, show that they do not meet the requirements in terms of the inter-story drift in a building that exceed the height limit based on SNI 1726-2019, and larger axial forces for braces.
3. The structures obtained from MRSA have adequate performance levels, i.e., Immediate Occupancy (IO), and no element reaches the Collapse Prevention (CP) condition. The distribution of the plastic hinges in the three models has also been as expected: the braced element yields earlier than the other elements. Thus, the BRBF designed with MRSA has good performance and can be utilized, even when the height limit is exceeded.
4. The BRBF structure with a height exceeding the limit in SNI 1726-2019 must be designed with LRHA, so that the inter-story drift of the structure obtained meets the requirements, to improve the performance of the structure and obtain design results that meet all the requirements of SNI 1726-2019.

References

[1] Naghavi, M., & Malekinejad, M. 2019 *A Numerical Study of the Behavior of Steel Frame with Concentric Buckling Restrained and Conventional Braces*. Journal of Building Material Science, 1(1), 18.

[2] Lopez, W. A., &Sabelli, R. 2004. *Seismic Design of Buckling-Restrained Braced Frames*. Steel TIPS Report, Structural Steel Education Council, Chicago, IL.

[3] Kersting, R. A., Fahnestock, L. A., & López, W. A. 2015 *Seismic Design of Steel Buckling-Restrained Braced Frames A Guide for Practicing Engineers About the Review Panel*. NEHRP Seismic Design Technical Brief, 11

[4] Badan Standarisasi Nasional 2019 *Tata Cara Perencanaan Ketahanan Gempa untuk Struktur Bangunan Gedung dan Nongedung*. SNI 1726:2019

[5] American Institute of Steel Construction 2016 *Seismic Provinsions for Structural Steel Buildings*.ANSI/AISC 341-16

[6] American Society of Civil Engineers 2012 *Seismic Evaluation and Retrofit of Existing Building*. ASCE/SEI 41

Seismic Performance of Ordinary and Damage-tolerant Concrete Beam-column Joints under Reversed Cyclic Loading

P Suprobo[1*], A Tambusay[1,2], B Suryanto[2], J.J.M Nelson[2]

[1]Department of Civil Engineering, Faculty of Civil, Planning and Geo Engineering, Sepuluh Nopember Institute of Technology, Surabaya, East Java, Indonesia.
[2]Institute of Built Environment, School of Energy, Geoscience, Infrastructure and Society, Heriot-Watt University, Edinburgh, United Kingdom.

*Corresponding author's e-mail: priyo@ce.its.ac.id

Abstract. The adoption of the capacity design principle in modern codes of practice allows for multi-storey frame structures to exhibit ductile behaviour in a major earthquake without collapsing. Whilst the risk of shear failure in regions of large inelastic deformation can be largely reduced, the detailing requirements in capacity design often leads to reinforcement congestion that adds difficulty during the construction process, due to the high amount of transverse reinforcement required to provide shear resistance and confinement. The work presented in this paper forms a part of a wider research programme that aims to simplify the detailing requirements of transverse reinforcement in the joint and plastic hinge regions of moment-resisting frame structures. To this end, damage-tolerant concrete known as engineered cementitious composite (ECC) was used to substitute concrete in these critical regions. In this article, the results of reversed cyclic tests on two half-scale exterior beam-column joints, constructed with ordinary concrete and ECC, are presented. Key features of the response such as the reversed cyclic hysteresis loops, stiffness degradation and cumulative energy dissipation are discussed, along with crack development under reversed cyclic loading in the form of digital strain maps.

Keywords: beam-column joint; damage tolerant; digital image correlation; engineered cementitious composite; moment-resisting frame, plastic hinge.

1. Introduction

The seismic performance of a reinforced concrete frame relies on the ability of its structural elements to dissipate energy without undergoing a significant loss of strength and stiffness. In a frame structure, critical regions that are ill-fit for energy dissipation, such as the beam-column connections, must be designed to withstand high levels of load reversals without losing their structural integrity. This is normally achieved through the provision of closely spaced, transverse reinforcement, which not only serves to improve the joint shear strength but also to provide confinement to the joint core [1]. From major seismic events that have taken place around the world, however, it has been reported that joint failure is still a pressing problem and has been identified as one of the most frequently reported mechanisms responsible for partial or even full collapse of a structure [2]. In some cases, joint failure can be traced back to the high detailing requirements in the beam-column joint region, which result in congestion issues that adds difficulties in the construction process. This often leads to construction errors such as honeycombing in the joint core or inadequate anchorage of the beam longitudinal reinforcement, making the joint more vulnerable to earthquakes. In other critical regions, where the plastic hinge is expected to occur (i.e., at the beam end), the issue of reinforcement congestion is generally less problematic but may still pose some difficulties during the construction process due to the high amount of transverse reinforcement required. Moreover, due to the brittle nature of concrete, there is a strong likelihood that under a moderate to large earthquake, damage in the plastic hinge and joint regions would become too costly to repair which, in some cases, requires complete demolition of the entire structure.

The work presented in this article describes an attempt to substitute concrete throughout the joint and plastic hinge regions with a form of damage-tolerant concrete, known as an engineered cementitious composite (ECC) [3,4]. ECC are a group of cementitious structural materials which are particularly known for their tensile strain-hardening properties (see Figure 1). In this article, the effectiveness of ECC in reducing reinforcement congestion issues in the joint and plastic hinge regions is investigated. Emphasis is placed on comparing the damage progression in the ECC under cyclic loading to ordinary concrete.

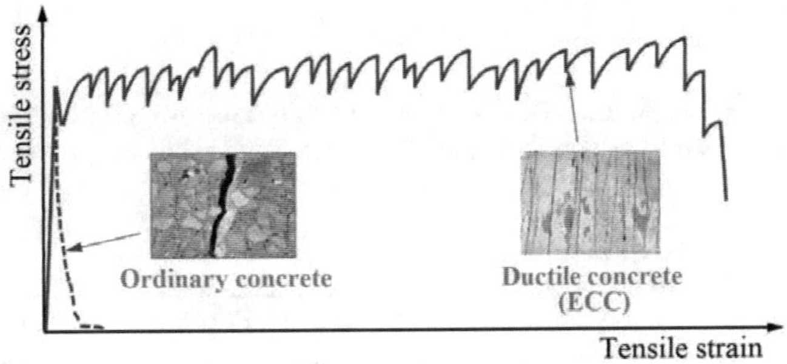

Figure 1. Schematic comparison of ECC and concrete under tensile loading.

2. Experimental Programme

The test programme involved the testing of two exterior joints, each comprising a 1.2m long column and 1.15m long beam. Both columns and beams had the same 200mm square cross-section. Figure 2 presents the dimensions, reinforcement detailing and material configuration for each specimen. In the beam, five 10mm longitudinal reinforcements were provided at the top and bottom of the beam, with 6mm transverse reinforcement of varying spacing. The column had eight 10mm longitudinal bars and 8mm transverse reinforcement throughout the column and the joint. Both specimens were designed according to the weak beam-strong column principle [5].

Each specimen was labelled based on the material utilised in its joint and plastic hinge region, as well as the spacing of the transverse reinforcement in the plastic hinge region. The first specimen, specimen RC-025d, was the control specimen and was constructed with ordinary Portland cement concrete. The numbers refer to the spacing of transverse reinforcement in the plastic hinge region of the beam (~0.25d or in this case, 40mm) where d is the effective depth to the centroid of the longitudinal reinforcement. The second specimen, ECC-050d, was constructed with ECC throughout the plastic hinge region and joint core (indicated by the shaded area on the right hand-side of Figure 1). In this specimen, the transverse reinforcement spacing in the plastic hinge region was ~0.50d (= 80mm). Outside of the plastic hinge region in the beam, both specimens had transverse reinforcement with a spacing of 80mm (i.e., s = ~0.50d) to fulfil the requirement of ACI 318-19 [6].

Regarding the transverse spacing in the joint, in the control specimen (i.e., RC-025d), a spacing of 50mm was used in the joint as well as throughout the entire column. In specimen ECC-050d, the transverse spacing was increased to 100mm to determine the effectiveness of ECC in providing confinement to the joint. Outside of the joint region, the ECC specimens had the same 50mm spacing for the transverse reinforcement, as in specimen RC-025d.

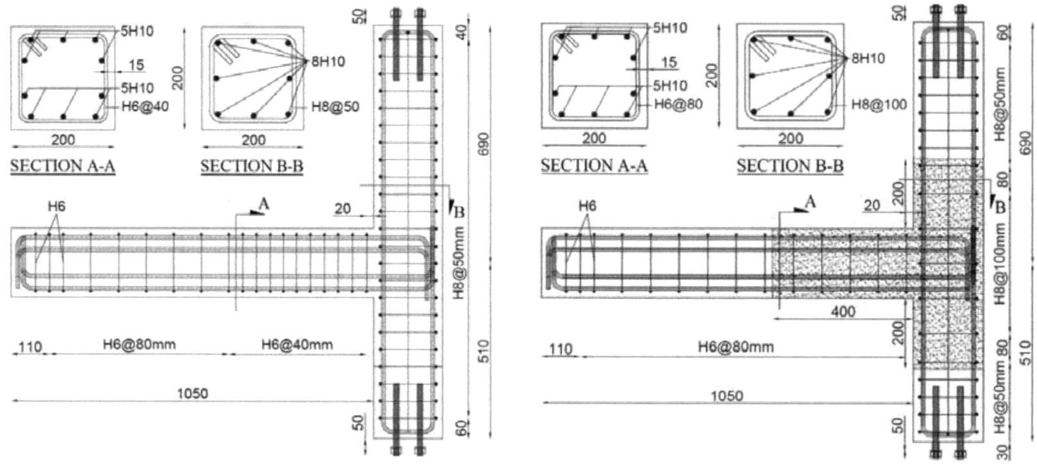

Figure 2. Dimensions, reinforcement details and material configuration for specimens: (a) RC-025d; and (b) ECC-050d (adapted from [5]). The shaded area in (b) is the ECC.

2.1. Materials and Fabrication

The concrete used in the specimens was Grade C30/37 ordinary concrete, provided by a local ready-mix supplier. Table 1 describes the mix used for the ECC specimen. It comprised a binder, which used a combination of CEM I 52.5N cement and fine fly-ash, fine silica sand (with a mean particle size of 120μm), a high-range water-reducing (HRWR) admixture to improve workability and aid fibre dispersion, and Polyvinyl Alcohol (PVA) fibres at an amount of 2% by volume. All were mixed using a 15-litre planetary motion mixer at a water/binder ratio of 0.28. The reinforcement used in all specimens was high-yield strength deformed reinforcement of Grade B500.

Table 1. ECC mix composition used in the experiment (kg/m3).

Cement	Fly-ash	Sand	Water	HRWR	PVA fibre
431	776	388	338	6.04	26

The fabrication of the ECC-050d specimen was undertaken in two main stages. The first stage involved the casting of the ECC precast segment which formed the joint and plastic hinge regions (the shaded region in Figure 2). In the regions separating the ECC and concrete, shear keys spanning the width of the specimen were provided to aid in the mechanical bonding between the ECC and the concrete. The second stage of the fabrication process was undertaken the next day and involved the casting of concrete for the remaining parts of the beam and columns. On the same day, specimen RC-025d was fabricated, using the same batch of concrete used for the other specimen. Both specimens were removed from the formwork after two days and wrapped with wet burlap and several layers of cling film to ensure that they remained fully saturated.

During the curing process, both specimens were left in the laboratory (10±5°C) until they were required for testing. Specimen ECC-050d was tested around 3.5 weeks after casting and specimen RC-025d a few days thereafter. The average cube compressive strengths for the concrete and ECC on the day of testing were 38.7MPa and 49.5MPa (determined on 100mm and 50mm cubes, respectively).

2.2. Test Setup and Instrumentation

Figure 3(a) presents a schematic diagram of the setup used in the experiment. Each specimen was connected to a reaction frame by a pinned support at the bottom of the column and a roller at the end of

IOP Conf. Series: Earth and Environmental Science 1244 (2023) 012011 doi:10.1088/1755-1315/1244/1/012011

the beam, at a distance of 1.03m from the column centreline. These support conditions were used to simulate points of contra flexure in a frame structure subjected to lateral loading (see the shaded area in Figure 3(a)). Lateral loading was induced by a 200kN servo-hydraulic actuator that was positioned horizontally at a height of 1.2m from the bottom pinned support. This actuator was mounted to the loading frame and connected to the specimen at the top of the column by means of a flexible (pin) connection. The pin was connected to the specimen using four tie rods and a thick plate placed on the opposite face of the column. Using this special loading configuration, it was possible to push the specimen laterally regardless of the load direction (e.g., forward (+ve) or reverse (–ve) loading).

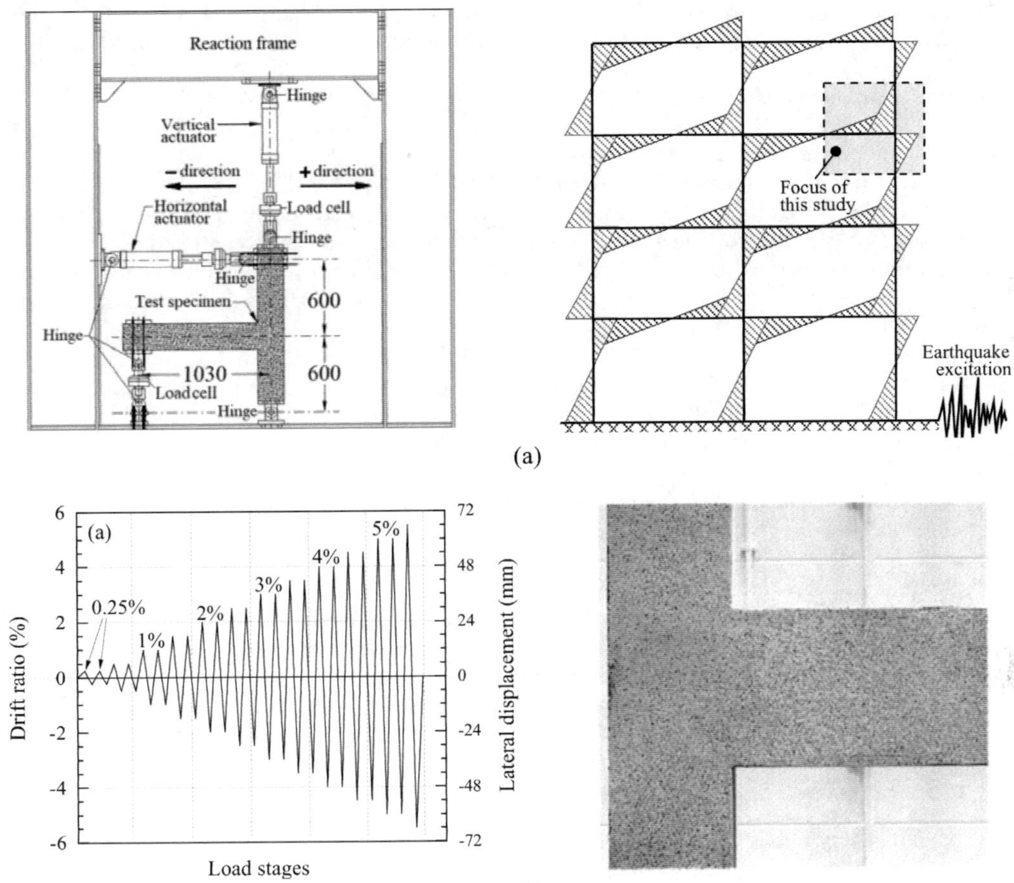

Figure 3. (a) Schematics of test setup and expected bending moment diagram in a frame under lateral loading, and (b) load protocol and speckled pattern for automated crack mapping (from [5]).

Another 200kN actuator was used to apply an axial load of 80kN to the top of the column, to simulate gravity loads. During testing, this load was first applied and then held constant throughout, in a load-controlled manner. Lateral displacements were then applied as two full loading cycles in both the "positive" and "negative" directions in accordance with the loading protocol presented in Figure 3(b). At each cycle, the lateral displacement is expressed as the drift ratio, which is defined as the ratio of the lateral displacement applied at the point of load application to the distance to the bottom hinge (e.g., 1.2m). Testing started at a drift ratio of 0.25% and this was then increased by 0.5% after the 0.5% drift cycle, ending at a drift ratio of 5% or when a significant strength degradation (> 40%) was observed.

To provide detailed observations on crack development, a combination of manual and automated crack mapping was utilised during testing. For the manual crack mapping, one side of the specimen was painted white and 50mm gridlines were drawn throughout the specimen. At the end of each cycle (both in the positive and negative directions), new crack formations were marked and photographed. For the automated (digital) crack mapping, the other side of the specimen was painted white and then sprayed with a random black speckle pattern (see Figure 3(b)). A digital single-lens reflex (DSLR) camera was setup on a rigid tripod at a distance of ~1m from the specimen and used to capture images throughout each stage of loading (24 images in total at each cycle). Once testing had concluded, the images were uploaded to an open-source Digital Image Correlation (DIC) software, Ncorr [7,8], and analysed to produce strain maps for the specimens. In this article, the crack maps obtained at 1% drift increments are presented to compare the damage progression in the ECC under cyclic loading to ordinary concrete.

3. Test Results and Discussion

3.1. Hysteresis Response and Crack Pattern

The hysteretic loops for specimens RC-025d and ECC-050d are presented in Figures 4(a) and (b). In these Figures, the lateral loads have been corrected for the horizontal component of the force applied on the vertical actuator (i.e., the horizontal component of the applied load is removed) and as such, they include the remaining vertical component of the force (which introduces P-delta effects) and the restoring force of the specimen. For reasons of clarity, the hysteretic responses at 1% drift increments are presented in Figures 5(a) and (b) in the form of exploded hysteresis plots, with each loop plotted and offset on the horizontal axis to highlight their distinctive shape at each drift level. Moreover, to provide insights into the extent and location of damage and crack development in the joint and plastic hinge regions at the same drift levels, the longitudinal strain maps obtained from the DIC analysis are displayed in Figures 6 and 7 for specimens RC-025d and ECC-050d, respectively.

With reference to the hysteretic response of specimen RC-025d, it is apparent from Figures 4(a) and 5(a) that the initial response is predominantly flexural (see the crack patterns displayed in Figure 6) and is characterised mainly by narrow hysteretic loops, which is indicative of low energy dissipation. This trend continues until approximately the 2% drift level (i.e., 24mm lateral displacement), at which point the hysteresis starts to develop more of a pinched response. This can be attributed to the yielding of reinforcement in the plastic hinge of the beam, which started from ~1.5% drift, and inelastic deformation of the joint due to bar slippage and early development of shear cracking [9]. The evidence of which could be obtained from the DIC strain maps presented in Figure 6, which shows that beyond 1% drift, there is a significant increase in strain at the face of column and damage is seen to have progressed into the joint core, as indicated by the initial development of X-shaped diagonal shear cracks. As the drift level increases (see Figures 4(a)), the lateral resistance continues to increase and reaches a maximum load of 31.1kN in the positive direction, during the first loading cycle of 2.5% drift. In the negative direction, the load reaches its maximum of 28.5kN at a slightly lower, 2% drift. Beyond these peak loads, a significant reduction in load-carrying capacity is evident. From 2.5% to 4% drift, for example, the peak lateral load decreases by ~20% and this continues to decrease by a further ~25% from 4% to 5% drift alone (see Figures 4(a) and 5(a)), which indicates rapid strength degradation. This is as a result of the combined effects of P-delta (see the insets in Figures 4(a) and (b)) and damage accumulation in the joint core. The evidence of which can be found from the DIC strain maps presented in Figure 6. During the post-peak regime (e.g., drift >2.5%), for example, it is evident that the strain values in the joint and plastic hinge regions remain high after reversed loading, indicating that the regions have developed significant accumulation of damage and plasticity due to the yielding and slippage of reinforcement, which prevent full closure of the cracks upon unloading. In the hysteretic response displayed in Figures 4(a) and 5(a), this is manifested in the increasing magnitude of the residual displacements of the specimen, which increases rapidly beyond ~2% drift. Moreover, from the DIC strain maps displayed in Figure 6, it is apparent that during the final few load cycles, the X-shaped cracks have propagated along the longitudinal reinforcement into the column sub-assemblage, creating a wedge-shaped mechanism of failure [10].

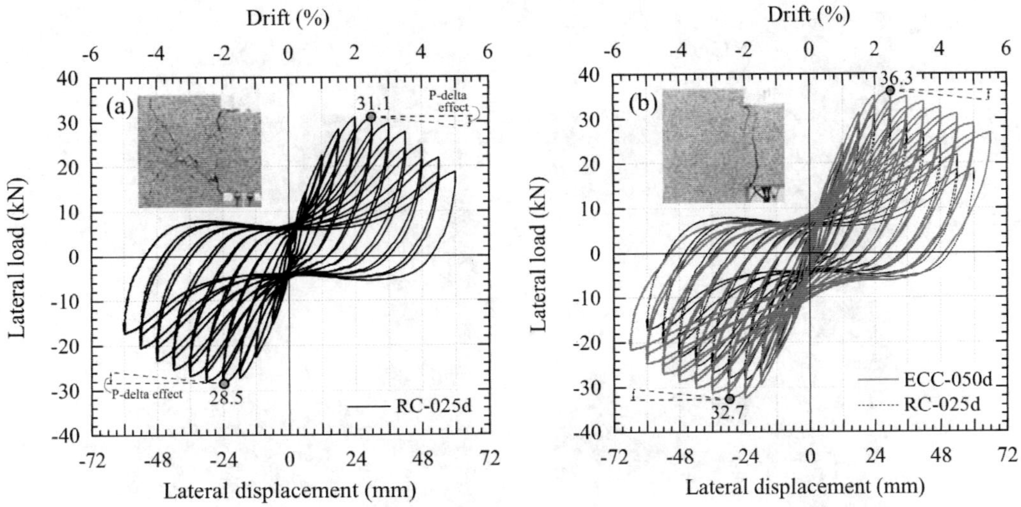

Figure 4. Lateral load versus lateral displacement: (a) RC-025d and (b) ECC-050d [5].

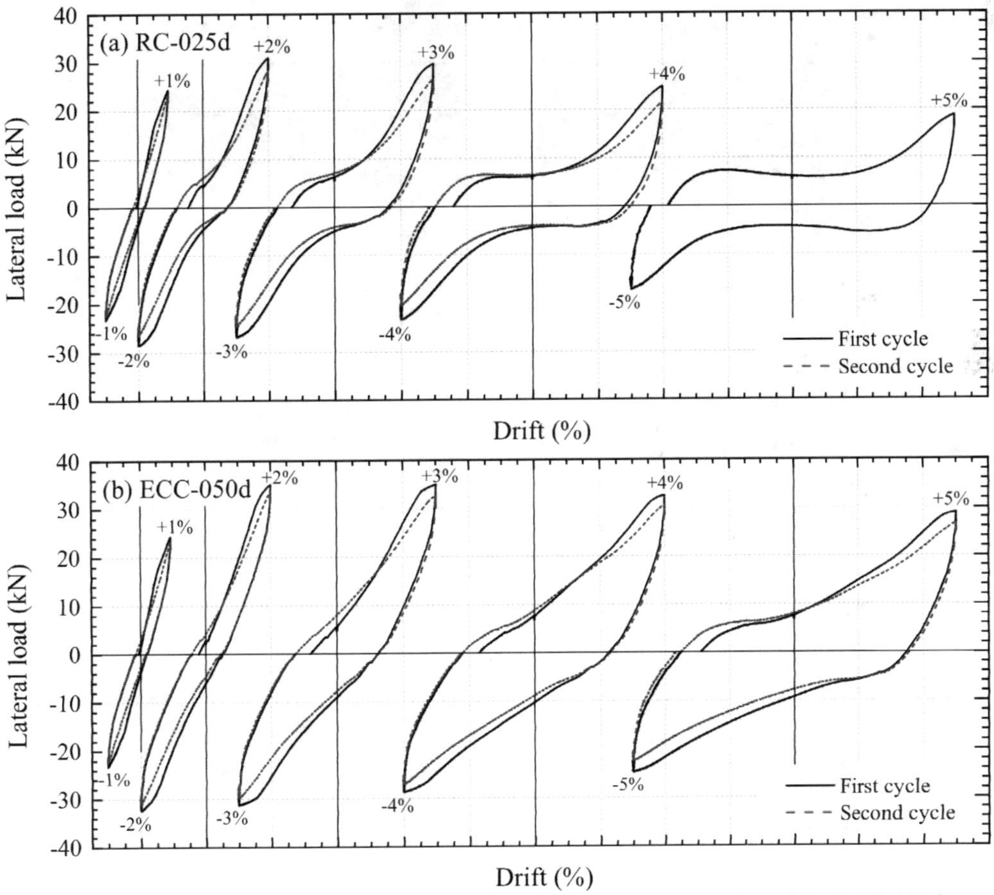

Figure 5. Exploded hysteresis diagrams for specimens: (a) RC-025d; and (b) ECC-050d.

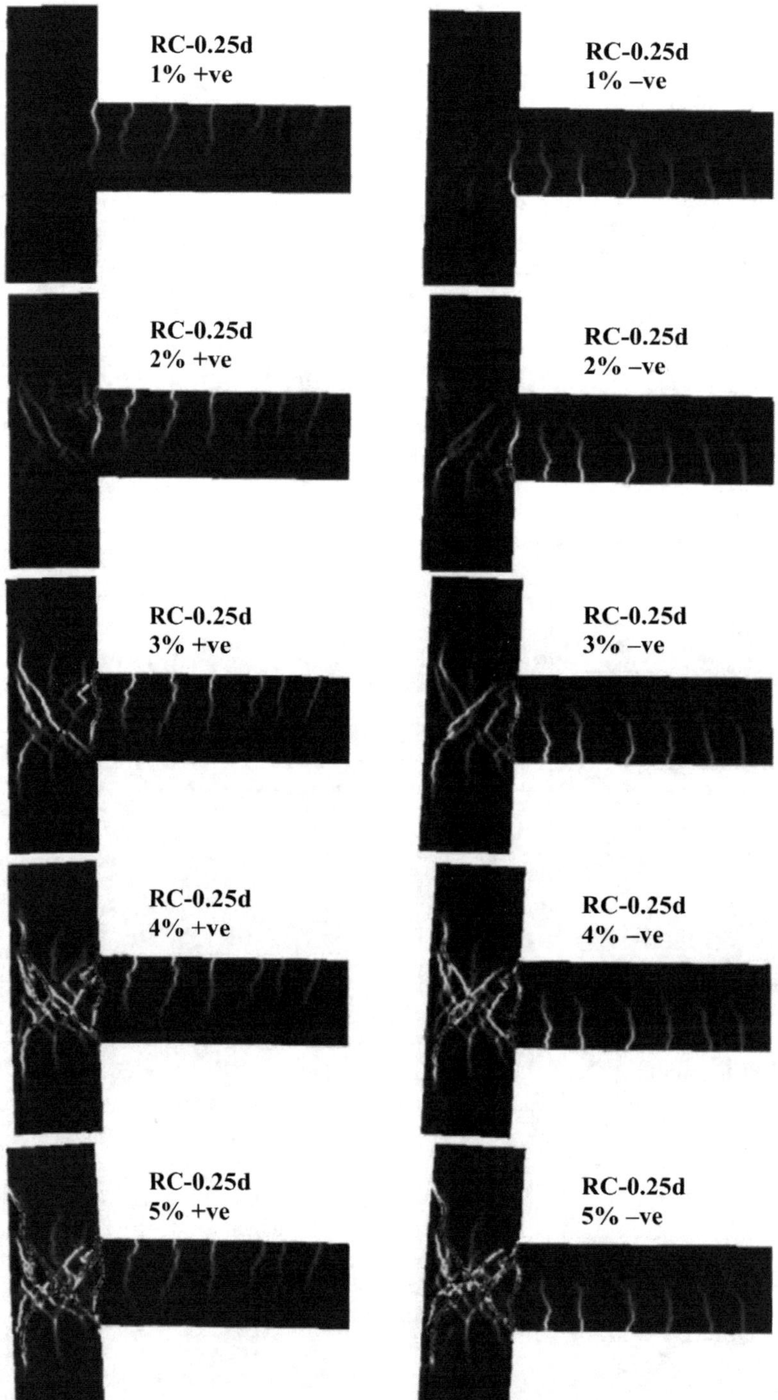

Figure 6. DIC strain maps of specimen RC-025d at 1% drift increments.

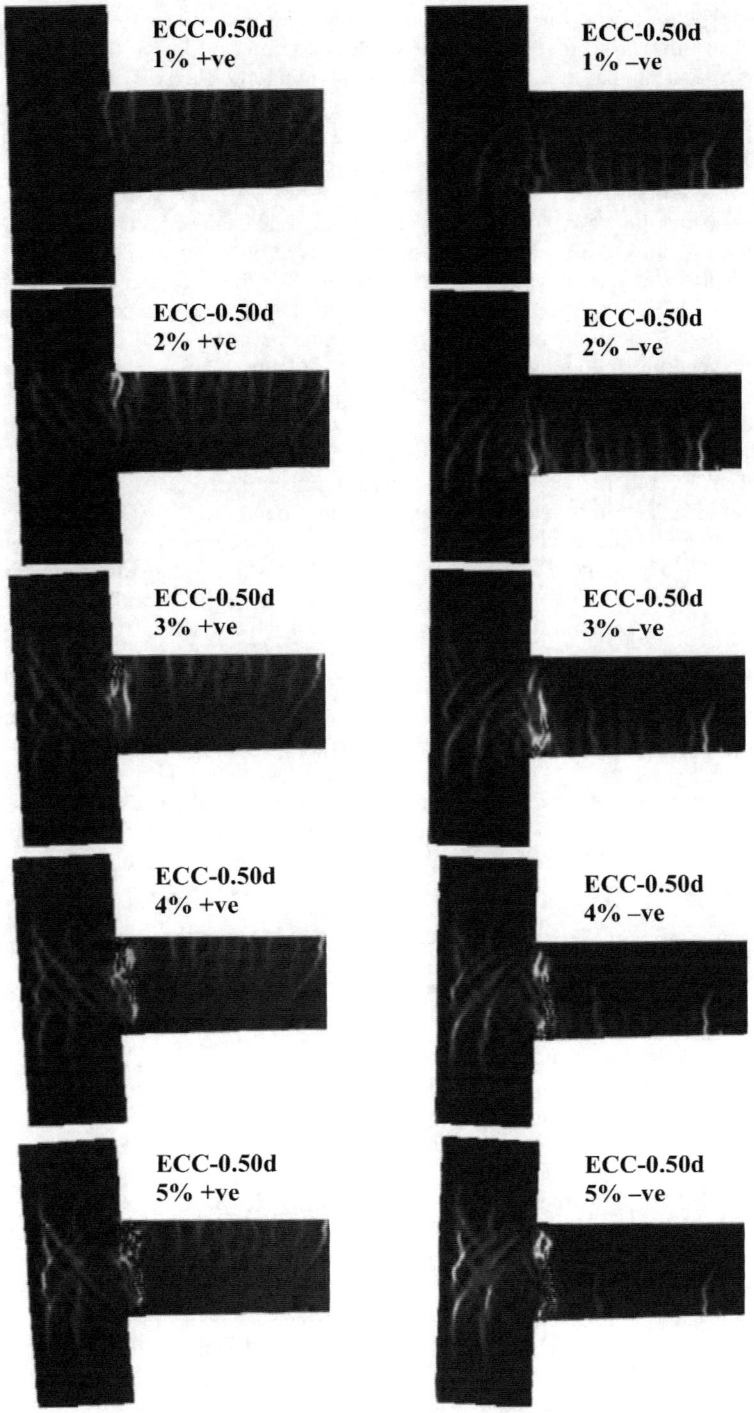

Figure 7. DIC strain maps of specimen ECC-050d at 1% drift increments.

Concerning the response of specimen ECC-050d displayed in Figures 4(b) and 5(b), a stable hysteresis response is evident throughout reversed loading cycles, which is similar to the control specimen. The initial response of specimen ECC-050d is also predominantly flexural and accompanied by the formation of numerous (barely visible) flexural cracks in the plastic hinge region of the beam (see Figure 7). As the drift level increases beyond 2%, the ECC specimen begins to develop some notable differences to the RC-025d specimen. The most notable difference can be seen from the shape of the hysteresis response, which does not begin to pinch from 2% drift, but instead displays a much more rounded response (see Figure 5(b) for clarity). Accordingly, the enclosed area under one full load cycle is greater, which is indicative of improved energy dissipation characteristics. As with the control specimen, as the drift level increases, the lateral resistance of the specimen increases and reaches its maximum value of 36.3kN at 2.5% drift in the positive direction, and a maximum value of 32.7kN in the negative direction of the same drift level. Both values outperformed specimen RC-025d by ~15% in both positive and negative directions and can be attributed mainly to an increase in the flexural capacity of the beam (this is as a direct result of the tensile strain-hardening properties of ECC [11]). With reference to the response beyond the peak load, the ECC specimen displays a smaller reduction in load-carrying capacity than specimen RC025d. Between 2.5% and 5% drifts, for example, specimen ECC-050d displayed an overall reduction in load-carrying capacity of ~25% while specimen RC-025d displayed an overall reduction of over 40% (both include P-delta effects). This can be attributed to the differences in crack and damage patterns sustained by the two specimens, particularly in the joint core regions (compare Figures 6 and 7). During the intermediate stages of loading, specimen ECC-050d displayed initial formation of diagonal cracks in the joint core region (see Figure 7); however, these cracks displayed little progression until the end of the test which is clearly in contrast to the major damage found in the joint regions of specimen RC-025d. This would indicate that ECC is effective in increasing the joint shear strength, even with a relatively large transverse reinforcement spacing in the joint region. In specimen ECC-050d, the most apparent accumulation of damage occurred near the beam-column interface, where a full-depth, localised macro crack formation is evident.

To provide a further comparison of the performance of the two joints, the peak-to-peak stiffness and cumulative energy dissipation for both specimens are displayed in Figures 8(a) and (b), respectively. The peak-to-peak stiffness of the sub-assemblage in each load cycle was obtained from the gradient between the peak loads in the positive and negative directions, as given by

$$k_{\mathrm{pp}} = \frac{F_{+\mathrm{ve}} - F_{-\mathrm{ve}}}{\delta_{+\mathrm{ve}} - \delta_{-\mathrm{ve}}} \tag{1}$$

where $F_{+\mathrm{ve}}$ and $F_{-\mathrm{ve}}$ are the maximum restoring forces in the two loading directions during one full cycle, respectively, and $\delta_{+\mathrm{ve}}$ and $\delta_{-\mathrm{ve}}$ are the corresponding values of lateral displacements during the same cycle. The amount of energy dissipation at each cycle is represented by the enclosed area of each hysteresis loop (see the inset displayed in Figure 8(b)).

Concerning the peak-to-peak stiffnesses of the two specimens presented in Figure 8(a), it is evident that both specimens exhibit an almost identical behaviour, with the ECC specimen consistently outperforming the RC specimen. From 0.25% to 1% drift, the two specimens display rapid degradation of peak-to-peak stiffness, which is a direct result of the formation of new cracks in the concrete or in the ECC, and the propagation of pre-existing cracks. This is then followed by a reduced rate of decay of stiffness, as the cracks in the concrete or in the ECC begin to stabilise. It is interesting to note that between 1% and 2% drift, the RC specimen exhibited a faster reduction in overall stiffness, thereby causing a larger offset. This can be associated with the development of X-shaped cracks in the joint region, which reduces the overall stiffness of the sub-assemblage.

With reference to the cumulative energy dissipation, it is apparent from Figure 8(b) that the ECC joint exhibited an improved energy dissipation to the concrete joint, particularly at higher drift levels. During the initial stages of loading, both specimens exhibit a gradual increase in energy dissipation. The initial energy dissipation of the ECC specimen is slightly lower due to a reduction in the residual drift at low drift levels; the initial values for the two specimens are, however, very close and hence the curves are superimposed on one another. At the 1.5% drift level, due to the yielding of longitudinal reinforcement in the beam, the energy dissipation rises rapidly, with values at 2.5% drift increasing by a factor of more than three relative to those at 1.5% drift. Beyond the 3% drift level, the curves begin to diverge as the ECC specimen exhibits greater energy dissipation than the RC specimen. At 5% drift, this culminates in cumulative energy dissipation of almost 18kNm for the ECC specimen and just slightly over 14kNm for the RC specimen (~20% difference).

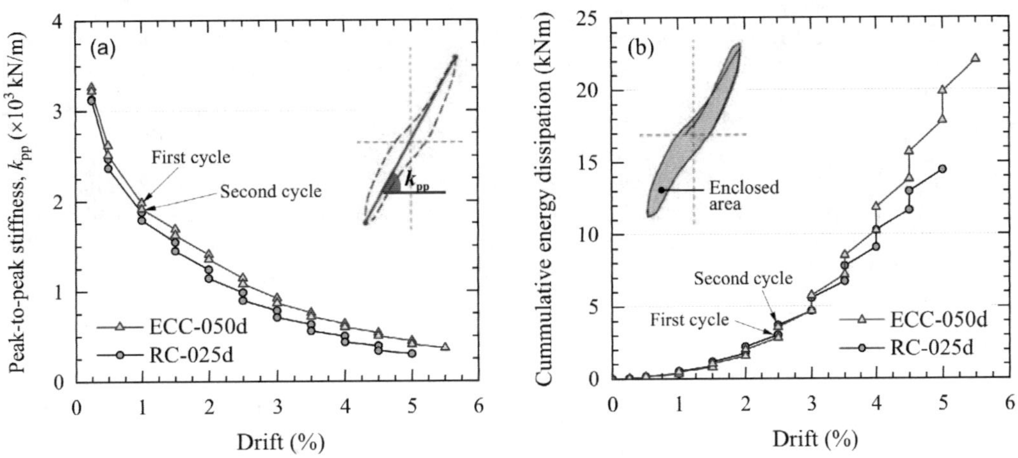

Figure 8. (a) Degradation in stiffness, and (b) cumulative energy dissipation (adapted from [5]).

4. Concluding Remarks and Future Outlook

From the work presented, it is shown that the ECC specimen designed with a reduced amount of transverse reinforcement, with the aim of reducing congestion of reinforcement, was able to exhibit better seismic performance and structural integrity than the seismically designed concrete specimen. In particular, the utilisation of the ECC in the sub-assemblage is shown to be successful in adding to the seismic resistance of the joint. Damage accumulation at the beam-column interface was evident when the two specimens were loaded well beyond their load-carrying capacity and future work will be directed at spreading this local damage over the plastic hinge region in a large-scale testing.

Acknowledgements

This work was supported by a Newton Fund Impact Scheme Grant, ID 624577537 and the Ministry of Research, Technology and Higher Education of the Republic of Indonesia partnership. The grant is funded by the UK Department for Business, Energy and Industrial Strategy and the Indonesian Ministry of Education, Culture, Research, and Technology (Grant No. 342/E4.1/AK.04.PT/2021 and 084/E5/PG.02.00.PT/2022) and delivered by the British Council.

References

[1] Paulay T and Priestley M J N 1992 Seismic design of reinforced concrete and masonry buildings *John Wiley & Sons* New York United States

[2] Takagi J and Wada A 2019 Recent earthquakes and the need for a new philosophy for earthquake-resistant design *Soil Dyn. Earthq. Eng.* **119** 499-507

[3] Li V C 2008 Engineered cementitious composite (ECC): material, structural, and durability performance *In: Nawy E.G., eds., Concrete Construction Engineering Handbook* CRC Press 24.1-24.46.
[4] Suryanto B, Nagai K and Maekawa K 2010 Smeared-crack modeling of R/ECC membranes incorporating an explicit shear transfer model *J. Adv. Concr. Technol.* **8** 3 315-26
[5] Suryanto B, Tambusay A, Suprobo P, Bregoli G and Aitken M W 2022 Seismic performance of exterior beam-column joints constructed with engineered cementitious composite: comparison with ordinary and steel fibre reinforced concrete *Eng. Struct.* **250** 113377
[6] ACI Committee 2019 ACI-318-19: Building code requirements for structural concrete and commentary. Farmington Hills: American Concrete Institute
[7] Tambusay A, Suryanto B and Suprobo P 2020 Digital image correlation for cement-based materials and structural concrete testing *Civ. Eng. Dimens.* **22** 1 6-12
[8] Suryanto B, Tambusay A and Suprobo P 2017 Crack mapping on shear-critical reinforced concrete beams using an open source digital image correlation software *Civ. Eng. Dimens.* **19** 2 93-8
[9] Pan Z, Guner S and Vecchio F J 2017 Modeling of interior beam-column joints for nonlinear analysis of reinforced concrete frames *Eng. Struc.* **142** 182-91
[10] Pampanin S, Calvi G M and Moratti M 2002 Seismic behavior of RC beam-column joints designed for gravity only *Proc., 12th Euro. Conf. Earthq. Eng.* Paper 726 London United Kingdom
[11] Suryanto B, Reynaud R and Cockburn B 2018 Sectional analysis of engineered cementitious composite beams. *Mag. Concr. Res.* **70** 22, 1135-1148.

The behavior of slab on pile structure with a variation of support end interconnection under earthquake load according to the SNI 2833:2016

T S M Aritonang[1, 2*] and A F Setiawan[1]

[1] Department of Civil and Environmental Engineering, Gadjah Mada University, Yogyakarta, Indonesia
[2] Department of Civil Engineering, Cenderawasih University, Jayapura, Indonesia

*Corresponding author: tobok.sm.aritonang@mail.ugm.ac.id

Abstract. This paper discusses the behaviour of the slab-on-pile (SOP) structure with variations support end interconnection under earthquake loads. Analysis of the SOP structure using macro modelling of numerical analysis. The SOP structure model is developed with three variations of the model with different support conditions, namely Model 1 with the end of the structure attached to the pile head and spun pile, Model 2 with the end of the structure attached to roller support, and Model 3 with the end of the structure not being supported, or the same as the cantilever system. Earthquake load is assigned using two methods according to SNI 2833:2016, namely the Single-mode Spectra Method and the Response Spectrum Method. Based on the analysis results, the vibration period of the SOP structure calculated by the single mode spectra method is 0.60 seconds for the first mode and 0.34 seconds for the second mode. Based on the numerical analysis results, the vibration period of Model 3 is greater than that of other models. The maximum internal forces and the most significant displacement occur in Model 3, while the most considerable base reaction occurs in Model 1. The magnitude of the base shear force occurs because Model 1 uses more piles, thus increasing the stiffness of the structure. This results in an increase in seismic forces but reduces displacement of the structure. Furthermore, for the analysis and design of SOP structures, Model 3 is the most reliable because this model produces the maximum load effect and the most significant displacement of the structure.

1. Introduction

A bridge is a traffic infrastructure connecting two parts of the road cut off by obstacles such as rivers, valleys, or other buildings underneath. One type of bridge built today is the slab-on-pile (SOP) structure, which consists of piles, pile heads, and slabs, as shown in Figure 1. The slab is a precast element supported by the pile head. Pile heads function as slab supports, connecting piles in one row and stiffeners between piles in the transverse direction. The piles serve as the foundation and also as the pillars that support the floor slab.

The spun pile is a type of pile that is currently widely used for bridge piers [1, 2]. The advantages of using spun piles are high strength and quality, longer durability, and more economical and practical in its implementation. In addition to the benefits, spun pile also has drawbacks in its use, that as low energy absorption and failure patterns due to seismic loads in the form of brittle failure that occurs suddenly [3]–[6]. Efforts to increase the strength and ductility of the spun pile have been carried out by researchers, among others, by filling the spun pile hole with concrete, then wrapped with carbon fiber [7], and by external restraint using fiber-reinforced polymer (FRP) [8].

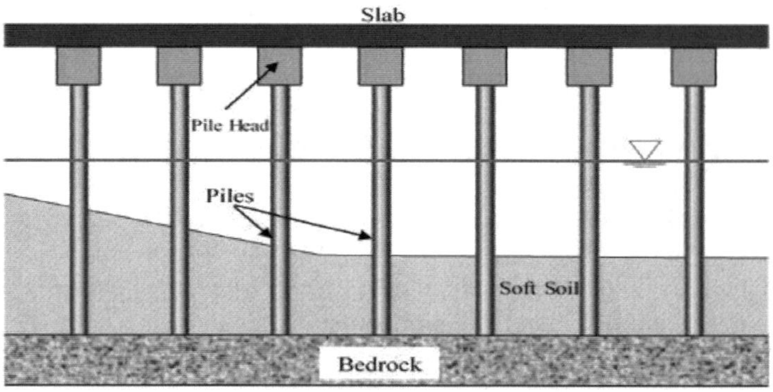

Figure 1. Slab-on-pile structural components.

On the other hand, many bridges are built in areas frequently hit by earthquakes. Earthquakes can cause severe damage to the bridge structure, which can take many forms depending on ground motion, site conditions, and structure configuration [9]. This damage is certainly not easy to repair and requires relatively expensive costs. When the earthquake load works, the pile will receive a high lateral force which will cause a high bending moment, which can cause significant damage to the pile [10, 11].

The length of the existing SOP structure can reach hundreds to thousands of meters and consists of connected segments. However, in design, the SOP structure is often designed and analyzed only per SOP segment, so the supports at the ends of the SOP segments need to be modelled. This study aims to determine the behavior of the SOP structure with variations of support end interconnection of the SOP segment due to earthquake loads and the effect of soil-structure interaction on the SOP structure. SOP behavior reviewed includes internal forces on piles, displacements, and base reactions.

1.1. Earthquake load analysis method

The loading applied in the bridge structure design refers to the SNI 1725:2016 [12] regarding bridge loading standards and SNI 2833:2016 [13] regarding seismic bridge design. These standards specify the loading requirements and actions to be used in the design of highway bridges. Based on [13], each bridge must be located in one of the four seismic zones, where seismic analysis is not required for bridges located in seismic zone 1 and for single-span bridges in all seismic zones. Seismic analysis is needed for multiple-span bridges, where the available analysis types are Uniform Load Method, Single-mode Spectra Metho, Multi-mode Spectra Method, and Time History Method.

1.2. Single-mode spectra analysis method

The single-mode spectra method should be based on the variation of fundamental vibration in the bridge's longitudinal or transverse directions. This variety of vibrations can be obtained by applying a uniform horizontal force to the structure to obtain a change in the structure's shape. The natural period of the bridge, T_f, can be calculated using the equation [13, 14]:

$$T_f = 2\pi \sqrt{\frac{\gamma}{p_0 g \alpha}} \tag{1}$$

and the equivalent static earthquake load, $p_e(x)$, is calculated using the equation:

$$p_e(x) = \frac{\beta C}{\gamma} w(x) v_s(x) \tag{2}$$

where p_0 is the uniform load (equal to 1 kN/mm), $v_s(x)$ is the resultant of deformation (mm), $w(x)$ is the not factored dead load on superstructure and substructure (kN/mm), g is the acceleration due to gravity (mm/s^2), and C is the elastic earthquake response coefficient. Factors α, β and γ are calculated using the equation:

$$\alpha = \int v_s(x)dx \quad (3)$$

$$\beta = \int w(x)v_s(x)dx \quad (4)$$

$$\gamma = \int w(x){v_s}^2(x)dx \quad (5)$$

1.3. Fixity point

Hardiyatmo [15] describes the procedure for calculating the depth of the fixity point, z_f, as shown in Figure 2. On the pile works horizontal loads H and vertical loads Q. The pile is considered to be pinched at a depth of z_f. The depth z_f for soil that has a subgrade modulus that increases linearly with depth is:

$$z_f = 1.8\,T \quad (\text{for } I_{max} = L/T > 4) \quad (6)$$

$$T = \sqrt[5]{\frac{E_p I_p}{n_h}} \quad (7)$$

where T is the stiffness factor for the non-constant soil modulus, L is the penetration depth of the pile, E_p is the modulus of elasticity of the pile, I_p is the moment of inertia of the pile, and n_h is the coefficient of the modulus of variation. Furthermore, the equivalent pile length (L_e) is calculated by the equation:

$$L_e = e + z_f \quad (8)$$

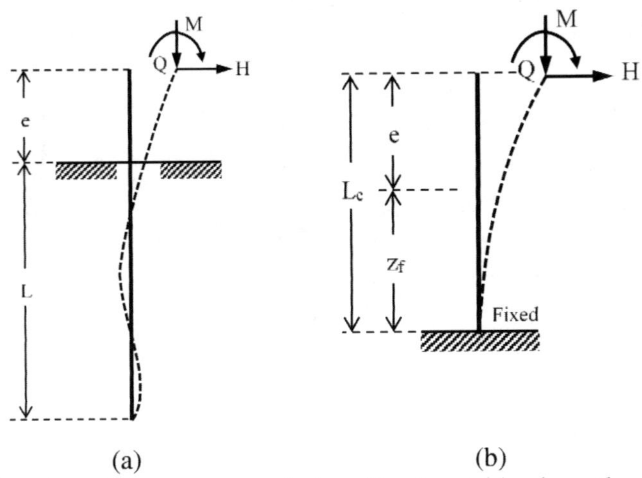

Figure 2. Pile bending that supports vertical and horizontal loads on the pile head, (a) Pile installed partially embedded in the ground, (b) Equivalent pile length (L_e) [15].

2. Research methods

2.1. Slab on pile structural model

In this study, the SOP bridge structure is considered in Jayapura City. The bridge category is the essential bridge. The length of the bridge segment is 25 m, with the distance between the spun piles in the longitudinal direction is 5 m. The pile height is 34 m, where 30 m of the pile is embedded in the ground, then the remaining 4.0 m is above ground level. The spun pile property refers to the research of Irawan et al. [16]. The outer diameter of the spun pile is 400 mm, and the wall thickness is 100 mm. The SOP structure and the cross-section of the spun pile are shown in Figure 3.

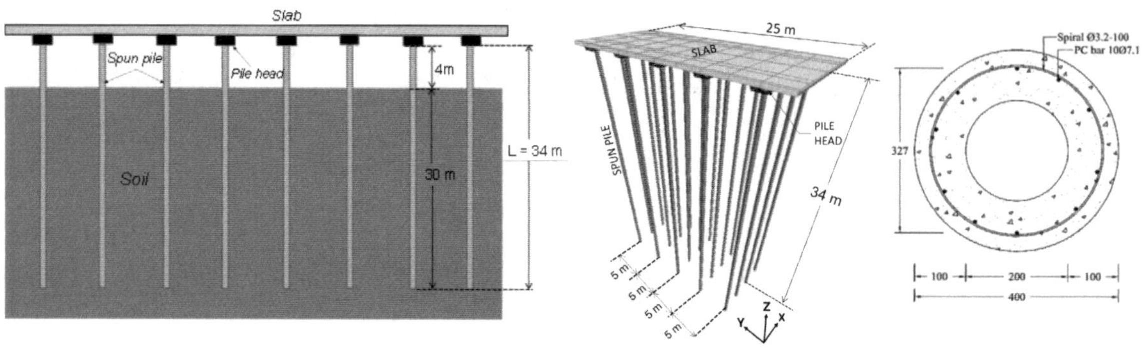

(a) (b) (c)

Figure 3. Analyzed SOP structure, (a) Longitudinal section, (b) Isometric view, and (c) Spun pile cross-section [16].

Furthermore, to model the support conditions of the end interconnection of the SOP structure, three variations of the structural model were made with different support conditions at the end interconnection of the SOP structure. The three structural models are: (1) Model 1, with the end of the structure attached to the pile head and spun pile; (2) Model 2, with the end of the structure attached to roller support; and (3) Model 3, with the end of the structure not being supported, or the same as the cantilever system. The modelling results of the three SOP structure models in SAP2000 are shown in Figure 4. To consider the soil-structure interaction, the modelling of each SOP model has defined the spring stiffness values in both the horizontal and vertical directions. The definition is carried out along the spun pile embedded in the ground, with the distance between the spring supports is 0.5 m.

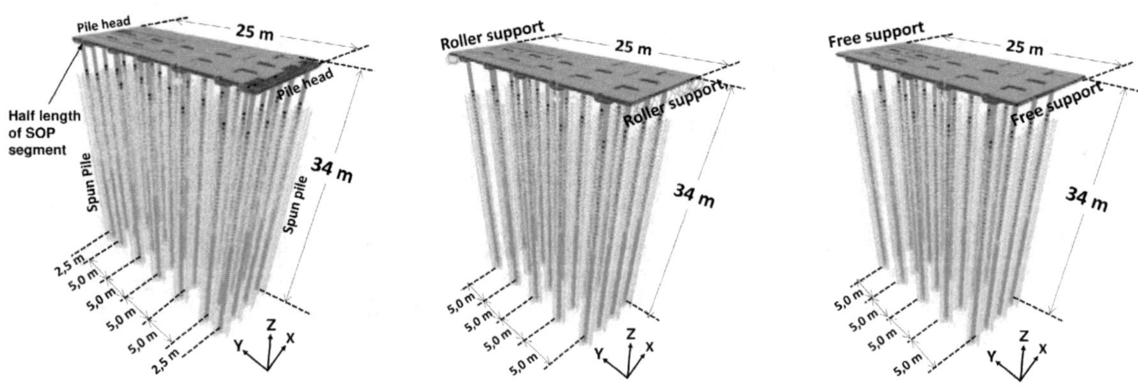

(a) Model 1 (b) Model 2 (c) Model 3

Figure 4. Variation of SOP structure model.

2.2. Material properties

The analyzed spun pile property data was obtained from the research of Irawan et al. [16]. Concrete compressive strength (f_c') = 54.5 MPa. Weight of concrete (w_c) = 25 kN/m ^{3}. Modulus of elasticity of concrete (E_c) = 39644 MPa. The prestressed steel reinforcement in the spun pile uses PC bar 10 ∅7.1 mm, while the non-prestressed reinforcement uses spiral ∅3.2-100. The idealized stress-strain curve of ∅7.1 mm PC bar and ∅3.2 mm spiral reinforcement can be seen in Figure 5. The compressive stress-strain relationship curve for concrete is based on the equation proposed by Tsai [17], while for the stress-strain relationship in tensile concrete, the equation proposed by Vecchio and Collins is used [17]. The stress-strain relationship curve for concrete can be seen in Figure 6.

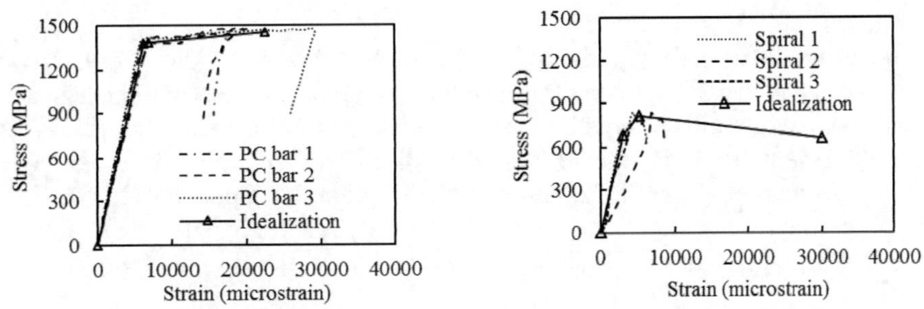

(a) PC bar ∅7.1 mm (b) Spiral ∅3.2 mm

Figure 5. Stress-strain idealization curve [16].

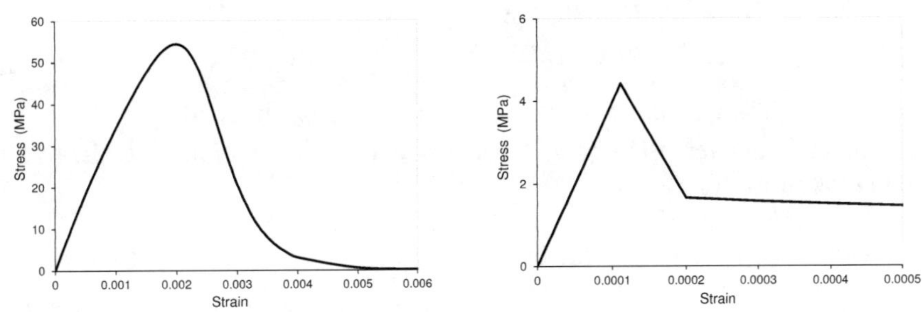

(a) Compression (b) Tension

Figure 6. Concrete stress-strain relationship curve for f' $_c$ = 54.4 Mpa.

2.3. *The spun pile modelling on the SAP2000 program*

Spun pile modelling in the SAP2000 is carried out using the Section Designer, where the spun pile cross section is made using Caltrans Shape. The prestressing force added to the tendon force is the maximum prestressing force following SNI 2847:2019 [18], which should not exceed 0.94 f_{py}, but not be greater than less than 0.80 f_{pu} and the maximum value recommended by prestressing steel manufacturers.

2.4. *Soil data, site class, and response spectra design*

The soil data used for analysis is standard penetration test (SPT) data taken from one location in Jayapura City, Indonesia. The results of the site class investigation based on SPT data found that the soil at the research site was soft soil or SE site class. Response spectra design in Jayapura City for the SE site class is shown in Figure 7.

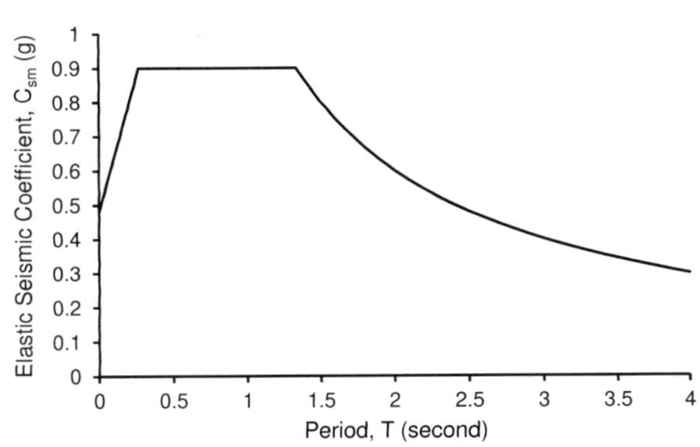

Figure 7. Response spectra design in Jayapura City for SE site class.

2.5. Structure loading

This study calculates the loads acting on the SOP structure based on [12] and [13]. These loads include dead loads, additional dead loads, traffic loads, brake loads, pedestrian loads, wind loads, and earthquake loads. Specifically for calculating earthquake loads, several assumptions are used, that is: (a) the bridge is categorized as the essential bridge, and (b) the bridge is considered a regular bridge. The earthquake load is then calculated using two methods, namely: (1) Single-mode Spectra Method (SM), where the magnitude of the earthquake load is calculated based on SNI 2833:2016 [13]; and (2) Response Spectrum Method (RS), where earthquake loads are defined using the SAP2000 program.

2.6. Calculation of earthquake load

The stiffness factor for the non-constant soil modulus (T) is calculated using equation (7), where $E_p = 39644041\ \mathrm{kN/m^2}$, $I_p = 0.00116\ \mathrm{m^4}$, and $n_h = 11779\ kN/\mathrm{m^3}$. The stiffness factor for the non-constant soil modulus (T) is calculated as follows:

$$T = \sqrt[5]{\frac{\mathrm{E_p I_p}}{\mathrm{n_h}}} = \sqrt[5]{\frac{39644041.17 \times 0.001163}{11779}} = 1.314\ \mathrm{m}.$$

The fixity point depth is calculated using equation (6), it is obtained $z_f = 1.8 \times 1.314 = 2.4$ m. Therefore, the equivalent pile length is: $H = 4 + 2.4 = 6.4$ m. Calculation of the factors α, β and γ and structure period (T) is shown in Table 1.

Table 1. Calculation of the factors α, β and γ and structure period (T).

Variable	Unit	Longitudinal direction	Transverse direction
Factor α	m^2	0.0146	0.00316
Factor β	kN. m	2.2162	12.004
Factor γ	$\mathrm{kN.\,m^2}$	0.00129	0,00009
Period (T)	second	0.60	0.34

Furthermore, based on the calculation results, the longitudinal shear force for one pile is 64.0 kN, while the transverse shear force for one pile is shown in Table 2. The application of earthquake loads to the SOP structure in the numerical model is shown in Figure 8.

Table 2. Calculation of earthquake loads in the transverse direction.

Location	$P_e(x)$ [a] (kN/m)	V_T [b] (kN)	V_i [c] (kN)
End of structure-1	0.000	0.00	0.00
Pile 1	6,980	21.14	5.28
Pile 2	20,940	63.42	15.85
Pile 3	34,899	105.70	26.42
Pile 4	20,940	63.42	15.85
Pile 5	6,980	21.14	5.28
End of structure-2	0.000	0.00	0.00

[a] The equivalent static earthquake load.
[b] Transverse shear force for one line of pile.
[c] Transverse shear force for one pile.

Earthquake loads were applied to the SOP structure model using a predefined response spectral function (Figure 7). Then the scale factor (SF) for the transverse direction (X-axis) is calculated, where the important factor is 1.0, and the response modification factor (R) is 3.5 (considered as a double column). Then obtained SF = 1 x 9.81 / 3.5 = 2.8. The earthquake load in the transverse direction is combined with the longitudinal direction, which is 30 percent of the earthquake load. So, the SF for combination with longitudinal direction is 0.3 × 2.8 =0.84. For the longitudinal direction (Y-axis), the response modification factor is 2.0 (considered as a single column), so SF = 1 x 9.81 /2 = 4.91. Furthermore, the SF for combination with the transverse direction is 1.47.

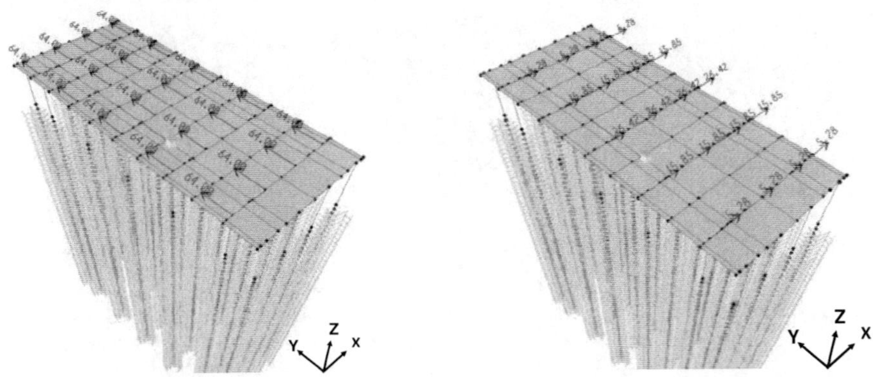

(a) Longitudinal direction (b) Transverse direction

Figure 8. Application of earthquake loads on numerical models.

2.7. Load combination
The design loads calculated above must be combined to obtain a predetermined load combination [12]. In this study, the load combination combines Service-1, Strength-1, Strength-2, Strength-3, Strength-5, and Extreme-1. Especially for the variety of Extreme-1, this combination is distinguished by the direction of earthquake loading, namely Extreme-1 (EQY) for earthquake loads in the longitudinal direction and Extreme-1 (EQX) for the transverse direction.

3. Result

3.1. Structure vibration period
Table 3 presents a comparison of the structural vibration period for all models. The vibration period obtained from the Single-mode Spectra Method is the same for all structural models, i.e. 0.60 seconds for the first mode (longitudinal translation) and 0.34 seconds for the second mode (transverse translation). Meanwhile, the structural motion pattern for the first mode of each model obtained from the numerical analysis is shown in Figure 9. The dominant motion pattern for the first mode is translational in the longitudinal direction.

Table 3. Comparison of the structural vibration period for all models.

Structure model	Calculation method	Period (Second)	
		Mode 1	Mode 2
Model 1	Single-mode Spectra Method	0.60	0.34
	Numerical Analysis	0.687	0.676
Model 2	Single-mode Spectra Method	0.60	0.34
	Numerical Analysis	0.757	0.104
Model 3	Single-mode Spectra Method	0.60	0.34
	Numerical Analysis	0.765	0.757

3.2. Structural Analysis Results on Spun Pile
The comparison of the maximum internal forces for the most critical spun pile is shown in Table 4. Compared to the highest internal forces of the three structural models, the most significant internal forces are generally generated by Model 3. The recapitulation of the largest maximum internal forces on the spun pile is as follows: $P^{-}_{(max)}$ = 1897.83 kN; $V^{+}_{(max)}$ = 91.94 kN; $V^{-}_{(max)}$ = –91.84 kN; $M^{+}_{(max)}$ = 347.73

kN.m; and $M^{-}_{(max)}$ = −353.87 kN.m. In addition, from the analysis results, the spun pile does not experience a tensile force.

(a) Model 1 T_1 = 0.687 seconds (b) Model 2 T_1 = 0.757 seconds (c) Model 3 T_1 = 0.765 seconds

Figure 9. Structure motion pattern result of numerical analysis for the first mode.

Table 4. Comparison of the structural analysis results for the most critical spun pile.

Structure model	Calculation method	$P^{-}_{(max)}$ (kN)	$V^{+}_{(max)}$ (kN)	$V^{-}_{(max)}$ (kN)	$M^{+}_{(max)}$ (kN.m)	$M^{-}_{(max)}$ (kN.m)
Model 1	SM[a]	-1263.8	50,095	-32.24	92.95	-187.918
	RS[b]	-1263.8	70.92	-70.81	260.16	-260.73
	SM/RS Ratio	1.00	0.71	0.46	0.36	0.72
Model 2	SM	-1304.41	73.774	-45.47	130.79	-284.69
	RS	-1304.41	91.94	-91.835	347.72	-348.24
	SM/RS Ratio	1.00	0.80	0.50	0.38	0.82
Model 3	SM	-1897.83	88.03	-50.68	216.21	-353.87
	RS	-1304.41	91.94	-91.84	347.73	-348.24
	SM/RS Ratio	1.45	0.96	0.55	0.62	1.02

[a] Single-mode Spectra Method.
[b] Response Spectrum Method.

Furthermore, the internal forces on the spun pile due to a load combination involving earthquake loads are discussed. In this case, the spun pile number 19 is chosen, located at the end of the structure, as shown in Figure 10.

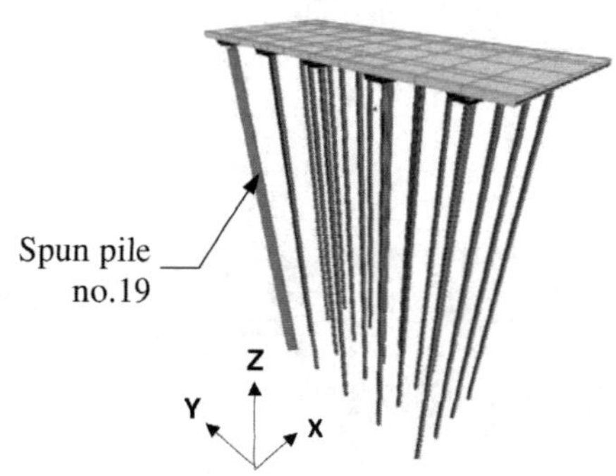

Figure 10. *Spun pile* position under review.

The diagram of the bending moment and shear force that occurs in the spun pile number 19 due to the load combination of Extreme-1 can be seen in Figure 11 and Figure 12. Based on Figure 11, it can be seen that the fixity point on the spun pile is at a depth of -2.5 m, which is at a location where the positive moment reaches its maximum value. Meanwhile, the maximum negative moment is at the end of the pile at an elevation of +4.0 m above ground level.

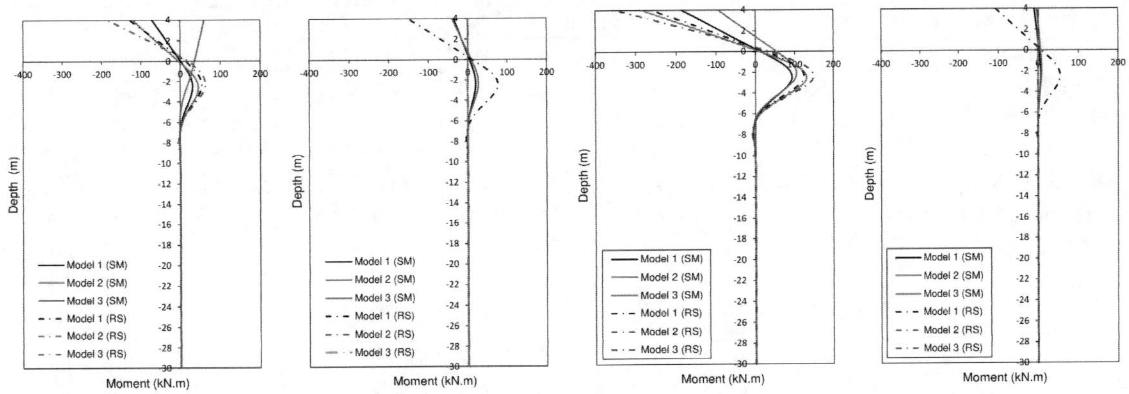

(a) Moment-M2 Ekstreme-1 (EQX) (b) Moment-M3 Ekstreme-1 (EQX) (c) Moment-M2 Ekstreme-1 (EQY) (d) Moment-M3 Ekstreme-1 (EQY)

Figure 11. Moment diagram on spun pile no.19 due to load combination of Extreme-1.

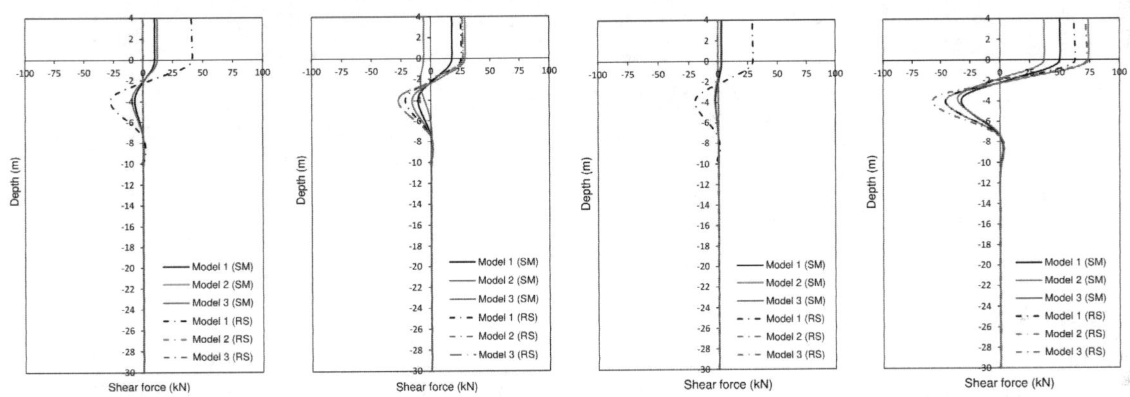

(a) Shear Force-V2 Ekstreme-1 (EQX) (b) Shear Force-V3 Ekstreme-1 (EQX) (c) Shear Force-V2 Ekstreme-1 (EQY) (d) Shear Force-V3 Ekstreme-1 (EQY)

Figure 12. Shear force diagram on spun pile no.19 due to load combination of Extreme-1.

3.3. Displacement

Table 5 shows the displacement values in the SOP structure due to the load combination involving earthquake loads. Based on the table, it is found that the largest transverse displacement (U1) is obtained from Model 3 (RS) due to the load combination of Extreme-1 (EQX), i.e. 44.673 mm. The most significant displacement in the longitudinal direction (U2) is obtained from Model 3 (RS) due to the load combination of Extrem-1 (EQY), i.e. 69.92 mm. While the largest vertical displacement (U3) is obtained from Model 3 (RS) due to the load combination of Extreme1-EQY, i.e. -42.724 mm.

Figure 13 shows the displacement in the spun pile number 19 due to the load combination of Extreme-1 (EQX) and Ekstreme-1 (EQY). The largest transverse displacement (U1) in the spun pile was obtained from Model 1 (RS) due to the Extreme-1 (EQX) combination, i.e. 35.4 mm. The largest longitudinal displacement (U2) was obtained from Model 2 (RS) and Model 3 (RS) due to the Extreme-1 (EQY) combination, i.e. 68.5 mm. Meanwhile, the most significant displacement in the direction of U3 (vertical) is obtained from Model 3 (SMT) due to the combination of Ekstrem-1 (EQY), i.e. -14.5 mm.

Table 5. Displacement in the SOP structure due to a load combination involving earthquake loads.

Structure model and calculation method	Displacement (mm)					
	Extreme-1 (EQX)			Extreme-1 (EQY)		
	U1	U2	U3	U1	U2	U3
Model 1 (SM)	9.365	11.754	-10.173	3.670	39.187	-10.335
Model 2 (SM)	0.000	16.091	-10.385	0.000	53.630	10.385
Model 3 (SM)	13.529	16.423	-38.702	5.557	54.789	-41.520
Model 1 (RS)	35.600	25.341	1.155	25.726	56.558	-10.734
Model 2 (RS)	0.000	30.675	-10.385	0.000	68.455	-10.385
Model 3 (RS)	44.673	31.319	-39.837	32.455	69.920	-42.724

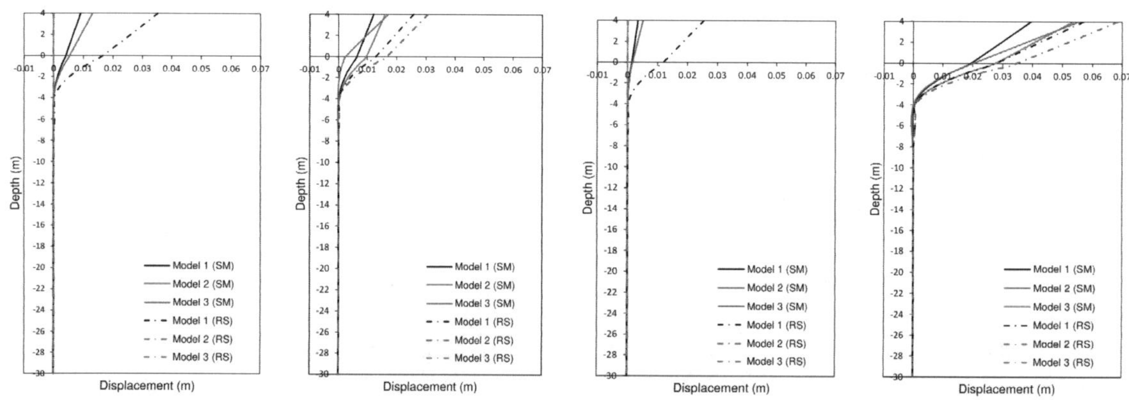

(a) Direction U1 Ekstreme 1-EQX
(b) Direction U2 Ekstreme 1-EQX
(c) Direction U1 Ekstreme 1-EQY
(d) Direction U2 Ekstreme 1-EQY

Figure 13. *Displacement* on *spun pile* no. 19 due to load combination of Extreme-1.

3.4. Base reaction

A comparison of the base reaction results of numerical analysis for all structural models is shown in Table 6. Based on the table, the most significant base reaction in the vertical direction (FZ) is 27491.11 kN, resulting from the analysis of Model 1. The largest base reaction in the vertical direction of Model 2 and Model 3 is the same, i.e. 26164.59 kN. Compared to the three models, the largest base reaction in the vertical direction of Model 1 is 5.07% greater than that of Model 2 and Model 3.

Table 6. Comparison of the base reaction results for all structural models.

Structure model	Load combination	Single-mode Spectra (SM)			Response Spectrum (RS)		
		FX (kN)[a]	FY (kN)[b]	FZ (kN)[c]	FX (kN)	FY (kN)	FZ (kN)
Model 1	Strength-1	0.00	0.00	27491.11	0.00	0.00	27491.11
	Extreme-1 (EQX)	-274.72	-384.00	18236.26	1179.94	841.62	18236.26
	Extreme-1 (EQY)	-82.42	-120.00	18236.26	840.99	1878.28	18236.26
Model 2	Strength-1	0.00	0.00	26164.59	0.00	0.00	26164.59
	Extreme-1 (EQX)	-274.72	-384.00	16909.74	0.36	742.15	16909.74
	Extreme-1 (EQY)	-82.42	-120.00	16909.74	0.26	1656.27	16909.75
Model 3	Strength-1	0.00	0.00	26164.59	0.00	0.00	26164.59
	Extreme-1 (EQX)	-274.72	-384.00	16909.74	1039.74	741.51	16909.75
	Extreme-1 (EQY)	-82.42	-120.00	16909.74	741.07	1654.86	16909.75

[a] Base reaction in the transverse direction.
[b] Base reaction in the longitudinal direction.
[c] Base reaction in the vertical direction.

4. Conclusion

Based on the results of the analysis and discussion, some conclusions can be drawn as follows:

1. The SOP vibration period for the first mode based on the Single-mode Spectra Method is 0.60 seconds. Meanwhile, the vibration period for the first mode based on numerical analysis is 0.687 seconds for Model 1, 0.757 seconds for Model 2, and 0.765 seconds for Model 3. The dominant motion pattern for the first mode in all SOP structural models is a translation in the longitudinal direction.
2. The fixity point based on the results of theoretical calculations is at a depth of -2.4 m, while the fixity point based on numerical analysis is at a depth of -2.5 m.
3. The maximum internal force and the most significant displacement of the SOP structure are obtained from the analysis results of Model 3, while the largest base reaction was obtained from Model 1. The magnitude of the base shear force occurs because Model 1 uses more piles, thus increasing the stiffness of the structure. Increasing the stiffness of the structure raises seismic forces but reduces displacement.
4. Model 3 is the most reliable model for the analysis and design of SOP structures because this model produces the maximum load effect and the most significant displacement of the structure.

References

[1] Gerwick B C 1968 *PCI Journal* pp 68-93
[2] Cheon J H, Kim T H, Lee B J, Lee J H and Shin H M 2012 *Magazine of Concrete Research*, **64**, pp 919–930
[3] Budek A M, Benzoni G, Priestley M J N 1997 Experimental Investigation of Ductility of In-ground Hinges in Solid and Hollow Prestressed Piles (San Diego: University of California)
[4] Kishida S, Horii M, Kuwabara F and Hayashi S 1998 *Transactions of AIJ* **63** pp 123–130
[5] Hoshikuma J and Priestley M J N 2000 *Flexural Behavior of Circular Hollow Columns with A Single Layer of Reinforcement Under Seismic Loading* (San Diego: University of California)
[6] Setiawan A F, Darmawan M F, Ismanti S, Mukhlis S and Muria A G 2020 *E3S Web Conf.* **156** p 03007
[7] Akiyama M, Abe S, Aoki N and Suzuki M 2012 *Engineering Structures* **34** pp 259–270
[8] Lignola G P, Prota A, Manfredi G and Cosenza E 2007 *Analysis of the confinement of RC hollow columns wrapped with FRP* p 9
[9] Chen W F and Duan L 2003 *Bridge engineering, Seismic design* (Boca Raton: CRC Press)
[10] Joen P H 1987 *Seismic Performance of Prestressed Concrete Piles and Pile-Pile Cap Connections* (Christchurch: University of Canterbury) chapter 1 pp 1-2
[11] Budek A M, Priestley M J N and Benzoni G 2000 *J. Struct. Eng.* **126** pp 510–517
[12] BSN 2016 *SNI 1725:2016 Pembebanan untuk Jembatan* (Jakarta: BSN)
[13] BSN 2016 *SNI 2833:2016 Perencanaan Jembatan Terhadap beban Gempa* (Jakarta: BSN)
[14] FHWA 2014 *LRFD Seismic Analysis and Design of Bridges Reference Manual* (Washington: U.S. Department of Transportation)
[15] Hardiyatmo H C 2020 *Analisis dan Perancangan Pondasi II* (Yogyakarta: Gadjah Mada University Press) pp 314-317
[16] Irawan C, Djamaluddin R, Raka I G P, Faimun, Suprobo P and Gambiro 2018 *International Journal on Advanced Science, Engineering and Information Technology* **8** p 501
[17] Chang G A and Mander J B 1994 *Seismic Energy Based Fatique Damage Analysis of Bridge Columns : Part I – Evaluation of Seismic Capacity* (New York: State University of New York at Buffalo)
[18] BSN 2019 *SNI 2847:2019 Persyaratan Beton Struktural untuk Bangunan Gedung* (Jakarta: BSN)

Pushover analysis in slab-on-piles bridge using reinforced concrete-filled spun piles

C A S Riyono[1], I Satyarno[1*], A F Setiawan[1] and A Awaludin[1]

[1]Department of Civil and Environmental Engineering, Universitas Gadjah Mada, Yogyakarta, Indonesia

*Corresponding author: imansatyarno@ugm.ac.id

Abstract. Slab-on-piles is a bridge structure consisting of a continuous pile as a pier, pile head, and slab. Generally, the spun pile is chosen as the pile pier because of the quick construction and low maintenance. However, due to the hollow section with prestressed concrete, without filler, it has weakness in the form of brittle failure under bending moment and shear. Some studies mention that the initial failure occurs due to an explosion in the inner surface of the hollow spun pile or on the surface of the hole. The implementation spun pile for the slab-on-piles bridges should consider that behavior. Therefore, some practical engineers usually use concrete-infill treatment, however, the performance improvement in the nonlinear behavior needs to be known. This study investigates the use of hollow spun piles filled with concrete and normal reinforcement for a slab-on-piles bridge structure. This type of spun pile will be referred to as a reinforced concrete-filled spun pile (RCFSP). The location of slab-on-piles bridge is assumed to be built in Aceh, Indonesia. The height of the RCFSP on the slab-on-piles bridge is 6 m relative to the ground surface. A preliminary design using the linear elastic method is conducted using the response modification factor of 3.5. The slab-on-piles bridge requires 25 RCFSP for each bridge segment. Based on the pushover analysis method, this structure could not reach a performance point. A redesign is conducted using a response modification factor of 1.5. The redesign result shows that the structure requires 60 RCFSP and a performance point is found at a displacement of 85 mm in the longitudinal direction and 81 mm in the transversal direction. This research concludes that a slab-on-piles bridge is not efficient to be designed with a high value of response modification factor even though using the concrete-infill treatment.

1. Introduction

The slab-on-piles structure is one type of railway bridge structure that is being developed in Indonesia. Some railway bridges with the slab-on-piles type that have been built recently are the railway bridges of Soekarno-Hatta International Airport, Adi Soemarmo International Airport, and Yogyakarta International Airport. This structure consists of a slab, pile head, and spun pile [1]. The advantages of the slab-on-piles bridge structure, among others, are more stable for roads that are above water-saturated soil because the foundation uses piles with a certain depth to hard soil, more guaranteed quality because it is produced at the factory, faster construction time, and the costs for construction and maintenance are more efficient than conventional methods. The problem is that the construction of a slab-on-piles bridge in a high earthquake zone causes the structure to be prone to earthquake loads. On the other hand, previous studies concluded that the failure pattern of a spun pile with a hole in the center (hollow spun pile) as a result of earthquake loads is in the form of brittle failure. The initial failure occurred due to an explosion on the inner surface of the hollow spun pile or the surface of the hole [2]. Therefore, several methods to increase the ductility of a slab-on-piles structure have been studied, i.e., the application of a shear panel damper system [3], the use of composite steel pipes for the support piles [4], or the method of spun pile concrete-infilling [5].

This study investigates the use of a hollow spun pile (HSP) filled with concrete and normal reinforcement for slab-on-piles bridge structures. This type of spun pile is hereinafter referred to as reinforced concrete-filled spun pile (RCFSP). The structure of the slab-on-piles bridge in this study is modeled in the form of one segment bridge in three dimensions. Nonlinear structure analysis was carried out using a static pushover method in the SAP2000 software. The location of the bridge is determined to be in Aceh, Indonesia. The free height of each RCFSP used on the bridge is 6 m, while the total length is 30 m.

2. Theoretical background

2.1. Reinforced concrete-filled spun pile

A reinforced concrete-filled spun pile (RCFSP) is a hollow spun pile (HSP) that is added with concrete and normal reinforcement by filling it in the hole. The cross-sectional comparison of HSP and RCFSP is shown in Figure 1. A recent study has been conducted to investigate the flexural behavior of a spun pile filled with concrete. The experiment result shows that infilled concrete increases the cracking moment resistance and increases the ductility of a spun pile by 38% [6]. RCFSP will be used in this study as a support pier for a railway bridge with a slab-on-piles structure type. The structural configuration of one bridge segment is shown in Figure 2.

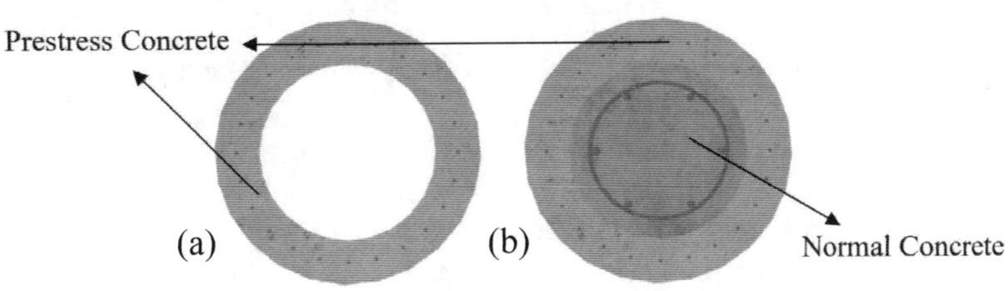

Figure 1. Cross-sectional comparison: (a) HSP; (b) RCFSP.

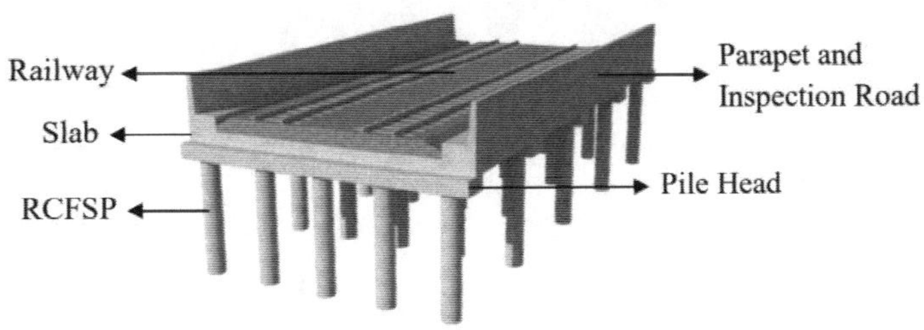

Figure 2. Structural configuration of railway slab-on-piles bridge.

2.2. Performance based-design

Performance-based design is based on the understanding that the performance of a structure can be predicted and evaluated properly and measurably so that in the planning process it can provide certainty of performance targets regarding the designed structure [7]. During its development, there were several documents containing performance-based design methodology [8]. Determining the performance of a structure can use various nonlinear analysis methods, including the static method and dynamic method. The static method is also called pushover analysis. Pushover analysis is a simplified tool that permits the estimation of the capacity of a structure subjected to a given external lateral load [9].

Based on ATC-40, pushover analysis can be done by the capacity spectrum method. The capacity spectrum method shows a graphical representation of the force and displacement relationship (pushover curve) which is then compared with demand in the acceleration-displacement response spectra format [10]. Two important things in this analysis are capacity and demand. Capacity is the ability of a structure to resist seismic demand. The value of structure capacity depends on the strength and deformation ability of each component of the structure. Demand is earthquake load in the form of elastic spectrum response at 5% damping. Capacity spectrum and demand are depicted in a single graph, as shown in Figure 3, to find the performance point. The performance point is the intersection of the capacity spectrum with the appropriate demand spectrum.

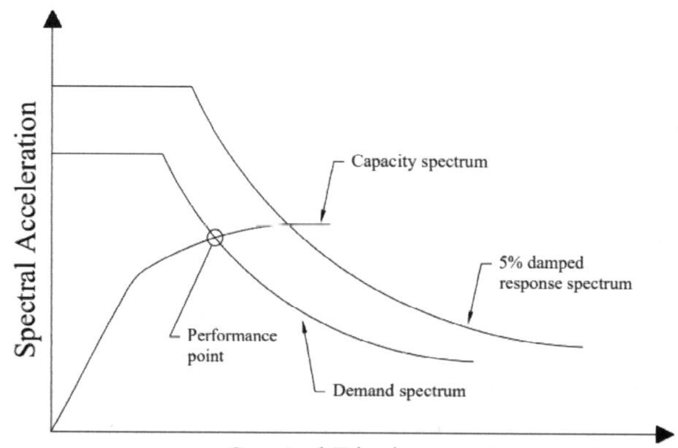

Figure 3. Intersection of the capacity and demand spectrum [10].

FEMA 356 sets the pushover analysis method by finding the target displacement of a structure. Target displacement is an estimate of the maximum expected displacement of the roof of a building calculated for the design earthquake [11]. The target displacement shall be calculated following Equation (1) after the force-displacement curve is obtained.

$$\delta_t = C_0\, C_1\, C_2\, C_3\, S_a \frac{T_e^2}{4\pi^2} g \tag{1}$$

where:

- C_0 = modification factor to relate spectral displacement of an equivalent SDOF system to the roof displacement of the building MDOF system.
- C_1 = modification factor to relate expected maximum inelastic displacements to displacements calculated for linear elastic response.
- C_2 = modification factor to represent the effect of pinched hysteretic shape, stiffness degradation, and strength deterioration on maximum displacement response.
- C_3 = modification factor to represent increased displacements due to dynamic P-Δ effects.
- S_a = response spectrum acceleration, at the effective fundamental period and damping ratio of the building in the direction under consideration, g.
- T_e = effective fundamental period of the building in the direction under consideration, sec.
- g = acceleration of gravity.

Performance determination of the slab-on-pile bridge structure is based on ASCE 61-14 because the structure is similar to the wharf structure. The performance is divided into three levels, i.e., minimal damage, controlled and repairable damage, and life safety protection [12]. The strain limits for each level of performance are described in Table 1.

Table 1. Strain limits for each level of performance.

Level	Component	Hinge strain location	
		Top of pile	In ground
Minimal damage (MD)	Concrete	$\varepsilon_c \leq 0.005$	$\varepsilon_c \leq 0.005$
	Steel	$\varepsilon_s \leq 0.015$	-
	Tendon	-	$\varepsilon_p \leq 0.015$
Controlled and repairable damage (CRD)	Concrete	$\varepsilon_c \leq 0.005 + 1.1\rho_s \leq 0.025$	$\varepsilon_c \leq 0.005 + 1.1\rho_s \leq 0.008$
	Steel	$\varepsilon_s \leq 0.6\varepsilon_{smd} \leq 0.06$	-
	Tendon	-	$\varepsilon_p \leq 0.025$
Life safety protection (LSP)	Concrete	No Limit	$\varepsilon_c \leq 0.005 + 1.1\rho_s \leq 0.012$
	Steel	$\varepsilon_s \leq 0.8\varepsilon_{smd} \leq 0.08$	-
	Tendon	-	$\varepsilon_p \leq 0.035$

where:

- ε_{smd} = strain at peak stress of the reinforcement steel.
- ρ_s = volumetric ratio of confining steel.

3. Research method

3.1. Linear elastic design

The study started with a preliminary design using the linear elastic method. Structural elements for the slab-on-piles bridge that is used in this study consist of RCFCP, pile heads, and a slab. The pile heads are divided into two groups, i.e., the T-type which is located on the edge of the structure, and the L-type which is located in the middle of the structure. Each element's cross-section is shown in Figure 4.

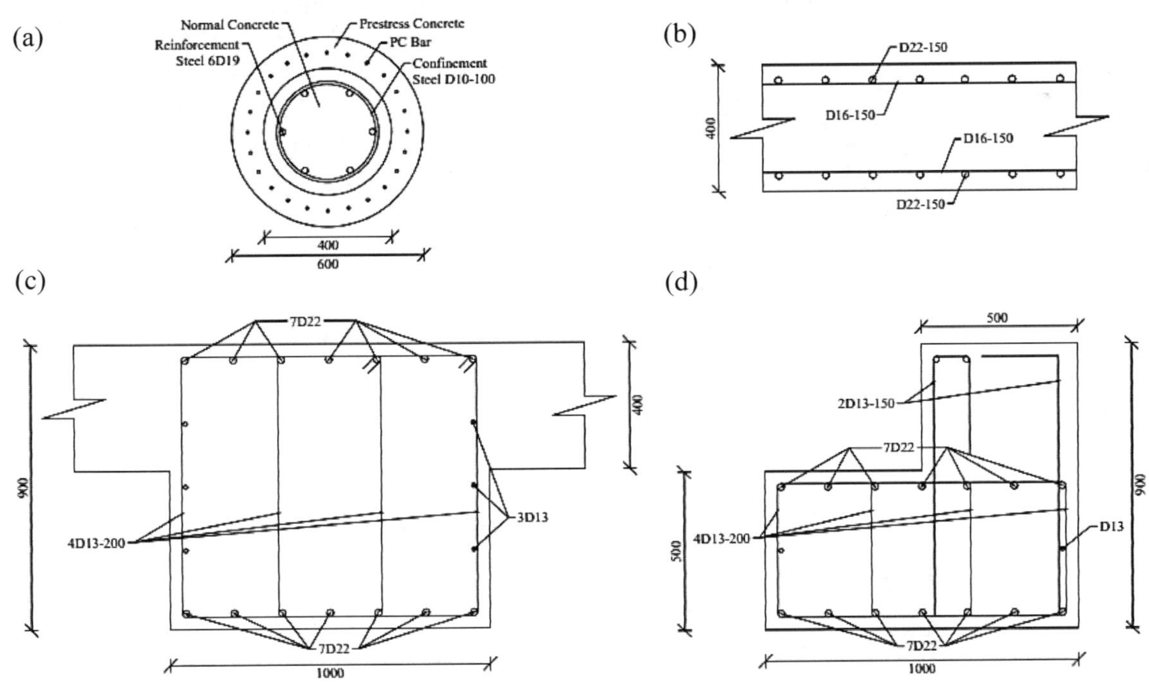

Figure 4. Cross-section: (a) RCFSP; (b) slab; (c) T-type pile head; (d) L-type pile head (dimensions in mm).

The slab dimensions are 20 m x 10.3 m and the free height of RCFSP is 6 m. The bridge uses concrete, reinforcement steel, and prestress steel (PC bar) for its material. Each of them has a property as shown in Table 2.

Table 2. Material properties.

Property	Spun pile concrete	Spun pile filler concrete	Pile head and slab concrete	PC bar	Reinforcement steel
Unit weight, γ (kN/m^3)	24	24	24	78.438	78.5
Compressive strength, f_c' (MPa)	60	K-300 (24.9 MPa)	K-350 (29.05 MPa)	-	-
Yield stress, f_y (MPa)	-	-	-	1275	390
Ultimate stress, f_u (MPa)	-	-	-	1420	500
Elasticity modulus, E (MPa)	36406.044	23452.953	25332.084	230000	200000
Effective prestress force (kN)	-	-	-	78.438	-

Soil-type data to a depth of 30 m is obtained from the standard penetration test (SPT). The soil type is classified into soft soil. The fixity depth of 6 m RCFSP is obtained at 3 m from the ground level, based on a lateral loading analysis of a single RCFSP where the soil bearing capacity is determined by the modulus of subgrade reaction, so the effective length of RCFSP is 9 m. The type of earthquake analysis used is response spectrum analysis. The response spectrum design for the planned earthquake is constructed based on SNI 2833:2016 [13], where the location of the bridge is in Aceh, Indonesia. Linear elastic bridge design is carried out with a response modification factor, R, of 3.5. The design result shows that the number of RCFSP needed in one bridge segment is 25 pieces that are divided into 5 lines (5 pieces per line). The design structure configuration is shown in Figure 5 and Figure 6.

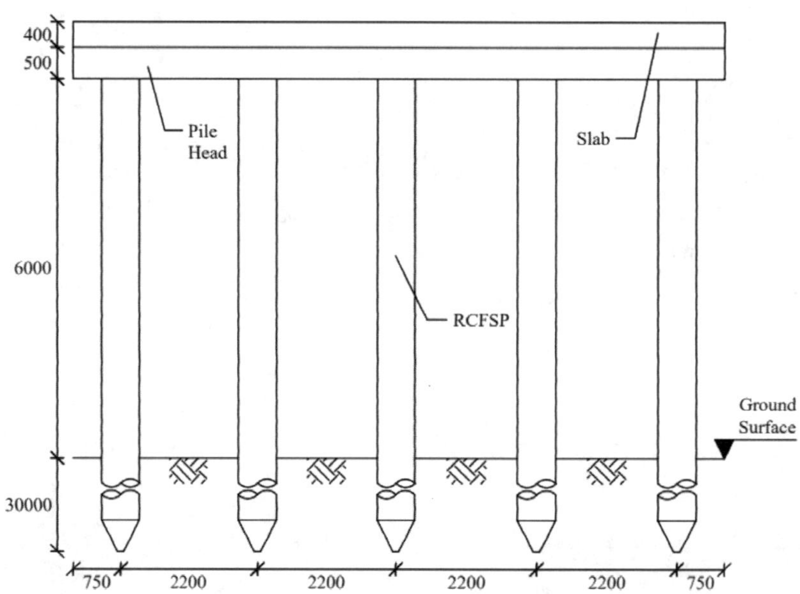

Figure 5. Transversal view of design using R = 3.5 (dimensions in mm).

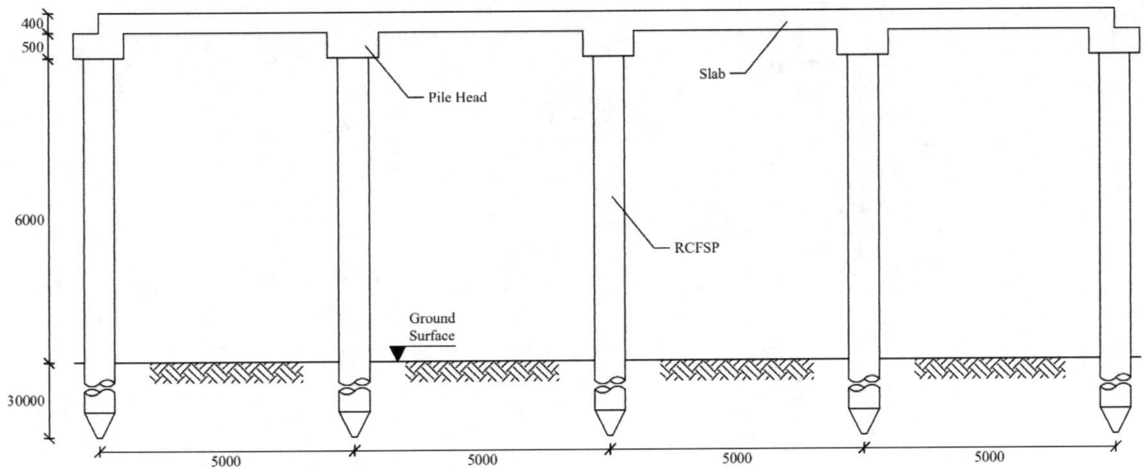

Figure 6. Longitudinal view of design using $R = 3.5$ (dimensions in mm).

3.2. *Nonlinear static pushover*

The first step in this section is the determination of plastic hinge properties. Plastic hinges are assumed to be formed only at the points where the maximum moment occurs. The points are located at the top and the fixity point of each RCFSP. The plastic hinge properties are defined in the form of a moment-curvature diagram. Moment-curvature diagram calculations are carried out under considerations of axial load which consist of structural dead load, additional dead load, and half of a live load. After each moment-curvature diagram is defined, the nonlinear static pushover analysis can be done. Unit lateral loads are assigned at all top nodes of the RCFSP to perform the pushover analysis in the longitudinal and transversal directions. The nonlinear geometric parameter of P-Δ is applied in all pushover analyses. The analysis is performed until the monitored displacement reaches the displacement control. The displacement control used in this analysis is a 5% drift ratio of the effective length of RCFSP. In this study, ATC-40 and FEMA 356 are used as a reference for the nonlinear static pushover analysis method.

4. Result and discussion

4.1. *Plastic hinge definition*

The axial load that occurs on each RCFSP in the design using $R = 3.5$ is shown in Table 3. Each axial load is then applied in the calculation of the moment-curvature diagram. The plastic hinge length is defined as $0.5D$ [14], where D is the diameter of RCFSP. Since the diameter of RCFSP is 0.6 m, the plastic hinge length is 0.3 m. The result of the moment-curvature diagram could be seen in Figure 7.

Table 3. Relation of moment-curvature depends on different axial load.

RCFSP	Axial load (kN)	RCFSP	Axial load (kN)	RCFSP	Axial load (kN)	RCFSP	Axial load (kN)	RCFSP	Axial load (kN)
1	229	6	378	11	371	16	378	21	229
2	183	7	304	12	295	17	304	22	183
3	230	8	351	13	341	18	351	23	230
4	183	9	304	14	295	19	304	24	183
5	229	10	378	15	371	20	378	25	229

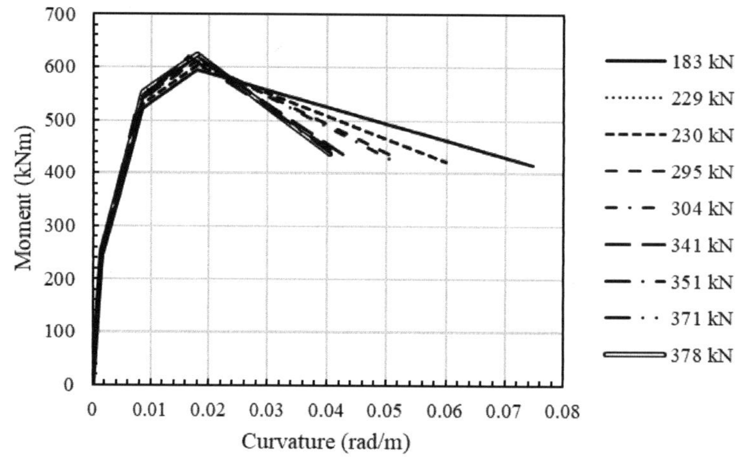

Figure 7. Moment-curvature diagram of RCFSP for design using $R = 3.5$.

The comparison shows that in a lower axial load condition, the ductility is higher, but the flexural strength is lower. The plastic hinges that are applied on the bridge structure are divided into two groups, i.e., the top which is located at the connectivity between RCFSP and pile head, and the in-ground which is located at the fixity point. The strain limits at each performance level, i.e., minimal damage, controlled and repairable damage, and life safety protection, are defined on every plastic hinge.

4.2. Pushover analysis

The result of nonlinear static pushover analysis in longitudinal and transversal directions is shown in Figure 8 and Figure 9. The result from the ATC-40 method shows that the performance point could not be determined because there is no intersection between the capacity and demand curve. It is validated by the result from FEMA 356 calculation where the target displacement could not be reached by the capacity of the structure. Both results indicate that the bridge structure could not resist the designed earthquake load.

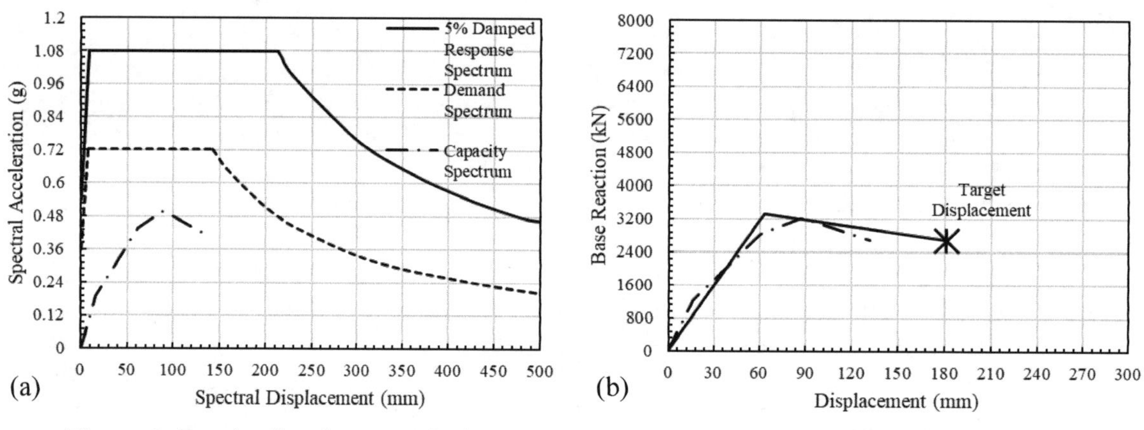

Figure 8. Result of pushover analysis: (a) ATC-40; (b) FEMA 356 in longitudinal direction.

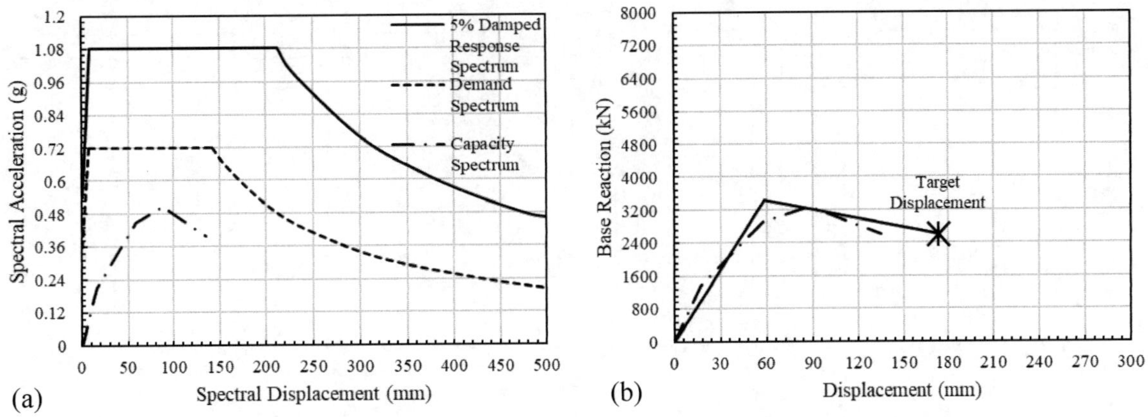

Figure 9. Results of pushover analysis: (a) ATC-40; (b) FEMA 356 in transversal direction.

This condition is not in accordance with the result of the linear elastic design, where the design process has considered the earthquake load. After being reviewed, this happens because the value of *R* is 3.5. The *R*-value in the linear elastic design process is determined at 3.5 because it assumes the bridge structure with RCFSP has higher ductility, but in fact, the ductility of this structure is lower so its capacity is not sufficient to resist the designed earthquake load. Therefore, the structure is redesigned with a smaller *R*, which uses a value of 1.5. The redesign result shows that the number of RCFSP needed in one bridge segment is 60 pieces that are divided into 5 groups (12 pieces per group). RCFSPs in each group are arranged into two lines. The redesign structure configuration is shown in Figure 10 and Figure 11.

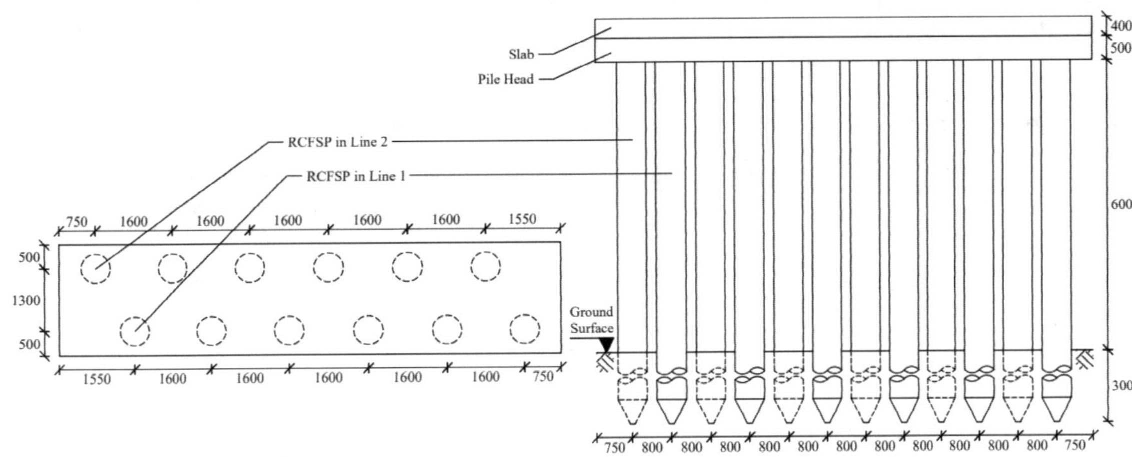

Figure 10. Transversal view of redesign using $R = 1.5$ (dimensions in mm).

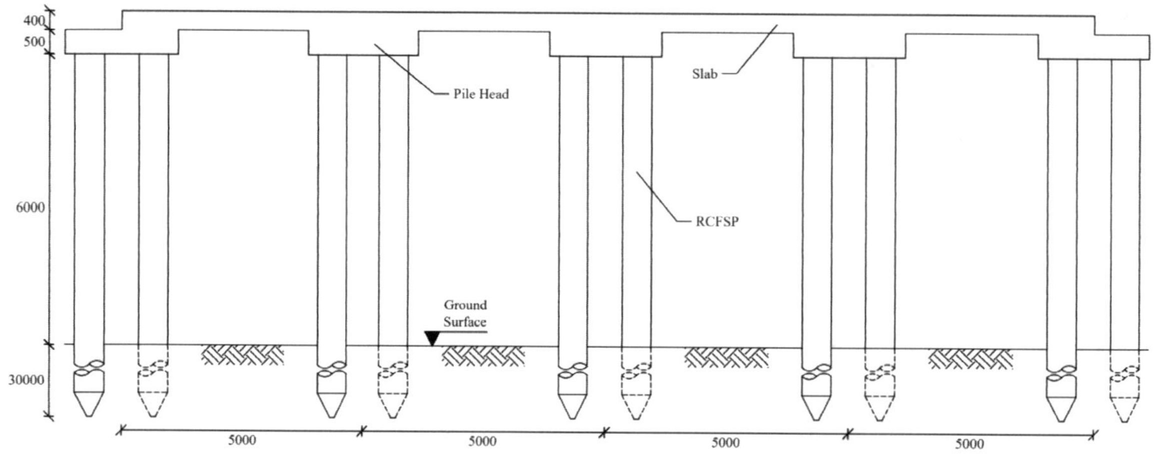

Figure 11. Longitudinal view of redesign using $R = 1.5$ (dimensions in mm).

The result of nonlinear static pushover analysis for the redesigned bridge structure in longitudinal and transversal directions is shown in Figure 12 and Figure 13. Based on the graph, the performance point of the redesigned bridge structure could be determined at the displacement of 85 mm in the longitudinal direction and 81 mm in the transversal direction. The target displacement of 65 mm in the longitudinal direction and 60 mm in the transversal direction also could be reached by the redesigned bridge structure capacity. All of the results indicate that the redesigned bridge structure capacity is sufficient to resist the designed earthquake load. The comparison of performance point and target displacement is shown in Table 4. The comparison shows that the different calculation methods used produce different base shear and displacement values. The maximum strain at the top and in-ground hinge based on ATC-40 and FEMA 356 could be seen in Table 5 and Table 6. All strains show that the plastic hinge is at a minimal damage level, so globally, the redesigned bridge structure performance also falls into this category. The bridge structure performance is at the same level for both ATC-40 and FEMA 356, even though the base shear and displacement results are quite different.

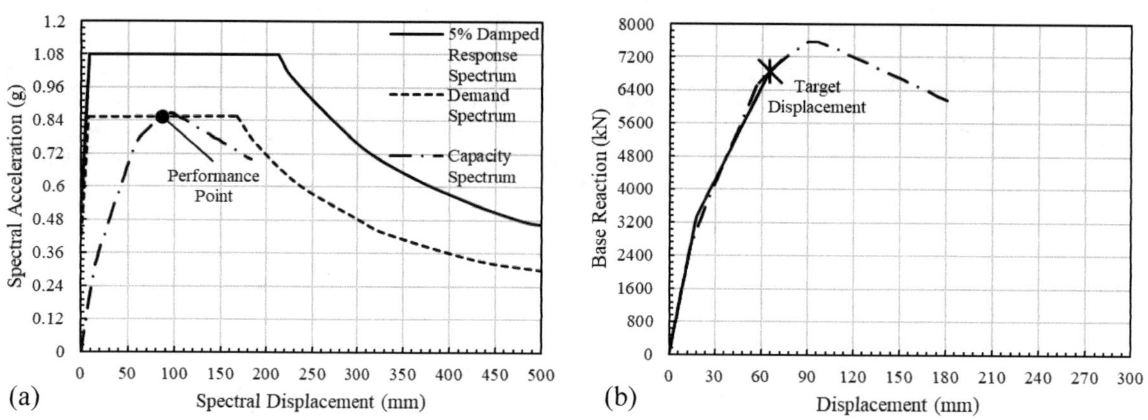

Figure 12. Results of redesigned structure pushover analysis: (a) ATC-40; (b) FEMA 356 in longitudinal direction.

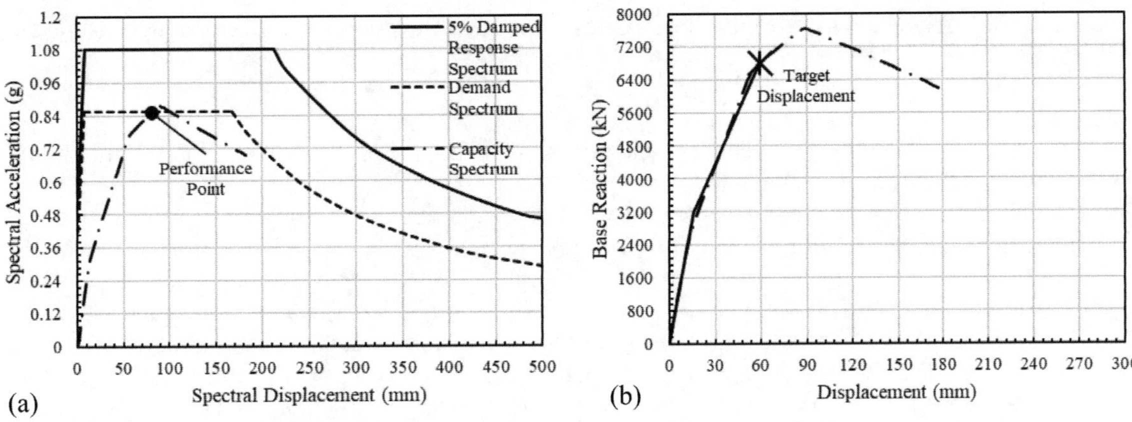

Figure 13. Results of redesigned structure pushover analysis: (a) ATC-40; (b) FEMA 356 in transversal direction.

Table 4. Comparison of performance point and target displacement.

Direction	Performance point (ATC-40)		Target displacement (FEMA 356)	
	Base shear, V (kN)	Displacement, δ (mm)	Base shear, V (kN)	Displacement, δ (mm)
Longitudinal	7461.7	85	6860.702	65
Transversal	7430.563	81	6805.092	60

Table 5. Maximum strain result of redesigned structure pushover analysis using ATC-40.

Top hinge				In-ground hinge				Level
Concrete		Steel		Concrete		Tendon		
Strain	Limit	Strain	Limit	Strain	Limit	Strain	Limit	
	0.005		0.015		0.005		0.015	MD
0.003	0.014	0.005	0.06	0.003	0.008	0.011	0.025	CRD
	No limit		0.08		0.012		0.035	LSP

Table 6. Maximum strain result of redesigned structure pushover analysis using FEMA 356.

Top hinge				In-ground hinge				Level
Concrete		Steel		Concrete		Tendon		
Strain	Limit	Strain	Limit	Strain	Limit	Strain	Limit	
	0.005		0.015		0.005		0.015	MD
0.002	0.014	0.003	0.06	0.002	0.008	0.008	0.025	CRD
	No limit		0.08		0.012		0.035	LSP

5. Conclusion

The design of a slab-on-piles bridge with RCFSP, based on a linear elastic method using an R of 3.5, requires 25 RCFSP for each bridge segment, but this design could not reach a performance point based on a nonlinear static pushover analysis because RCFSP has low ductility. When the slab-on-piles bridge is redesigned using an R of 1.5, the requirement of RCFSP increases to 60 RCFSP for each bridge segment. The redesigned structure could achieve a performance point at a displacement of 85 mm in the longitudinal direction and 81 mm in the transversal direction. At these displacements, the performance of the structure is included in the minimal damage level. The results of pushover analysis on a slab-on-piles bridge with RCFSP, using R of 3.5 and 1.5, show that a slab-on-piles bridge is not efficient to be designed with a high value of response modification factor even though using the concrete-infill treatment.

References

[1] Kurniadi A, Rosyidin I F, Indarto H and Atmono I D 2015 Desain struktur slab on pile *Jurnal Karya Teknik Sipil* **4** (4) 57-68

[2] Budek A M, Benzoni G and Priestley M J N 1997 Experimental investigation of ductility of in-ground hinges in solid and hollow prestressed piles *Report No. SSRP-97/17* University of California

[3] Darmawan M F 2021 Seismic performance investigation of pile-supported slab viaduct structure braced with shear panel damper *MEng Thesis* Universitas Gadjah Mada, Yogyakarta

[4] Ramadhanti D 2020 Perancangan struktur viaduct tipe slab on pile menggunakan tiang pipa baja komposit pada berbagai level zona gempa di Indonesia *BEng Thesis* Universitas Gadjah Mada, Yogyakarta

[5] Riyono C A S 2021 Perancangan jembatan kereta api tipe slab on pile menggunakan reinforced concrete filled spun pile pada berbagai level zona gempa di Indonesia *BEng Thesis* Universitas Gadjah Mada, Yogyakarta

[6] Irawan C, Djamaluddin R, Raka I G P, Faimun, Suprobo P and Gambiro 2020 The effect of the presence of infilling concrete on flexural performance of spun pile – an experimental study *Jurnal Teknologi (Science & Engineering)* **82** (1) 85-94

[7] Suryadi T 2011 Analisa riwayat waktu perilaku nonlinear pilar jembatan beton gelagar box bentang menerus *MEng Thesis* Universitas Indonesia, Depok

[8] Zhang Q 2015 Performance based design and evaluation of reinforced concrete bridges *MAppSc Thesis* The University of British Columbia, Vancouver

[9] Cimellaro G P and Marasco S 2018 *Introduction to Dynamics of Structures and Earthquake Engineering* (Cham: Springer International Publishing AG)

[10] ATC 1996 *ATC-40: Seismic Evaluation and Retrofit of Concrete Buildings* (Applied Technology Council)

[11] ASCE 2000 *FEMA 356: Prestandard and Commentary for the Seismic Rehabilitation of Buildings* (Federal Emergency Management Agency)

[12] ASCE 2014 *ASCE/COPRI 61-14: Seismic Design of Piers and Wharves* (American Society of Civil Engineers)

[13] BSN 2016 *SNI 2833:2016 Perencanaan Jembatan terhadap Beban Gempa* (Jakarta: Badan Standardisasi Nasional)

[14] Corley G 1966 Rotational capacity of reinforced concrete beams *ASCE Journal of the Structural Division* **92** (ST5) 121-46

Analysis Of RC Precast Modular Building with Frame Element Approach

C Johan[1], I Satyarno[*], A Awaludin[2], R Fikri[3], and A F Setiawan[2]

[1] Master Student in Structural Engineering, Department of Civil and Environmental Engineering, Universitas Gadjah Mada, Yogyakarta, Indonesia.
[2] Department of Civil and Environmental Engineering, Universitas Gadjah Mada, Yogyakarta, Indonesia.
[3] Department of Civil Engineering, Universitas Syiah Kuala, Banda Aceh, Indonesia.

*Corresponding author: imansatyarno@ugm.ac.id

Abstract. Concrete buildings with modular systems are still not widely used for high-rise buildings. One of the main reasons is the lack of knowledge regarding the structural design of concrete modular buildings. In Indonesia itself, there are not even many concrete modular buildings because the majority of the area is surrounded by earthquake zones. This study aims to develop a frame analysis method using a section designer for structural analysis of concrete modular buildings. The function of using frame elements is to simplify the calculations that should be modelled with shell elements in order to ensure that the connection between segments occurred only vertically. The analysis was carried out for an 8-storey apartment modular building composed of 4 module segments (M-24A, M-24B, M-36A and M-36B). Seismic analysis was carried out using the parameters of the city of Kutai with S_{ds} = 0.27 and S_{d1} = 0.3. All module segments are capable of enduring the maximum moment caused by seismic loading. The M-36B segment has a maximum moment capacity of 62,318.73 kNm and can withstand an ultimate moment of 2,227.55 kNm. The combined effect of the earthquake due to the response spectrum created maximum building story drift of 0.11% does not exceed the required limit of 2%.

1. INTRODUCTION

Modular system is increasingly used because they can simplify the building process. Modular buildings are commonly constructed by arranging factory-printed modules [1] and the fabrication module can be a partition or an entire room [2]. Furthermore, modular construction is typically used for buildings which consisted of the same room. This system is also generally employed for hotels, schools, apartments or other buildings that have the typical floor plan at each level of the building [3].

Nevertheless, previous studies have predominantly focused on steel buildings, such as box connection modeling [4], module corner connection performance due to seismic loads [5], and the collapse of steel module buildings caused by fire [6]. There has not been much research conducted on modular concrete. The most recent research on modular concrete construction modeling is structural analysis using the MEFM method (Module Equivalent Frame Method). The MEFM method analogizes the module wall into frame elements in the form of beams and columns [7].

In contrary to the MEFM analysis, this study utilized the section designer to model the module segments. The section designer is a feature of the SAP2000 application to draw cross sections of frame elements. The depth analysis is performed globally for each module segment. The significance of using section designer is the automatic calculation of the moment of curvature, torque and moment of inertia [8].

2. THE SCHEME OF MODULAR STRUCTURE

Throughout this research, a modular building comprises four crucial components: a foundation, a concrete module, a shear wall, and a diaphragm floor. The module wall will transmit the influence of the gravity load to the foundation, while the diaphragm will transmit the lateral load to the shear wall. Diaphragms can be created through the reinforced concrete material that is casted in place or concrete with toppings on the deck [10]. The diaphragm in this study employed the diaphragm topping method. The diaphragm topping has a thickness of 75 mm. Unlike factory-molded module walls, shear walls and floor diaphragm used the on-site casting method. The modular building appearance showed in figure 1.

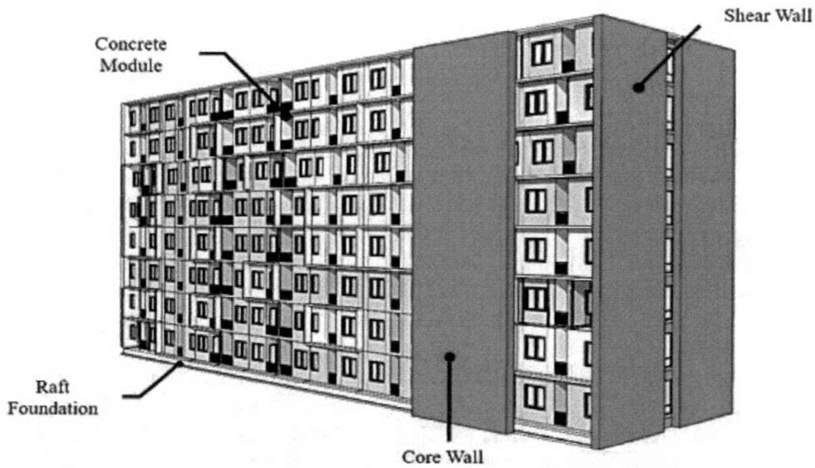

Figure 1. The Modular Building with Three Dimension viewpoint

2.1. The Scheme of Modular Construction

The module size is limited by the transportation capacity to the construction site. In addition, the maximum module dimensions of the segment are 12 m length, 3.4 m width and 4.5 m height [9]. The precast module is divided into 4 segments, namely the M-24A, M-24B, M36- A, and M-36B segments. Each segment will be arranged into two types of rooms (Type 24 and Type 36), see figure 2. The Segment Modules will be arranged to form an 8-storey apartment according to the floor plan shown in figure 3.

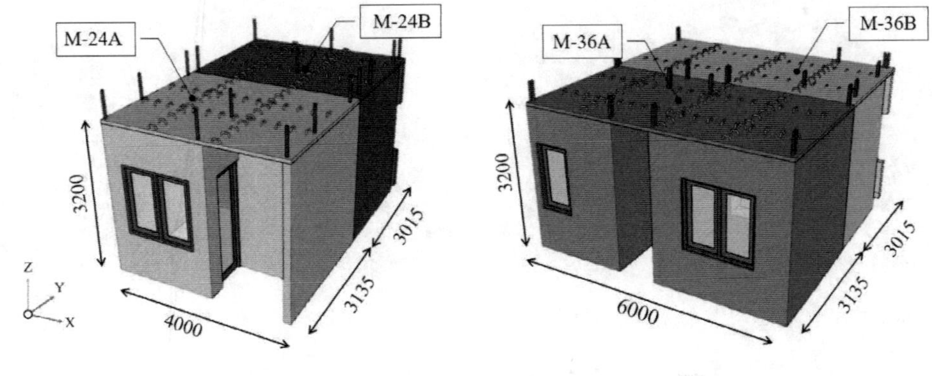

(a) (b)

Figure 2. Module segments (a) Type 24 and (b) Type 36 (unit in mm)

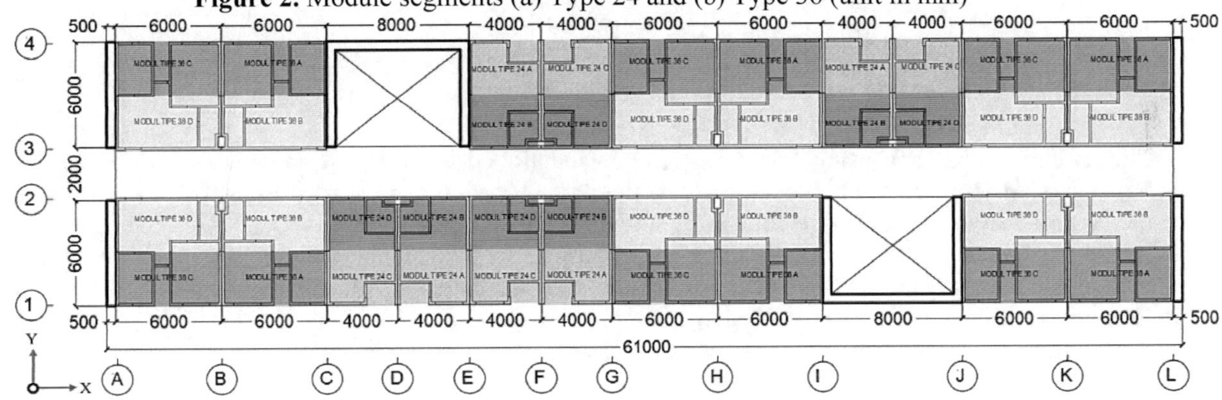

Figure 3. Floor plan (unit in mm).

2.2. The Connection Scheme

Firstly, each module segment was transported from the factory and then it was arranged on the foundation. The Inter-module reinforcement is also used to connect each segment to the foundation. As shown in figure 4(a), on each module segment there is a concrete slab equipped with a shear connector. Then, as depicted in figure 4(b) the floor diaphragm reinforcement is rafted and connected to the shear connector. After the floor diaphragm reinforcement raft is completed, the diaphragm floor is casted as shown in figure 4(c). Besides transmitting the lateral loads to the shear wall, the diaphragm also functions to bind each segment horizontally. Furthermore, second floor module segments are set up on a diaphragm plate which has been casted and then it is connected to the grouted inter-module reinforcement, see figure 4(d).

The connection between each module is crucial since it determined the integrity and stability of the building [11]. The flexibility of the building increases and affects the distribution of lateral loads, module shifts and deformation as a result of a discontinuous structural system [12].

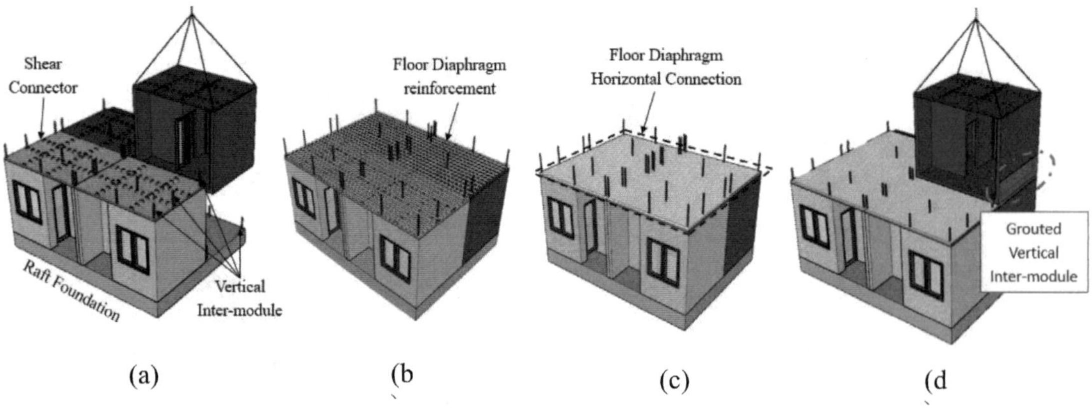

Figure 4. The construction of connection scheme (a) Module connected to foundation, (b) The diaphragm reinforcement installation, (c) The diaphragm casting, (d) The second level of module segment pile.

3. STRUCTURE MODELING

3.1. Opening wall segment

The modeling approach using frame elements is functioned to define segment walls into columns. Segment walls are required for opening access points such as doors or the need for air circulation. Then, the voltage that occurred is a function of the module size and the length of the opening segments [13]. In the SAP2000 designer section, no wall openings are available, therefore an approach should be employed. The approach taken is to make walls with openings have a thinner thickness. The wall thickness of the opening (t_o) is calculated based on the wall height (H), opening height (H_0) and module thickness (t_m) using the equation (1) as demonstrated in the following equation.

$$t_o = \frac{(H-H_o)\times t_m}{H} \quad (1)$$

The coefficient of the opening height of the module in equation (1) is shown in figure 5. The calculation of wall thickness for Doors and Windows is displayed in table 1 with opening wall code shown in figure 6.

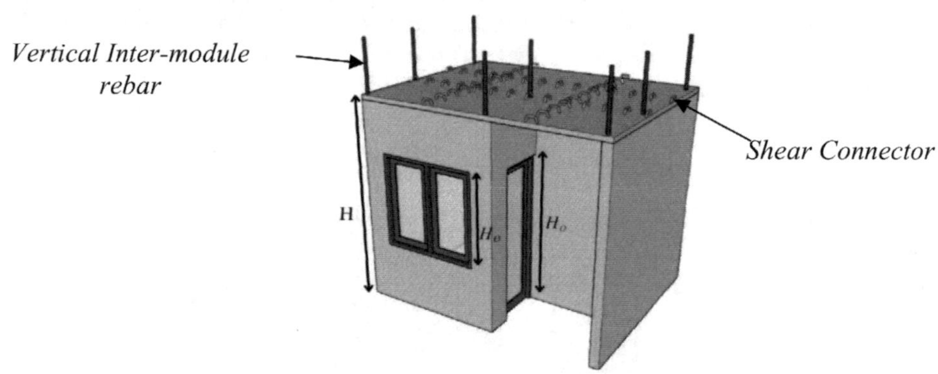

Figure 5. Module Segment M-24A and Module Height H=3,200 mm.

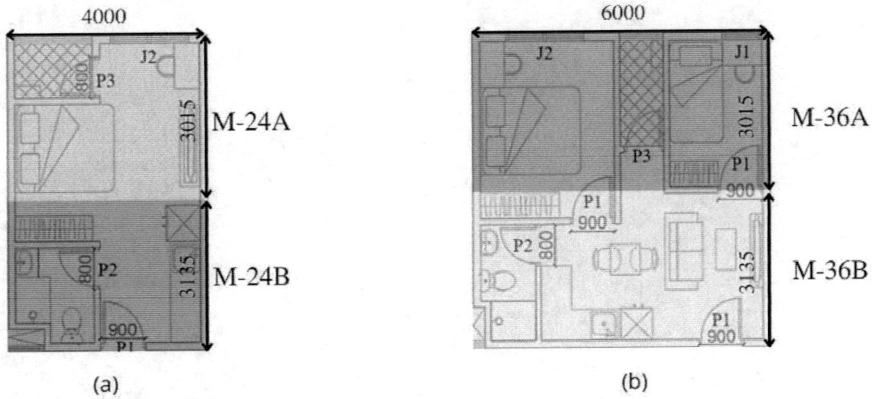

Figure 6. Opening wall code for (a) Type 24 (b) Type 36.

Table 1. Wall thickness in opening area.

Opening Type	Opening Wall Code	t_m (mm)	H (mm)	H_o (mm)	t_o (mm)
Door	P1	150	3,200	2,150	49.22
	P2	120	3,200	2,150	40.31
	P3	120	3,200	2,400	37.5
Window	J1	150	3,200	1,550	77.34
	J2	150	3,200	1,550	77.34

3.2. The modelling of segment module wall using section designer

Each module segment in the form of a wall is modeled as a column using a frame element. Figure 6 showed how the section designer describes the frame element's design in accordance to the segment wall's shape. The function of using frame elements is to simplify the calculations that should be modeled with shell elements in order to ensure that the connection between segments occurred only vertically. Vertical connection using grouted steel rebar as Inter-module connection Inter-module connection is one-point reinforcement with an area equivalent to 5 D32 rebar's area is 4,022.85 mm^2. Figure 7 portrays the process of defining the frame element.

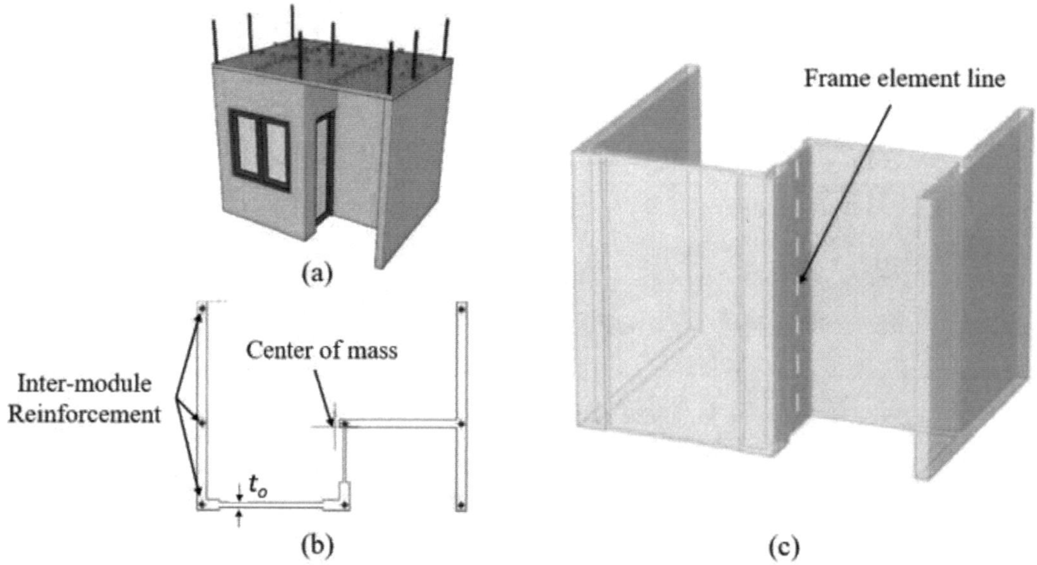

Figure 7. (a) 3D View Segment M-24A, (b) Section Designer Cross Section of Segment M-24A wall, (c) Frame Element SAP2000 3D view for Segment M-24A.

3.3. *The Model Connection of SAP2000*

The Modeling using frame elements is only applied for module segments while the shear walls and floor diaphragms are modeled using shell elements. The issue of this model is that the frame elements will only be connected at one point to the floor slab. The floor slab has 150mm thickness and is made up of 75mm for topping floor diaphragm and 75mm for precast module deck plate. Figure 8 depicts the modeling connection between the modular system's components.

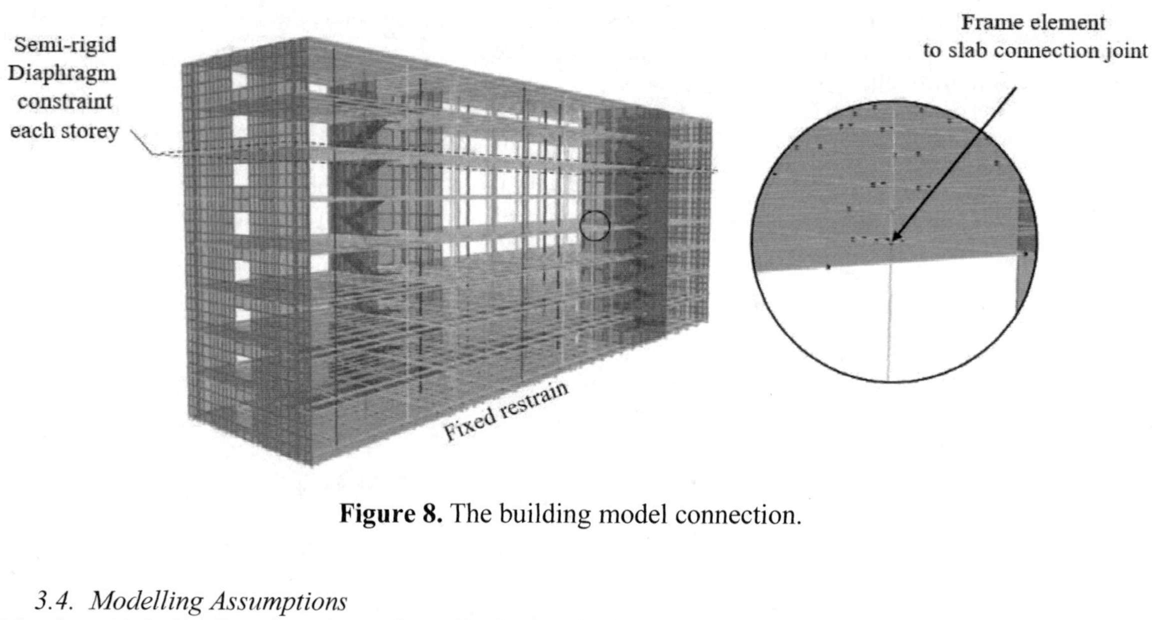

Figure 8. The building model connection.

3.4. *Modelling Assumptions*

The floor slab that functioned as a floor diaphragm is imposed a semi-rigid diaphragm constraint. The shear walls and module walls are assumed to be cracked with an in-plane stiffness reduction factor of

0.35. The out-of-plane wall is not taken into account therefore it was given a factor of 0.001. Then, the floor slab is modeled as a crack with an in-plane stiffness reduction factor of 0.25[14].

The module will only endure the compressive force while the inter-module connection should hold the tensile force. This underlies the need to modify the material properties of the module segment. The tensile stress of the concrete material for the module wall is made to 0 MPa so that it aligns with the characteristics of modular system. In the following table, the characteristics of the concrete material are presented in Table 2, and the characteristics of the reinforcement material are listed in Table 3. Reinforcement material used are elastic perfectly plastic as shown in figure 9.

Table 2. Concrete Material Properties.

	Concrete	f_c (MPa)	ε_c	f_{ct} (MPa)
Wall	Unconfined	50	0.003	0
Floor Slab	Unconfined	25	0.003	3.1
Shear Wall	Unconfined	25	0.003	3.1

Table 3. Steel Material Properties

Material	f_y (MPa)	E (MPa)	ε_{min}	ε_{max}
Steel	400	200,000	-0.20	0.20

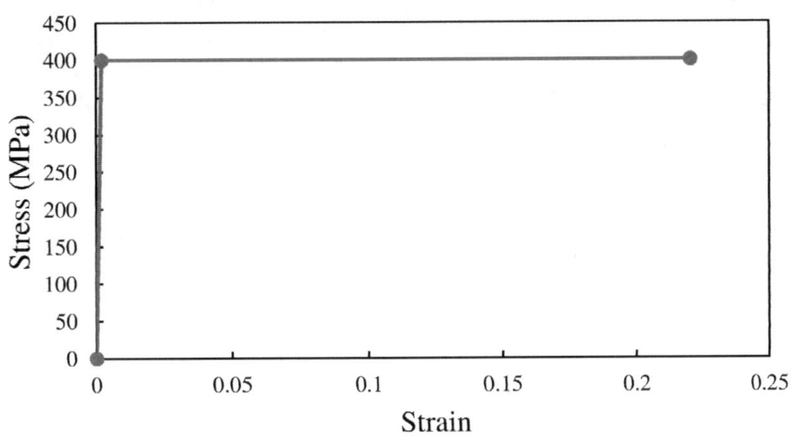

Figure 9. Stress strain curve of steel materials.

3.5. *Mode Shape*

Based on the modeling steps that has been carried out, the mode shape is obtained for the analysis as depicted in figure 10. In Mode 1 there is a translation of the Y-axis direction with a period of 0.478. Then, Mode shape in Mode 2 occurred rotation with a period of 0.351 seconds. Meanwhile, the shape mode in mode 3 is translated in the X direction with a structure period of 0.343 seconds. The fundamental period of the structure is still below the fundamental period hinted at by SNI. The fundamental period is taken from the multiplication of the Cu coefficient and the fundamental period of the approach (Ta) which is T= 0.77 seconds.

IOP Conf. Series: Earth and Environmental Science 1244 (2023) 012014 doi:10.1088/1755-1315/1244/1/012014

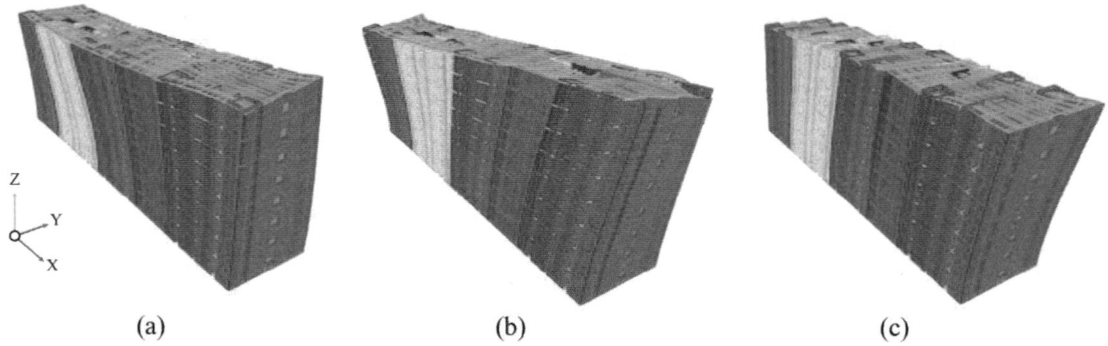

Figure 10. Mode Shape for (a) Mode 1, (b) Mode 2 and (c) Mode 3

4. SEISMIC ANALYSIS

This research employed equivalent static and response spectrum for seismic analysis. Scaled response spectrum analysis using to simplify static equivalent. In this study, the earthquake data used was taken from Cipta Karya Spectrum Response Application for one of Indonesia's regions, Kutai. Table 4 displays the seismic parameters and the system used to transport lateral loads is a special moment-resisting frame structure with coefficients as shown in table 5.

Table 4. The Seismic Parameters.

Site Class	PGA	S_S	S_1	S_{DS}	S_{D1}	T_S	T_0	T_L
SE	0.085	0.171	0.109	0.27	0.3	1.11	0.22	16

Table 5. Factor R, Cd, and Ω_0 for seismic force resisting system.

Seismic force resisting system	Coefficient of response modification (R)	Over strength factor (Ω_0)	Deflection magnification factor (C_d)
Special reinforced concrete shear walls	5	2.5	5

4.1. *Equivalent Static Analysis*

4.1.1. *Manual calculated static equivalent base force*

In accordance to SNI 1726:2019, this building is included in category D with shear walls as seismic force-resisting system. Buildings with special moment resisting frame are taken as response modification $R = 5$ [15]. Since this structure functions as an apartment, the utilized value $Ie = 1$. Based on the parameter value, the seismic response coefficient (Cs) is calculated using equation (2). Maximum response coefficient should not exceed the equation (3) and (4). If the fundamental period of the structure (T) does not exceed long period (T_L) then the response coefficient is calculated based on the equation (3) and if period (T) exceed (T_L) then calculated based on the equation (4).

$$C_s = \frac{S_{DS}}{\left(\frac{R}{I_e}\right)} \tag{2}$$

$$C_s = \frac{S_{D1}}{T\left(\frac{R}{I_e}\right)} \tag{3}$$

$$C_s = \frac{S_{D1} T_L}{T^2\left(\frac{R}{I_e}\right)} \quad (4)$$

For T<TL then the maximum limit of Cs is 1,255.The value of *Cs* = 0.054 is obtained from equation (2). After getting the value of *Cs*, then determined the total design force at the base of the structure in the direction using equation 3. In this case *W* is the total weight of the structure, and from the manual calculation, the structure weight *W* = 90,145.07 kN. From Equation (3) the value of total base force *V* = 4,867.83 kN.

$$V = C_s \times W \quad (5)$$

4.1.2. Equivalent Static Analysis Input

The analysis of static equivalents used an automatic lateral load pattern. The earthquake parameters and coefficients in table 4 and table 5 are input into the SAP2000 load pattern. The load pattern is generated with each earthquake direction having an eccentricity of 0.05 in the X and Y directions. For each direction of the seismic force is added 30% in its perpendicular direction. The period value for static load was taken from the modal analysis period. Then, the static force in the X direction used a period of *T* = 0.343 seconds, while the Y direction used *T* = 0.478 seconds.

4.2. Response Spectrum Analysis

The response spectrum design used curves from the Spectra Cipta Karya Response Application. Since no soil data were obtained for planning, SE (soft soil) was used in the soil site class. The response force spectrum is input by using the Load Pattern with a curve as illustrated in figure 11.

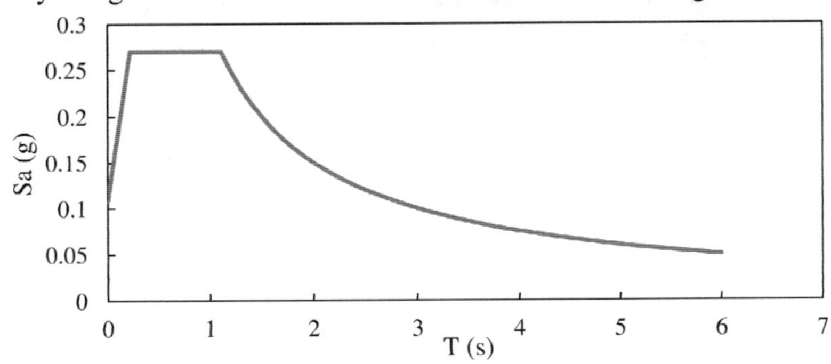

Figure 11. The curve of spectrum response.

The Base shear caused by the spectrum response must be scaled to at minimum 90% of the static load's base shear force. Base shear for static and dynamic earthquakes can be seen from the output of the SAP2000 table. In the load pattern, the response spectrum is used a scale of 1.19 so that the base shear is more than 90% of a static earthquake. Table 6 provides the base shear force's magnitude.

Table 6. Base shear output SAP2000.

	Automatic static equivalent X direction (kN)	Response spectrum X direction (scaled 1.19) (kN)	Automatic static equivalent Y direction (kN)	Response spectrum Y direction (scaled 1.19) (kN)
Base Shear	4,273.37	4,264.23	4,273.37	3,993.33

4.3. Assigned Load and Combination

The earthquake analysis is combined with several forces due to dead and live loads. The structure self-weight is known as the dead load (*D*). Meanwhile, the Super dead load (*SDL*) is the weight of extra accessories, such as ceilings, ceramics, installations, and others. Then, the live load (*L*) acts on the floor slab with the amount determined based on the function of the room. Table 7 shows the value of the horizontal load. The earthquake load (*E*) is then combined to the horizontal load, as illustrated in table 8. On the application of static earthquake load an unexpected torque effect of 5% is added for each direction and 30% lateral force is added in its perpendicular direction.

Table 7. Load assign on slab and roof.

Super Dead Load, *SDL* (kN/m2)	Live Load on Corridor, *L* (kN/m2)	Apartment Room Live Load, *L* (kN/m2)
1.132	4	2.4

Table 8. Load combination.

	Load Combination
Combination 1	$0.9\,D + 0.9\,SDL + E_{\text{static}}$
Combination 2	$1.2\,D + L + E_{\text{static}}$
Combination 3	$0.9\,D + 0.9\,SDL + E_{\text{Response spectrum}}$
Combination 4	$1.2\,D + L + E_{\text{Response spectrum}}$

5. RESULTS AND DISCUSSION

5.1. The Shear Force of Module segment

Lateral forces in the modular system caused shear forces in each segment of the module. The shear force which occurred on each floor had a different value. In order to see the variation of the shear force occurrence, a section cut is made for all segments each storey. The shear inspection for each storey is performed in both the X and Y directions as shown in figure 12. The first floor has the highest floor segment shear force. The highest shear force is 3,657.69kN for the X direction and 2,417.74kN for the Y direction.

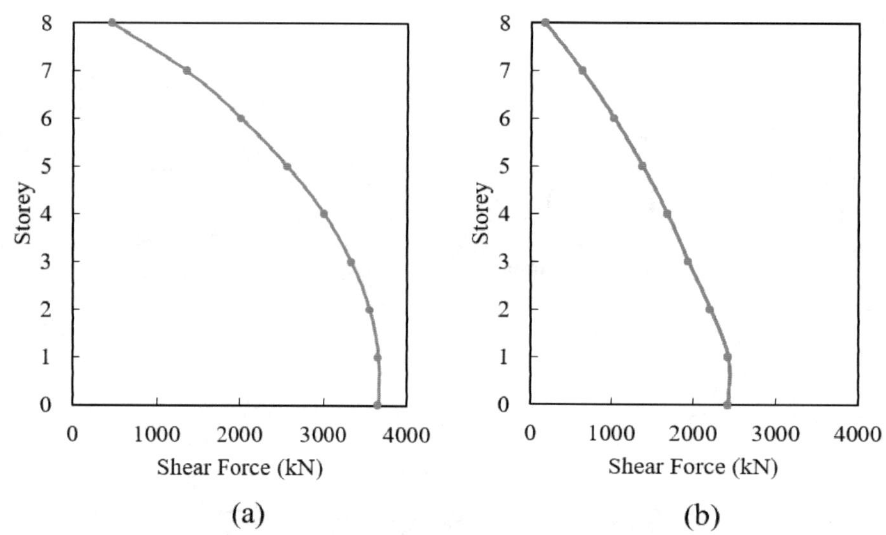

Figure 12. Comparison of shear forces on each floor (a) X direction, (b) Y direction.

The ultimate shear inspection for each segment was carried out on the first floor. The ultimate shear force in each segment must be able to be resisted by the shear capacity of joint. The shear capacity of the inter-module connection is a function of the area of the reinforcement multiplied by the yield strength of the steel material used. The number of inter-module connections determines the capacity of each segment. The calculation of the joint shear capacity of each module is provided in table 9.

Table 9. Shear capacity of each segment

	Amount of Inter-module	Inter-module area (mm^2)	Total area (mm^2)	Shear capacity (kN)	Ultimate shear fore (kN)
M - 24A	8	4,022.857	32,182.86	12,873.14	205.30
M - 24B	9	4,022.857	36,205.71	14,482.29	258.55
M - 36A	10	4,022.857	40,228.57	16,091.43	773.54
M - 36B	10	4,022.857	40,228.57	16,091.43	637.15

5.2. Moment Capacity

Each module segment should be able to endure the moments due to seismic forces. The maximum ultimate moment for the module segment occurred on the first floor. The ultimate moment of each segment must be able to be resisted by the moment capacity of the segment which is analyzed by using the section designer. To check the moment due to the earthquake is carried out for the X and Y directions. Since it uses the same inter-module connection, the moment capacity of the segment for each story is also the same. As a result, it attempted to facilitate the manufacture and implementation in the field.

The segment capacity is determined based on the shape of the wall in the section designer. The module segment capacity is exaggerated because the wall segment is regarded as the same as the column. The maximum capacity of the M-36B segment is 62,318.73 kNm for X direction and the M-36B segment is 27,382.32 kNm for Y direction. The comparison between the moments caused by the earthquake and the moment capacity of the segment shown in the figure 13.

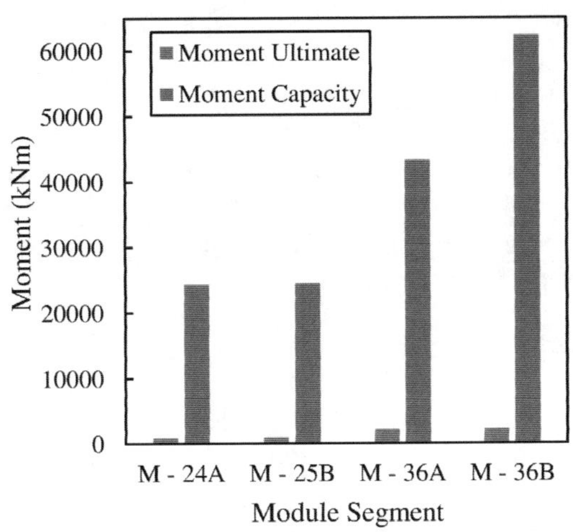

(a)

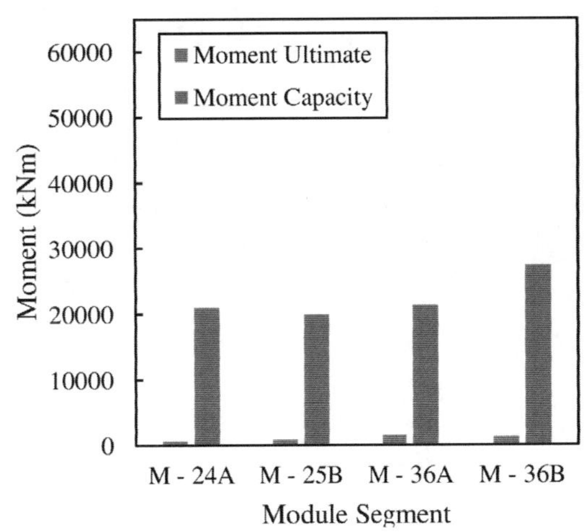

(b)

Figure 13. Comparison of moment capacity and ultimate moment (a) X direction, (b) Y direction

5.3. Drift Ratio

In accordance with SNI 1726:2019 for modular building structures, it is limited to less than 2%. The drift between floors should be checked for the response of the structure due to the combined seismic load. The drift ratio value is the deflection of the location under consideration multiplied by the deflection amplification factor (Cd). The value of the deflection factor for buildings with shear walls is Cd=5.

Based on the drift ratio calculation, the larger story drift is actually caused by the earthquake response spectrum. Then, the largest maximum story drift is in the Y direction of 0.11% due to the combination of load response spectrum as shown in figure 14(b). Figure 14 depicts each floor's story drift as a result of the equivalent static load and response spectrum.

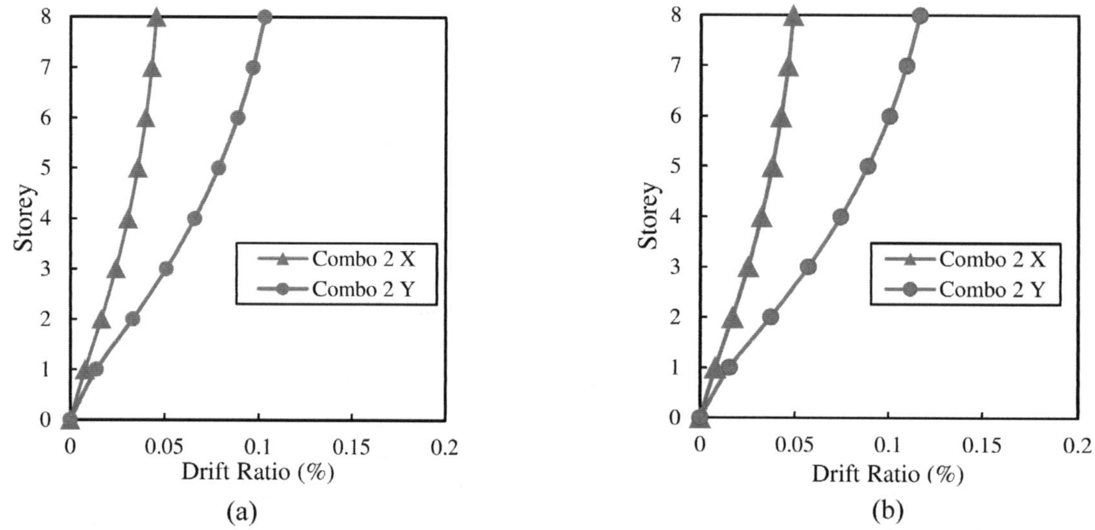

Figure 14. Maximum story drift using a combination of loading (a) Static earthquake, (b) Response Spectrum.

5.4. *P-delta effect*

The lateral movement that is influenced by axial loads and horizontal displacement will cause secondary moments that cause an increase in story drift. P-delta effect needs to be examined to see the stability of the structural system. P-delta effect can be ignored if the stability coefficient (θ) is below 0.1. Stability coefficient calculated by equation (6) where (P_x) is the total vertical design load on floors x and above, (Δ) is the design story drift, utilized value of the building $I_e = 1$, (V_x) seismic shear force at storey x, (h_{sx}) story height at storey x and deflection amplification factor (Cd).

$$\theta = \frac{P_x \Delta I_e}{V_x h_{sx} C_d} \tag{6}$$

The calculation of stability coefficient for the X and Y direction is shown in Tables 8 and 9. From the calculation results, the stability coefficient is still below the value of 0.01 so that the P-delta effect can be ignored in this case.

Table 10. Stability coefficient along X direction

Storey	*Δ* (mm)	*Px Dead* (kN)	*Px Live* (kN)	*Px* total (kN)	*V* (kN)	h_{sx} (mm)	θ
1	0.33265	90145.07	15287.04	105432.1	746.1856	3200	0.003
2	0.917035	78333.918	13376.16	91710.08	1029.45	6400	0.003
3	1.273355	66522.767	11465.28	77988.05	883.4853	9600	0.002
4	1.503835	54711.615	9554.4	64266.02	736.2377	12800	0.002
5	1.62908	42900.464	7643.52	50543.98	588.9902	16000	0.002
6	1.672125	31089.312	5732.64	36821.95	441.7426	19200	0.001
7	1.677745	19278.161	3821.76	23099.92	294.4951	22400	0.001

8	1.59571	7481.7126	1910.88	9392.593	147.2475	25600	0.001

Table 11. Stability coefficient along Y direction

Storey	*Δ* (mm)	*Px Dead* (kN)	*Px Live* (kN)	*Px* total (kN)	*V* (kN)	h_{sx} (mm)	*θ*
1	0.64834	90145.07	15287.04	105432.1	746.1856	3200	0.006
2	1.64641	78333.918	13376.16	91710.08	1029.45	6400	0.005
3	2.282005	66522.767	11465.28	77988.05	883.4853	9600	0.004
4	2.686405	54711.615	9554.4	64266.02	736.2377	12800	0.004
5	2.909955	42900.464	7643.52	50543.98	588.9902	16000	0.003
6	2.98797	31089.312	5732.64	36821.95	441.7426	19200	0.003
7	2.964795	19278.161	3821.76	23099.92	294.4951	22400	0.002
8	2.92242	7481.7126	1910.88	9392.593	147.2475	25600	0.001

6. Conclusion

The analysis of modular system building using section designer as a frame element is an approach method. The conclusions obtained from this analysis are as follows:

1. The approach of using the frame element section designer can be employed for the analysis of high-rise buildings. The use of the section designer can also be applied to various forms of module segments. Furthermore, section designer modeling helps to ensure that each wall is connected only at joints.

2. The joint shear capacity of each segment could resist the ultimate force. The M-36A and M-36B segments have the largest moment capacity with the same amount of inter- module connections. M-36A and B segment connection has a shear capacity of 16,091.43 kN and can withstand an ultimate shear force of 773.54 kNm.
3. All module segments are capable of enduring the maximum moment caused by earthquake loading. The M-36B segment has a maximum moment capacity of 62,318.73 kNm and can withstand an ultimate moment of 2,227.55 kNm.
4. The combined effect of the earthquake due to the response spectrum created a larger story drift between floors. The maximum building story drift of 0.11% does not exceed the required limit of 2%.

Acknowledgments

The authors would like to gratefully acknowledge the Department of Civil and Environmental Engineering Gadjah Mada University and PT. Waskita Karya Precast for the data support for this study.

References

[1] Gunawardena T. *Behaviour of Prefabricated Modular Buildings Subjected to Lateral Loads*.2016.
[2] Lawson M, Ogden R, Goodier C. *Design in Modular Construction*. 2014.
[3] Lawson RM, Ogden RG, Bergin R. *Application of Modular Construction in High-Rise Buildings*.

J Archit Eng 2012; 18: 148–54. https://doi.org/10.1061/(asce)ae.1943- 5568.0000057.

[4] Qin J, Tan P. *Design method of innovative box connections for modular steel constructions*. J Build Eng 2022;57: 104820. https://doi.org/10.1016/j.jobe.2022.104820.

[5] Lian JY, Deng EF, He JM, Cai LM, Gao SC, Zhou JJ. *Numerical analysis on seismic performance of corner fitting connection in modular steel building*. Structures 2021;33:1659–76. https://doi.org/10.1016/j.istruc.2021.05.044.

[6] Shan S, Pan W. *Collapse mechanisms of multi-story steel-framed modular structures under fire scenarios*. J Constr Steel Res 2022;196:107419. https://doi.org/10.1016/j.jcsr.2022.107419.

[7] Pan W, Wang Z, Zhang Y. *Module equivalent frame method for structural design of concrete high-rise modular buildings*. J Build Eng 2021;44.

[8] CSI. SAP2000 *Integrated Solution for Structural Analysis and Design*. Comput Struct Inc 2011:556.

[9] Liew JYR, Chua YS, Dai Z. *Steel concrete composite systems for modular construction of high-rise buildings*. Structures 2019;21:135–49.

[10] Moehle JP, Hooper JD, Kelly DJ, Meyer TR. *NEHRP Seismic Design Technical Brief No. 3 Seismic Seismic Design of Cast-In-Place Concrete Diaphragms, Chords, and Collectors*. NEHRP Seism Des Tech Br No 3 2010.

[11] Thai HT, Ngo T, Uy B. *A review on modular construction for high-rise buildings*. Structures 2020;28:1265–90. https://doi.org/10.1016/j.istruc.2020.09.070.

[12] Srisangeerthanan S, Hashemi MJ, Rajeev P, Gad E, Fernando S. *Numerical study on the effects of diaphragm stiffness and strength on the seismic response of multi-story modular buildings*. Eng Struct 2018;163:25–37.

[13] Wenke JM, Dolan CW. *Structural integrity of precast concrete modular construction*. vol. 66. 2021.

[14] Badan Standardisasi Nasional. *Persyaratan Beton Struktural untuk Bangunan Gedung dan Penjelasan* (SNI 2847:2019). 2019.

[15] Badan Standarisasi Nasional. SNI 1726-2019 tentang *Tata cara perencanaan ketahanan gempa untuk struktur bangunan gedung dan nongedung*. 2019.

IOP Conf. Series: Earth and Environmental Science 1244 (2023) 012015 doi:10.1088/1755-1315/1244/1/012015

The effect of floor slabs in a structure with frame shear walls modeling in pushover analysis

A G Y Sweiga[1], I Satyarno[1], B Supriyadi[1]

[1] Department of Civil and Environmental Engineering, Universitas Gadjah Mada, Yogyakarta, Indonesia

E-mail: grafita.yans@mail.ugm.ac.id, imansatyarno@ugm.ac.id, bambang.supri@ugm.ac.id

Abstract. Pushover is a nonlinear static analysis method that provides information on nonlinear structures' behavior due to earthquakes. Subsequently, it has been recommended as a reliable tool to help know the capacity of structures. Another consideration is its easiness of interpretation than the nonlinear dynamic analysis. Numerical modeling of the structure plays an important role in this analysis since every detail of modeling could give different results. The objective of this study is to analyze the effect of the existence of floor slabs in a structure with frame shear walls modeling, which considers the frequent attempt to simplify structural modeling. The effect of the position of components assignation is going to be analyzed too, especially floor slabs and beams, whether it was assigned with centroid or top center insertion point in pushover analysis. In this study, an existing hospital building with vertical irregularity has been modeled into four various models using SAP2000. As the result of pushover analysis, every model gives its pushover curves which reflect its behavior. Those results show that the models with floor slabs have higher capacities. In addition, the combination of the assignation of floor slabs and the top center insertion point gives the highest capacity compared to the other models.

1. Introduction

Pushover is a nonlinear static analysis method that is often used to find out the post-yield behavior of structures. Therefore, this analysis is expected to provide information on the structure's characteristics with higher accuracy which could not be provided from a linear static or dynamic analysis, also the results from this form of analysis are easier to interpret than nonlinear dynamic analysis [1,2]. Numerical modeling for pushover analysis can be done with the help of computer software. SAP2000 is one of the most popular for its compatibility to analyze any kind of structure which was developed by Computer and Structure, Inc. (CSI). As an important note, every modeling detail could give a different analysis result. Specifically, this study will discuss the effect of the existence of floor slabs in pushover analysis considering the frequent attempts to simplify structural modeling. Furthermore, the thing that is often overlooked especially when using SAP2000 for modeling is all components will be assigned with centroid insertion point. Thus, this paper will also give its pushover analysis result if the floor slabs and beams of the structure are being modeled with the top center insertion point following the real condition of the building.

2. Methodology

An existing hospital building located at DI Yogyakarta, Indonesia is used for this numerical modeling analysis. This building consists of eight storey with vertical irregularity because of the two separated towers in the middle of its height. Firstly, the structure with shell shear walls is being compared with the structure with frame shear walls for their modal analysis results considering that the running process for pushover analysis takes a longer time for the structure with shell shear walls than the structure with frame shear walls. The frame shear walls were created using section designer as shown in Figure 1. Table 1 shows the time period comparison between those two structures then decided that the model with frame shear walls in Figure 2 as the base model (Model 0). Furthermore, there are four various models which have been prepared and presented in Table 2. Model 0 represents how the real structure is being built while the other three are the variations of the floor slabs are either being modeled or not and as well the assignation of the top center or centroid insertion point. The plastic hinges were assigned at both edges of the beams using the moment-curvature parameter and at the bottom of columns and shear walls using the default from section parameter. Meanwhile, all the slabs are still modeled as elastic components. The pushover analysis method and its parameters from ASCE 41-17 code are used for this study.

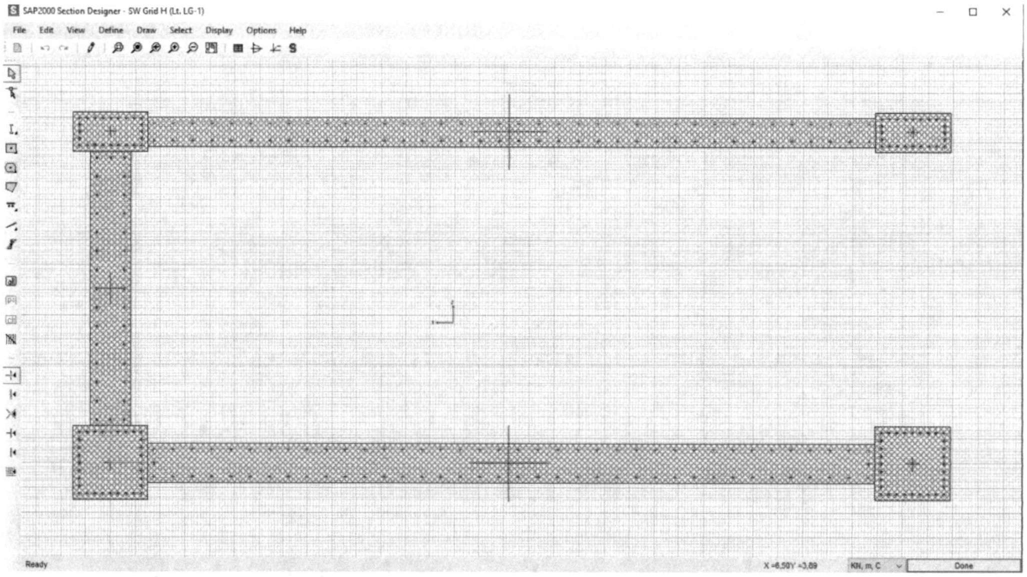

Figure 1. Frame shear wall component.

Table 1. Time period comparison for structures with shell SW and frame SW.

Mode	Time period (s)	
	Structure with shell SW	Structure with frame SW
1	1.13	1.13
2	0.77	0.86
3	0.61	0.61

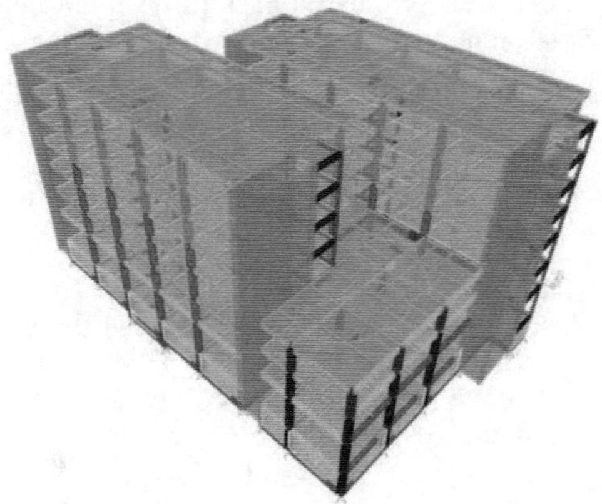

Figure 2. Structure with frame shear wall (Model 0).

Table 2. Modeling variations.

Model	Description	Detailed picture
Model 0	Structure with floor slabs (insertion point: top center)	Figure 3 (a)
Model 1	Structure with floor slabs (insertion point: centroid)	Figure 3 (b)
Model 2	Structure without floor slabs (insertion point: top center)	Figure 3 (c)
Model 3	Structure without floor slabs (insertion point: centroid)	Figure 3 (d)

(a) (b) (c) (d)

Figure 3. Detail of modeling variations.

3. Nonlinear static procedure (NSP)

Nonlinear static procedure (NSP) is also known as pushover analysis. If this method is being used, numerical modeling directly incorporating the nonlinear load-deformation characteristics of individual components of the building shall be subjected to monotonically increasing lateral loads representing inertia forces in an earthquake until a target displacement is exceeded [3]. The target displacement, δ_t, should be calculated based on Equation (1), where C_0 is a modification factor to relate the spectral displacement of an equivalent single-degree-of-freedom (SDOF) system to the roof displacement of the building multiple-degree-of-freedom (MDOF) system, C_1 is a modification factor to relate expected maximum inelastic displacements to displacement calculated for the linear elastic response, C_2 is a modification factor to represent the effect of pinched hysteresis shape, cyclic stiffness degradation and strength deterioration on the maximum displacement response, S_a is response spectrum acceleration, T_e is effective fundamental period, and g is acceleration of gravity [3]. Furthermore, its pushover result will be shown as idealized force-displacement curves with the control node shall be located at the center of mass on the roof of its building [3].

$$\delta_t = C_0\, C_1\, C_2\, S_a\, \frac{T_e^2}{4\pi^2}\, g \tag{1}$$

For new existing buildings, there are two seismic hazard levels to be evaluated, BSE-1N and BSE-2N [3]. BSE-1N seismic hazard level shall be taken as two-thirds of the values of the parameters for the BSE-2N, where BSE-2N itself shall be taken as values from response spectrum parameters where the probability of exceedance is 2% in 50 years or 2,475 years of mean return period [3].

4. Result and discussion

4.1. Modal analysis

The fundamental parameter in the static force-based design of a structure is its vibration characteristic [4,5]. Table 3 presents each model's periods and mode shapes from the first to the third mode. Furthermore, its details are shown in Figure 4 for the periods, and from Figure 5 to Figure 8 for the mode shapes. As can be seen from the results, Model 0 has the smallest period compared to other models. It shows that the assignation of floor slabs and the top center insertion point will increase the structure's stiffness. The moment of inertia is higher if the floor slabs and beams were assigned with the top center than the centroid insertion point. All the models have the translation in the Y-direction, translation in the X-direction, and rotation mode shapes sequentially from the first to the third mode.

The mass participation ratio of structure in numerical modeling, especially in the dynamic analysis is very important and needs to achieve 90% of it for the least. Table 4 presents each model's mass participation results up to the 2500th mode. However, it shows that the two models with floor slabs have not reached 90% of mass participation in both X and Y directions. Since this study specifically only discusses pushover analysis, mass participation will not affect the result.

Table 3. Vibration characteristic comparison.

Mode	1	2	3
Model 0			
Period (s)	1.13	0.86	0.61
Mode shape	Trans. Y	Trans. X	Rotation
Model 1			
Period (s)	1.35	0.98	0.69
Mode shape	Trans. Y	Trans. X	Rotation
Model 2			
Period (s)	1.24	1.05	1.02
Mode shape	Trans. Y	Trans. X	Rotation
Model 3			
Period (s)	1.69	1.55	1.24
Mode shape	Trans. Y	Trans. X	Rotation

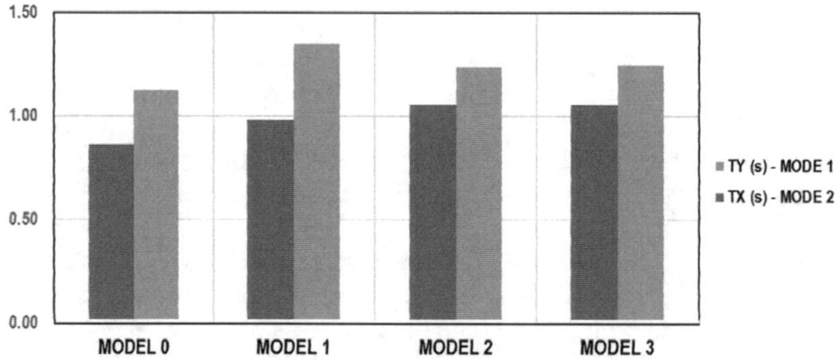

Figure 4. Fundamental period (s).

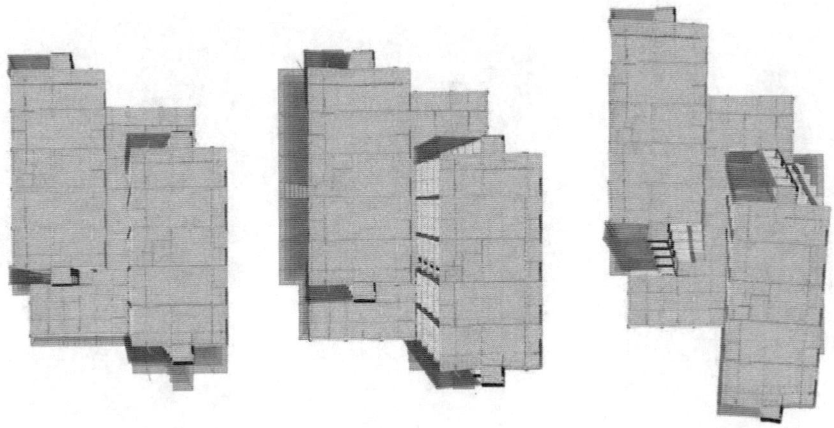

(a) Mode 1 (a) Mode 2 (a) Mode 3

Figure 5. First three shape modes of Model 0.

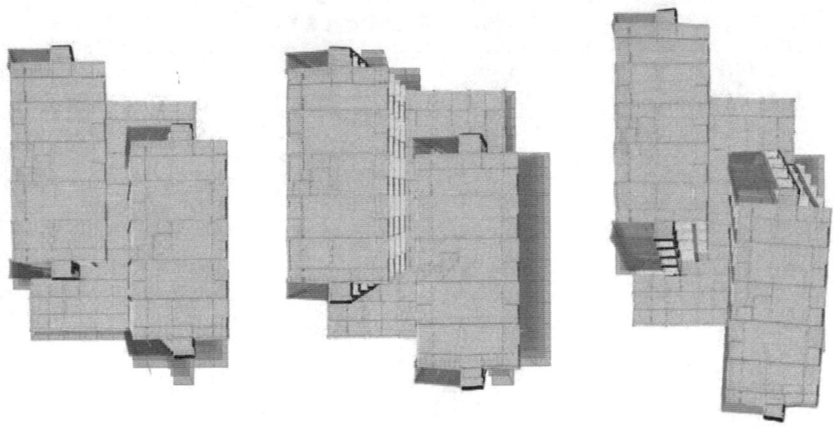

(a) Mode 1 (a) Mode 2 (a) Mode 3

Figure 6. First three shape modes of Model 1.

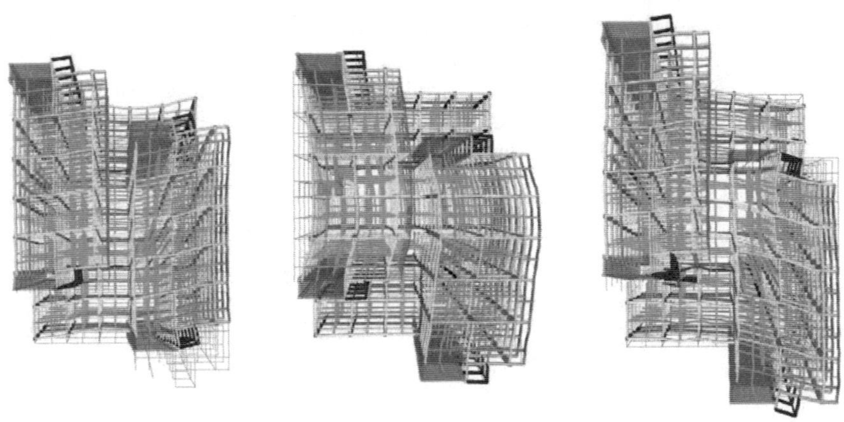

(a) Mode 1 (a) Mode 2 (a) Mode 3

Figure 7. First three shape modes of Model 2.

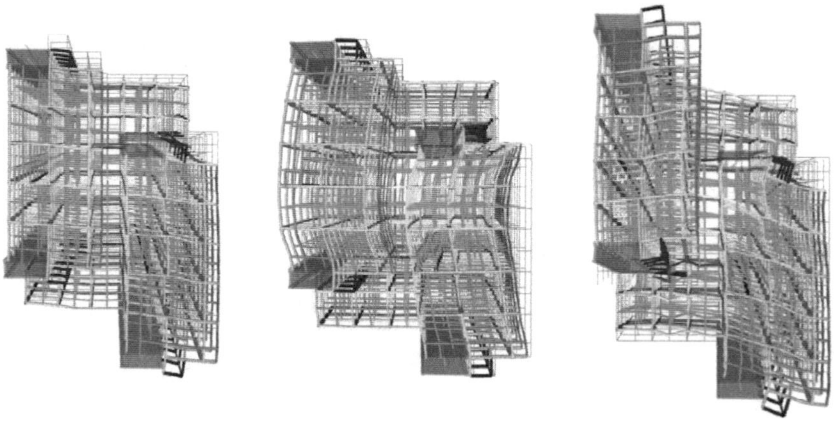

(a) Mode 1 (a) Mode 2 (a) Mode 3

Figure 8. First three shape modes of Model 3.

Table 4. Mass participation.

Mode no.	Model 0		Model 1		Model 2		Model 3	
	SumUX	SumUY	SumUX	SumUY	SumUX	SumUY	SumUX	SumUY
1	0.01%	59.86%	0.01%	57.85%	0.04%	55.86%	67.56%	0.28%
2	58.32%	59.87%	57.11%	57.86%	51.42%	55.95%	67.86%	67.23%
3	58.32%	59.87%	57.12%	57.87%	51.44%	61.48%	68.01%	67.49%
4	58.33%	59.87%	57.12%	57.87%	51.45%	61.48%	68.16%	67.52%
5	58.33%	60.06%	57.12%	59.80%	51.70%	61.49%	68.17%	67.61%
-	-	-	-	-	-	-	-	-
-	-	-	-	-	-	-	-	-
2499	87.84%	88.27%	87.60%	88.13%	92.01%	90.90%	92.14%	90.93%
2500	87.84%	88.27%	87.60%	88.13%	92.03%	90.90%	92.15%	90.93%

4.2. Pushover analysis

As stated before, the pushover analysis result can be shown by its idealized force-displacement curve and every direction has its target displacement. The Y-direction results will be shown as the representative results for this analysis. Table 5 presents all models' target displacement for both BSE-1N and BSE-2N seismic hazard levels. Those target displacements will be plotted together with the pushover curves. Figure 9 shows the pushover curves for BSE-1N seismic hazard level, while Figure 10 is for BSE-2N. In addition, the results of this analysis also be presented with ATC-40 curve as shown from Figure 11 to Figure 14 for each model.

Table 5. Target displacement (Y direction).

Model	Seismic hazard level	
	BSE-1N	BSE-2N
	Target disp., δ_t (m)	Target disp., δ_t (m)
Model 0	0.24	0.36
Model 1	0.28	0.42
Model 2	0.26	0.39
Model 3	0.26	0.39

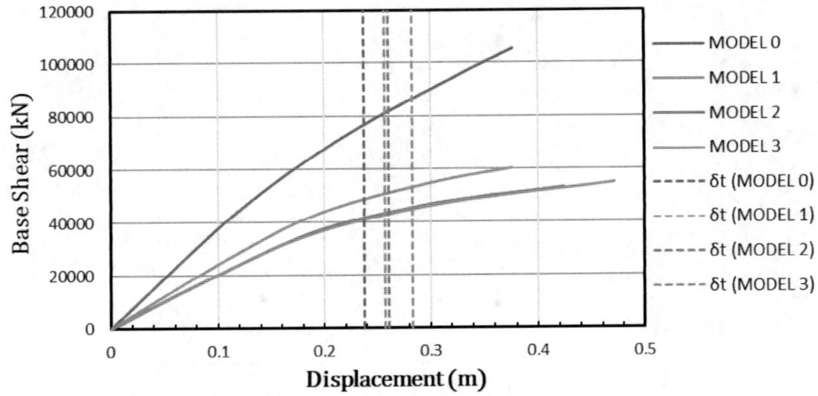

Figure 9. Pushover curve in Y-direction for BSE-1N.

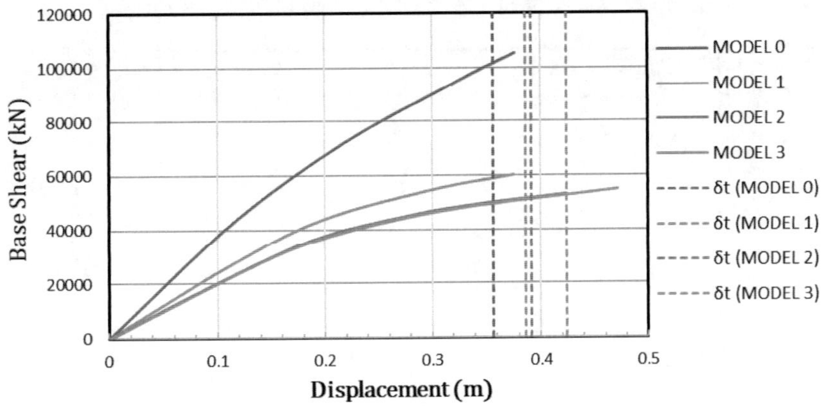

Figure 10. Pushover curve in Y-direction for BSE-2N.

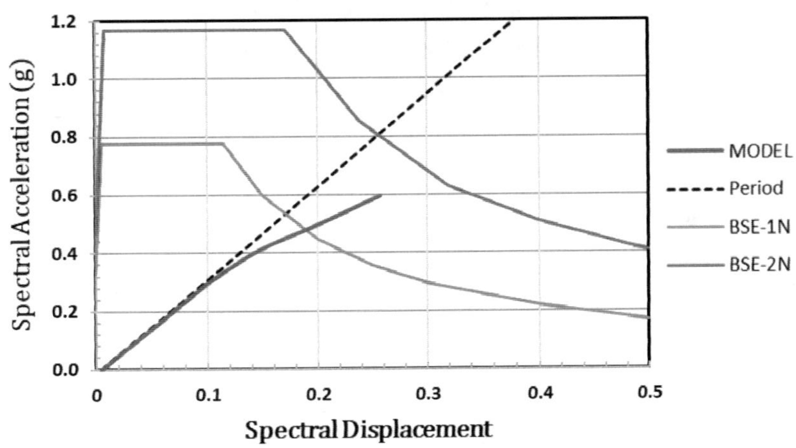

Figure 11. Model 0 ATC-40 Curve in Y-direction.

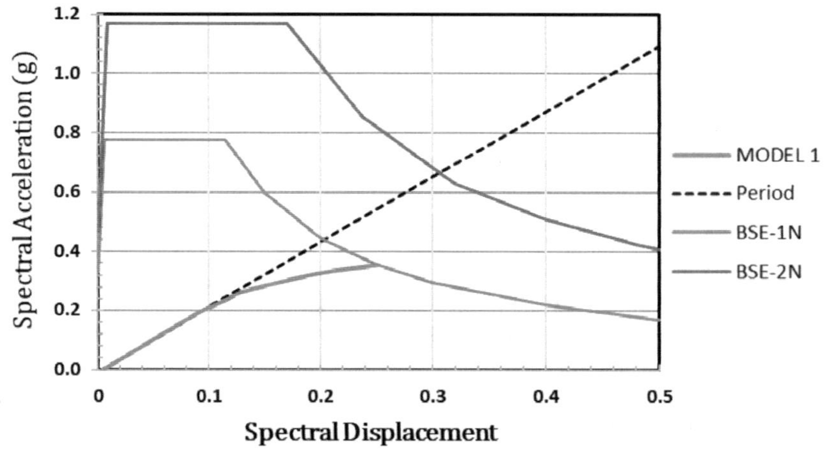

Figure 12. Model 1 ATC-40 Curve in Y-direction.

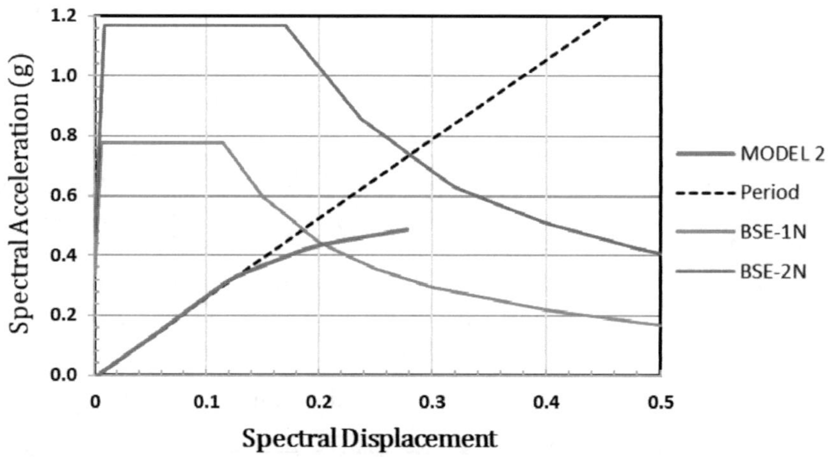

Figure 13. Model 2 ATC-40 Curve in Y-direction.

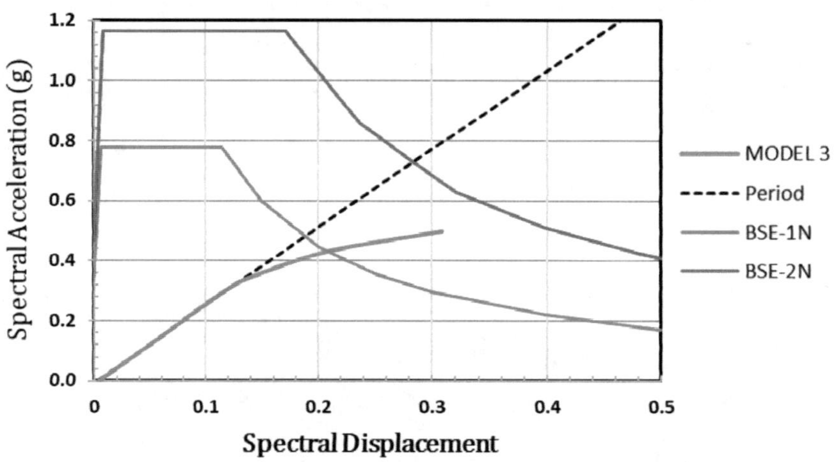

Figure 14. Model 3 ATC-40 Curve in Y-direction.

As can be seen from the pushover curves, Model 0 has the highest curve and is followed by Model 1, while Model 2 and Model 3 are the lowest. It shows that floor slabs have a significant effect by increasing their capacities. The insertion point does not give much effect when the floor slabs are not

IOP Conf. Series: Earth and Environmental Science 1244 (2023) 012015 doi:10.1088/1755-1315/1244/1/012015

being modeled, it is shown by the similarity of Model 2 and Model 3's curves. The top center insertion point will increase the pushover capacity by its higher moment of inertia if it is being modeled along with the floor slabs. All the models' pushover curves successfully achieved their target displacement for the BSE-1N seismic hazard level, while Model 1's pushover curve is not long enough to achieve its target displacement for the BSE-2N seismic hazard level.

CSI America declares on their website that the analysis procedure between the assignation of centroid and other insertion points in SAP2000 are different. Internal constraints will be generated because of the vertical offset from the top or bottom center since they use the centroid insertion point as the default modeling procedure along with its object boundary conditions [6]. Table 6 presents the free body diagram comparison between Model 0 with the top center insertion point and Model 1 with the centroid insertion point. Therefore, it is really important to take notice and consider every detail when doing numerical modeling of a structure because it will affect the analysis result.

Table 6. Free body diagram comparison.

FBD	Model 0	Model 1
BMD	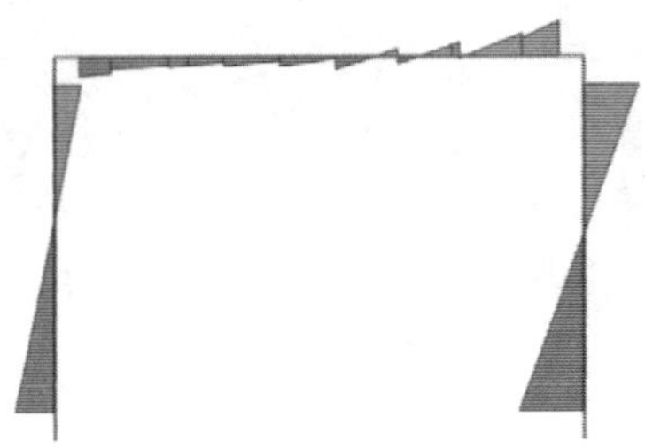	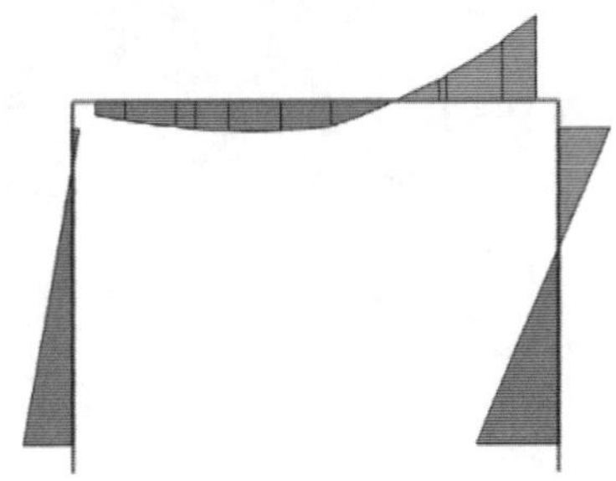
SFD	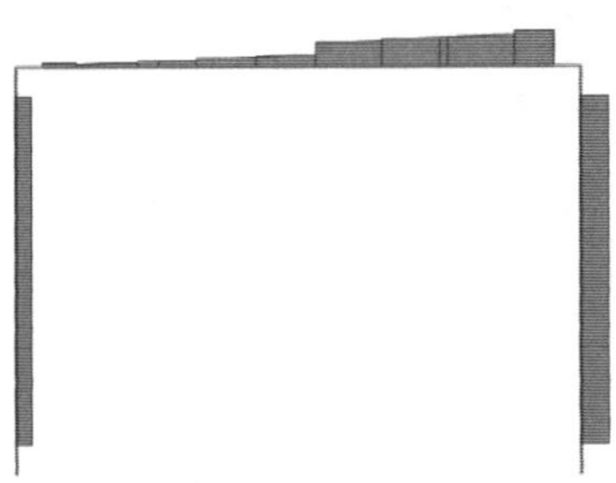	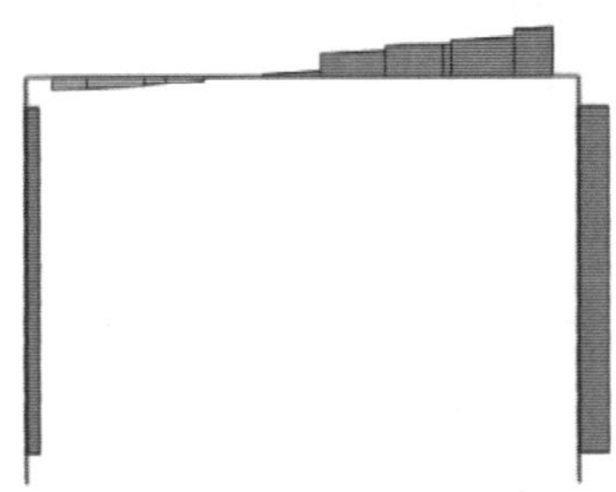

5. Conclusion

This paper evaluated the effect of floor slabs and also the position of floor slabs and beams assignation in a structure with frame shear walls modeling in pushover analysis. According to the results, the main conclusions are summarized as follows:

1. The assignation of floor slabs in the modeling structure significantly affects its pushover analysis results.
2. The combination of floor slabs assignation and the top center insertion point gives the highest capacity of structure compared to the other model variations.

References

[1] Pannirselvam, N., Sreelekshmi, M. (2022). "Study on Irregular Tall RC Structure and Composite Structure by Pushover Analysis." *In: Loon, L.Y., Subramaniyan, M., Gunasekaran, K. (eds) Advances in Construction Management. Lecture Notes in Civil Engineering*, Vol. 191, 179-194.

[2] Mohod, M.V. (2015). "Pushover Analysis of Structures with Plan Irregularity." *IOSR-JMCE*, Vol. 12(4), 46-55.

[3] ASCE (2017). *ASCE 41-17: Seismic Evaluation of Retrofit of Existing Buildings*, American Society of Civil Engineers, Virginia.
[4] Ahmed, M.M.M, Abdo, M.A., Mohamed, W.A.E. (2021). "Vertical Geometric Irregularity Effect on Performance-Based Seismic Design for Moderate Rise RC Moment Resisting Frame Buildings." *Arabian Journal for Science and Engineering*.
[5] Liu, H., Sun., S, Xu, M., Peng, Y. (2020). "Dynamic Nonlinear Analysis of an Irregular Office Building." *E3S Web of Conferences*, Vol. 173.
[6] CSI America (2019, March 27). CSI Knowledge Base. *Effect of insertion point on beam reaction*. Retrieved from https://wiki.csiamerica.com/display/tp/Effect+of+insertion+point+on+beam+reactions.

Numerical comparative study on seismic response of half-span and full-span scaled models of a floating type cable-stayed bridge

S Yadi[1], B Suhendro[2*], H Priyosulistyo[2] and A Aminullah[2]

[1] Civil Engineering Study Program, Faculty of Engineering, Universitas Muhammadiyah Yogyakarta, Yogyakarta, Indonesia
[2] Department of Civil and Environmental Engineering, Universitas Gadjah Mada, Yogyakarta, Indonesia

*Corresponding author: bsuhendro@ugm.ac.id

Abstract. This study compares the results of numerical simulations of half-span and full-span cable-stayed bridge models based on the seismic response of the pylons and main girders subjected to earthquake loads. The half-span model represents the most critical condition of the bridge during the construction phase, while the full-span model represents the bridge in operational phase. The three-dimensional finite-element modeling of the two bridge models uses data from the longest existing cable-stayed bridge in Indonesia which was scaled down geometrically to 1/160, using Midas CIVIL software. The Kobe earthquake record from the PEER Strong Motion Database was used in this study by applying the spectral matching method to modify its acceleration according to the location of the existing bridge being analyzed. This study also examines the influence of the directional effects of earthquake load relative to the longitudinal axis of the bridge, namely 0, 45, and 90 degrees. The results of the analysis show that for the half-span bridge model, the maximum acceleration occurs at the top of the pylon due to earthquake excitation at 90 degrees direction, with an amplification factor of 4.658, while the amplification factor in full-span bridge conditions is only 2.44. In addition, in both the half-span and full-span bridge models, the largest acceleration amplification factor occurs in the main girder that is caused by earthquake excitation at 90 degrees direction. The largest transverse displacement occurs at the free ends of the main girder, but not in the full-span model. The results of the study indicate that during the detailed design process the designer has to consider the effect of earthquake on the strength and stability of the bridge during construction especially if the balanced cantilever method applies.

1. Introduction

Several long span bridges have been constructed in Indonesia over the last few decades, including the Suramadu bridge in East Java, the Tengku Agung Sultanah Latifah bridge in Siak-Riau, the Merah Putih bridge in Ambon, the Ir. Soekarno bridge in Manado-North Sulawesi, the Barito bridge in South Kalimantan, and the Mahakam Ulu bridge in East Kalimantan. Furthermore, these long bridges are typically cable-stayed bridges. Cable-stayed bridges are still considered the best choice for medium- to long-span bridges with main spans ranging from 200 m to 1000 m [1, 2, 3].

Cable-stayed bridge construction in Indonesia takes a long time, with an estimated average time of three years or more. A major earthquake is very likely during this time period for the cable-stayed bridge. This issue has not yet been addressed in the bridge design code. During the construction phase, bridge girders are at their least stable, so it is crucial that bridge designers and contractors comprehend and account for these conditions [4].

In the past several decades, both numerical and experimental research has been conducted on the dynamic response of cable-stayed bridges. Few studies have, however, examined the dynamic response of cable-stayed bridges during construction stage.

Simulations of the Sutong cable-stayed bridge's construction using the ANSYS software to identify stresses and deformations in the steel box girder, the concrete tower, and the cable tensile stress have been performed by [5]. Wang et al. [6] analyzed cable-stayed bridges under various erection conditions during the construction process using the balanced cantilever method. The finite element computation procedure for the construction process employs forward process analysis and backward process analysis, as well as linear and nonlinear computational procedures. In the meantime, Jadhav et al. [7] presented an analysis of the response of cable-stayed bridge structures during construction using Midas/Civil software. In this study, the balanced cantilever method was utilized for construction. The parameters of the study include the ratio of side span to main span, the ratio of pylon height to main span, and the geometry of the pylon. The entire study did not compare the response of the bridge during construction to response of the bridge after construction was completed.

On the basis of the factors discussed, a comparative study between cable-stayed bridges during construction (half span bridges) and full span bridges is considered necessary. In light of the fact that this study is a continuation of an experimental study conducted by [8], the scaled bridge model utilized in the experiment is utilized in this study. A 1/160 scale model of a cable-stayed bridge was subjected to a variety of earthquake excitation directions, including in the direction of the bridge's longitudinal axis, at 45 degrees, and at 90 degrees to the longitudinal axis using the Midas/Civil software. The objective of comparing the results of the analyses of the two numerical models is to determine which of the cable-stayed bridge model's two conditions elicits a more severe response.

2. Prototype of the bridge

The prototype of the bridge used in this study was the Suramadu Bridge. The main structural type is a cable-stayed bridge. The span configuration is 192+434+192 m using twin pylons with double cable planes and composite girders. The width of the bridge deck is 30 m consisting of two directions, and each direction consists of 4 lanes.

The general description of the main bridge and the typical cross-section of the bridge was described in detail by [8] as shown in Figure 1 and Figure 2.

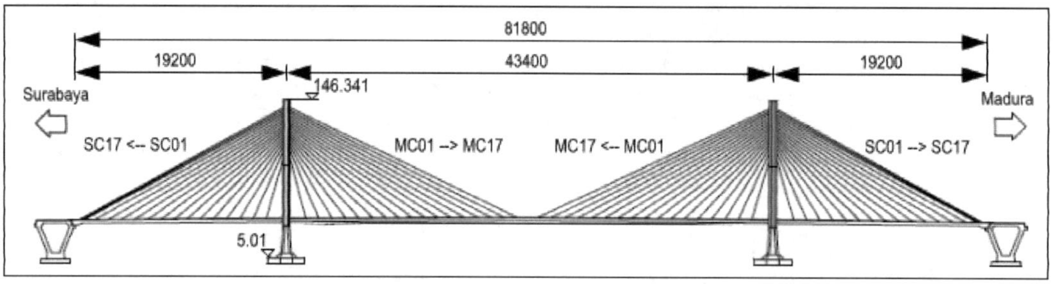

Figure 1. Suramadu bridge layout (unit:milimeters) [8].

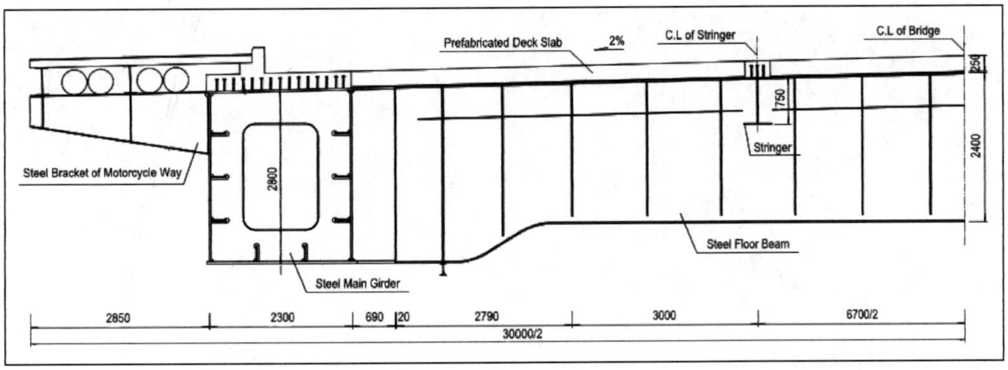

Figure 2. Typical cross section of Suramadu bridge [8].

3. Procedure of numerical study

This study compared the numerical simulation of the half-span and full-span cable-stayed bridge models based on the seismic response of the pylons and main girders subjected to earthquake loads. The half-span model represented the most critical bridge condition during the construction phase, while the full-span model represented the bridge in the operational phase. The three-dimensional finite element modeling of these two bridge models used data on the longest cable-stayed bridge in Indonesia which was reduced geometrically to 1/160 following the similarity requirements clearly explained by [8]. The details of the similarity scale in the scaled-bridge model are shown in Table 1.

Table 1. Similitude requirements of the scaled model.

Variabel	Notation	Value
Geometry scale	S_L	160
Pylon material scale	S_{Epilon}	12.778
Girder material scale	$S_{Egirder}$	3
Cable material scale	S_{Ekabel}	1.062
Gravity acceleration scale	S_g	1
Displacement scale	S_δ	160
Frequency scale	S_f	0.079
Time scale	S_t	12.649
Force scale on cable	S_{Fkabel}	27191.71
Strain scale	S_ε	1

According to Table 1, there were material distinctions between pylons, girders, and cables of the bridge model. The pylon was made of acrylic (modulus of elasticity about 2,7 GPa). The girder was made of aluminum (modulus of elasticity about 70 GPa). The material of cable was stainless steel wire type 304 soft with a diameter of 1.0 mm, 1.2 mm, and 1.5 mm, according to the group, where the value of the modulus of elasticity about 193 GPa.

The idealization of the numerical model of the Suramadu Bridge in this study was modeled using finite element-based software, namely Midas/Civil (Modeling, Integrated Design and Analysis Software). The numerical model of the full-span Suramadu scaled-bridge included 4074 elements and 2083 nodes. The total number of frame elements was 2238, with 1696 plate elements and 140 cable elements. The idealization of the numerical model is shown in Figure 3.

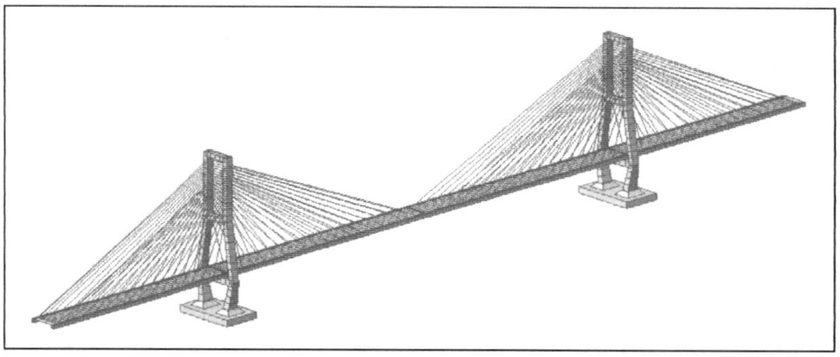

Figure 3. Full-span numerical scaled model of the Suramadu bridge on Midas/Civil software.

4. Design of ground motion

The ground motion of the Kobe earthquake at The Takatori station was used in this paper (peak ground acceleration and time duration 0.62g and 40.95 seconds) [9]. The acceleration time history of the Kobe earthquake was modified by performing a "response spectrum matching" which refers to the Indonesian earthquake regulation [10] which has been explained by [8]. Response spectrum matching refers to the method proposed by [11], which is based on the least square method. The graph of the earthquake time history that has been adjusted to the Suramadu bridge region and which has been time-scaled can be seen in Figure 4.

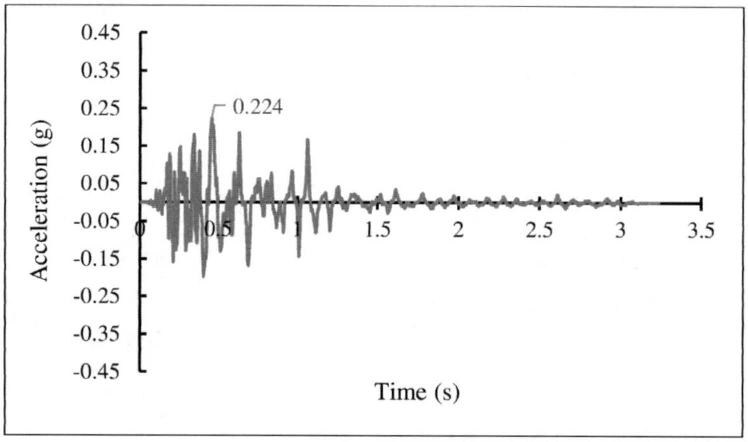

Figure 4. Time-scaled graph of the Kobe earthquake time history that has been adjusted to the Suramadu bridge region.

This study also examines the influence of the directional effects of earthquake load relative to the longitudinal axis of the bridge as the direction of the bridge's fundamental mode, namely 0 degrees, 45 degrees, and 90 degrees, respectively. The image below depicts the direction of 45 degrees of earthquake excitation applied to the half-span scaled bridge model shown in Figure 5.

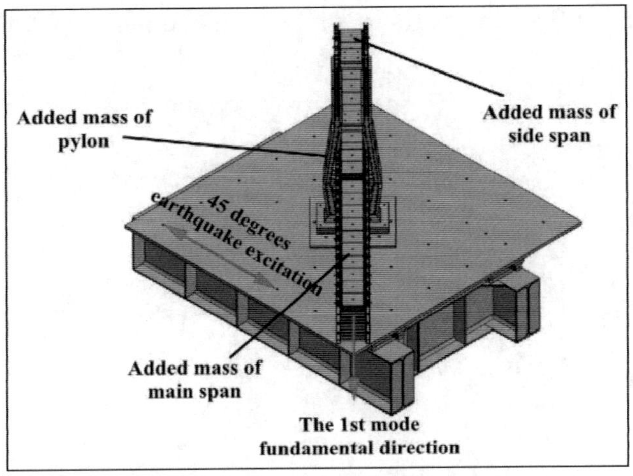

Figure 5. The earthquake excitation direction of 45 degrees was applied to the half-span scaled bridge model.

5. Results and discussion
This study investigates the effect of the angle of incidence of earthquake excitation on the bridge's longitudinal axis in both half-span and full-span bridge models. The acceleration and displacement responses of the top pylon and the bridge girder were investigated. As a result, the amplification factor $AF = R_{output}/R_{input}$ must be determined, where R_{output} is the maximum peak acceleration response of the bridge model at a specific position and R_{input} is the peak acceleration of the seismic excitation input applied to the pylon foundation.

5.1. The top pylon acceleration response
Figure 6 show the acceleration response of the top pylon of the half-span scaled bridge model in Y-direction due to the Kobe earthquake excitation at 0 degrees, 45 degrees, and 90 degrees to the bridge's longitudinal axis.

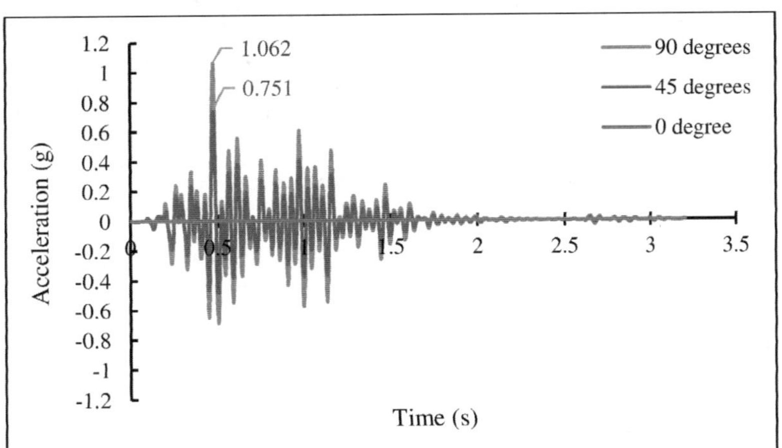

Figure 6. Acceleration response at the top of the pylon on a half-span scaled bridge model in the Y direction for various Kobe earthquake excitation directions.

Furthermore, the acceleration response at the top of the pylon in the full-span scaled bridge model can be seen in Figure 7.

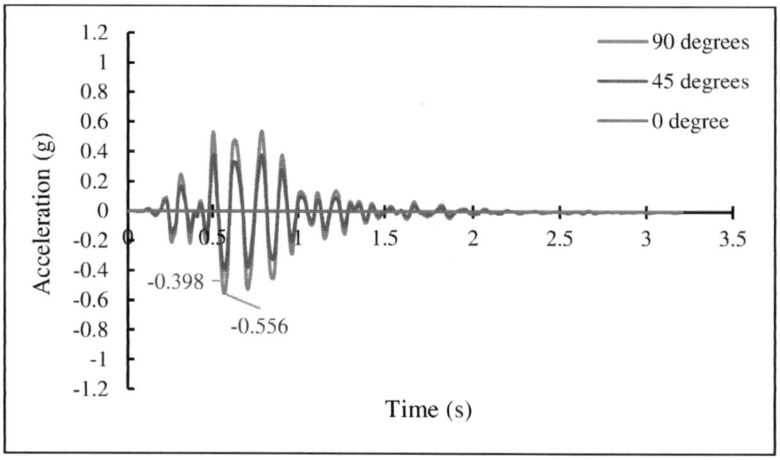

Figure 7. Acceleration response at the top of the pylon on a full-span scaled bridge model in the Y direction for various Kobe earthquake excitation directions.

Based on Figure 6 and Figure 7, the maximum acceleration response at the top of the pylon in the transverse direction (perpendicular to the bridge's longitudinal axis) tends to increase as the angle of the earthquake excitation direction increases, whereas the maximum acceleration response in the longitudinal direction tends to decrease as the angle of the earthquake excitation direction increases. The maximum amplification factor table clarifies the preceding statement.

The amplification factor for the maximum acceleration response at the top of the pylon on the half-span and full-span scaled bridge model is shown in Table 2 below.

Table 2. The amplification factor of the maximum acceleration response at the top pylon of scaled bridge model caused by different directions of Kobe earthquake excitation.

The direction of the earthquake excitation about the longitudinal axis	AF of maximum acceleration response			
	Half-span model		Full-span model	
	X-direction	Y-direction	X-direction	Y-direction
0°	0.817	-	0.354	-
45°	0.582	3.338	0.253	1.768
90°	-	4.658	-	2.440

The value AF 4.658 denotes that the maximum acceleration response at the top of the pylon in the Y direction is 4.658 times greater than the input earthquake acceleration at the bottom of the pylon with an earthquake excitation direction of 90 degrees to the bridge's longitudinal direction.

Based on Table 2, it can also be concluded that the largest maximum amplification factor at the top of the pylon occurs due to the 90-degree earthquake excitation about the longitudinal axis of the bridge, with an amplification factor of 2.440 for the full-span scaled bridge model. This value is much smaller than the amplification factor for the half-span scaled bridge model, which is 4.658.

5.2. The top pylon displacement response

The displacement response at the top of the pylon on the scaled model of the half-span bridge in the X-direction due to the excitation of the Kobe earthquake at 0 degree, 45 degrees, and 90 degrees to the longitudinal axis of the bridge can be seen in Figure 8 below.

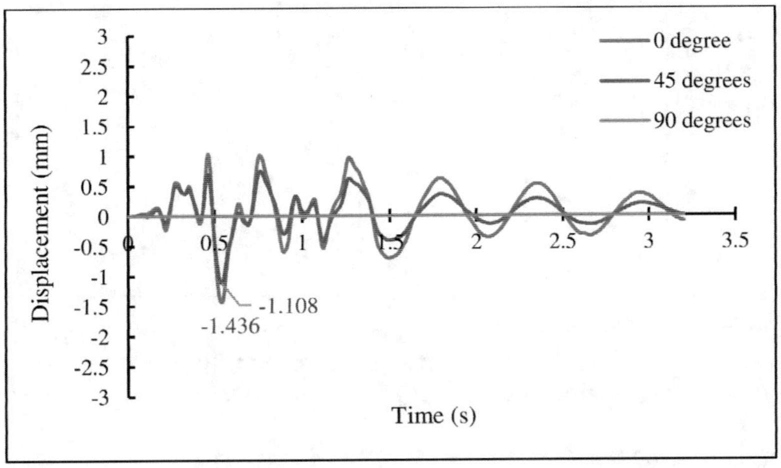

Figure 8. Displacement response at the top of the pylon on a half-span scaled bridge model in the X direction for various Kobe earthquake excitation directions.

Furthermore, the displacement response at the top of the pylon on a scaled model of a half-span bridge in the Y direction can be seen in Figure 9 below.

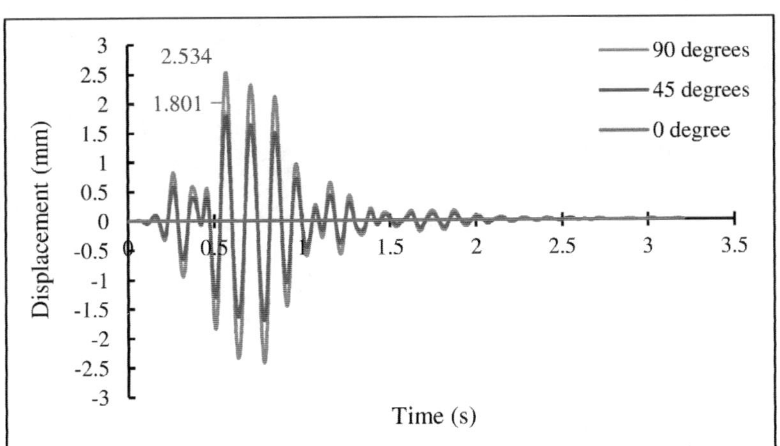

Figure 9. Displacement response at the top of the pylon on a half-span scaled bridge model in the Y direction for various Kobe earthquake excitation directions.

Meanwhile, Figure 10 and Figure 11 show the displacement response at the top of the pylon on the full-span scaled bridge model, both in the X-direction and in the Y-direction, due to the excitation of the Kobe earthquake in various directions to the bridge's longitudinal axis.

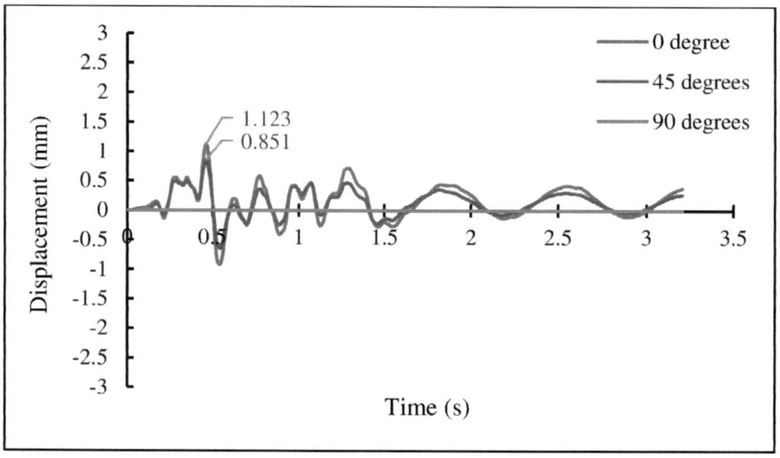

Figure 10. Displacement response at the top of the pylon on a full-span scaled bridge model in the X direction for various Kobe earthquake excitation directions.

Based on Figure 8 and Figure 10, it can be seen that the displacement response at the top of the pylon in the X direction is similar in both the half-span and full-span scaled bridge models. The response to the half-span scaled bridge model, however, is still greater than the response to the full-span scaled bridge model.

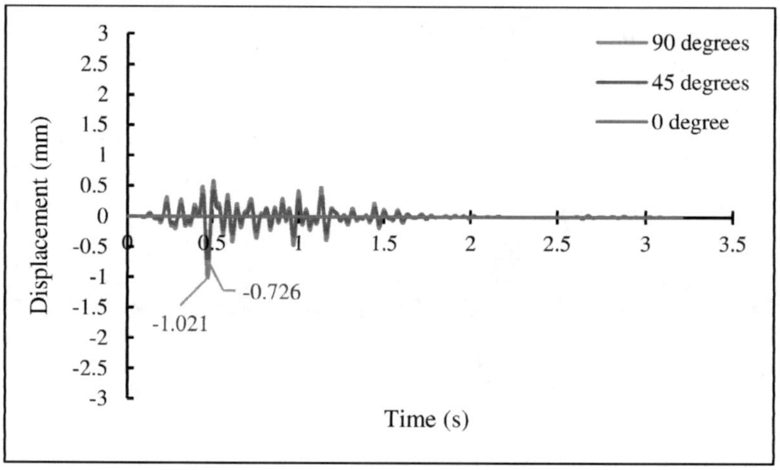

Figure 11. Displacement response at the top of the pylon on a full-span scaled bridge model in the Y direction for various Kobe earthquake excitation directions.

In contrast to the displacement response of the top pylon in the X direction, which is nearly identical, Figure 9 and Figure 11 show a significant difference in the displacement response of the top pylon in the Y direction between the half-span scaled bridge model and the full-span scaled bridge model.

5.3. The girder acceleration response

The acceleration response of the girder in both the half-span and full-span scaled bridge models in the Y-direction due to the Kobe earthquake excitation at 0 degree, 45 degrees, and 90 degrees to the bridge's longitudinal axis can be seen in Figure 12 and Figure 13.

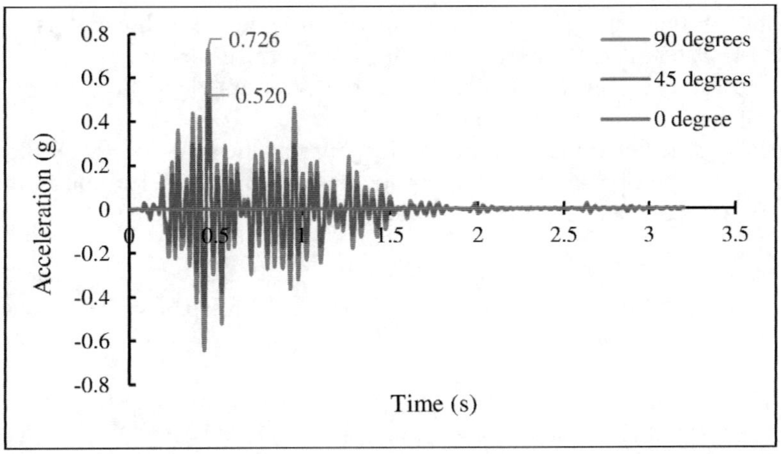

Figure 12. Acceleration response at the girder on a half-span scaled bridge model in the Y direction for various Kobe earthquake excitation directions.

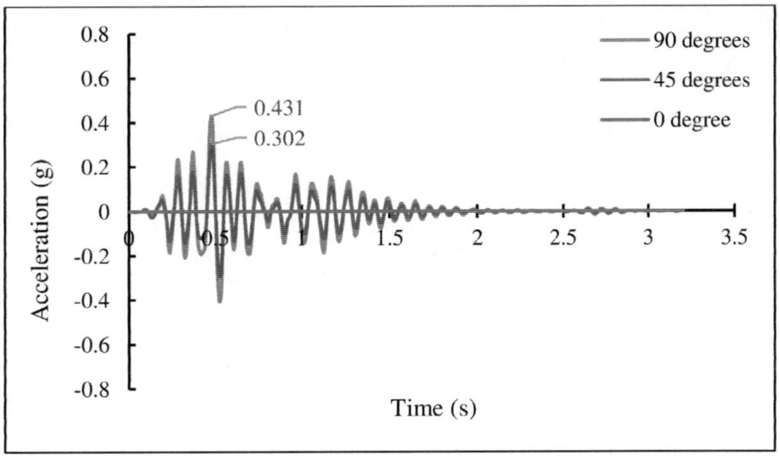

Figure 13. Acceleration response at the girder on a full-span scaled bridge model in the Y direction for various Kobe earthquake excitation directions.

Meanwhile, the amplification factor for the maximum acceleration response of the girder for both the half-span and full-span scaled bridge models for various directions of earthquake excitation about the longitudinal axis of the bridge can be seen in Table 3.

Table 3. The amplification factor of the maximum acceleration response at the girder of scaled bridge model caused by different directions of Kobe earthquake excitation.

The direction of the earthquake excitation about the longitudinal axis	AF of maximum acceleration response			
	Half-span model		Full-span model	
	X-direction	Y-direction	X-direction	Y-direction
0°	0.068	-	0.063	-
45°	0.051	2.311	0.047	1.344
90°	-	3.186	-	1.889

Table 3 indicates that the response of the girder acceleration in the X direction is very small in both the half-span and full-span scaled bridge models. Meanwhile, the girder acceleration response in the Y direction indicates that the half-span scaled bridge model is more vulnerable than the full-span scaled bridge model, where the AF ratio is around 1.70.

Table 3 also demonstrates that the acceleration response of the girder in the X direction decreases as the angle of arrival of the earthquake excitation increases, whereas the response to the acceleration of the girder in the Y direction exhibits the opposite phenomenon.

5.4. The girder displacement response

The response of the girder displacement on the half-span and full-span scaled bridge models can be seen in Table 4 below.

Table 4. The displacement response at the girder of scaled bridge model caused by different directions of Kobe earthquake excitation.

The direction of the earthquake excitation about the longitudinal axis	Displacement response (mm)			
	Half-span model		Full-span model	
	X-direction	Y-direction	X-direction	Y-direction
0°	0.975	-	1.066	-
45°	0.721	-0.356	0.809	-0.476
90°	-	-0.500	-	-0.677

According to Table 4, it is clear that the girder displacement response does not exceed the specified displacement requirement of 3 mm.

6. Conclusion

This study aims to compare the numerical simulation of cable-stayed half-span and full-span bridge models subjected to earthquake loads considering the effects of earthquake directions, focusing on the seismic response of the pylons and main girders. The three-dimensional finite element modeling of these two bridge models took the use of data from the longest cable-stayed bridge in Indonesia, which was reduced geometrically to 1/160 in accordance with the similarity requirements.

The results of the analysis show that for the half-span bridge model, the maximum acceleration occurs at the top of the pylon due to earthquake excitation at 90 degrees direction,with an amplification factor of 4.658, while the amplification factor in full-span bridge conditions is only 2.440. In addition, in both the half-span and full-span bridge models, the largest acceleration amplification factor occurs in the main girder that is caused by earthquake excitation at 90 degrees direction. The largest transverse displacement occurs at the free ends of the main girder, but not in the full-span model. The results of the study indicate that during the detailed design process the designer has to consider the effect of earthquake on the strength and stability of the bridge during construction especially if the balanced cantilever method applies.

Acknowledgments

I would like to thank Mr. Budi Santoso for supporting the MIDAS/Civil 2014 software license in order to assist in the numerical study of this research.

References

[1] Chang F K and Cohen E 1981 *J. Struct. Div.* **107(7)** 1145–60

[2] Gimsing N J and Georgakis C T 2012 *Cable-Supported Bridges: Concept and Design* vol 3, ed N J Gimsing and C T Georgakis (West Sussex: John Wiley & Sons Ltd)

[3] Svensson H 2012 *Cable-Stayed Bridges: 40 years of Experience Worldwide* (Berlin: Wilhelm Ernst & Sohn)

[4] McPheron D, McCullouch B and Bowman M D 2012 *Structural Impact of Construction Loads* Publication FHWA/IN/JTRP2012/18 Joint Transportation Research Program (West Lafayette, Indiana: Department of Transportation and Purdue University)

[5] Chengfeng X, Laijun L, Fangwen W and Caofang Y 2015 Numerical analysis of long-span cable-stayed bridge in the construction phase *Open Civ. Eng. J.* **9** 896–905

[6] Wang P H, Tang T Y and Zheng H N 2004 Analysis of cable-stayed bridges during construction by cantilever methods *Comput. Struct.* **82(4-5)** 329–46

[7] Jadhav P, Ganesh G M and Vinayagamoorthy M 2017 Erection stage dynamic behavior of cable stayed bridge using construction stage analysis *IJCIET* **8(4)** 252–64

[8] Yadi S, Suhendro B, Priyosulistyo H and Aminullah A 2018 Shake table test of floating cable-stayed bridge under earthquake excitation during construction with balanced cantilever method *IJCIET* **9(11)** 2063–81

[9] PEER Ground Motion Database. Kobe Earthquake Ground Motion at Takatori Station. https://ngawest2.berkeley.edu/spectras/160674/searches/148164/edit

[10] Badan Standardisasi Nasional 2016 *Design of Bridges for Earthquake Load* SNI 2833:2016 (Jakarta: National Standardization Agency of Indonesia)

[11] Kalkan E and Chopra A K 2010 *Practical Guidelines to Select and Scale Earthquake Records for Nonlinear Response History Analysis of Structures* (Virginia: USGS–EERI)

Dynamic field load testing of a horizontally curved long-span LRT Kuningan bridge using eccentric mass vibrator and experimental modal analysis

B Santoso[1*], A Pramudito[1] and M K Zamad[1]

[1]PT. Pratamadaya Cahyamanunggal, Topaz Commercial TC-A09, Summarecon Bekasi, 17142, Indonesia

*Corresponding author: budi@pdcm.co.id

Abstract. Light Rail Transit (LRT) infrastructure in Indonesia is developing at a rapid pace. The LRT challenging construction is the Kuningan-Longspan Curved bridge with a span of 148 meters and the radius of curvature R = 115 meter. Therefore, Kuningan LRT Bridge is considered as one of the longest concretes curved balanced cantilever LRT Bridge in the world. The implementation of bridge regulations in Indonesia requires a functional feasibility test before the bridge is operational. In the case of the Kuningan LRT bridge, functional feasibility tests were carried out in the form of static and the dynamic tests. The dynamic load test aim is finding the structure natural frequency in the vertical, Longitudinal and Transverse direction. There are 3 methods that being used to measure the dynamic parameters of the train bridge structure in Indonesia. The first method is the Impulsive method using the moving train with the speed of 25 Km/hour as a load. The second method is the usage of The Experimental Modal Analysis and the third method is the combination method of Experimental Modal Analysis with a vibrating instrument in the form of an Eccentric Mass Vibrator (Exciter). The experiment results and the modal fem from the designer is compared and presented in this paper.

1. Introduction

LRT Kuningan bridge is a Light Train bridge that crosses the Gatot Subroto Highway in the city of Jakarta, Indonesia. This bridge was awarded by Indonesia National Record Museum (MURI). The bridge is internationally recognized as an achievement in structural engineering due to its sharp curvature as a concrete cantilever Bridge. Before it is opened to traffic, the new bridge is subjected to static and dynamic load tests. All test should be complied to the Bridge Safety Committee Recommendation [1]. The traditional dynamic test for simple bridge usually is enough as cited in [2], [3], but for more complex bridge structure, more dynamic test is required like EMA, OMA, or The Mass Exciter Testing. After a brief description of the structure, this paper presents the dynamic testing procedures and the experimental equipment used. Some experimental results are presented and compared with the analytical values evaluated using RM BRIDGE software in designer report as cited in [4].

LRT Kuningan Bridge is a concrete-balanced cantilever bridge with a main span of 148 m and a curvature radius of 115 m. Besides the main span, the bridge has two lateral spans on the left and right side of the main span, with lengths of 86.5 m. A general view of the bridge is presented in Figure 1.

The main span is sharply curved supported by 2 piers on each side of the span. The bridge pier has a height of 24.9 m above ground and a clearance of 9.45 m from Gatot Subroto elevated Highway below the bridge main span, material used in this bridge is reinforced concrete with compressive strength of f'c 50 MPa for the bridge superstructure and f'c 40 MPa for the bridge piers. The bridge girders are precast segments of box girders that have a cross-section of 9.15 m with a total depth that varies from 8 m over the piers to 3.5m at the midspan of the Bridge (Figure 2).

Figure 1. General View of The LRT Kuningan Bridge.

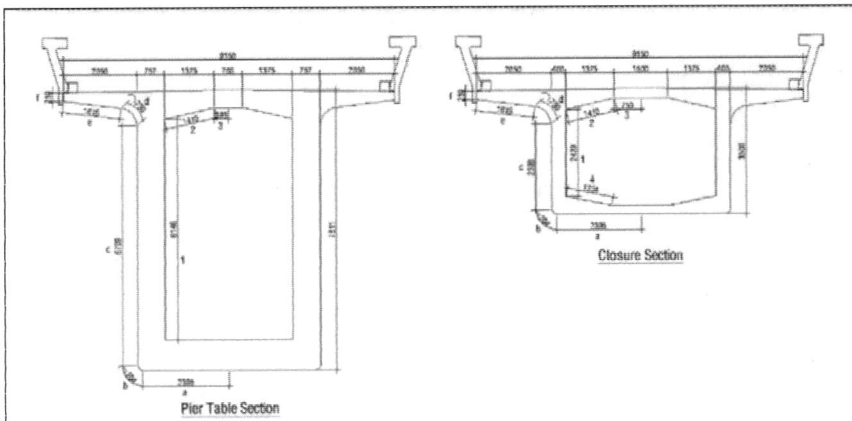

Figure 2. Girder Segments of LRT Kuningan Bridge

2. The Experiment Method

In traditional modal analysis, an excitation force produced then the vibration responses were measured. This response is used to produce frequency response cited in [5], and [6]. The Operational modal analysis as cited in [7], are based on noise excitation for exciting all modes of vibration. Several real case examples on some bridge structures can be found detail in Cantieni papers in [8],[9],[10]. The test was using the excitation by force (exciter) and ambient (moving train) to check all modes in the bridge.

The structural monitoring system was installed during the loading test of the bridge. The structure acceleration measurement was using 18 3-axis wireless Accelerometer (Figure 3). The vibrations of the structure in vertical, tangential and radial directions were measured respectively.

The placement of these accelerometers is planned according to the bridge geometry which accounts for the most likely points of vibrations in the bridge model, these nodes are then translated into the actual bridge structure. Accelerometers placed on the bridge main span and side span, with the point of placements of the accelerometer being at 1/4, 1/2, and 3/4 of the length of the bridge main span and half side spans. The Accelerometer used for EMA data capture is LORD Microstrain 2G G-Link-200 Wireless Triaxial Accelerometers with LXRS protocol that allows lossless data collection, scalable network size, and node synchronization of ±50µs. (Figure 4). The LORD accelerometers accompanied by Exciter monitoring accelerometer that produced by Silicon Design Inc.

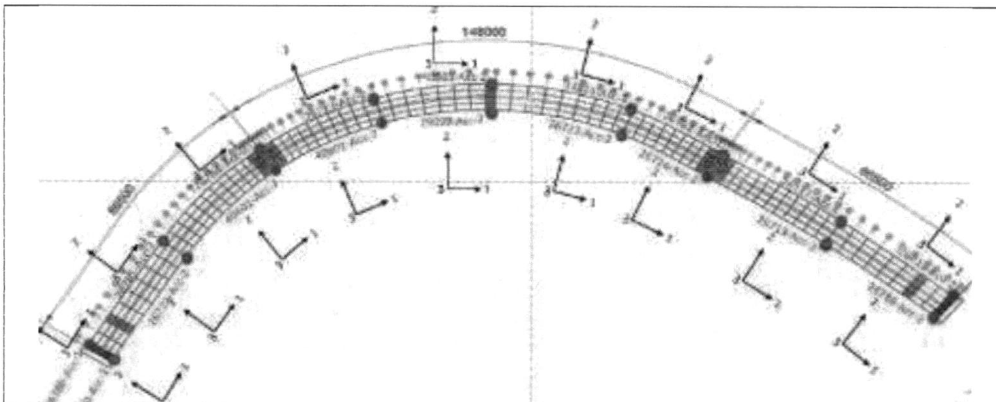

Figure 3. Accelerometer Sensors Location

Figure 4. Wireless Accelerometers

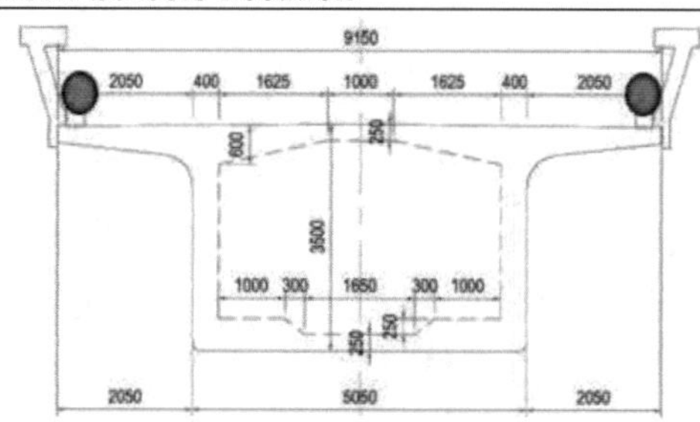

Figure 5. Accelerometers Location in Cross Section

2.1. Testing Procedure

For dynamic testing of LRT Kuningan Bridge there are 3 methods of testing; Impulse testing using staged train load as an impact force, forced vibrations using Eccentric Mass Vibrator, and EMA testing using Eccentric Mass Vibrator.

2.1.1. Impulse Testing.

Accelerometers are placed at the middle section of the main span (½ L), and then a train starts moving past the accelerometer point at a speed of 25 km/hour.

These trains introduced the vibrations to the bridge which would be translated into vertical vibration. There were 3 schemes for impulse testing. The first scheme was the passing train at the accelerometer at speed of 25 km/hour. The second scheme was the train passing at the accelerometers at speed of 25 km/hour then braking gradually to stop at the middle span. The third scheme was the one train stopped at the second track while another train was passing at the speed of 25 Km/hour then braking gradually to stop at the middle span.

2.1.2. EMA Testing.
For EMA test accelerometers were placed with the configurations of Figure 5, in order to create vibrations on the structure, exciter machine was used in this case for generating vibration. Every vibration created the exciter were recorded. The collected vibration data used as the data input in ARTeMIS program. The program analysed the vibrations data that caused by exciter for modal analysis for modal analysis.

2.1.3. Exciter Testing.
Eccentric mass vibrators or more commonly known as exciter are a machine capable of introducing vibrations into the structure being tested with the frequency of vibrations controlled by the user. The schematics of Exciter can be seen in (Figure 6).

Accelerations were measured at the same points as EMA testing and the frequencies of vibrations are increased gradually by the exciter machine from 0 Hz up to 10 Hz. Finally, the vibrations that were caused by the exciter were recorded. For finding the resonance excitation vibrations from the exciter, these data were combined from exciter vibration data reading and accelerometer data reading were analysed.

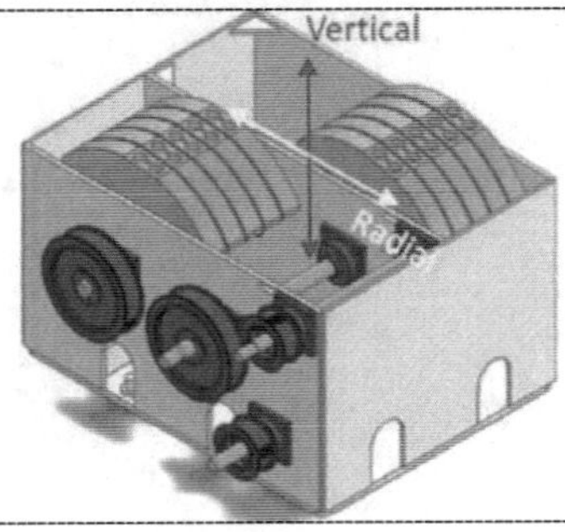

Figure 6. Eccentric Mass Vibrator Machine

3. The Experiment Results

3.1. Impulse Test Results
The measurement of impulse test were the vertical vibrations of the bridge spans due to the impact load of trains; the resulting vibrations can be seen in (Figure 7a, b, and c).

The free vibration reading from accelerometer were transformed into frequency reading using Fourier Transformation to find the dominant frequency of the structure, these dominant frequencies were identified by the highest peak in the vibrations reading of the accelerometer.

There were 3 schemes for the impulse testing, each scheme corresponds to higher load on the bridge by adding breaking force of 1 train and by adding a second train stopping at the loading area. These loads introduced additional mass to the bridge, therefore as the mass increased, theoretically the bridge natural frequency should decrease due to the bridge added mass, lowering the vibrations for the structure.

From the accelerometer vibrations reading as seen in Figure 8, these theories are proven as the mass increased in scheme 2 by the stopping of the train resulted in the frequency value lower than the frequency result obtained from scheme 1 as shown in the results.

The natural frequency for vertical vibrations obtained in scheme 1 were 0.87 Hz, for the added weight of 1 train by stopping in scheme 2 the vertical natural frequency obtained were 0.84 Hz, and for the added weight for 2 trains by first stopping a train at the middle span of the bridge and then a second train was moving to loading area and breaking to a stop next to the first train, the natural frequency obtained were 0.79 Hz.

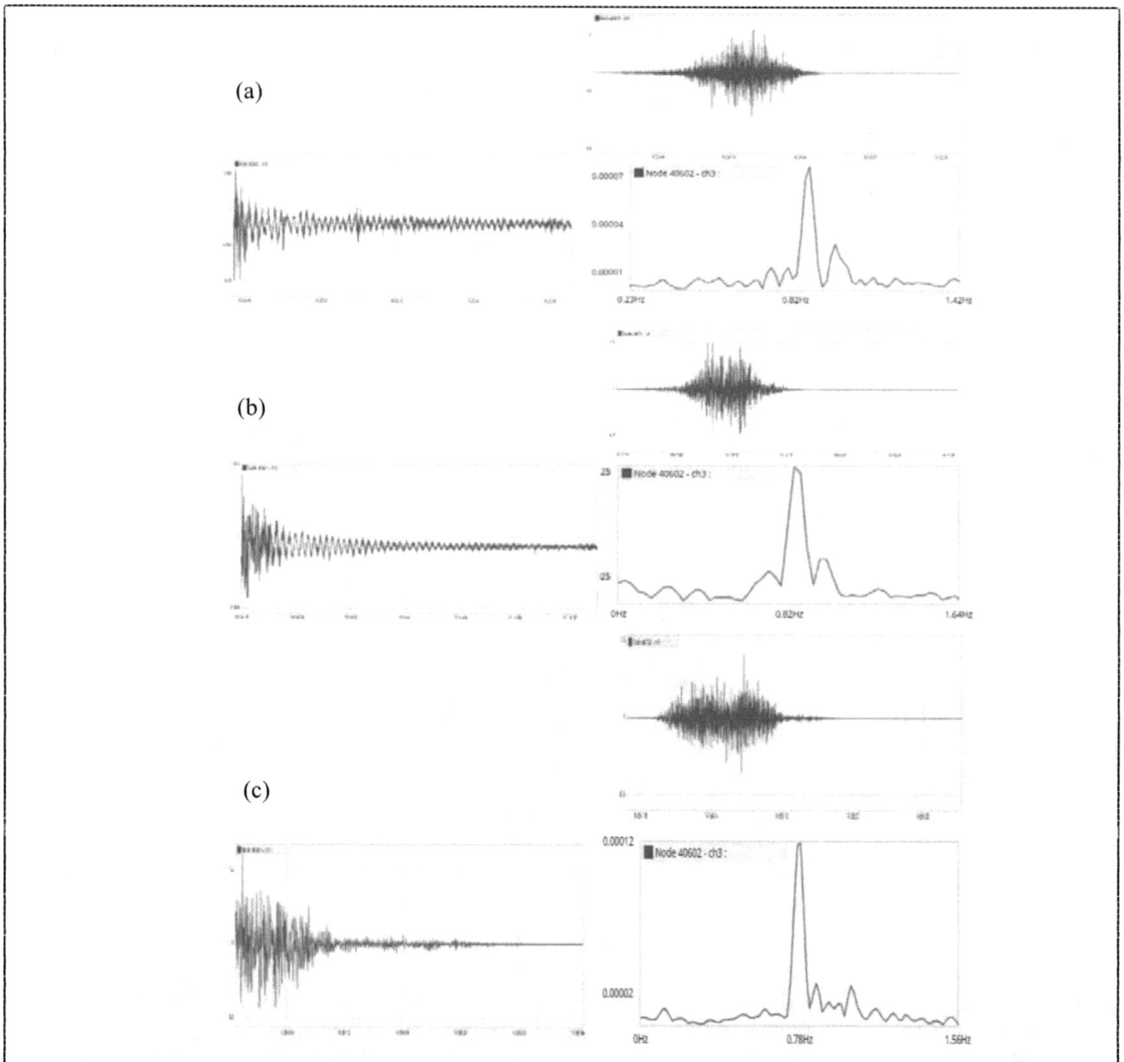

Figure 7. Accelerometer Reading Scheme 1(top), Scheme 2 (middle), and Scheme 3 (bottom)

3.2. Exciter Test Results

The eccentric mass vibrator (exciter) vibrated the structure in the vertical and transversal directions, the vibration given by the machine were transferred into the structure by a set of anchor bolt connected into the exciter machine. The exciter vibrated the structure by rotating a mass on an axis. The range of frequencies the exciter can generate are proportional to the amount of mass put on the axis and the operator input. The results from the POVDAR ANCO Exciter system are presented in figure 8.

IOP Conf. Series: Earth and Environmental Science 1244 (2023) 012017 doi:10.1088/1755-1315/1244/1/012017

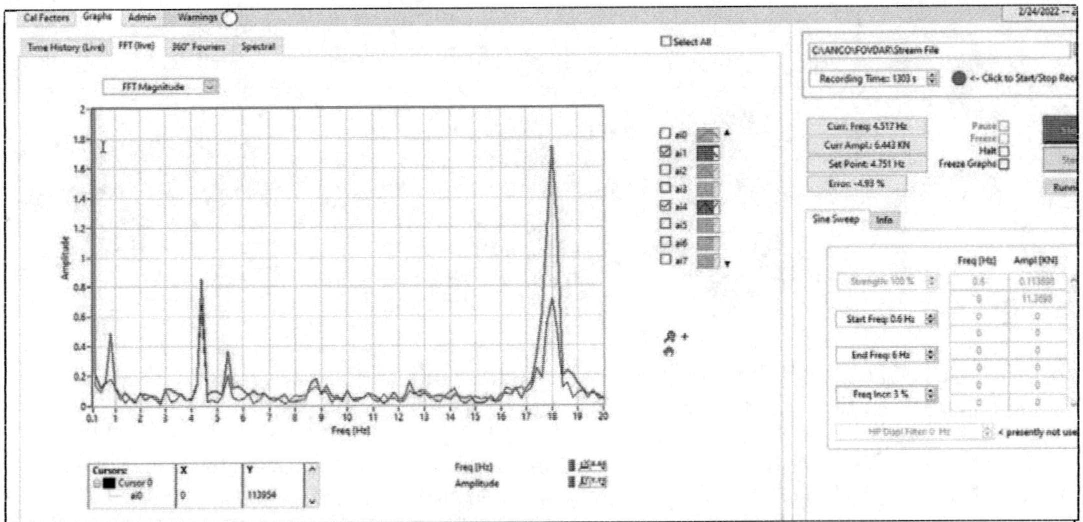

Figure 8. Eccentric Mass Vibrator Machine Exciter frequency reading resonance peaks during excitation

3.3. EMA Test Results

The measurement of EMA test was capturing the natural frequency of bridge in all 3 axes of vibration, and the vibration in the vertical vibrations of the bridge spans could be compared to impulse testing result obtained from previous reading.

Due to the bridge curvature the reading transversal and longitudinal vibrations in global axis were then translated into tangential and radial reading of each Node, these direction on local axis of node can be seen in (Figure 5).

As the exciter vibrated the structure, the accelerometer recorded the vibration of bridge in each axis and those data were plugged into ARTeMIS software which would read the data stored in the accelerometer. The vibration reading of each accelerometer converted into a global vibrations of combined accelerometer data at each node. The natural frequency analysed by check its complexity value and the mode shape animation results shown in ARTeMIS software.

These vibration simulations from ARTeMIS software were compared to the mode shape from designer's analytical approach using RM BRIDGE software. EMA test result in the vertical direction and its mode shape compared analytical approach are shown in (Figure 8a) and (Figure 8b), the result shown are only the vertical directions of the bridge. The vertical vibrations shown were the first order of vertical vibration and the second order of vertical vibration. These orders were sorted from mass participation appearance order on the analytical approach.

Natural frequency result from EMA test in the vertical directions obtained by the experiment yielded 0.79 Hz for the first order of vertical vibrations, this result have a complexity value of 3.42 %, due to the small result of complexity and the similar movement simulation, these result shows that analytical approach using RM BRIDGE model on Kuningan LRT Bridge are validated. The experimental frequency result coincided with analytical model. Another point of interest is the different peaks of vibration from the simulation generated by ARTeMIS. The peak displacement of animation simulated ARTeMIS coincides with the model vertical vibration. From the vibration simulation on the structure presented Figure 9, the first order of vertical vibration (Figure 9a) only generates one peak in the middle of the span, these results are further similar to the analytical model by RM BRIDGE which also shown that the peak displacement of node due to vibration happened at the middle of the span of the bridge. For the 2nd order vertical vibration shown in Fig. 9b, there are slight differences between analytical model and EMA measurements, although the difference is quite negligible, (the difference of frequency value is >10%). The peak on 2nd order vertical vibration results similar to the model generated by analytical method and the simulation generated by EMA. There are 3 peaks on each of the span with each peak of displacement located on the middle span of the bridge.

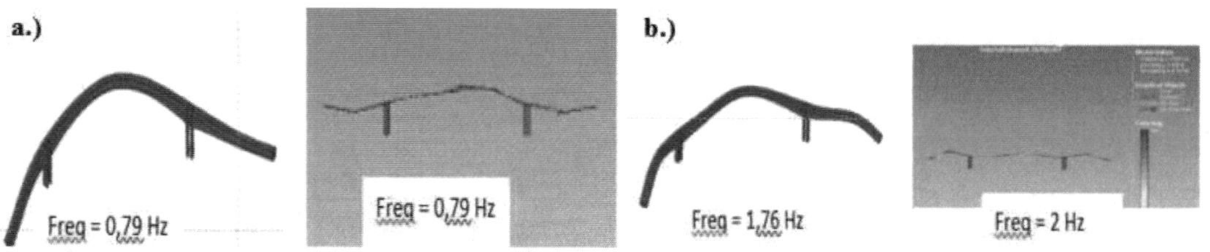

Figure 9. Modelling (left) vs EMA Vibration Simulation for Vertical 1st Order (right) (a), Modelling (left) vs Vibration Simulation for Vertical 2nd Order (right) (b)

EMA test shows that the bridge vibrated according to the simulation generated by analytical model from RM BRIDGE software. Both vertical frequency of the first order and second order simulation are similar between EMA and model. The similarity between the result of EMA and modelling by RM BRIDGE justifies that the model is correct and the model can confidently represent the actual condition of the bridge, especially for the vibrations behaviour due to ambient loading or impulse loading at the structure of the bridge.

4. Conclusion

The LRT Kuningan bridge was subjected to dynamic load tests, performed at the end of its construction and before it was opened to traffic.

During these tests, the existing structural health monitoring system and the additional equipment installed for these tests allowed the measurement of vertical vibration of the structure. Dynamic Load tests were carried out with 3 different methods which is Impulse testing using Train Load, EMA testing, and Exciter testing.

The experimental results acquired during Dynamic load tests were used to validate the numerical model and to assess the structural behaviour of the bridge, which proved to be linear for all the Dynamic load methods conducted. The vertical natural frequencies of the bridge using impulsive test are around 0.79-0.87 Hz, Exciter result is 0.8 Hz and EMA result is 0.78 Hz, these vertical natural frequencies value are compared to the model vertical frequency of 0.78 Hz which shows all of the test resulted in a similar frequency relative to the model for the vertical vibration natural frequencies. This similar value from the test result and from the modelling conclude that the model confidently represents the structure. This information is invaluable to validate the modelling and as a record of the bridge natural frequencies. Tangential and Radial vibration at Kuningan LRT Bridge is still being analysed as this paper is being written.

5. Acknowledgements

The authors would like to acknowledge the help and guidance of the Indonesia National Committee for Safety of Road and Bridge (KKJTJ), and all technicians and engineers of PT. Pratamadaya Cahyamanunggal (PT.PDCM) and other parties that participated in the loading tests of the Kuningan LRT Bridge.

References

[1] Wardana PK, 2002 Pembahasan Penyelenggaraan Keamanan Jembatan Khusus (Jakarta: Ministry of Public Works and Housing)
[2] Trisnanto L, and Sunardi 2002 Penilaian kondisi jembatan untuk bangunan atas dengan cara uji getar (Bandung: Pusjatan - Ministry of Public Works and Housing)
[3] Trisnanto L, and Sunardi 2002 Penilaian kondisi jembatan untuk bangunan atas dengan cara uji getar (Bandung: Pusjatan - Ministry of Public Works and Housing)
[4] Delitriana A 2021 Line2-Kuningan Bridge; Report SuperStructure Analysis (Jakarta: Adhi Karya)

[5] Ewins D 2000 Modal Testing – Theory, Practice and Application (Baldock: Research Studies Press, UK)
[6] Maia N, and Silva J 1997 Theoretical and Experimental Modal Analysis (Baldock: Research Studies Press, UK)
[7] Peeters B, De Roeck G, Hermans L, Wauters T, Krmer C, and Smet C 2000 Comparison of System Identification Methods Using Operational Data of a Bridge Test
[8] Deger Y, Cantieni R, Pietrzko S, Rücker W, and Rohrmann R 1995 Modal Analysis of a Highway Bridge: Experiment, Finite Element Analysis and Link
[9] Ietrzko S, and Cantieni R 1996 Proc. 14th Int. Modal Analysis Conference. Modal testing of a steel/concrete composite bridge with a servo-hydraulic shaker p 91-98
[10] Cantieni R 2004 Experimental methods used in system identification of civil engineering structures.

The comparison of X-braced frame and moment frame on structural building (case study: government office of North Toraja, South Sulawesi, Indonesia)

J Tanijaya[1*], R Kwandou[2] and N S Indranata[3]

[1]Professor in Structural Engineering, Civil Engineering Department, Universitas Kristen Indonesia Paulus, Makassar, Indonesia
[2]Alumni, Universitas Hasanuddin, Makassar, Indonesia
[3]Alumni, Universitas Atmajaya, Makassar, Indonesia

*Corresponding author: jonie.tanijaya@gmail.com

Abstract. The seismic force can damage the building and harm the occupants. Therefore, its structure needs to be designed seismic-resistant. One of the essential things in seismic-resistant building is the effect of lateral force to its structural rigidity. It can be improved by using the lateral resistant element such as bracing and shear wall. The lateral element used in this study is concentric X-bracing. This study conducted an optional structural model of the office of North Toraja Regent, Panga, South Sulawesi. The building was originally analyzed as a whole unit structure, but it has already rotated in its first mode shape due its irregularity. The comparison analysis is conducted by using the same existing structure configuration, except with the addition of X-bracing. The analysis results of the dynamic response are the fundamental building period, mode shape, base shear, lateral displacement and P-delta effect. The fundamental building period using X-bracing is smaller than moment frame building. It indicates that the use of X-bracing lead to the stiffer building. The use of the X-bracing results in a larger base shear value than the moment-frame building. The difference in base-shear between X-braced frame with moment frame is about 211.77% and 212.10% for seismic force with X-direction and Y-direction respectively.

1. Introduction

The seismic force can damage the building and harm the occupants [1]. Therefore, its structure needs to be designed seismic-resistant. One of the essential things in seismic-resistant building is the effect of lateral force to its structural rigidity. It can be improved by using the lateral resistant element such as bracing and shear wall. Khairiansyah [2] conducted a research of steel frame structures with and without bracing due to the effect of seismic lateral force. The rigidity of the braced structure is dependent on the rigidity of diagonal bracing in the center of the span. The variables studied include lateral displacement, base shear, and fundamental building period. It was concluded that the braced-frame has significant stiffness compared to the steel frame structure. Viswanath et al.[3], comparing different types of bracing in simple structures. The use of bracing can transfer the lateral forces so that the structure becomes more stable. Structure with X-braced frame has the smallest bending moment value compared to other types. The addition of X-bracing also reduces the lateral displacements that occur in the structure. Windah [4], stated that bracing can reduce the maximum lateral displacement about approximately 60% compared to moment frame. The maximum lateral displacement of X-braced frame and Chevron bracing is smaller than the moment frame.

Fajri [5], conducted a study of 10 types of bracing configuration, namely the bracing layout and different dimensions of bracing. It was concluded that the right selection of the building structure configuration can have a significant effect besides the bracing dimension. The bracing location on the edge portal both in X-direction and Y-direction apparently still exhibit symmetrical structural behavior. While, the bracing location on the central building, even though the shape and location are symmetrical, but do not show symmetrical structural behavior. This can be seen from the results of the analysis where the direction of the mode shape rotates. Sri Haryono et al. [6], conducted research to evaluate the structural behavior before and after the addition of X-bracing. Evaluation is carried out on the internal forces and displacement between structures before and after the addition of X-bracing. It shows that the drift limit on moment frame without bracing is not safe. The use of bracing on the right location can significantly improve the structural performance. Thoeng et al. [7], examines the extent to which concentric and eccentric bracing are able to provide lateral resistance to structures. The study was conducted by analyzing and comparing the lateral resistant element such as Concentrically Braced Frame (CBF) and Eccentrically Braced Frame (EBF). The increase in shear resistant for CBF was 74.0% against moment frame, whereas EBF was 52.0%, so it can be concluded that CBF has higher elastic stiffness compared to EBF. While the EBF is more ductile than CBF. The lateral element used in this study is concentric X-bracing. This study conducted an optional structural model of the office of North Toraja Regent, Panga, South Sulawesi, Indonesia. The comparison analysis is conducted by using the same existing structure configuration, except with the addition of X-bracing. The analysis results of the dynamic response are the fundamental building period, structural mode shape, base shear, lateral displacement and P-delta effect.

2. Methodology

The building consists of 4 storeys + 1 roof story. It is located on Panga, North Toraja, South Sulawesi, Indonesia. The building location can be seen in Figure 1. The design was conducted on 2017. The building function is generally for offices equipped with many assembly areas. Structural model can be seen in Figure 2 and Figure 3. It can be seen that the building model is irregular with T-shape. The unsymmetrical plan of building or irregular building shape would tend to have rotational (torsional) mode shape. Torsional motion is produced by the eccentricity existing between the center of mass and the center of rigidity. The torsional mode shape could be serious because it leads to increased displacement at the extremes and may cause suffering in the lateral load-resisting elements located at the edges, particularly in the systems that are flexible. Consequently, torsional motion has been the cause of major damage to buildings vibrated by strong earthquakes, ranging from visible distortion of the structure to structural collapse.

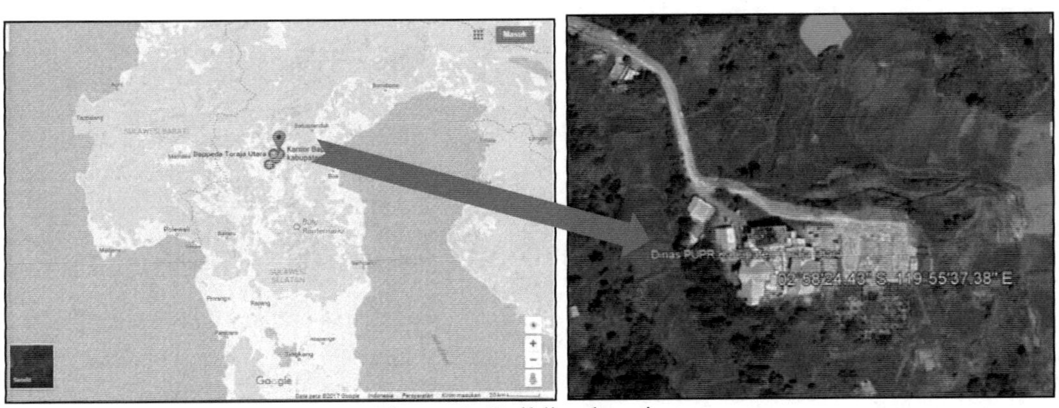

Figure 1. Building location

The building was originally analyzed as a whole unit structure, but it has already rotated in its first mode shape due its irregularity. This is contrary to the requirement of first and second mode-translations. This is what underlies the authors to examine the building by using alternative models with X-braced frame. The bracing is commonly used due to its effect to improve the building stiffness. Steel profile yield strength is 240 MPa while the concrete compression strength $f'_c = 25$ MPa. Building height is 19.60 m. The load specification is based on SNI 1727:2013 [8] while the steel structures is based on SNI 1729:2015 [9]. Based on the site soil properties, the site is classified as Site Class C (SC). Seismic design parameter can be seen in Table 1. Table 2 shows the design coefficient and factors for the structural system used in this study. The seismic evaluation criteria, such as structural period, minimum base shear value, displacement and P-delta effects, is based on SNI 1726:2012 [10].

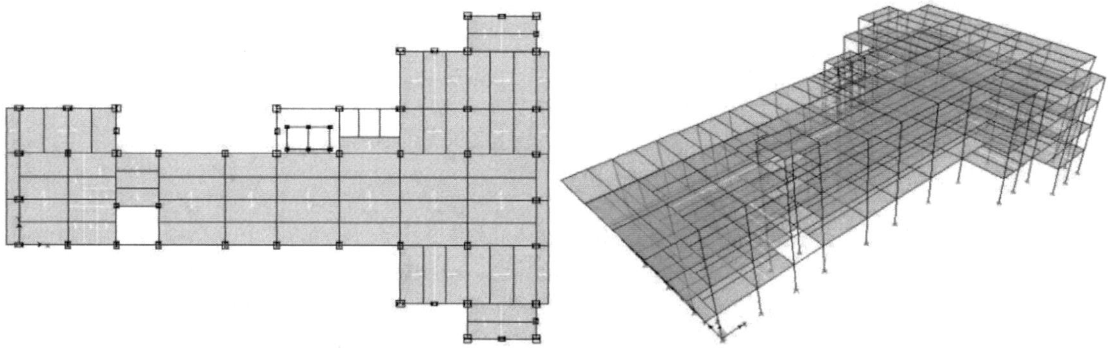

Figure 2. Model for moment frames.

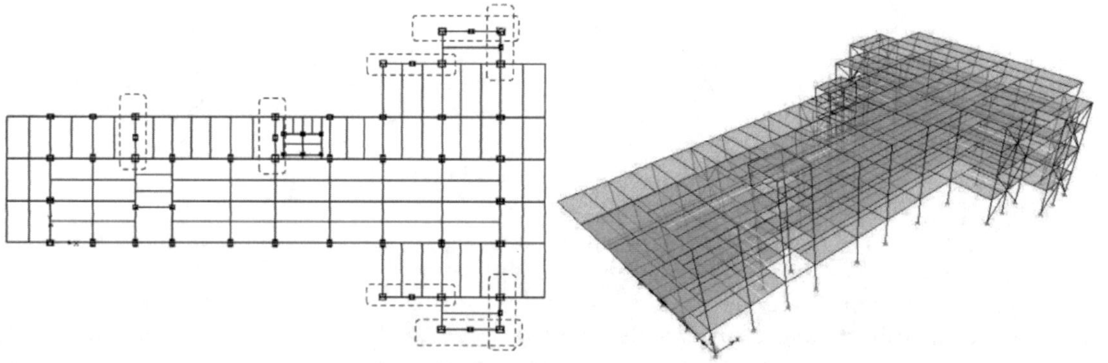

Figure 3. Model for X-braced frames.

Table 1. Seismic design parameter.

Parameter	Value
Risk categories	II
Importance factor (I)	1.00
MCEr response acceleration parameter at short period (S_s)	1.00
MCEr response acceleration parameter at 1-s period (S_1)	0.30
Site class	SC
Short period coefficient (F_a)	1.10
Long period coefficient (F_v)	1.80
Seismic design cathegory	D

Table 2. Design coefficient and factor for seismic force-resisting system.

Structural system	Response modification coefficient (R)	Overstrength factor (Ω_o)	Deflection amplification factor (C_d)
Steel and concrete composite special moment frame	8	3.00	5.50
Steel and concrete composite special concentrically braced frame	6	2.50	5.00

3. Result and discussion

The fundamental building period using X-bracing is smaller than moment frame building. It indicates that the use of X-bracing lead to the stiffer building. It can be seen that the fundamental building period for X-braced frame building is smaller the maximum period limit while it is larger than the maximum period for moment-frame building. Hence, the fundamental building period is taken as the maximum period limit for moment-frame building (Table 3). For the first mode, the fundamental building period of X-braced frame is around 2.97 times the fundamental building period of moment frame. While in second mode, it is around 2.73 times (Table 4 and Table 5). A minimum number of modes to obtain a combined modal mass participation should be at least 90%, as in SNI 1726:2012 sec 7.9.1 [10]. It can be seen that the requirement has been fulfilled. The modal mass participation for moment-frame building is about 97.22% and 90.16% at fourth mode for X-direction and Y-direction respectively. While for X-braced frame is about 96.15% dan 97.56% at seventh mode for X-direction and Y-direction respectively.

Table 3. Fundamental building period.

Type	T_x (s)	T_y (s)	T_{max} (s)	$T_{x\ use}$ (s)	$T_{y\ use}$ (s)
Moment frame	1.16	0.86	0.64	0.64	0.64
X-braced frame	0.34	0.39	0.64	0.34	0.39

Table 4. Modal mass participation for moment frame.

Mode	Period (s)	Ux (%)	Uy (%)	Rz (%)	sum Ux (%)	sum Uy (%)
1	1.16	87.42	0.04	0.91	87.42	0.04
2	0.93	0.69	37.98	51.16	88.11	38.02
3	0.86	0.21	52.13	38.46	88.32	90.15
…15	0.04	0.13	0.08	0.00	99.99	99.99

The moment frame building show rotational mode for the second mode (Table 4) while the building mode shape of X-braced frame tends to be translational mode for the first and second mode (Table 5). The combined response for the modal base shear (V_t) should be at least 85% of the calculated base shear (V) using the equivalent lateral force procedure (SNI 1726:2012 sec 7.9.4.1). If the modal base shear (V_t) is smaller than 85% V then it is permitted to multiply the forces by 0.85V/Vt. The modal base shear (V_t) before scaling can be seen in Table 6. While the comparison with the base shear (V) can be seen in Table 7. The use of the X-bracing results in a larger base shear value than the moment-frame building. It indicates that the stiffer frame tends to attract more base-shear. The difference in base-shear between X-braced frame with moment frame is about 211.77% and 212.10% for seismic force with X-direction and Y-direction respectively (Table 8). The scale factor for moment-frame is larger than X-braced frame.

Table 5. Modal mass participation for X-braced frame.

Mode	Period (s)	Ux (%)	Uy (%)	Rz (%)	sum Ux (%)	sum Uy (%)
1	0.39	0.00	85.45	0.30	0.00	85.45
2	0.34	85.08	0.00	0.12	85.08	85.45
3	0.33	0.12	0.21	84.42	85.20	85.65
…15	0.03	0.27	0.04	0.00	99.97	99.97

Table 6. Unscaled base shear.

Moment frame		X-braced frame	
RSX (ton)	RSY (ton)	RSX (ton)	RSY (ton)
49.57	47.04	178.00	178.91

Table 7. Comparison of base shear to determine the scale factor.

Moment frame		X-braced frame	
$V_{staticX=Y}$ (ton)	$0.85V_{staticX=Y}$ (ton)	$V_{staticX=Y}$ (ton)	$0.85V_{staticX=Y}$ (ton)
239.85	203.87	508.84	432.51

Table 8. Scale factor for base shear.

Moment frame				X-braced frame			
Seismic-X	$V_{x\text{-after scale}}$ (ton)	Seismic-Y	$V_{y\text{-after scale}}$ (ton)	Seismic-X	$V_{x\text{-after scale}}$ (ton)	Seismic-Y	$V_{y\text{-after scale}}$ (ton)
4.12	204.24	4.34	204.14	2.43	432.53	2.42	432.96

The building height is 19.60 m. Inelastic displacement limit is taken as 0.02 H (= 392 mm). The inelastic displacement using deflection amplification factor can be seen in Table 9 for moment frame and X-braced frame respectively. It can be seen that the inelastic displacement values do not exceed the limit. Hence both the model has already fulfilled the requirements. P-delta effects are not required to be considered when the stability coefficient (θ) is equal to or less than 0.10. From Table 10 and Table 11, it can be seen that the P-delta effect need not to be considered in analysis due to its θ value that is lower than the minimum requirement ($\theta < 0.10$).

Table 9. Inelastic displacement

Storey	Moment frame		X-braced frame	
	UX (mm)	UX (mm)	UX (mm)	UY (mm)
Roof	266.46	167.78	48.23	66.25
4th	218.57	140.93	38.58	54.27
3rd	162.57	107.47	24.43	36.99
2nd	86.83	58.86	11.59	18.02
Base	0.00	0.00	0.00	0.00

Table 10. Check P-delta effect for moment-frame building.

Storey	X-direction			Y-direction		
	θ	θ_{limit}	Note	θ	θ_{limit}	Note
Roof	0.0081	0.0909	Ok	0.0075	0.0909	Ok
4th	0.0100	0.0909	Ok	0.0091	0.0909	Ok
3rd	0.0148	0.0909	Ok	0.0141	0.0909	Ok
2nd	0.0198	0.0909	Ok	0.0210	0.0909	Ok

Table 11. Check P-delta effect for X-braced frame building

Storey	X-direction			Y-direction		
	θ	θ_{limit}	Note	θ	θ_{limit}	Note
Roof	0.0013	0.1000	Ok	0.0016	0.1000	Ok
4th	0.0014	0.1000	Ok	0.0018	0.1000	Ok
3rd	0.0014	0.1000	Ok	0.0020	0.1000	Ok
2nd	0.0015	0.1000	Ok	0.0024	0.1000	Ok

4. Conclusion

The fundamental building period using X-bracing is smaller than moment frame building. For the first mode, the fundamental building period of X-braced frame is around 2.97 times the fundamental building period of moment frame. While in second mode, it is around 2.73 times. It indicates that the use of X-bracing lead to the stiffer building. The building mode shape of X-braced frame tends to be translational mode for the first and second mode while the moment frame building show rotational mode for the second mode. It indicates that the use of X-bracing can change the rotational mode shape of moment-frame building. The use of the X-bracing results in a larger base shear value than the moment-frame building. It indicates that the stiffer frame tends to attract more base-shear. The difference in base-shear between X-braced frame with moment frame is about 211.77% and 212.10% for seismic force with X-direction and Y-direction respectively. P-delta effect need not to be considered in analysis due to its θ value that is lower than the minimum requirement.

5. References

[1] Youssef M A, Ghaffarzadeh H and Nehdi, M 2007 *Seismic Performance OF RC Frames with Concentric Internal Steel Bracing*

[2] Khairiansyah S 1999 *Analysis of frame stiffness due to the effect of stories number on steel structure using seismic lateral force*

[3] Viswanath K G, Prakash K B and Desai, A 2010 *Seismic Analysis of Steel Braced Reinforced Concrete Frames*

[4] Windah R S 2011 *The use of braced frames element as seismic resistance element on multi-stories building*

[5] Fajri 2015 *Configuration analysis of the shape of bracing for designing steel building using CBF*

[6] Haryono S 2015 *The use of X-braced frame to improve the structural performance*

[7] Thoeng S, Tanijaya J and Kalangi H T 2016 *The comparison performance of CBF and EBF (unpublished thesis)*

[8] National Standardization Agency 2013 *SNI 1727:2013 Minimum Load for Building Design and Other Structures* (Jakarta: National Standardization Agency)

[9] National Standardization Agency 2015 *SNI 1729:2015 Specifications for Building Structural Steel Buildings* (Jakarta: National Standardization Agency)

[10] National Standardization Agency 2012 *SNI 1726:2012 Earthquake Resistance Design Procedures for Structural and Non-Structural Building* (Jakarta: National Standardization Agency)

Performance-based Analysis of Building Structures in Various Earthquake Zones

S A Nurjannah[1*], Rosidawani[1], Saloma[1], Hanafiah[1], and M Rafly[1]

[1]Department of Civil Engineering, Universitas Sriwijaya, Indralaya, Indonesia

*Corresponding author: sitiaisyahn@ft.unsri.ac.id

Abstract. The structural design of an eight-story building was examined to determine its performance in resisting ultimate loads. This was necessary for applying the building design in Indonesian regions with low to high seismic hazard levels. Studies have indicated that pushover analysis is suitable to assess the capacity of building structures against earthquake loads. However, as the structural design of each building produces its nominal strength, performance analysis must always be conducted. This study aimed at comparing the structural performance of buildings in regions with low, medium, and high seismic hazards, represented by the cities of Palembang, Jakarta, and Palu, respectively. A performance-based analysis was conducted using the pushover method to obtain the performance points. The responses of the building structures were the base shears, floor drifts, and inter-story drifts. The structural performance levels in the three regions were between immediate occupancy and life safety or were categorized as damage control, indicating minor damage. Because the structures exhibited high stiffness and ductility, they resisted the ultimate forces up to the damage control performance level in all regions. The greatest lateral drifts of 79.744 mm and 51.656 mm in the X- and Y-axis directions, respectively, occurred in Palu's high seismic hazard region. The lowest lateral drift occurred in Palembang's low seismic hazard region, in the X- and Y-axis directions of 59.401 mm and 38.378 mm, respectively.

1. Introduction

Security is a significant factor in designing and developing high-rise buildings, hence, adequate planning is necessary to ensure optimal structural performance and resistance to natural disasters, such as earthquakes [1]. The lateral loads of earthquakes can cause damage and lower the structural performance of a building. The elastic structures are burdened with combined loads in the design process to obtain the ultimate condition (boundary). However, plastic failure behavior occurs during earthquakes [2], and therefore, the structures should also be analyzed under plastic conditions. Nonlinear static or pushover analysis is used to analyze the earthquake resistance of structures. It is a part of performance-based seismic (PBS) design that is widely used for new structures or for evaluating existing buildings [3]. Static loads are applied to each floor, resulting in a demand curve representing the structural behavior [4]. This method is used to assess the collapse behavior by indicating the types of plastic hinges formed on the structure. The load value is increased until the structure reaches a plastic condition or exhibits a certain deviation value.

Structures can be designed to experience damage under certain conditions according to the seismic design loads [5]. The determination of structural performance is based on the function of the building and is related to the economic factors of post-earthquake repair. The building structure should perform adequately under life safety (LS) and collapse prevention (CP) conditions owing to the seismic design loads and maximum severity of earthquakes considered, respectively [6]. Under these conditions, it is hoped that earthquakes will not cause any casualties. However, the structural design of each building provides its own nominal capacity and a performance analysis is always necessary. This study aimed to analyze the performance of an existing eight-story building structure in various (low, intermediate, and high) seismic-hazard regions using pushover analysis.

2. Materials and method

A prototype building of an eight-story apartment was analyzed. The structure had previously been designed for all the seismic-hazard regions in Indonesia. The general data structure and dimensions of the structural members are presented in Table 1 and Table 2, respectively. The structural model constructed using SAP2000 software is shown in Figure 1. Figures 2 and 3 show the plans of the building.

Table 1. Data of the building

Item	Data	Item	Data
Function	Apartment	Height of building	31 m
Number of floors	8	Concrete strength	30 MPa (beam) 35 MPa (column & shear wall)
Length of building	60 m	Steel strength	420 MPa
Width of building	14.8 m	Seismic Design Criteria	D

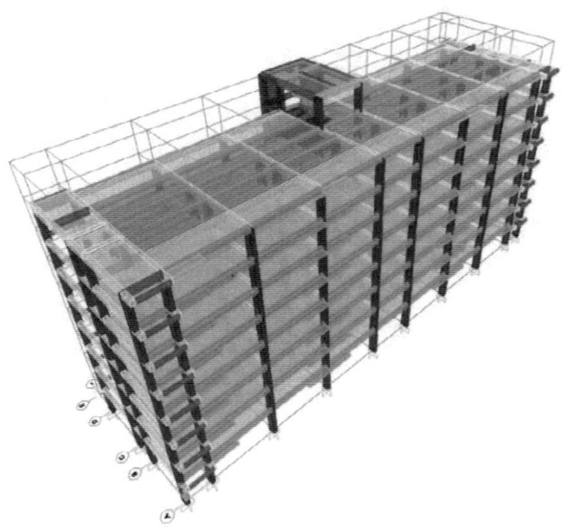

Figure 1. Structural model of the building.

The live load was applied according to the codes [7], as well as the dead load [7,8] and earthquake load [9]. The building was included in hazard category II and had a priority factor value (I_e) of 1.00. In each reviewed region, the building structure was assumed to be on soft soil (SE). The type of soil layer influences seismicity parameters. The long-period transition (TL) value was assumed to be 20 s. The response spectra and other earthquake parameters of Palembang, Jakarta, and Palu are shown in Figure 4 and Table 3, respectively [10]. These cities represent regions of low, medium, and high seismic hazards in Indonesia. The seismic parameters and load combinations used to design earthquake-resistant structures followed the code [9].

IOP Conf. Series: Earth and Environmental Science 1244 (2023) 012019 doi:10.1088/1755-1315/1244/1/012019

The capability of the structures was determined using base shear, response spectrum force scale, lateral drift, and inter-story drift obtained from pushover analysis. The analysis was conducted according to the provision in [11]. he code requires the building to be designed at least under LS conditions which ensures that the building suffers moderate damage but does not collapse after the earthquake. The pushover analysis yields a capacity curve for the base shear and roof drift [12]. The response spectra [10] of the seismic data were used in the PBS. Both the capacity curves and response spectra are converted into $S_a - S_d$ based curves [11,13], namely capacity spectrum curves and demand spectra, respectively. The performance point is the intersection of the capacity spectrum curve and demand spectra [14]. The performance point indicates the performance level of the structure due to the designed earthquake load, ranging from light to heavy damage to collapse. The structural performance levels were determined based on the ratio of the reference point displacement at the time of the performance point achieved to the total height of the building. The ratio was categorized into the maximum values of drift and inelastic drift [11].

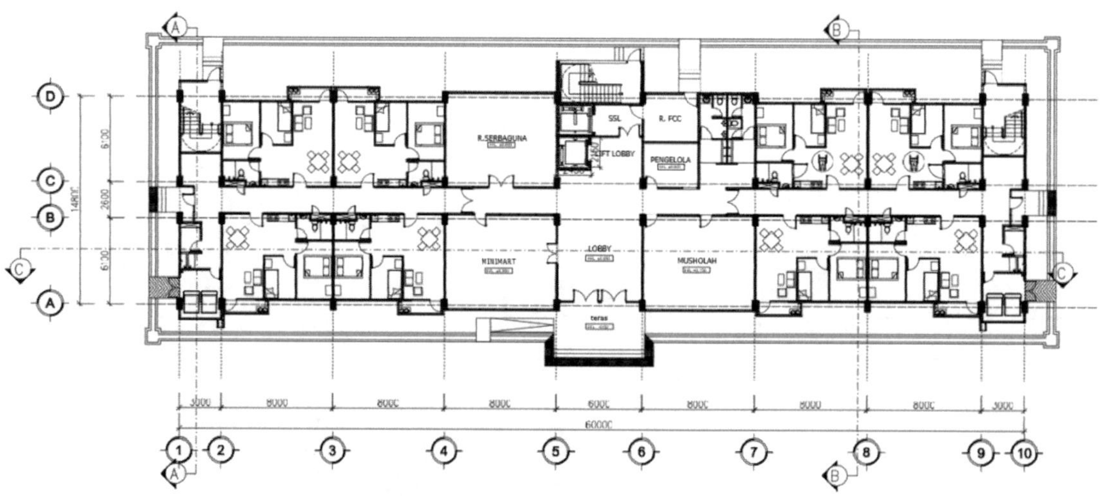

Figure 2. Plan of the ground floor.

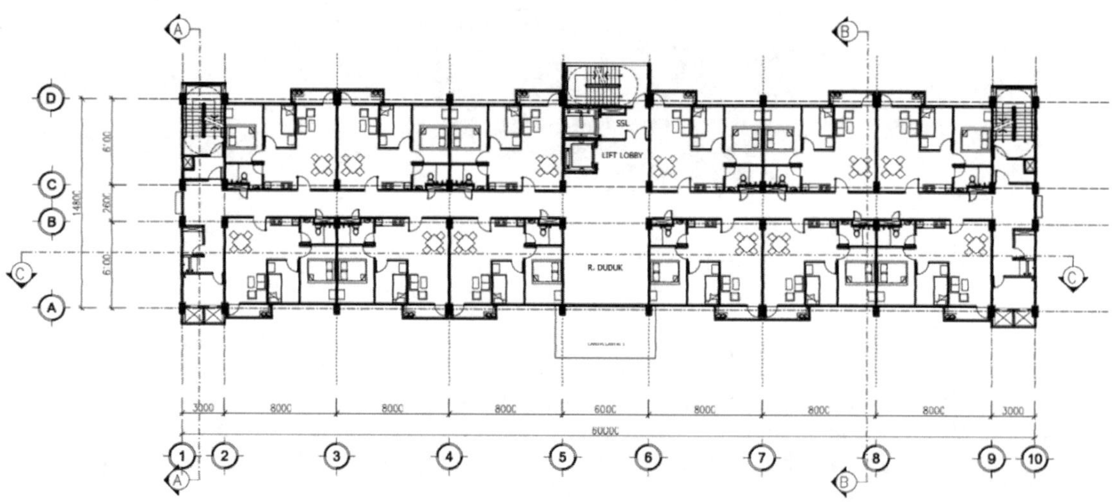

Figure 3. Plan of the 1st – 8th floors.

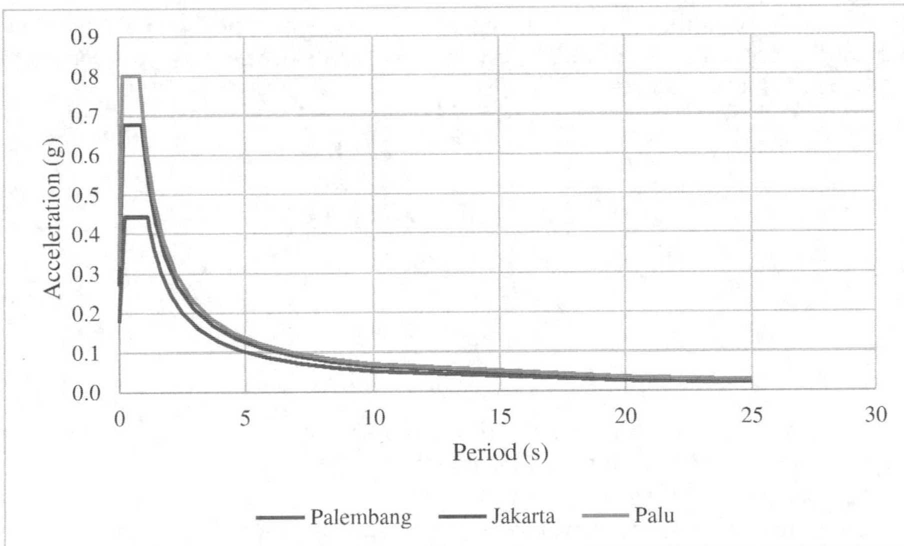

Figure 4. Response spectra.

Table 2. Dimensions of the structural members

Type	Dimension ($b \times h$) in mm	Color	Type	Dimension (t) in mm	Color
C48	400 × 800	brown	S10	100	light green
B24	200 × 400	red	S13	130	cyan
B24A	250 × 400	dark red	S15	150	light red
B36	300 × 600	green	S20	200	yellow
B36A	150 × 300	orange	CL	200	red
B46	400 × 600	yellow	SW1	200	magenta
B13A	150 × 300	grey			
B47	400 × 700	blue			
B57	500 × 700	cyan			

Note: C: column; B: beam, b = width; h = height; t = thickness; S: slab; CL: L shaped column; SW: shear wall

Table 3. Earthquake parameters of Palembang, Jakarta, and Palu

Seismic parameter	Palembang	Jakarta	Palu
S_{ss}	0.29 (g)	0.78 (g)	1.5
S_1	0.25 (g)	0.38 (g)	0.6
F_a	2.28	1.28	0.8
F_v	3.06	2.47	1.7
S_{MS}	0.67	1.02	1.2
S_{M1}	0.76	0.95	1.02
S_{DS}	0.44	0.68	0.80
S_{D1}	0.51	0.63	0.68
T_0	0.23	0.19	0.17
T_S	1.14	0.93	0.85

In the PBS, the roof displacement and the base shear correlation forms the capacity curve (Figure 5). The points on the curve indicate the structural damage to the building, which occurs gradually. To

IOP Conf. Series: Earth and Environmental Science 1244 (2023) 012019 doi:10.1088/1755-1315/1244/1/012019

connect the capacity curve to the seismic demand curve to produce a performance point [11], it must be converted into a capacity spectrum (Figure 6) using the acceleration displacement response spectrum (ADRS) method and Equations 1 – 4 [11].

$$PF_1 = \left[\frac{\sum_{i=1}^{N}(w_i\emptyset_{i1})/g}{\sum_{i=1}^{N}(w_i\emptyset_{i1}^2)/g}\right] \quad (1)$$

$$\alpha_1 = \frac{[\sum_{i=1}^{N}(w_i\emptyset_{i1})/g]}{[\sum_{i=1}^{N}(w_i/g)][(w_i\emptyset_{i1}^2)/g]} \quad (2)$$

$$S_a = \frac{V/W}{\alpha_1} \quad (3)$$

$$S_d = \frac{\Delta_{roof}}{PF_1\phi 1_{1,roof}} \quad (4)$$

where:

PF_1 : modal participation factor for the first natural mode
α_1 : modal mass coefficient for the first mode
w_{ii}/g : mass assigned to story – i
$\emptyset_{1,ii}$: amplitude of first mode at story – i
$\emptyset_{1,roorr}$: amplitude of first mode at the roof
N : number of story
V : base shear
WW : weight of the building (including self and life loads)
$\Delta\Delta_{roorr}$: roof drift
S_a : spectral acceleration
S_d : spectral displacement

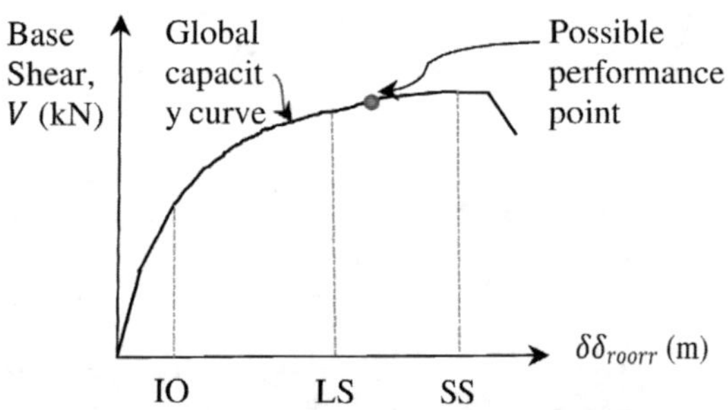

Figure 5. Building structure capacity curve [11].

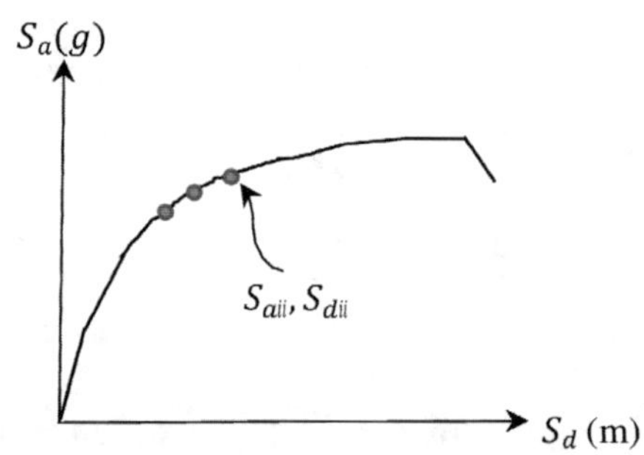

Figure 6. Building structure capacity spectrum curves [11].

The response spectrum relates the spectral acceleration (S_a) and period (T) (Figure 7). To correlate the response spectrum with the capacity spectrum, it must be converted into a demand spectrum in the ADRS format (Figure 8) using Equations 5-7 [11].

$$S_{di} = \frac{T_i^2}{4\pi^2} S_{ai}g \quad (5)$$

$$S_{ai}g = \frac{2\pi}{T_i} S_v \quad (6)$$

$$S_{di} = \frac{T_i}{2\pi} S_v \quad (7)$$

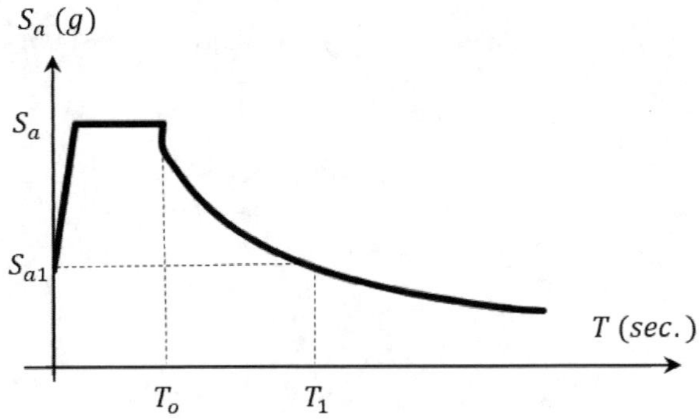

Figure 7. Response spectrum [11]

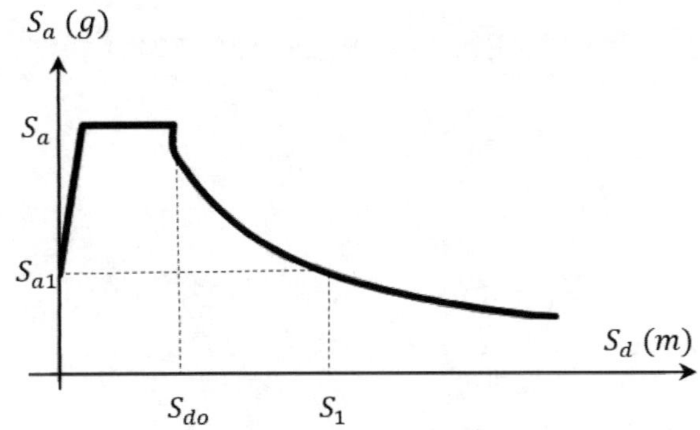

Figure 8. Demand spectrum [11]

3. Results and discussion

The static base shear (V_S), dynamic base shear (V_D), and structural volume were calculated using SAP2000 software and are presented in Table 4. The different base shear values for each area were influenced by the parameters of the response spectra and the seismic coefficients (S_S, S_1, and C_S). The Palu region had the highest base shear, followed by Jakarta and Palembang. The structural weight also affected the base shear. The original design of the building had different lengths along the X- and Y-axes which caused the stiffnesses in the X- and Y-axis directions to be different. This condition resulted in a difference of base shear in the X-direction ($\Delta\Delta$) compared with that in the Y-direction ($\Delta\Delta$) of 16.05%. The increased stiffness in one direction resulted in increased base shear, and vice versa. The structural stiffness in the Y-axis direction was greater than that in the X-axis direction.

Table 4. Structure base shear and volume

Base Shear	Axis	Region					
		Palembang (kN)	$\Delta\Delta$	Jakarta (kN)	$\Delta\Delta$	Palu (kN)	$\Delta\Delta$
V_S	X	3122.122	16.05%	3881.493	16.05%	4191.381	16.05%
V_S	Y	3718.972		4623.511		4992.639	
V_D	X	3122.066	16.05%	3881.64	16.05%	4191.465	16.05%
V_D	Y	3718.979		4623.537		4992.664	
Volume		56937.14		56937.14		56937.14	

The story drift is a deviation owing to the base shear which is distributed to each floor as lateral forces [11]. As shown in Figures 9 and 10, the lateral drifts increased with building height on the X- and Y-axes, respectively. In both the X- and Y-axes, the greatest story drift occurred on the ninth floor of the Lower Machine Room (LMR). The story drifts at the xth floor ($\delta\delta_{ex}$) were determined by elastic analysis owing to the floor design seismic load using the SAP software. Here, x denotes the floor number. The enlarged drifts of the center of mass at the xth floor ($\delta\delta_x$) were obtained using Equation (8) [9]. The values of $\delta\delta_{ex}$ and $\delta\delta_x$ for each floor in the X- and Y-axis directions are presented in Tables 5 and 6, respectively.

$$\delta_x = \frac{C_d \delta_{ex}}{I_e} \quad (8)$$

where the lateral drift enlargement factor (C_d) and important seismic factors (I_e) are 5.5 and 1.0, respectively [9].

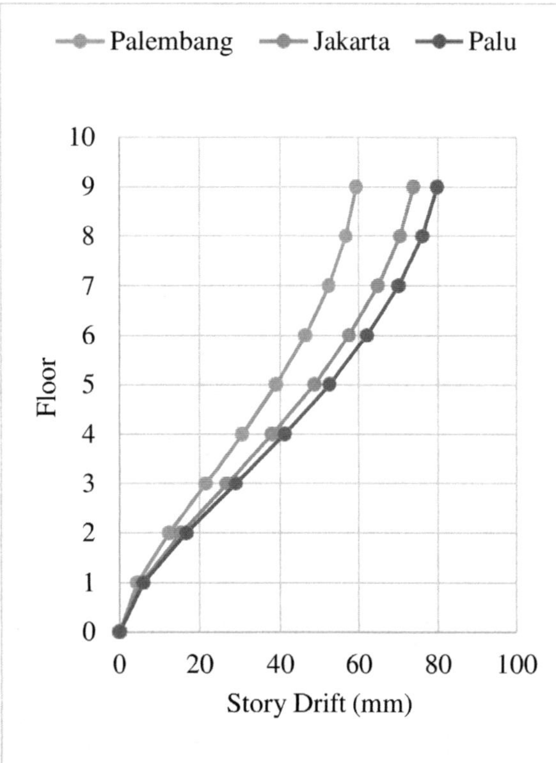

Figure 9. Story drifts in the X-axis.

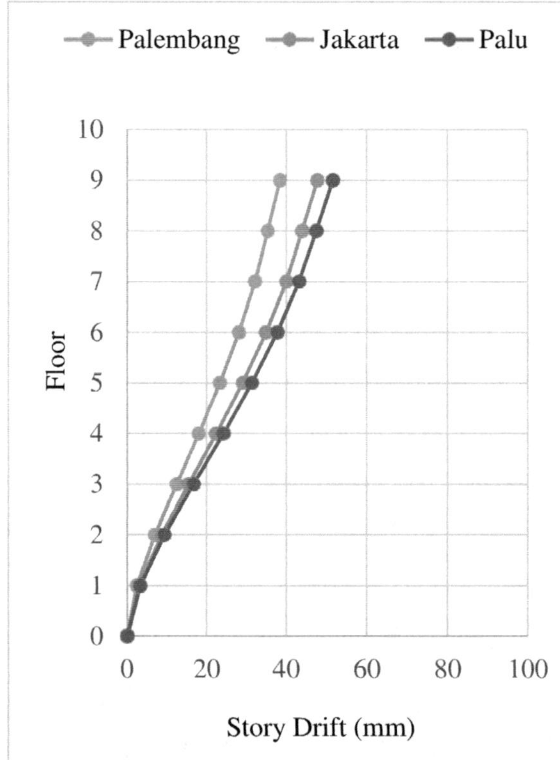

Figure 10. Story drifts in the Y-axis.

Table 5. Story drifts in the X-axis direction

Floor	Height	Palembang		Jakarta		Palu	
		$\delta\delta_{ex}$	$\delta\delta_{x}$	$\delta\delta_{ex}$	$\delta\delta_{x}$	$\delta\delta_{ex}$	$\delta\delta_{x}$
	(m)	(mm)	(mm)	(mm)	(mm)	(mm)	(mm)
LMR	3.6	59.40	326.70	73.85	406.16	79.74	438.59
Roof	3.4	56.66	311.65	70.45	387.45	76.07	418.38
7	3.4	52.20	287.08	64.89	356.90	70.07	385.39
6	3.4	46.29	254.57	57.54	316.49	62.14	341.76
5	3.4	38.99	214.42	48.47	266.57	52.34	287.85
4	3.4	30.59	168.26	38.03	209.18	41.07	225.88
3	3.4	21.55	118.50	26.79	147.33	28.93	159.09
2	3.4	12.47	68.61	15.51	85.30	16.75	92.11
1	3.6	4.48	24.65	5.57	30.65	6.02	33.09
Base	0	-	-	-	-	-	-

Table 6. Story drifts in the Y-axis direction

Floor	Height	Palembang		Jakarta		Palu	
		$\delta\delta_{ex}$	$\delta\delta_{x}$	$\delta\delta_{ex}$	$\delta\delta_{x}$	$\delta\delta_{ex}$	$\delta\delta_{x}$
	(m)	(mm)	(mm)	(mm)	(mm)	(mm)	(mm)
LMR	3.6	38.48	211.63	47.84	263.10	51.66	284.11
Roof	3.4	35.42	194.80	44.03	242.18	47.55	261.51
7	3.4	32.20	177.11	40.03	220.18	43.23	237.76

Floor	Height	Palembang		Jakarta		Palu	
		$\delta\delta_{ex}$	$\delta\delta_x$	$\delta\delta_{ex}$	$\delta\delta_x$	$\delta\delta_{ex}$	$\delta\delta_x$
	(m)	(mm)	(mm)	(mm)	(mm)	(mm)	(mm)
6	3.4	28.17	154.95	35.03	192.64	37.82	208.02
5	3.4	23.40	128.69	29.09	159.99	31.41	172.76
4	3.4	18.06	99.33	22.45	123.48	24.24	133.34
3	3.4	12.45	68.49	15.48	85.15	16.72	91.95
2	3.4	7.01	38.56	8.72	47.94	9.41	51.77
1	3.6	2.43	13.35	3.02	16.60	3.26	17.93
Base	0	-	-	-	-	-	-

The inter-story drifts ($\Delta\Delta_x$) on each floor were checked using Equation (9) [9] to ensure that the values remained within the acceptable interstory drift ($\Delta\Delta_a$). Tables 7 and 8 show that all inter-story drifts were within acceptable values.

$$\Delta_x = \frac{(\delta_{ex} - \delta_{e(x-1)})\ C_d}{I_e} \leq \Delta_a = 0.025 h_{sx} \tag{9}$$

where h_{sx} is the floor height below the xth floor.

Table 7. Inter-story drifts in the X-axis direction

Floor	Height	$\Delta\Delta_a$	Region					
			Palembang		Jakarta		Palu	
			$\Delta\Delta_x$	Note	$\Delta\Delta_x$	Note	$\Delta\Delta_x$	Note
	(m)	(mm)	(mm)		(mm)		(mm)	
LMR	3.6	90	15.05	OK	18.72	OK	20.21	OK
Roof	3.4	85	24.57	OK	30.55	OK	32.99	OK
7	3.4	85	32.50	OK	40.41	OK	43.64	OK
6	3.4	85	40.15	OK	49.92	OK	53.91	OK
5	3.4	85	46.16	OK	57.39	OK	61.97	OK
4	3.4	85	49.75	OK	61.85	OK	66.79	OK
3	3.4	85	49.90	OK	62.03	OK	66.98	OK
2	3.4	85	43.96	OK	54.65	OK	59.01	OK
1	3.6	90	24.65	OK	30.65	OK	33.09	OK
Base	0	0	0	-	0	-	0	-

Table 8. Inter-story drifts in the Y-axis direction

Floor	Height	$\Delta\Delta_a$	Region					
			Palembang		Jakarta		Palu	
			$\Delta\Delta_x$	Note	$\Delta\Delta_x$	Note	$\Delta\Delta_x$	Note
	(m)	(mm)	(mm)		(mm)		(mm)	
LMR	3.6	90	16.83	OK	20.93	OK	22.60	OK
Roof	3.4	85	17.69	OK	21.99	OK	23.75	OK
7	3.4	85	22.15	OK	27.54	OK	29.74	OK
6	3.4	85	26.26	OK	32.65	OK	35.26	OK
5	3.4	85	29.36	OK	36.50	OK	39.42	OK

Floor	Height	Δ_a	Region					
			Palembang		Jakarta		Palu	
			Δ_x	Note	Δ_x	Note	Δ_x	Note
	(m)	(mm)	(mm)		(mm)		(mm)	
4	3.4	85	30.83	OK	38.33	OK	41.39	OK
3	3.4	85	29.93	OK	37.21	OK	40.18	OK
2	3.4	85	25.21	OK	31.34	OK	33.84	OK
1	3.6	90	13.35	OK	16.60	OK	17.93	OK
Base	0	0	0	-	0	-	0	-

The greatest story drift occurred in the Palu region because the lowest period (T_S) in Palu resulted in the greatest base shear and seismic multiplier factor compared to those in the other regions. The story drift in the X- and Y-axis directions was significantly different because of the moment of inertia of the structure members along those directions. Figures 11 and 12 show that the greater inertia moment along the Y-axis provides stiffer structure members, which results in lower story drifts along the Y-axis. The nonlinear story drift on each floor was influenced by the acceleration function of Newton's second law [15]. The greatest inter-story drift (ISD) along the X- and Y-axis directions occurred on the third and fourth floors, respectively. If the ISD (Δ_x) is less than the acceptable ISD (Δ_a), the structure is safe.

The performance points of the models in the three cities are shown in Figures 13–18, representing the ability of the structure to perform LS under seismic loads along the X and Y axes. The structural model in the lowest seismic risk of Palembang achieved less spectral acceleration and spectral displacement than the other models in the medium- and high-seismic-risk regions of Jakarta and Palu, respectively. This behavior shows that the performance points provide information regarding the performance of structures under different earthquake loads [16]. Pushover analysis using response spectrum data can produce ISD ratios that nearly match the analysis results using the nonlinear response time history analysis (NTHA) method. The NTHA method provides the most accurate analysis results compared to the three other methods: equivalent static method, pushover analysis, and response spectrum method [17, 18].

According to the code [11], the structural performance levels are shown by three plastic hinge conditions: the first yield, performance point, and ultimate in Tables 9 – 11. As shown in Table 9, the structural model in the Palembang region experienced the greatest drift when plastic joints were first formed. The performance points exhibited the opposite behavior, where the structural model in the highest seismic hazard region of Palu drifted greater than its counterparts in other regions, as shown in Table 10. The performance points were used to assess the structural performance level. A greater drift in this condition indicates more critical structural performance level or structural damage. The structural performance level was determined using the ratio of the roof drift to the total height of the building [11]. The data was obtained at the performance point. In the ultimate state in Table 11, the drifts of the three models in all the regions are nearly the same. Table 12 lists the maximum total drifts. All the models in the low, medium, and high seismic hazard regions reached the condition of damage control (DC) owing to earthquakes in the directions of the X- and Y-axes. At this stage, the building structure was damaged but within the safe limit and could be easily repaired [11]. The maximum inelastic drifts are listed in Table 13. The model for the low seismic hazard in the Palembang region only experienced immediate occupancy (IO) on the X-axis and did not suffer significant damage. Although it could still be used [11], it reached the DC performance level in the Y-axis direction. The models in the medium and high seismic hazard regions of Jakarta and Palu reached the DC structure performance level owing to earthquakes along the directions of the X- and Y-axes. Thus, the structural model can be applied to the low-to-high seismic-hazard areas of Indonesia. However, optimizing the dimensions of the structural elements according to the earthquake loads of the building locations is necessary, especially in low-seismic hazard regions.

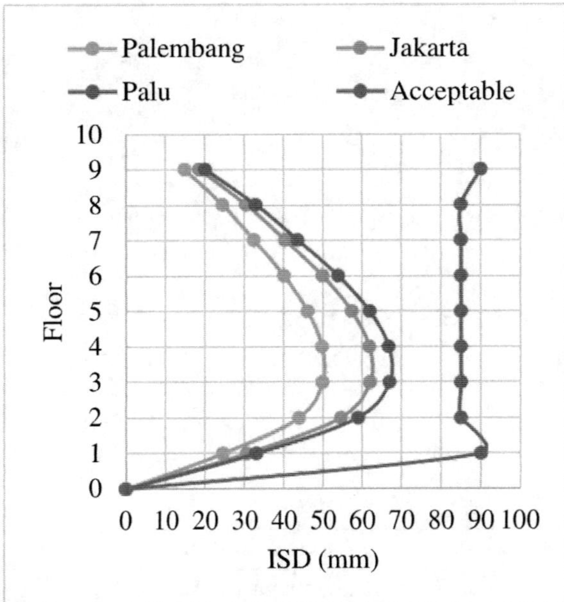

Figure 11. The ISD along the X-axis direction.

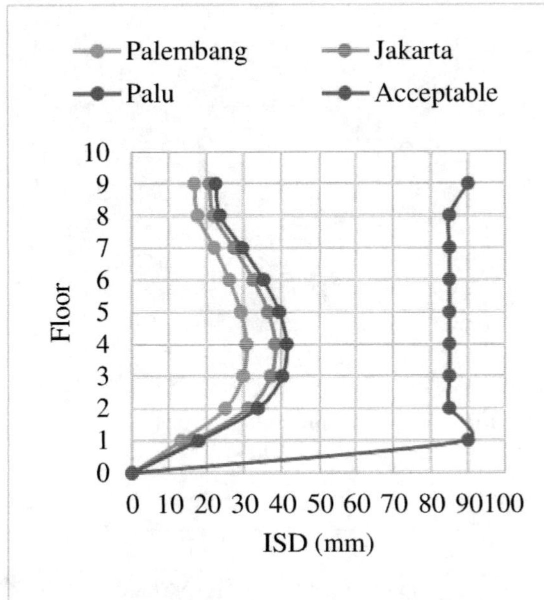

Figure 12. The ISD along the Y-axis direction.

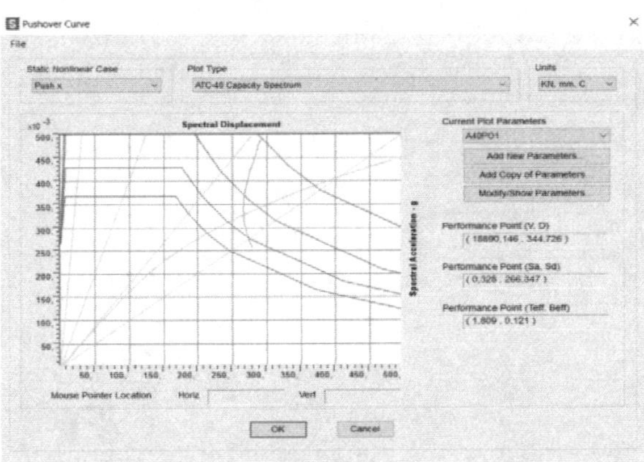

Figure 13. Performance point for the X-axis in Palembang.

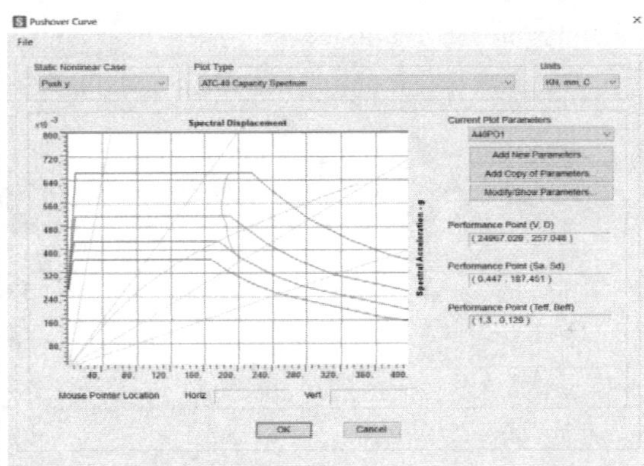

Figure 14. Performance point for the Y-axis in Palembang.

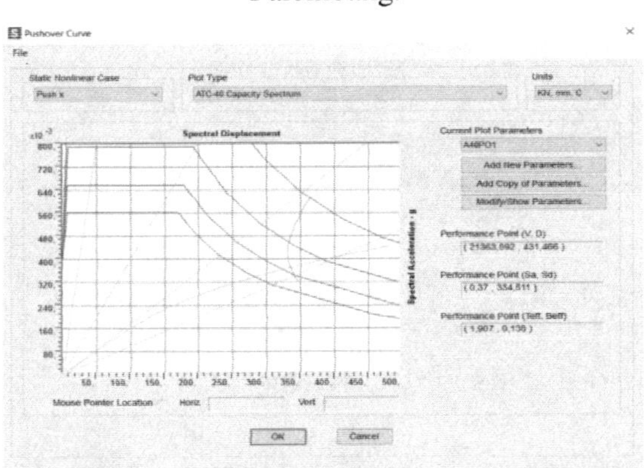

Figure 15. Performance point for the X-axis in Jakarta.

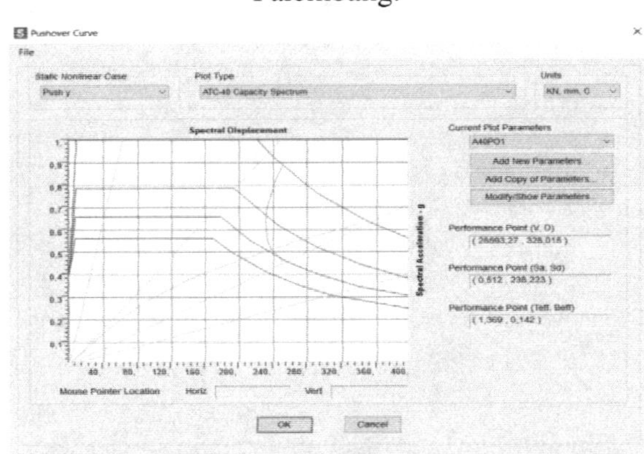

Figure 16. Performance point for the Y-axis in Jakarta.

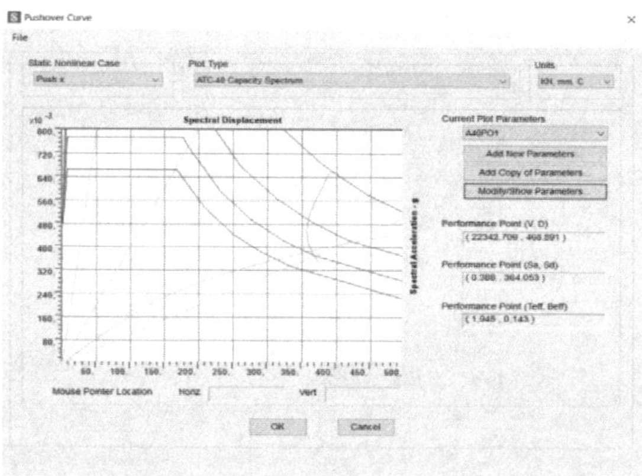

Figure 17. Performance point for the X-axis in Palu.

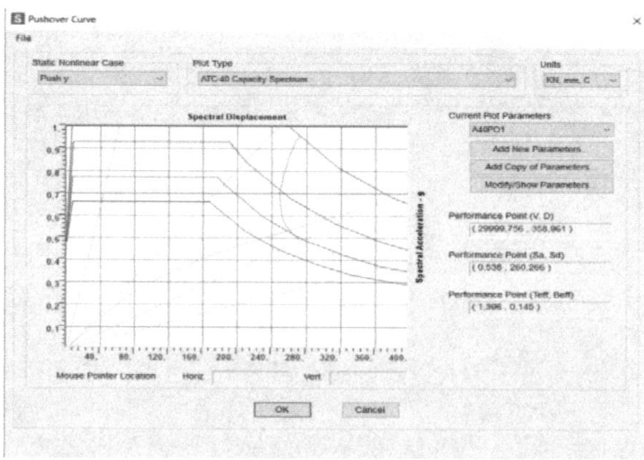

Figure 18. Performance point for the Y-axis in Palu.

Table 9. Plastic hinges in the first yield condition

Region	X-axis direction			Y-axis direction		
	Step	Base Shear (kN)	Drift (mm)	Step	Base Shear (kN)	Drift (mm)
Palembang	17	8395.22	118.03	12	10745.1	77.39
Jakarta	13	5912.61	81.83	8	6911.33	47.56
Palu	11	4840.54	66.76	6	5043.72	34.06

Table 10. Plastic hinges in the performance point condition

Region	X-axis direction			Y-axis direction		
	Step	Base Shear (kN)	Drift (mm)	Step	Base Shear (kN)	Drift (mm)
Palembang	56	18890.1	344.73	38	24967.1	257.05
Jakarta	69	21363.6	431.47	50	28593.3	328.02
Palu	73	22342.7	468.89	57	29999.8	358.96

Table 11. Plastic hinges in the ultimate condition

Region	X-axis direction			Y-axis direction		
	Step	Base Shear (kN)	Drift (mm)	Step	Base Shear (kN)	Drift (mm)
Palembang	100	25553.38	620.24	72	34654.2	471.19
Jakarta	100	25553.38	620.24	72	34654.2	471.19
Palu	97	25548.49	620.24	75	34615	470.83

Table 12. Structure performance levels based on the total maximum drift

Region	X-axis direction		Y-axis direction	
	Max. total drift	Performance level	Max. total drift	Performance level
Palembang	0.0111	DC	0.0083	DC
Jakarta	0.0139	DC	0.0105	DC
Palu	0.0151	DC	0.0116	DC

Table 13. Structure performance level on maximum inelastic drift

Region	X-axis direction		Y-axis direction	
	Max. inelastic drift	Performance level	Max. inelastic drift	Performance level
Palembang	0.0083	IO	0.0058	DC
Jakarta	0.0113	DC	0.0091	DC
Palu	0.0129	DC	0.0105	DC

4. Conclusions

Building structures designed to be applied in various seismic regions of Indonesia were modeled using performance-based method analysis. The following conclusions were drawn from this study:

1. The structural model exhibited the greatest base shear in the high seismic hazard region of Palu in both the X- and Y-axis directions. The structural models followed it in the medium and low seismic hazard regions of Jakarta and Palembang'. The percentage difference of the base shear in the X- and Y-axis directions was 16.05% for all structure models owing to the different lengths of the building along both axes, which resulted in different stiffnesses and ductility values in these directions. The base shear increased with the stiffness in each direction. The structure exhibited stiffer behavior in the X-axis direction than in the Y-axis direction, and the base shear of the X-axis was higher than that of the Y-axis.

2. The difference in the base shear resulted in different floor drifts. The greatest drifts of 79.74 mm and 51.66 mm, respectively, in the X and Y-axis directions, occurred in the Palu region. The lowest drifts of 59.40 mm and 38.48 mm in the X and Y-axis directions occurred in the Palembang region. Greater floor drifts occurred with greater earthquake loads exerted on the structure.

3. The inter-story drifts of all the models were less than the acceptable inter-story drift, hence, building structures were declared safe. The greatest inter-story drifts in the X- and Y-axis directions occurred on the third and fourth floors, respectively.

4. The structural model in all regions had performance levels between IO and LS or were categorized as DC. In these states, minor damage occurred, and the building could still be used without repair. The relatively similar structural performance levels in all regions resulted from the high stiffness and ductility behavior. Then, the structural models could resist seismic loads up to the DC performance level. However, structural elements are needed to optimize the dimensions and details to adapt the design to earthquake loads, especially in low-seismic-hazard regions.

5. References

[1] Hasan M Saidi T Afifuddin M and Setiawan B 2021 The assessment and strengthening proposal of building structure after the Pidie Jaya earthquake in December 2016 *Journal of King Saud University – Engineering Sciences* pp 1-12 https://doi.org/10.1016/j.jksues.2021.02.007

[2] Dewobroto W 2006 Performance Evaluation of Earthquake Resistant Steel Buildings Using SAP2000 *Jurnal Teknik Sipil* 3(1) pp 7-24 (text in Indonesian)

[3] Daei A and Poursha M 2021 On the accuracy of enhanced pushover procedures for seismic performance evaluation of code-conforming RC moment-resisting frame buildings subjected to pulse-like and non-pulse-like excitations *Structures* 32 pp 929–945 https://doi.org/10.1016/j.istruc.2021.03.035

[4] Kheirollahi M Abedi K and Chenaghlou M R 2021 A new pushover procedure for estimating seismic demand of double-layer barrel vault roof with vertical double-layer walls *Structures* 34 pp 1507–1524 https://doi.org/10.1016/j.istruc.2021.08.090

[5] Lesmana Y 2019 *Concept and Design of Special Moment Frame System (SMFS) Earthquake Resistant Reinforced Concrete Based on SNI 2847:2013 and SNI 1726:2012* (Yogyakarta: Deepublish) (in Indonesian)
[6] Budiono B 2011 *Comparative Study of Earthquake Resistant Building Designs Using SNI 03-1726-2002 and RSNI 03-1726-201x* (Bandung: Publisher ITB) (in Indonesian)
[7] National Standardization Board 2020 *Minimum Design Loads and Associated Criteria for Buildings and Other Structures* SNI 1727:2020 (Jakarta: BSN) (in Indonesian)
[8] Department of Public Works 1987 *Guidelines of Load Planning for Houses and Buildings* SKBI - 1.3.3.1987 (Jakarta: Public Works publishing board foundation) (in Indonesian)
[9] National Standardization Board 2019 *Earthquake Resistance Planning Procedures for Building and Non-Building Structures* SNI 1726:2019 (Jakarta: BSN) (in Indonesian)
[10] Ministry of Public Works and Human Settlements 2021 Indonesian Spectral Design http://rsa.ciptakarya.pu.go.id/2021/
[11] Applied Technology Council 1996 ATC-40 *Seismic Evaluation and Retrofit of Concrete Buildings* vol 1 (Redwood: California Seismic safety Comission)
[12] Mamesah H Y Wallah S E and Windah RS 2014 Pushover Analysis on Buildings with Soft First Story *Jurnal Sipil Statik* 2(4) pp 214–224 (text in Indonesian)
[13] Nurjannah S A Megantara Y Lim E and Imran I 2020 Comparison of structural performance of open frame structures based on SNI 03-1726-2002 and SNI 03-1726-2012 *E3S Web of Conferences* 156 05006 pp 1-10 Padang: September 26-27, 2019 https://doi.org/10.1051/e3sconf/202015605006
[14] Nurjannah S A Megantara Y and Yudha C 2011 Evaluation of Performance of Six-Story Structures Using Pushover Analysis in Soft Soil and Medium Soil *Proceeding of the 2nd International Conference of Earthquake Engineering and Disaster Mitigation: Seismic Risk Reduction and Damage Mitigation for Advancing Earthquake Safety of Structure* pp F-79 to F-87 Surabaya: July 19-21, 2011
[15] Leet K.M Uang C -M Lanning J T and Gilbert A M 2018 *Fundamentals of Structural Analysis* 5th ed (New York: McGraw-Hill Education)
[16] Kalkbrennera P Pelàa L Sandovalb C 2019 Multi directional pushover analysis of irregular masonry buildings without box behavior Engineering *Structures* 201 109534 https://doi.org/10.1016/j.engstruct.2019.109534
[17] Liu Y Kuang J S Huang Q Guo Z Wang X 2020 Spectrum-based pushover analysis for the quick seismic demand estimation of reinforced concrete shear walls *Structures* 27 pp 1490–1500 https://doi.org/10.1016/j.istruc.2020.07.040
[18] Jalilkhania M Ghasemib S H Danesha M 2020 A multi-mode adaptive pushover analysis procedure for estimating the seismic demands of RC moment-resisting frames Engineering *Structures* 213 110528 https://doi.org/10.1016/j.engstruct.2020.110528

Acknowledgments

The authors thank their colleagues and Universitas Sriwijaya for their support.

Non-engineering house damage after the February 25th, 2022 West Pasaman Earthquake

Fauzan[1*], B Istijono[1], F A Ismail[1], Y Narny[2] and Y S Putra[1]

[1] Civil Engineering Department, Andalas University, Padang, Indonesia
[2] Faculty of Humanities, Andalas University, Padang, Indonesia

*Corresponding author: fauzan@eng.unand.ac.id

Abstract. The earthquake that occurred on February 25, 2022 with a magnitude of M6.2 in West Pasaman Regency, West Sumatera has resulted in many losses such as casualties and damages to infrastructure, especially residential buildings. To determine the level of damage to community houses due to the earthquake, a rapid visual assessment survey was conducted with the National Agency for Disaster Countermeasure (BNPB) and 100 Civil Engineering students. The Method used in this survey is based on the Public Work Minister Regulation, which categorizes the damaged houses into 3 categories, namely light, moderate and heavy damage. Based on the results of the assessment, the damage generally occurred on the walls and structural elements of the houses. The damages is caused by human error factors such as lack of planning, poor construction implementation, site selection and selection of quality building materials that were not in accordance with earthquake safe house building standards. One of the efforts to overcome the problem of the damage is to strengthen the house using a ferrocement layer. The level of buildings damage based on the distance from the earthquake center and the implementation of retrofitting on the damaged houses are also discussed in this paper.

1. Introduction

Indonesia is one of the countries located in the highest seismic area in the world, which has experienced many strong earthquakes over the last few decades [1]. A series of earthquakes have struck several areas in Indonesia, such as Aceh in 2004 and 2016 [2], Bengkulu in 2007 [3], Yogyakarta in 2006 [4], West Sumatera in 2009 [5,6], Lombok 2018, and Central Sulawesi in 2018. An earthquake measuring M6.2 struck the West Pasaman Regency area, West Sumatra on Friday, February 25, 2022. A preliminary earthquake (foreshock) occurred at 08:35 am Western Indonesian Time. The United States Geological Survey (USGS) released that the predecessor earthquake occurred at coordinates 99.984°E and 0.222°N, with a magnitude of M5.0 at a depth of 10 km. Meanwhile, the main earthquake occurred at 08:39 am local time of Indonesia. The United States Geological Survey (USGS) released that the main earthquake occurred at coordinates 100.106°E and 0.233°N, with a magnitude of M6.2 at a depth of 12.3 km [7]. The earthquake area that shook around the epicenter is shown in figure 1.

The Meteorology, Climatology and Geophysics Agency (BMKG) recorded 201 aftershocks that occurred in West Pasaman Regency, West Sumatra, after an earthquake measuring M6.2 took place. The earthquake incident has caused a lot of losses, both loss of life and damages to infrastructure,

IOP Conf. Series: Earth and Environmental Science 1244 (2023) 012020 doi:10.1088/1755-1315/1244/1/012020

especially community houses. A few days after the earthquake, the National Agency for Disaster Countermeasure (BNPB) in collaboration with various universities in West Sumatra deployed 100 students majoring in civil engineering as a damaged house assessment team to the earthquake area in West Pasaman Regency, West Sumatra. The damaged house assessment team was dispatched directly to the affected area to collect data on the houses affected by the earthquake.

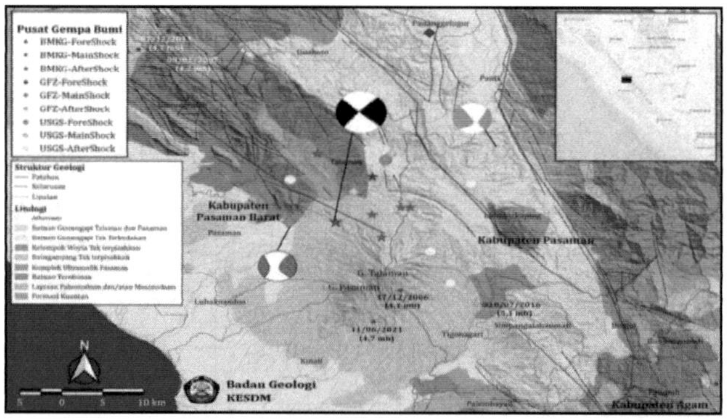

Figure 1. Map of the West Pasaman earthquake center.

This paper discusses the condition of damage to houses and analyze the damage to community housing structures after the earthquake disaster in West Pasaman Regency, as well as provide solutions to the problem of house damages in the form of a training on how to repair and strengthen earthquake-safe houses using the method of ferrocement layer to the local community, especially the local home builders.

2. Method

Method used in the implementation of this survey is field observations which are measured quantitatively and qualitatively. Qualitative measurement uses a check list that has been prepared, taking pictures and photo documents. Quantitative examination are in the forms of, namely; testing of concrete materials, building measurements, structural components and reinforcement dimensions. The data obtained from the field were evaluated based on the technical guidelines for assessing building damage due to earthquakes, with reference to predetermined criteria. The procedure afterwards is to conclude or determine the classification of damaged residential house condition [8].

The survey location for the assessment of damaged community houses is in the Nagari Kajai area, Talamau District, West Pasaman Regency, West Sumatra, as shown in figure 2.

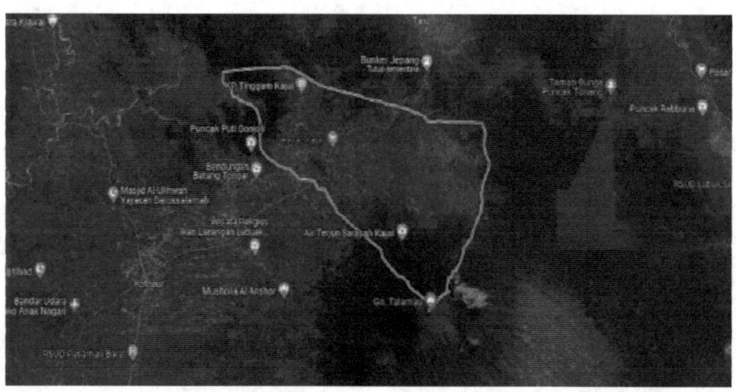

Figure 2. Map of the damaged house assessment survey location.

The implementation of the damaged house assessment started with the division of the assessment team. The team were divided into several groups to be distributed in the Nagari Kajai area, Talamau District, West Pasaman Regency. The data collection was carried out by visiting each of the resident house to perform a visual assessment of the damage to the house and immediately filling out the assessment form. In this study, the contents of the assessment form used referred to Technical Guidelines for Earthquake Resistant Houses and Buildings, Directorate General of Human Settlements, Indonesian Ministry of Public Works, 2006 [9]. The technical way to fill out the form was by filling in the identity of the house which includes the name of the house owner, the identification number, the address, the GPS coordinates, and the owner's phone number. The sketch of the house plan is made simply by describing the parts of the room in the house. After that, fill in the structural system used in the house and describe what potential hazards could occur around the location of the house. Furthermore, inspecting the general condition of the house, such as whether the building is still upright or has experienced a slope, inspecting the condition of structural damage, such as whether the foundation has settled, whether house walls and frames are cracked retaining walls, or whether the roof frame shifts or breaks, and finally non-structural damage, such as ceilings, roof coverings, doors, windows, and floor coverings. After the house damage checking, conclusions can be drawn on which categories of damage do the houses belongs to, such as whether heavy, moderate, light or no damage. Finally, a portrait of the building and detailing the damage as a whole is taken so that the evaluator gets a complete picture of the house being reviewed. The criteria for damage to residential buildings in the assessment form is shown in table 1.

Table 1. Criteria for damages to residential buildings due to earthquake [9].

No	Damage Category	Damage Criteria
1	Light Damage	a. The structure suffered minor damage < 30% b. Damages to building walls of 6 mm or frame crack smaller than 1 mm c. Non-structure component such as cracked walls not petrating, peeling plaster, celling and lisplank are slightly damage
2	Moderate Damage	a. The structure was damaged batween 30 – 50% b. Damages to the roof truss was more than 50%, and other criteria experienced minor damage > 50% c. Non-structure component such as plaster walls cracked or partially collapsed, plaster peeled off, ceiling and listplank were moderately damages
3	Heavy Damage	a. Global sloping building are more than 1° b. The structure is damaged more than 50% c. Non-structures component such as ceilings, doors, windows, roof coverings and floors are damaged a lot.

3. Results and discussion

3.1. Damages to comminunity house buildings

The survey on the damages to community houses due to the earthquake have been carried out, especially on the technical aspects of the damaged houses. Based on the survey results, the damages to community houses due to the West Pasaman earthquake are divided into non-structural and structural elements.

3.1.1. Non-structural element damages.

• Damages to the insulating/infilling brick walls. The survey results show that on average the houses affected by the West Pasaman earthquake, the bulkheads on the building structures including houses are made of non-reinforced brick walls and in some locations even non-reinforced brick columns are found, as shown in figures 3(a) and 3(b). Structural engineers generally ignore the contribution of infill walls in stiffening and bearing loads in designing the structure. Since they are not designed as structural elements, damage to infill or insulating masonry walls is generally unavoidable, especially in flexible and non stiffened frame systems with shear wall elements. When an earthquake occurs, most of the earthquake force is absorbed by the rigid wall structure because the structure have high stiffness. However, because the wall does not have sufficient strength to withstand the seismic forces it absorbs, the wall is mostly cracked. In some locations where the wall system was constructed without a complete and adequate bonding frame, the wall system was seen to be severely damaged, and even completely collapsed.

The survey result also found that there are many buildings whose roof level is not equipped with adequate roof beams. Sometimes the roof floor level is only given a ring of beams along the perimeter of the building, without being equipped with interior beams. As a result, there are many insulating walls between rooms that do not have tie beams above them, as shown in figure 3(c). Insulating walls like this are basically also prone to collapse. In addition, there are many conditions where the infill walls are completely detached from the binding frame or the frame system, namely practical beams and columns, which in fact are separated from the main frame of the building.

(a) (b) (c)

Figure 3. (a) Brick wall without reinforcement/binding, (b) non reinforced brick columns, (c) walls without tie beams.

• Damages to walls plaster and structural elements. Based on the survey results, on average, the damages to residential buildings generally occur in plastering walls and structural elements (columns and beams), as shown in figures 4(a,b,c). Damages to plastering walls or plastering beams/columns are basically unavoidable in a relatively flexible portal structural system. The damages that occur can also be exacerbated by the use of poor plastering materials and imperfect plastering methods. Damages to the stucco also occurred in several new houses, where the stucco reached a thickness greater than 7.5 cm. This is usually done to eliminate protrusions caused by the use of structural elements of different dimensions. Plaster of this thickness is basically very susceptible to shrinkage and cracking, especially if it is remembered that in usually plaster is not given steel reinforcement.

(a) (b) (c)

Figure 4. (a) Damage to wall plastering, (b) damage to column plastering, (c) damage to plastering beams.

3.1.2. Damage to structural elements. In general, the form of damage that occurs to the main structural elements due to earthquakes is mostly caused by inadequate strength, stiffness and ductility of the structural elements themselves in resisting earthquake loads as well as poor quality of materials and implementation [10]. The survey results found that in general, the average damage to structural elements occurred in the columns, beams and walls, as shown in figures 5(a,b,c). The damages are divided into 3 categories, namely light, moderate and heavy. For structural elements that suffered minor damages, the damages appear only in the form of cracks where the building was still standing firmly. Meanwhile, for structural elements with moderate damage, the damage was in the form of the destruction of the structural components. For the heavy or severe category, the damage was in the form of the destruction of all structural components, making the building collapsed.

(a) (b) (c)

Figure 5. (a) Damage to column, (b) damage to beam, (c) damage to walls.

3.2. The level of building damage based on the distance from the earthquake center

The results of the field survey show that the damage to community houses is more severe in areas that are close to the earthquake. Based on the results of a report by the National Agency for Disaster Countermeasure regarding data on damage to residential houses due to the 2022 West Pasaman earthquake, it was found that the total number of damaged residential buildings in West Pasaman Regency reached 6627 units, with details of 1075 units heavily damaged, 3447 units moderately damaged, and 2105 units lightly damaged. In order to investigate the correlation between the level of the house damages affected by the earthquake and the distance of the houses from the epicenter, the areas affected by the earthquake were divided into three areas covering the farthest (15 km), medium (10 km), to the closest (5 km) areas from the earthquake epicenter, as shown in figure 6.

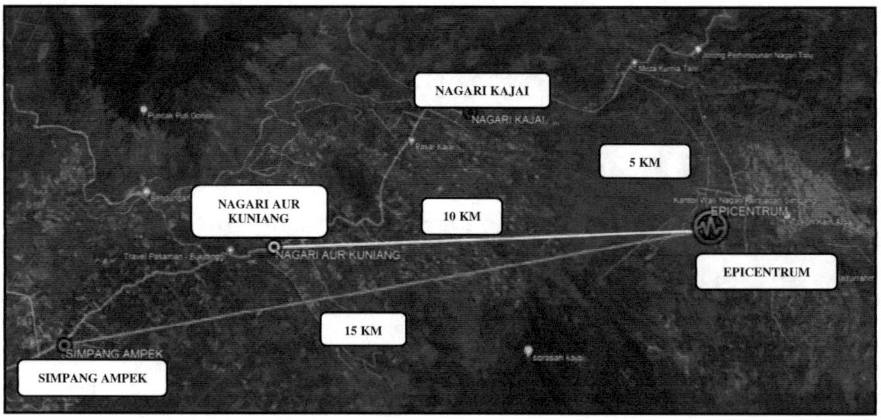

Figure 6. Map of the distance of the area from the epicenter of the earthquake.

3.2.1. Community house damage survey in Nagari Kajai. Nagari Kajai is the area closest to the epicenter of the earthquake. It is estimated that the distance of the area is about 5 km from the epicentre of the earthquake. Most of the heavily damaged houses were located in Timbo Abu Village. Based on observations, the condition of community houses due to the earthquake mostly suffered from heavy or severe damage, as shown in figure 7(a). The damages to the building occurs in all of its components which include damages to columns, beams, walls, ceilings, roofs and doors and windows (figure 7(b)). Most houses only have roof coverings that appear to cover the house while the other components are shattered. The condition of the foundation and floor of the house is in good condition with no damage.

(a) (b)

Figure 7. (a) Heavy damage building, (b) buildings in which all building components have been destroyed.

In addition to many heavily damaged community houses in areas close to the epicenter of the earthquake, there were also community houses that were still standing firmly and only suffered minor damage such as small cracks in the walls, as shown in figure 8(a). The quality of planning, construction implementation, site selection and selection of good building materials so that it is resistant to earthquake shocks are among the factors making these houses could resist the earthquake. As for the cause of the collapsed building, it is estimated that it did not pay attention to the four conditions above, as shown in figure 8(b).

(a) (b)

Figure 8. (a) Building community houses that are still standing strong, (b) collapsed community houses.

3.2.2. Community houses damage survey in Aur Kuniang. Nagari Aur Kuniang is an area located 10 km from the epicenter of the earthquake. Based on visual observations, the condition of community houses in the area is visibly quite severe, where most of the houses suffered moderate and light damage, and there were several houses that suffered heavy damage. Damages to the house are mostly found on the walls, ceiling and roof covering, as shown in figure 9. As for the condition of the house, the houses should be repaired and strengthened before it was reoccupied.

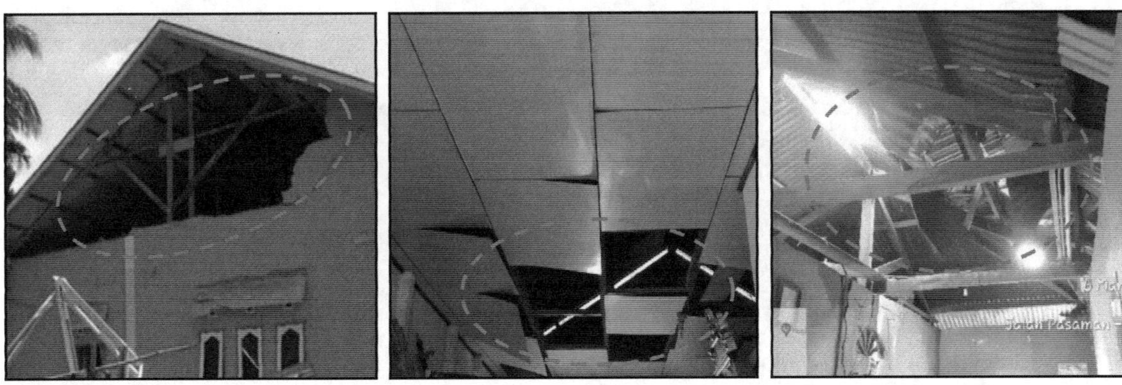

(a) (b) (c)

Figure 9. (a) Damage to walls, (b) damage to the ceiling, (c) damage to the roof covering.

3.2.3. Community houses damage survey in Simpang Ampek. Simpang Ampek is an area located about 15 km from the epicenter of the earthquake. Based on the results of observations, the houses in the area only experienced very small damage, even if they were not careful in looking at the condition of the house, no damage can be visible. Damages to the walls of the building marked by hair cracks is shown in figures 10(a) and 10(b). Repairs that can be done is patching the damaged wall. The houses in this lightly damaged category can be occupied directly by the owner even though the damaged components have not been repaired.

(a) (b)

Figure 10. (a) Small carck in the wall, (b) long crack in walls.

3.3. Solutions for strengthening and repairing damages to community house buildings due to the earthquake.

The solution that can be done to overcome the damaged house problems by the earthquake is repairing and strengthening using the ferrocement layer method. Ferrocement is a composite material made from a mixture of sand and cement using wire mesh which can form a unified whole and form a fairly strong construction. The advantages of using this method are the raw materials that can be easily obtained in all regions, relatively cheap processing costs, and very easy to do by the manual workers or home builders [11,12].

The procedure to repair and strengthen the crack wall using the ferrocement layer method is shown in figure 11:

a. The first step, before starting the repair and strengthen of houses, is to locate the point or location of the cracked wall. Draw a line about 20-30 cm from the crack as a benchmark. Then, peel off the old plaster with a concrete chisel or a drill with an iron chisel.
b. The cracked parts are poured with cement paste in order to strengthen the bond with the plaster later.
c. Cut the woven wire as needed, then attach the woven wire to the wall using roofing nails.
d. Finally, plaster the area that has been given woven wire until the whole surface is covered.

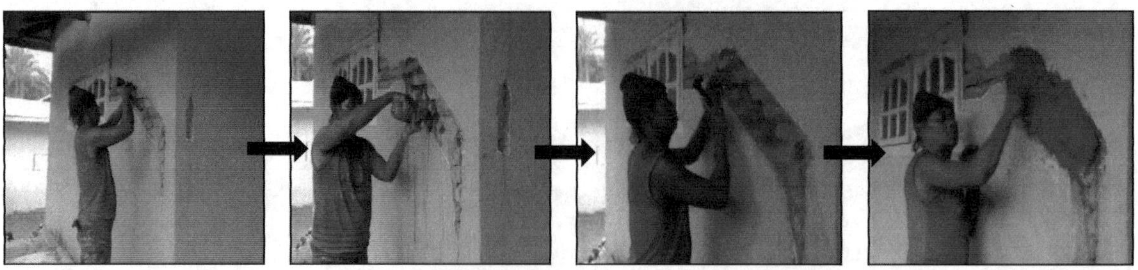

Figure 11. (a) Stripping old plaster, (b) filling of cracks with liquid cement, (c) laying of chicken wire, (d) plastering the repaired wall.

Before implementing the repair and strengthen the crack wall using ferrocement layer method, a training for the local community especially construction workers (builders) was carried out. The training was conducted in the Nagari Kajai area, Talamau District, West Pasaman Regency. This training activity begins with the explanation and presentation of theory and ends with the implementation of retrofitting practice, as shown in figure 12. Result of the training shows that local workers easily understood the method and were ready to implement the method to repair the damaged houses in West Pasaman Regency in order to speed up the rehabilitation and reconstruction post-earthquake.

(a) (b)

Figure 12. (a) Presentation of the training theory by the Andalas University PSB team, (b) implementation of retrofitting practices.

4. Conclusion

Based the results of assessment of the non-engineering house damage after the February 25th, 2022 West Pasaman Earthquake, it can be concluded that:

- The impact of the West Pasaman earthquake mostly occurred in community houses, in which many houses suferred from damage even collapse. The total number of damaged residential buildings in West Pasaman Regency reached 6627 units with details of 1075 heavily damaged units, 3447 moderately damaged units, and 2105 lightly damaged houses.
- There are two types of damage that occurred in community houses, namely the structural and non-structural damage. The damages to the building structure include damage to walls that have cracked or collapsed, damages to columns and beams that have been cracked or destroyed, and damages to foundations that have cracked or fallen. Meanwhile, non-structural damages to buildings include damages to ceilings, roof coverings, windows, doors and floors that were destroyed.
- The main causes of damages to community houses are due to human error factors such as lack of planning, construction implementation, site selection and the quality of building materials that are not in accordance with earthquake safe house building standards.
- The damages were affected by the distance from the earthquake epicenter, in which that the number of damaged houses close to the earthquake epicenter was greater than those that are far from the earthquake epicenter.
- Repair and retrofitting with the ferrocement layer method is one of the best solutions to overcome problems related to damage to community houses, because this method is easy to implement by workers or home builders using relatively cheap and available materials in the local area.

Acknowledgments
The authors gratefully acknowledge the financial support from Andalas University, which has funded this research. The authors also acknowledge National Agency for Disaster Countermeasure (BNPB) which has given an opportunity to conduct this research.

References

[1] Maidiawati, Tanjung J, Sanada Y, Nugroho F and Wardi S 2020 Seismic Analysis of Damaged Buildings Based on Post-Earthquake Investigation of the 2018 Palu Earthquake. Int. *J. of Geomate* **18**(70) 116-122

[2] Muzli, Umar M, Nugraha A D, Bradley K E, Widiyantoro S, Erbas K, Jousset P, Rohadi S, Nurdin I and Wei S 2018 The 2016 Mw 6.5 Pidie Jaya, Aceh, North Sumatra, Earthquake: Reactivation of an Unidentified Sinistral Fault in a Region of Distributed Deformation. *Seismological Research L* **20**(20)

[3] Maidiawati and Sanada Y 2008 Investigation and Analysis of Buildings Damaged during the September 2007 Sumatra, Indonesia Earthquakes, *J. of Asian Architecture and Building Engineering* **7**(2) 371-378

[4] Meireles H, Bento R 2013 Rehabilitation and strengthening of old masonry buildings, Technical Report ICIST DTC 02/13, ICIST-IST, Lisbon

[5] Choi H, Sanada Y, Kashiwa H, Watanabe Y, Tanjung J and Jiang H 2016 Seismic Response Estimation Method for Earthquake Damaged RC Buildings, *Earthquake Eng and Struc Dynamics* **45**(6) 999-1018

[6] Learning from Earthquakes 2009 The Mw 7.6 Western Sumatra Earthquake of September 30, 2009. *EERI Special Earthquake Report* (2009)1-12.

[7] Ministry of Energy and Mineral Resources of the Republic of Indonesia 2022 *Earthquake disaster report in West Sumatera* (Jakarta: Center for Volcanology and Geological Hazard Mitigation)

[8] Rakhman J 2008 *Pattern of Non-Structural Building Failure Due to Shallow Earthquake, 6.5 on the Richter Scale* (Bandung: Settlement Research and Development Center)

[9] Directorate General of Human Settlements, Indonesian Ministry of Public Works, 2006

[10] Antonius, Adhy D S, and Ruslan 2006 Study of Factors Affecting the Level of Building Damage Due to Earthquake (Case Study of Earthquake in *J. of Lecturer of Civil Engineering Department UNISSULA.*

[11] Rahman A, Bachtiar G, Daryati 2013 Study of the Flexural Strength of Ferrocement Plates with Aluminum Sheet Layers as Fixed Formwork on High-rise Building Floor Plates *J. of Menara Civil Engineering Department Faculty of Engineering* **8** 1

[12] Apriwelni S, Ismail F and Hakam A 2017 Numerical Study of Red Brick House Reinforcement Using Ferrocement Layers *In: Hidayat, B and Purnawan, P (Eds.) Proceeding of the 4 th Andalas Civil Engineering (ACE) Conference 2017, November 9, Andalas University, Padang, Department of Civil Engineering*, pp 583-593

IOP Conf. Series: Earth and Environmental Science 1244 (2023) 012021 doi:10.1088/1755-1315/1244/1/012021

Tunnel Stability Analysis Under Seismic Load Using Finite Element Method: A Case Study of Spillway Tunnel, Sidan Dam, Bali, Indonesia

Benny Arianto[1,2], I Gde Budi Indrawan[1], and I Wayan Warmada[1]

1. Gadjah Mada University, Geological Engineering Department, Yogyakarta, Indonesia
2. Ministry of Public Works and Housing of Indonesia

*E-mail: igbindrawan@ugm.ac.id

Abstract. The location of the Sidan Spillway Tunnel, as shown on the Earthquake Hazard Zone Map of Bali Island, is in a moderate hazard zone. The support system of the spillway tunnel of the Sidan Dam, Indonesia, was empirically designed based on rock mass classification systems. The objective of this research was to verify the performance of the proposed support system by a numerical method considering the uncertainty in using rock mass classification systems for designing a tunnel support system. In addition, being located in an active seismic region, the seismic load was not considered in the empirical tunnel support design. The proposed tunnel support design was numerically modeled in two dimensions using a finite element method and subjected to a seismic load obtained from probabilistic seismic hazard analysis. A numerical analysis without seismic load was also conducted for comparison. Phase2 software was used for the tunnel stability analysis. The input parameters in the numerical analysis were derived from field investigations (surface engineering geological mapping and evaluation of drilling cores) and laboratory tests. The spillway tunnel area consists of pyroclastic rocks, and the Generalized Hoek-Brown Failure Criterion was selected for the strength parameter of rocks. The suggested support designs are systematic rock bolts (20 mm diameter fully grouted), wired mesh, shotcrete, and steel ribs at specific locations. The numerical models determined that applying the proposed tunnel support system recommended based on the empirical method resulted in a roof strength factor of more than 1.5 under static load & 1.1 under seismic load and roof displacements lower than a 10 cm maximum displacement at every stage of construction. The results verify that the tunnel construction will be stable and fulfill the required safety standard, both under and without seismic loads.

1. Introduction

Tunnel construction for the spillway of Sidan dam positioned at Gianyar Regency, Bali Province, Indonesia. The lithology in this area is composed of pyroclastic rocks which are relatively weak rocks. According to the Earthquake Hazard Zone Map of Bali Island, the location of the tunnel is in a moderate hazard zone [1] (Figure 1). The tunnel has the potential to be impacted by quakes of magnitude VI on the Modified Mercalli Intensity Scale (MMI) and earthquake sources with moderate depth (30 to 80 km). Initially, this tunnel's design planning employs an empirical approach that relies on the Rock Mass Rating (RMR) classification system in the detailed designing stage, however, the problem is considerably more complicated, involving earthquake factors, changes in pore pressure, plasticity, lining deformations, and existing structures. Therefore, numerical analysis is required in order to evaluate the design and verify tunnel construction safety. This study aims to quantitatively evaluate the assessed tunnel support design based on an empirical methodology.

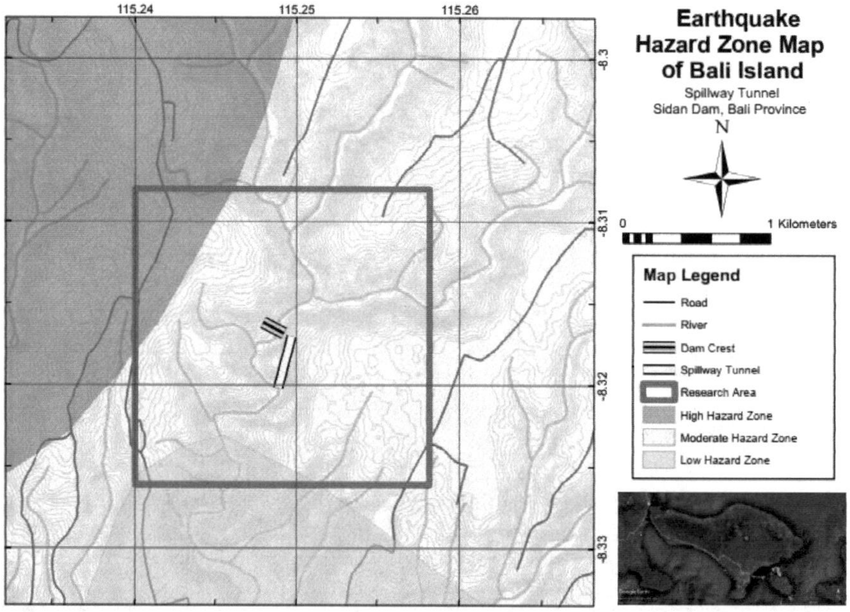

Figure 1. Hazard zone of research area map [1]

The finite element method is frequently used in tunnel construction analysis since it can analyze more challenging problems, such as: simulating construction sequences, dealing with varied ground conditions, modeling realistic soil behavior, handling complicated hydraulic conditions, accounting for adjacent services and structures, simulating intermediate and long-term conditions, and dealing with multiple tunnels [2]. In previous tunnel projects, this method has been applied in evaluating the effectiveness of the design and mitigating the earthquake effect on the road [3], rail [4], and dam [5] tunnels in Indonesia.

2. Methodology

During the tunnel design process, the rock mass classification and geological cross-section profile were obtained by processing conducted field observation and drilling reports. The tunnel support system was evaluated by the RMR [6], Q-System [7], RMi [8], GSI [9], and JSCE [10] classifications.

This research implements the 2D Finite Element Method (FEM) associated with the Phase2 software (Rocscience Inc.). The numerical analysis was conducted prior to the commencement of tunnel excavation. The Generalized Hoek-Brown parameters, deformation modulus, uniaxial compressive strength, Poisson's Ratio, and vertical and horizontal stresses (in situ) have been utilized to undertake a numerical analysis.

2.1. *The Generalized Hoek-Brown failure criterion*

The Generalized Hoek-Brown criterion to quantify rock mass strength is expressed as below formula (1). The criterion was chosen due to its applicability to extremely weak rock masses and the computation of comparable Mohr-Coulomb parameters from the Hoek-Brown failure envelope [11]. Phase2 simulates the material failure by initially yielding the intact strength and dropping to a lower residual state.

$$\sigma_1' = \sigma_3' + \sigma_{ci}\left(m_b \frac{\sigma_3'}{\sigma_{ci}} + s\right)^a \qquad (1)$$

Where σ_1' and σ_3' are the major and minor effective principal stresses at failure, σ_{ci} is the uniaxial compressive strength of the intact rock material, m_b is the decreased value of the material constant m_i that calculated by considering the Geological Strength Index (GSI) rating of the rock and disturbance factor (D) (2). D is a variable that depends on the degree of disturbance caused by blast damage and stress relaxation to the rock mass. It ranges from 0 for intact rock masses in situ to 1 for extremely disturbed rock masses. *s* and *a* are constants for the rock mass given by the following relationships.

$$m_b = m_i \exp\left(\frac{GSI-100}{28-14D}\right) \quad (2)$$

$$s = \exp\left(\frac{GSI-100}{9-3D}\right) \quad (3)$$

$$a = \frac{1}{2} + \frac{1}{6}\left(e^{-GSI/15} - e^{-20/3}\right) \quad (4)$$

The rock mass modulus of deformation that has been used for numerical analysis was derived by the Generalized Hoek-Diederichs [12] method, expressed in equation 5. The residual strength parameter of the rock was determined by the relationship between GSI residual/GSI and GSI (6) [13].

$$Em = Ei\left(0.02 + \frac{1-D/2}{1+e^{\left(\frac{60+15D-GSI}{11}\right)}}\right) \quad (5)$$

$$GSI\ residual = GSI\ x\ e^{(-0.0134\ GSI)} \quad (6)$$

2.2. Field Stress

Tunnel stability analysis necessarily requires the use of field stresses to characterize the in situ stresses that occur. This in situ stress is measured as horizontal (σ_h) and vertical (σ_v) two-way stresses in the rock. In determining σ_v, the depth of the overburden plays a role (7), whereas the stress ratio is applied to compare the K ratio (8) with Sheorey's model when determining σ_h [14]. It is presumed that the vertical stress increases linearly with depth due to the weight of the overburden.

$$\sigma_v = \gamma \times H \quad (7)$$

$$K = 0.25 + 7Em\ (0.001 + \frac{1}{H}) \quad (8)$$

2.3. The Seismic Load

The seismic load coefficient of the site study was derived using pseudo-static analysis based on SNI 8460: 2017 [15] and a peak acceleration map in bedrock (PGA) for a 7% probability of exceeding 75 years by the Ministry of Public Works and Public Housing of Indonesia [16]. Peak Ground Acceleration (PGA) is the maximum acceleration of the ground caused by earthquake shaking at a particular location. PGA is obtained by identifying seismic zones according to the Indonesian seismic code and ground types based on the average standard penetration test. However, the maximum considered earthquake geometric mean PGA adjusted for site class effects (PGA_m) is the value that is applied in generating the stability model of the tunnel (9). Site coefficients (F_{PGA}) were empirically determined from the site class of the research area and PGA value.

$$PGA_m = F_{PGA}\ x\ PGA \quad (9)$$

3. Results & Discussions

3.1. Engineering Geology Characteristics and Rock Mass Classifications

Based on engineering geology surface mapping and core analysis, the tunnel is located in the pyroclastic breccia rock unit (Figure 1), which is described as follows: blackish, lapilli to block-sized (2-30 cm) angular lithic and scoria fragments, poorly sorted, matrix ash sized supported and fining upward. As the matrix fraction increases, the upper portion of this unit can be identified as sub-unit tuff breccia. The results of the rock mass classification are included in Table 1. The rock properties utilized in numerical modeling are the results of laboratory tests conducted on samples from the tunnel portal, which are interpreted to represent the rock units' properties (Table 2).

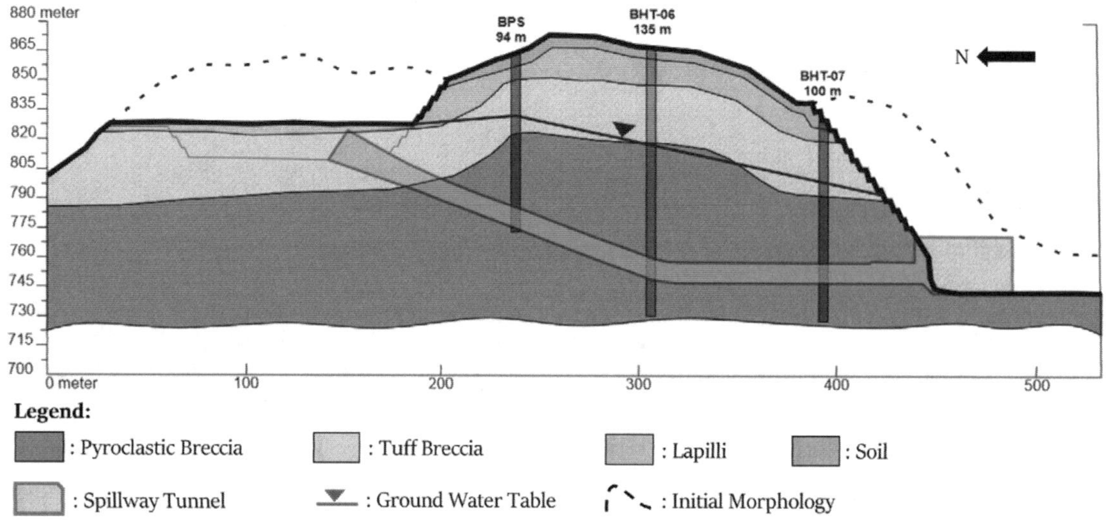

Figure 2. Spillway tunnel lithology cross-section

Table 1. Rock mass classification identifications at tunnel elevation

No.	Borehole	RMR Class	Q-System Class	RMi Class	JSCE Class	GSI Rating
1	BPS	Fair	Poor	Medium	D II	46
2	BHT 6	Fair	Very Poor	High	C II	45
3	BHT 7	Fair	Poor	High	C II	51

Table 2. Rock units' properties

No.	Lithology	UCS (MPa)	Poisson Ratio	Modulus Intact MPa	Cohesion MPa	Friction Angle (^{0})	Unit Weight Dry MN/m^3	Unit Weight Saturated MN/m^3	mi	D
1	Soil	-	0.3	-	0.02	43.92	0.011	0.015	-	-
2	Lapilli	1	0.3	-	0.07	33.00	0.005	0.011	13	0
3	Tuff Breccia	7.56	0.07	213	0.24	38.66	0.010	0.015	19	0
4	Pyroclastic Breccia	20.58	0.03	2082	0.203	34.99	0.015	0.017	19	0

3.2. Tunnel Support Design

The recommendations for the tunnel wall and roof primary reinforcement system using shotcrete and rock bolts are presented in Tables 3 through 5. The crown shotcrete thickness of the tunnel varies between 60 and 200 millimeters, and the wall shotcrete thickness varies between 30 and 400 millimeters. In addition, the rock bolt spacing and length range from 2.6 to 6 meters and 1 to 2.5 meters, respectively, for the tunnel's roof and wall supports. Shotcrete reinforced with wire mesh is implemented in accordance with System Q, RMR, and RMi recommendations.

Table 3. Crown and wall shotcrete design

		Support							
No.	Borehole	Crown shotcrete thickness (mm)				Wall shotcrete thickness (mm)			
		RMR	Q-System	JSCE	RMi	RMR	Q-System	JSCE	RMi
1	BPS	50-100	100	250	100	30	60	400	55
2	BHT 6	50-100	100	150	60	30	60	400	40
3	BHT 7	50-100	90	150	60	30	60	400	40

Table 4. Roof rock bolt design

		Roof support							
No.	Location/ Borehole	Rock bolt length (m)				Rock bolt spacing (m)			
		RMR	Q-System	JSCE	RMi	RMR	Q-System	JSCE	RMi
1	BPS	4	4	6	3.3	1.5-2	1.7	1	1.25
2	BHT 6	4	4	4	3.3	1.5-2	1.6	1.2	1.5
3	BHT 7	4	4	4	3.3	1.5-2	1.8	1.2	1.5

Table 5. Wall rock bolt design

		Wall support							
No.	Location/ Borehole	Rock bolt length (m)				Rock bolt spacing (m)			
		RMR	Q-System	JSCE	RMi	RMR	Q-System	JSCE	RMi
1	BPS	4	2.6	6	2.6	1.5-2	1.9	1	1.5
2	BHT 6	4	2.6	4	2.6	1.5-2	1.9	1.2	2.5
3	BHT 7	4	2.6	4	2.6	1.5-2	1.9	1.2	2.5

The support properties used in the numerical analysis were derived from the software Phase2 database which partially has been modified based on the design plan, as shown in Table 6. In numerical modeling, the support is classified as an elastic material. On the basis of the evaluation of the JSCE system, it is suggested that H200 (DII) and H150 (CII) steel sets be used to reinforce the tunnel's arch and walls.

Table 6. Support properties

Shotcrete	Wire Mesh	Rock Bolt	Steel Rib H-200 (DII JSCE's Class)	Steel Rib H-150 (CII JSCE's Class)
Young's Modulus (Mpa): 24000	Diameter (mm): 12	Type: Fully Bonded	H-Beam (W): W200x100	H-Beam (W): W150x37.1
Poisson Ratio : 0.15	Moment Inertia (m^4): $1.02x10^{-9}$	Diameter (mm): 20	Moment Inertia (m^4): $1.13x10^{-4}$	Moment Inertia (m^4): $2.22x10^{-5}$
Compressive Strength (MPa): 25	Young's Modulus (MPa): 200000	Young's Modulus (MPa): 200000	Young's Modulus (MPa): 200000	Young's Modulus (MPa): 200000
Tensile Strength (MPa): 1.25	Poisson Ratio : 0.25	Tensile Capacity (MN): 0.178	Poisson Ratio : 0.25	Poisson Ratio : 0.25
	Compressive Strength (MPa): 400		Compressive Strength (MPa): 400	Compressive Strength (MPa): 400
	Tensile Strength (MPa): 400		Tensile Strength (MPa): 400	Tensile Strength (MPa): 400

IOP Conf. Series: Earth and Environmental Science 1244 (2023) 012021 doi:10.1088/1755-1315/1244/1/012021

3.3. Finite Element Analysis

Phase2 automatically calculates failure criterion properties based on intact rock properties, GSI values, and the disturbance factor (Table 7). K Ratio and PGAm were derived from formulas 8 and 9. Utilizing Mohr & Coloumb and Generalized Hoek Brown failure criteria, numerical modeling classifies soils and rocks as plastic materials.

Table 7. Failure criterion, field stress, and seismic load properties

No.	Borehole	Rock Mass GSI	GSI Residual	m_b	s	A	H (m)	E_m (MPa)	K Ratio	PGA_m	Material Type
1	BPS	45	24.6	2.665	0.0022	0.51	73	466	0.30	0.44	Plastic
2	BHT 6	41	23.7	2.310	0.0014	0.51	108	356	0.28	0.44	Plastic
3	BHT 7	45	24.6	2.665	0.0022	0.51	69	466	0.30	0.44	Plastic

A finite element analysis is conducted at different stages of implementation, including the head excavation and support stage, the bench excavation and support stage, and the post-construction stage under seismic load (Figure 3). The results and analysis of drilling data are used to determine the thickness of the soil and rock layers and the position of the groundwater table. The results of numerical analysis concentrate on the value of the total displacement (TD) and strength factor on the tunnel's roof (RS), as shown in Figure 4, and the results of each model are summarized in Table 8.

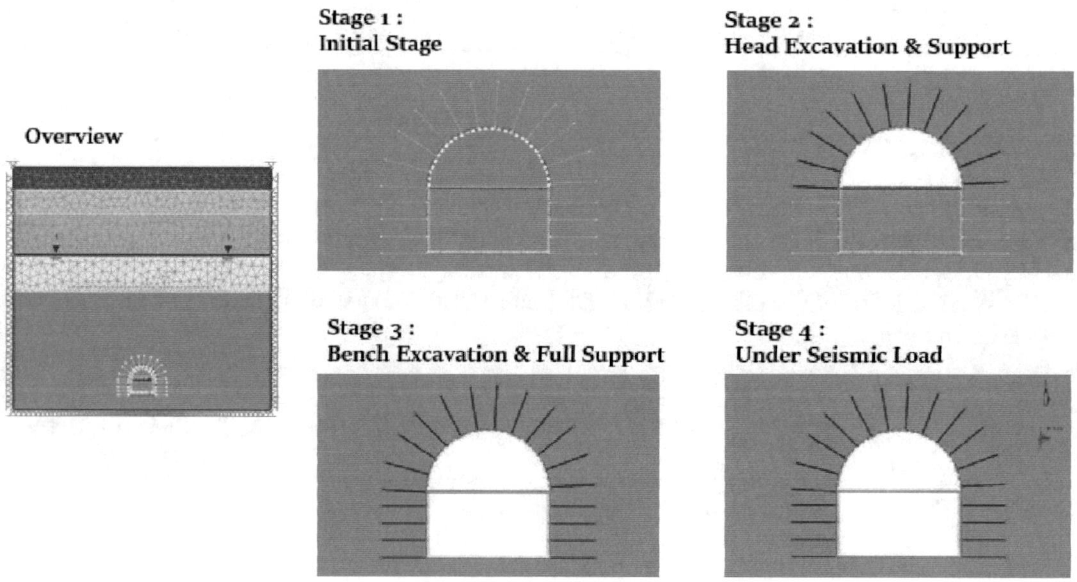

Figure 3. Modelling stages

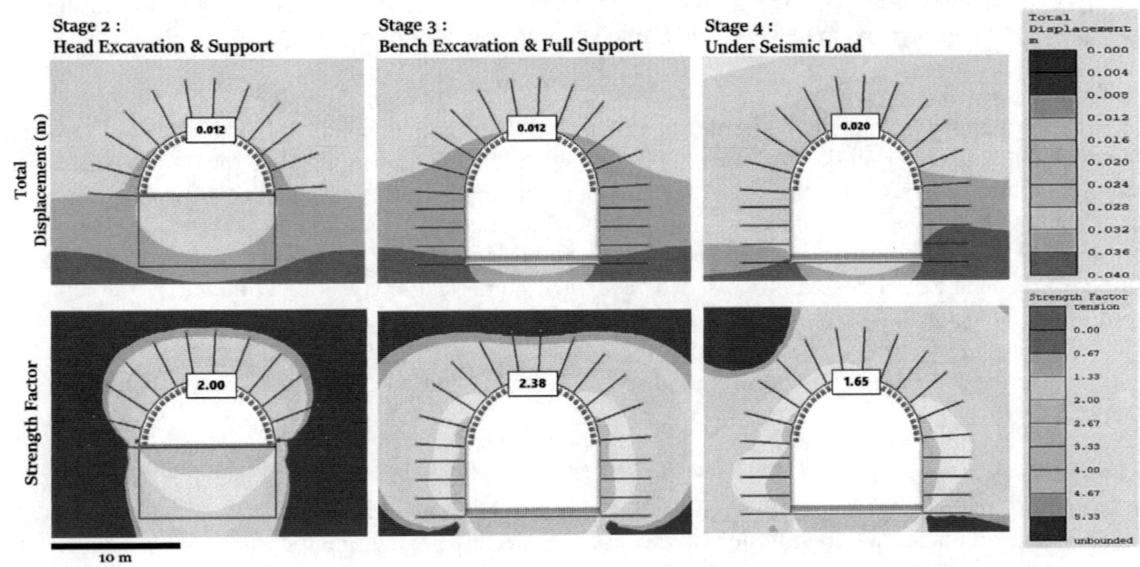

Figure 4. Total displacement (TD) and roof strength factor (RS) model (BPS section-RMR)

Table 8: Finite element analysis results

No.	Stages	Rock Mass Classification Support Design							
		RMR		Q-System		RMi		JSCE	
		TD (cm)	RS	TD (cm)	RS	TD (cm)	RS	TD (cm)	RS
BPS									
1	Head Excavation & Support	1.2	2.00	1.2	2.15	1.2	2.05	1.4	4.38
2	Bench Excavation & Full Support	1.2	2.38	1.2	2.52	1.2	2.44	1.5	9.32
3	Under Seismic Load	2.0	1.65	2.0	1.72	2.0	1.69	2.1	1.69
BHT-6									
1	Head Excavation & Support	2.7	2.58	2.7	2.63	2.5	1.80	3.0	4.23
2	Bench Excavation & Full Support	2.6	3.08	2.6	3.19	2.4	1.97	3.1	6.57
3	Under Seismic Load	3.0	2.79	3.0	2.89	2.9	1.92	3.4	4.20
BHT-7									
1	Head Excavation & Support	2.4	2.16	2.3	2.00	2.3	1.76	2.5	3.48
2	Bench Excavation & Full Support	2.3	2.62	2.4	2.41	2.3	1.99	2.6	5.12
3	Under Seismic Load	3.2	1.86	3.2	1.83	3.2	1.67	3.3	2.01

According to the assessments of the finite element analysis, several locations have a potential total displacement range of 1.2 to 3.4 cm, with the roof strength factor value varying between 1.65 and 9.32. The BHT6 bore with the highest overburden value has a higher potential for total displacement than the location with the lower overburden value in the static condition, however, the deeper the tunnel, the less impact the seismic load will have on the tunnel's stability. The use of a tunnel support system based on the JSCE recommendation yields a greater roof strength factor than other recommendations because this system provides a greater number of support systems than other systems. The influence of seismic load conditions has a negative effect on the total displacement value and roof strength factor by increasing the potential roof displacement by 10-67% and decreasing the roof strength factor by 3% up to 82%, relative to static load conditions.

4. Conclusions

In general, the analysis results determine that the total displacements below 10 cm and roof strength factor above 1.5 under static load and 1.1 under seismic load, indicating that the tunnel will be in stable condition, despite the fact that this tunnel is situated in an area that is moderately prone to earthquakes hazard. The numerical results validate that the empirical method for selecting reinforcement and excavation methods based on rock mass classification systems can be implemented safely during the tunnel design process. However, the numerical analysis illustrates that although the roof strength factor is still above the safety threshold, there is still an indication of the potential for displacement of the tunnel roof. Hence, the installation of additional reinforcement and further analysis during the excavation stage is required to minimize the potential for the displacement of the tunnel.

Acknowledgments

The authors express their gratefulness to Balai Wilayah Sungai Bali Penida for their assistance, management support, and permission to access the data. In addition, the first authors gratefully acknowledge the Indonesian Ministry of Public Works and Housing for granting the master's study program scholarship.

References

[1] Supartoyo, Suantika G and Djaja 2009 Peta Kawasan Rawan Bencana Gempabumi Pulau Bali *Kementerian Energi dan Sumber Daya Mineral Indonesia*
[2] Potts D M, Zdravković L, Addenbrooke T I, Higgins K G and Kovačević N 2001 *Finite element analysis in geotechnical engineering: application* vol **2** (Thomas Telford London)
[3] Nugraha J A N, Indrawan I G B and Karnawati D 2019 Numerical Evaluation of Earthquake Effect on Cisumdawu Tunnel Stability *Journal of Applied Geology* **4** 73–81
[4] Pramono R, Indrawan I G B and Junica M I 2021 Numerical evaluation of support design: A case study Jakarta–Bandung, Indonesia High-Speed Railway Tunnel 7 *IOP Conference Series: Earth and Environmental Science* vol **871** (IOP Publishing) p 012062
[5] Fatkhiandari I A, Indrawan I G B and Karnawati D 2020 Numerical Evaluation of Tunnel Portal Slope Stability at Bagong Dam Site, East Java, Indonesia *Journal of Applied Geology* **5** 40–52
[6] Bieniawski Z T 1989 *Engineering rock mass classifications: a complete manual for engineers and geologists in mining, civil, and petroleum engineering* (John Wiley & Sons)
[7] Barton N, Lien R and Lunde J 1974 Engineering classification of rock masses for the design of tunnel support *Rock mechanics* **6** 189–236
[8] Palmström A 1996 Characterizing rock masses by the RMi for use in practical rock engineering: Part 2: Some practical applications of the rock mass index (RMi) *Tunnelling and underground space technology* **11** 287–303
[9] Hoek E, Carter T G and Diederichs M S 2013 Quantification of the geological strength index chart *47th US rock mechanics/geomechanics symposium* (OnePetro)
[10] Japan Society of Civil Engineers 2007 Standard Specifications for Tunneling—2006: Mountain Tunnels *Tokyo, Japan: Japan Society of Civil Engineers*
[11] Hoek E, Carranza-Torres C and Corkum B 2002 Hoek-Brown failure criterion-2002 edition *Proceedings of NARMS-Tac* **1** 267–73
[12] Hoek E and Diederichs M S 2006 Empirical estimation of rock mass modulus *International journal of rock mechanics and mining sciences* **43** 203–15

[13] Cai M, Kaiser P K, Tasaka Y and Minami M 2007 Determination of residual strength parameters of jointed rock masses using the GSI system *International Journal of Rock Mechanics and Mining Sciences* **44** 247–65
[14] Sheorey P R 1994 A theory for in situ stresses in isotropic and transverseley isotropic rock *International journal of rock mechanics and mining sciences & geomechanics abstracts* vol **31** (Elsevier) pp 23–34
[15] Standar Nasional Indonesia 2017 Persyaratan perancangan geoteknik *SNI* 8460 2017
[16] Tim Pusat Studi Gempa Nasional 2017 Peta sumber dan bahaya gempa Indonesia tahun 2017 *Kementerian Pekerjaan Umum dan Perumahan Rakyat*

Identifying the potential development of HAZUS model as an earthquake disaster loss model for school buildings in Indonesia

R Milyardi[1,2*], **K S Pribadi**[1], **I Meilano**[3], **E Lim**[1]

[1]Civil Engineering, Faculty of Civil and Environmental Engineering, Institut Teknologi Bandung, Ganesha St. no.10, Bandung, 40132, West Java, Indonesia
[2] Civil Engineering, Faculty of Engineering, Universitas Kristen Maranatha, Surya Sumantri St. no. 65, Bandung, 40164, West Java, Indonesia
[3] Geodesy and Geomatics Engineering, Faculty of Earth Science and Technology, Institut Teknologi Bandung, Ganesha St. no.10, Bandung, 40132, West Java, Indonesia
*roimilyardi2@gmail.com

Abstract. Located in a complex tectonic zone, Indonesia is prone to earthquakes, which cause considerable damages to houses and buildings and lead to significant economic losses. One of the most significant damages to public facilities is the destruction of school buildings. In addition to hazard factor, poor building structure in most of Indonesia's school buildings have contributed to increased earthquake disaster losses as a vulnerability component. To reduce disaster losses, quantifying earthquake disaster losses was conducted as an initial process. In practice, Indonesian government and Indonesia's National Disaster Management Agency (BNPB) calculate disaster losses at the time of post-disaster using various methods. Such calculation method is generally time-consuming, which may impeded the disaster response process. Thus, pre-disaster disaster losses quantification method could be a solution to this problem, which have not been developed and widely applied in Indonesia. HAZUS model is used for estimating earthquake disaster losses in the pre-disaster phase and has been widely used based on the building performance developed by FEMA, USA. This study aims to identify the potential for developing the HAZUS model as a tool for estimating disaster losses in school buildings in Indonesia. The identification was conducted in four stages. The analysis of HAZUS models consisted of hazard analysis, capacity analysis, damage analysis, and losses analysis. A case study was conducted on disaster losses potential mapping in Cicendo District, Bandung City, Indonesia. On the hazard analysis, hazard data were applied on the basis of the Indonesian earthquake hazard and 2017 source map and building code of SNI 1726:2019 as response spectrum procedure. In the analysis of the building capacity, potential development was identified based on capacity building parameters (Ay, Dy, Au, Du) specific to the typical school building in Indonesia with a survey approach and analysis of existing buildings. Concrete Frame with infill masonry shear walls (C3) and Reinforced Masonry Wall with flexible diaphragms (RM1) were identified as the types of structures, which were close to the typical school buildings in Indonesia. The difference of local structure characteristics was identified as the potential to develop fragility curves of local structure on the damage analysis. In the losses analysis, four components of economic loss were applied, consisting of building replacement and repairment of losses, income losses, and building relocation losses. The potential and challenges lie in the use of limited taxonomic data on buildings and school administration data, as well as in the historical data approach to the costs and duration of school building repairs.

1. Introduction

Indonesia lies above a complex tectonic zone, where the Eurasian, Indo-Australian, Philippine and Pacific Plates meet, resulting in significant seismic hazards across the country. Earthquake records of in the last 30 years (2001-2020) show that earthquake occurrence was 1% of the total disaster event, but it could cause the most severe building damage and death toll [1]. Out of the damaged public facilities due to earthquakes, the number of damage school building was highest with 23.552 buildings from 2000-2021 [2]. Beside the significant seismic hazard, the school building condition contributed to the building damage and losses. In 2020, out of the 218,234 school buildings, only 13.59%-29.88% (from elementary to the high school level) were categorized in good condition by the Indonesian Ministry of Education and Culture (MEC), while the others were categorized as slightly damaged to extensively damaged buildings [3]. BNPB calculated disaster losses in Indonesia based on the qualitative method in the post disaster event. This practice gave under-estimated disaster losses and is time-consuming, which may impede disaster response process, particularly in terms of financial preparedness in dealing with disasters. Accurate disaster loss estimation has become one of the key factors of disaster resilience, specifically financial protection strategies against disaster [4,5].

Pre-disaster disaster losses quantification method could be a solution to obtain more accurate losses estimation, which has not been developed and widely applied in Indonesia. HAZUS model is one of the models for estimating earthquake disaster losses in the pre-disaster phase that has been widely used based on building performance developed by FEMA, USA. HAZUS model was developed by local condition in the USA with multidisciplinary area, in terms of geology, geotechnics, structures, architects, economists, social science, hydrology, emergency specialist, and GIS specialists [6]. The local characteristic on each disciplinaries area drive the adjustment on the losses model. So, when the losses model applied in another country, there is must be adjustment. The HAZUS earthquake loss estimation tools were adjusted to Indonesia's local condition, which could reduce the earthquake disaster losses, especially in school building. This study identifies the potential for developing the HAZUS model as a tool for estimating earthquake disaster losses in school buildings in Indonesia.

2. HAZUS earthquake loss estimation model

The HAZUS earthquake loss estimation model was developed by FEMA in 1992 to estimate earthquake loss in local, state, and regional level with state-of-the-art decision support software. The Geographic Information System (GIS)-based software was applied to small and large area with wide range of population and infrastructure (building, road, and lifeline infrastructures) database and can be implemented by users with varying expertise [6]. For general building stock, HAZUS model has four main stage analyses, made of hazard analysis, structure analysis, damage analysis, and losses analysis. The earthquake losses were calculated based on building performance (spectral displacement) and spectral acceleration, with three main replacement components, consisting of structural, non-structural drift sensitive, and non-structural acceleration sensitive as shown in Figure 1 with simplified building performance based on type of structure and building height.

Given its popular usage level of expertise users, HAZUS models are reliable to obtain width of earthquake loss estimation. The damage probability on those components was used to calculate replacement losses and operational losses. On the other hand, the non-simplified building performance method was conducted with more obstacles, such as limited building losses assessment that need detailed Engineering Demand Parameter (EDP) for each building [7]. FEMA has been officially used as HAZUS model for earthquake annual losses estimation in USA in the last 20 years. The following sub-section explains each main stage analysis in more detail.

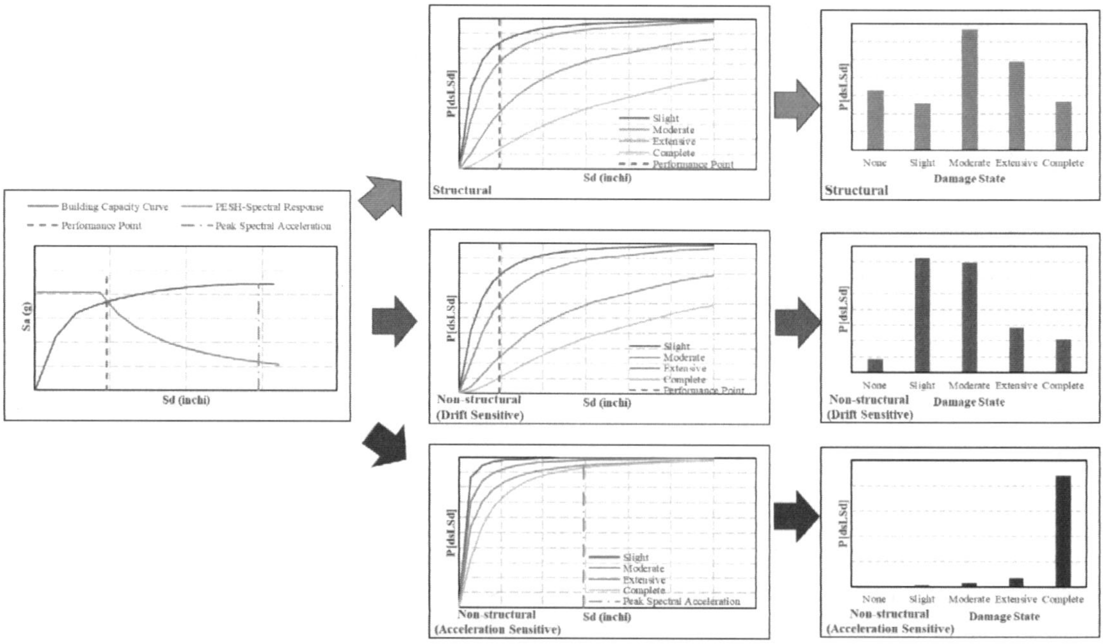

Figure 1. Conceptual Framework of HAZUS earthquake loss [8]

2.1. Hazard analysis

HAZUS model using demand spectra as a hazard on the observed location is shown in Figure 2a. Demand spectra was constructed based on response spectra parameter (Short period spectral acceleration, 1-second spectral acceleration, and specific site class) as presented in Figure 2b. The response spectra parameters were converted into demand spectra parameter (spectral displacement) using Equation 1, where S_D= spectral displacement (inch), S_A=spectral acceleration (g), and T= period for a given value of (seconds).

$$S_D = 9.8 \times S_A \times T^2 \quad (1)$$

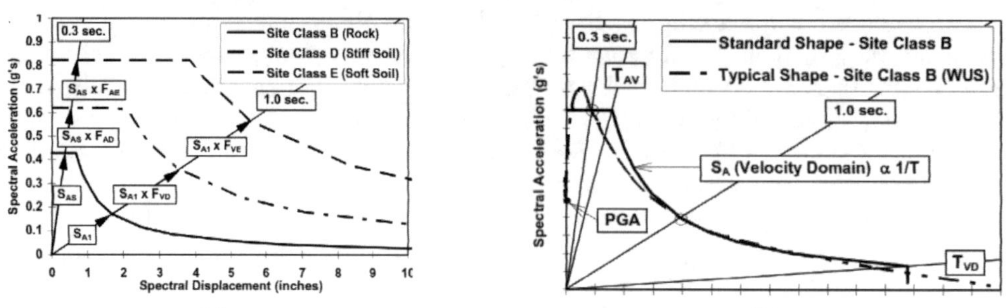

(a) (b)

Figure 2. Hazard calculations of HAZUS earthquake loss (a) Demand spectra construction[6] (b) Standardized Response Spectrum [6]

2.2. Structure analysis

In structure analysis, HAZUS uses simplified building capacity performance based on typical building type, level of code, and building height. In total, HAZUS has 16 types of structures consisting of wood, steel, and concrete structure, etc. The building capacity curve was superimposed to demand spectra to obtain building performance as shown in Figure 3a.

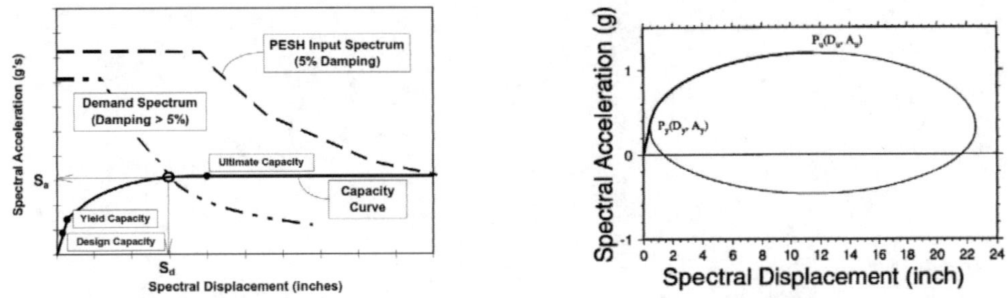

(a) (b)

Figure 3. Calculation of HAZUS structure analysis (a) Building capacity curve [6] (b) Ellipse approach to c yield point to ultimate point [9]

Building capacity curve was constructed using provided yield capacity point, and ultimate capacity can be found in HAZUS manual in Table 5-7 until Table 5-10 with different code levels (high code, moderate code, low code, and pre-code). Capacity curve was constructed by connecting two provided points as transits from elastic to nonlinear plastic states with ellipse form shown in Figure 3b and Equation 2 to Equation 3, where ellipse parameter (A,B,C) solve the existing parameters of Ay, Dy, Au, Du [9].

$$\left(\frac{SD-D_u}{C}\right)^2+\left(\frac{SA-A_x}{B}\right)^2=1 \tag{2}$$

$$\frac{dSA}{dSD}=-\left(\frac{SD-D_u}{SA-A_x}\right)\left(\frac{B^2}{C^2}\right) \tag{3}$$

2.3. Damage analysis

Fragility Curves of the building are conducted in damage analysis. Fragility curves are the lognormal probability functions for the structural and non-structural damage states (slight, moderate, extensive, and complete) to be reached or exceeded when the peak-responses displacement (SD) is given [9]. The form of Building Fragility Curves is shown in Figure 4. Each fragility curve is characterized by median ($\bar{S}_{d,ds}$) and lognormal standard deviation (β_{ds}) as displayed in Equation 4 [6]. From Equation 4, the equation can be rewritten in a more compact form by using the error function erf shown in Equation 5, where, *i*, refers to each damage state [9]. The value of $\bar{S}_{d,ds}$ and β_{ds} can be found in HAZUS manual Table 5-12 until Table 5-21 for structural and non-structural component with each damage state [6].

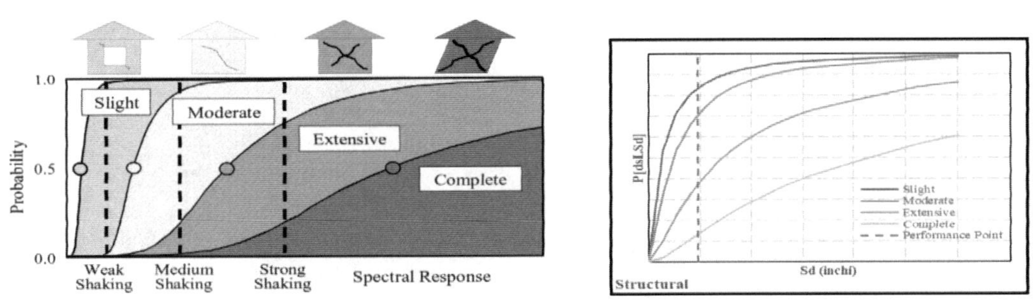

(a) (b)

Figure 4. Calculation of HAZUS damage analysis (a) Damage state in fragility curve [6] (b) Form of Building capacity curve

$$P\left[ds \| S_d\right] = \Phi\left[\frac{1}{\beta_{ds}} \times ln\left(\frac{S_d}{\overline{S}_{d,ds}}\right)\right] \tag{4}$$

$$P\left[ds \leq ds(i) \| S_d\right] = \frac{1}{2}\left(1 + erf\left(\frac{ln\left(SD / \overline{S}_{d,Sds}\right)}{\sqrt{2}\beta_{ds}}\right)\right), i = 1,2,3,4 \tag{5}$$

2.4. Losses analysis

HAZUS calculates Earthquake direct economic losses component, containing building repair costs, building content losses, building inventory losses, relocation expenses, income loss, rental income losses, and wage losses. All of the loses component are calculated based on previous stage analysis. For education facility, earthquake economic losses are limited to some components due to different activities, especially related to business activities. Earthquake economic losses relevant with education facility are calculated by the formula shown in Equation 6 to Equation 10. Structure repair cost is calculated in Equation 6, where BRC_i is Building replacement cost, $PMBTSTR_{ds,i}$ is probability of structure being in structural damage state, $RCS_{ds,i}$ structural repair cost ratio in damage state that is provided in HAZUS manual in Table 11-2. In equation 7 and 8, non-structural economic losses is calculated into non-structural drift sensitive component and acceleration sensitive component, where $PONSA_{ds,i}$ is probability of building being in non-structural acceleration with sensitive damage state, $RCA_{ds,i}$ is non-structural acceleration sensitive repair cost ratio in damage state that is provided in HAZUS manual in Table 11-3, $PONSD_{ds,i}$ is probability of building being in non-structural drift sensitive damage state, $RCD_{ds,i}$ is non-structural drift sensitive repair cost ratio in damage state that is provided in HAZUS manual in Table 11-4. Relocation expenses are conducted as losses that are considered when existing building is repaired and it is calculated based on time dependent function as shown in equation 9, where FA_i is floor area of the building, $\%OO_i$ is percentage of the building occupied (95% for school building), DC_i is disruption cost (1.16 $\$/ft^2$ for school building), $RENT_i$ is the rental cost (0.04 $\$/ft^2/day$ for school building), and RT_{ds} is recovery time state that is provided in HAZUS manual in Table 11-8. Income losses are considered as losses of income during building recovery time as shown in Equation 10, where INC_i is income per day, and LOF_{ds} is loss of function time for damage state that is provided in HAZUS manual in Table 11-8 and Table 11-9.

$$CS_{ds,i} = BRC_i \times \sum_{i=1}^{33} PMBTSTR_{ds,i} \times RCS_{ds,i} \tag{6}$$

$$CNSA_{ds,i} = BRC_i \times PONSA_{ds,i} \times RCA_{ds,i} \tag{7}$$

$$CNSD_{ds,i} = BRC_i \times PONSD_{ds,i} \times RCD_{ds,i} \tag{8}$$

$$REL_i = FA_i \times \left[\left(1 - \%OO_i\right) \times \sum_{ds=2}^{5}\left(POSTR_{ds,i} \times DC_i\right) + \%OO_i \times \sum_{ds=2}^{5}\left(POSTR_{ds,i} \times \left(DC_i + RENT_i \times RT_{ds}\right)\right)\right] \tag{9}$$

$$YLOS_i = \left(1 - RF_i\right) \times FA_i \times INC_i \times \sum_{ds=2}^{5} POSTR_{ds,i} \times LOF_{ds} \tag{10}$$

3. Methodology

The objective of this study is to identify the potential for developing the HAZUS model as a tool for estimating disaster losses in school buildings in Indonesia. The modification and adaptation of the initial model parameter are conducted in the analysis of each main stage with field assessment and Indonesian local data shown in Table 1. Earthquake losses are limited to structure repair cost, non-structural repair cost, relocation expense, and income losses. The hazard parameter adjusted with Indonesian earthquake source and hazard map of 2017 with each building coordinate [10] and local code (SNI 1726:2019) for response spectra procedure [11], with USGS Vs30 data as a site class reference [12]. Building capacity curve is constructed using HAZUS parameter with the building type structure closest to the FEMA Structure type criteria [13]. On the preliminary analysis, there is identified typical structure of school building in the Study case meet the 2 types of FEMA's structure criteria, Concrete Frame with infill masonry shear walls (C3) and Reinforced Masonry Wall with flexible diaphragms (RM1). The level of code is adjusted with Indonesian building code, where the year of construction after 2012 is classified as high-code (HC), 1991-2012 is classified as moderate code(MC), and before 1991 is classified as low-code(LC) [14].

Table 1. Model parameter used in case study

Stage Analysis	Parameters	Data Source
Hazard	S_s, S_1, F_a, F_v	Indonesia earthquake source and hazard map 2017 and SNI 1726:2019
	Site Class, V_{s30}	USGS Vs30 Map Viewer
Structure	A_y, D_y, A_u, D_u	HAZUS manual Table 5-7 to Table 5-10
	Type of structure	GIS tools
	Level Code	Building age base on Indonesia MEC data
Damage	$\bar{S}_{d,ds}$, β_{ds}	HAZUS manual Table 5-12 until Table 5-21
Losses	$PMBTSTR_{ds,i}$, $PONSA_{ds,i}$, $PONSD_{ds,i}$	From damage analysis for each building and damage states
	$RCS_{ds,i}$, $RCA_{ds,i}$, $RCD_{ds,i}$, LOF_{ds}, RT_{ds}, $\%OO_i$, $RENT_i$	HAZUS manual data for Education facility (EDU1)
	FA_i	GIS measurement
	BRC_i	State building standard unit price
	INC_i	School operational assistance form Indonesia's MEC data

The building replacement cost is calculated by multiple total floor area by State building standard unit price in Bandung in 2021 (IDR 6.6 million/m^2) [15]. Income losses are calculated by School operational assistance form Indonesia's MEC based on the number of students and teachers [16,17]. The other model parameter is adapted from HAZUS model. The identification process is led by a study on the assessment of elementary school building earthquake losses in Cicendo District, Bandung city, which have various school building types and high seismic hazard in the city [18]. A total of nineteen school buildings were assessed with field assessment and secondary data shown in Table 2. Field survey was done online due to pandemic condition in 2021, using GIS tools (QGIS, google earth, google maps) which acquired building data such as building coordinate, type of building structure, and building area.

Table 2. Elementary school building data in Cicendo District, Bandung City

Code	School	Location		Year of construction	Area (m^2)	Number of stories	Type of structure	Income (Million IDR/year)
		Latitude	Longitude					
S1	SDK 1 Paulus	-6.904	107.600	1953	10,101	3	C3	271.32
S2	SD Al Fatah	-6.893	107.559	1987	520	2	C3	210.82
S3	SD Angkasa 1	-6.906	107.586	1962	1,654	1	RM1	472.28
S4	SD Angkasa 3	-6.897	107.585	1962	793	1	RM1	414.73
S5	SD Industri	-6.913	107.595	1986	376	2	C3	194.73
S6	SDK 1 Bina Bakti	-6.912	107.594	1973	13,584	3	C3	450.28
S7	SDK 2 Bina Bakti	-6.912	107.594	1982	13,584	3	C3	574.49

Code	School	Location		Year of construction	Area (m^2)	Number of stories	Type of structure	Income (Million IDR/year)
		Latitude	Longitude					
S8	SDK BPPK	-6.907	107.594	1987	4,420	2	C3	197.47
S9	SDK Trimulia	-6.898	107.589	1988	19,468	4	C3	845.45
S10	SD Pandu	-6.905	107.593	1960	4,851	3	C3	466.80
S11	SD Pelita Jasa	-6.912	107.587	1987	1,624	2	C3	212.53
S12	SDN 016	-6.907	107.600	1948	4,706	2	C3	862.81
S13	SDN 045	-6.895	107.561	1998	818	2	C3	415.14
S14	SDN 155	-6.890	107.569	1979	2,992	2	C3	733.87
S15	SDN 156	-6.893	107.565	1978	1,024	2	C3	463.46
S16	SDN 157	-6.892	107.579	1975	2,314	2	RM1	201.89
S17	SDN 170	-6.891	107.559	1975	1,348	2	C3	413.11
S18	SDN 171	-6.894	107.583	1975	1,002	1	RM1	252.28
S19	SDN 187	-6.899	107.586	1974	2,787	1	RM1	442.15

4. Results and discussion

The results of earthquake losses assessment are shown in percentage of building cost (BRC) in Table 3. Earthquake losses are estimated for 50 years as the earthquake hazard map for probability exceeding 2% in 50 years. The annual losses could not be estimated due to limited seismic hazard parameter that needs minimum 8 seismic parameter for eight return periods. Non-structural repair costs dominate the total losses with average of 26.39%, followed by structural repair cost 15.50%, relocation expense of 3.33%, and income losses of 0.09% for different structure type (C3 and RM1). In C3 type, different code levels show Moderate Code (MC), which gives lower estimate than Low-Code (LC). Figure 5 shows the total estimated losses map, where the estimated loss value is more attributed by the size of buildings, with the hazard close to the same value at around S_S=1.20 and S_1=0.52g.

Identification of the potential development of HAZUS model as an earthquake disaster losses model for school buildings in Indonesia is shown in Table 4. Limited return period for seismic parameter that is provided by Indonesia earthquake source and hazard map 2017 (return period 1000 and 2500) makes it hard to estimate the annual losses. Annual loses in HAZUS model were estimated using eight return periods (2500, 2000, 1500, 1000, 750, 500, 250, 100). Estimated return period with provided 2 return period will give overestimated losses. Empirical earthquake parameter conversion was conducted using Eurocode equation, which could be one of alternatives. For structure analysis, building capacity curve could be developed with locally characterized school building structure.

Table 3. Earthquake losses of school buildings school in Cicendo District, Bandung City

School	Types of structure & Level of Code	BRC (Million IDR)	Earthquake losses component (% to BRC)				Earthquake losses in each damage state (% to BRC)			
			Structure repair cost	Non-structural repair cost	Relocation expense	Income losses	Slight	Moderate	Extensive	Complete
S1	C3L-LC	66,701	14.40%	23.91%	3.84%	0.01%	0.0%	0.6%	4.6%	36.9%
S2	C3L-LC	3,432	15.75%	25.24%	3.68%	0.04%	0.0%	0.4%	10.3%	77.2%
S3	RM1L-LC	10,922	17.51%	28.24%	3.42%	0.13%	0.0%	0.1%	2.7%	46.5%
S4	RM1L-LC	5,236	14.48%	24.07%	3.48%	0.20%	0.5%	3.8%	12.0%	57.2%
S5	C3L-LC	2,483	16.16%	28.72%	3.64%	0.22%	0.0%	0.3%	2.9%	45.5%
S6	C3L-LC	89,701	16.17%	28.73%	2.71%	0.01%	0.0%	0.3%	2.8%	44.5%

School	Types of structure & Level of Code	BRC (Million IDR)	Earthquake losses component (% to BRC)				Earthquake losses in each damage state (% to BRC)			
			Structure repair cost	Non-structural repair cost	Relocation expense	Income losses	Slight	Moderate	Extensive	Complete
S7	C3L-LC	89,701	16.17%	28.73%	2.71%	0.02%	0.0%	0.3%	2.8%	44.5%
S8	C3L-LC	29,187	16.19%	28.77%	3.71%	0.02%	0.0%	0.3%	2.9%	45.5%
S9	C3M-LC	128,555	12.94%	20.08%	3.44%	0.01%	0.1%	0.6%	5.1%	30.7%
S10	C3L-LC	32,033	16.02%	27.31%	3.49%	0.04%	0.0%	0.3%	3.4%	43.1%
S11	C3L-LC	10,724	16.18%	28.75%	3.71%	0.06%	0.0%	0.3%	2.9%	45.5%
S12	C3L-LC	31,076	16.18%	28.76%	3.71%	0.08%	0.0%	0.3%	2.9%	45.5%
S13	C3L-MC	5,402	14.77%	26.15%	3.61%	0.20%	0.0%	0.5%	3.9%	40.2%
S14	C3L-LC	19,757	14.80%	26.21%	3.61%	0.09%	0.0%	0.5%	3.9%	40.2%
S15	C3L-LC	6,762	14.80%	26.21%	3.03%	0.17%	0.0%	0.5%	3.8%	39.8%
S16	RM1L-LC	15,280	14.77%	26.16%	2.83%	0.03%	0.0%	0.5%	3.8%	39.4%
S17	C3L-LC	8,901	15.77%	25.28%	2.92%	0.13%	0.0%	0.3%	4.2%	39.6%
S18	RM1L-LC	6,617	15.71%	25.14%	2.85%	0.10%	0.0%	0.3%	4.2%	39.3%
S19	RM1L-LC	18,404	15.67%	25.02%	2.85%	0.06%	0.0%	0.3%	4.3%	39.0%
Average			15.50%	26.39%	3.33%	0.09%	0.06%	0.57%	4.39%	44.22%
Structure C3 Type Average			15.45%	26.63%	3.42%	0.08%	0.04%	0.41%	4.02%	44.20%
Structure RM1 Type Average			15.63%	25.73%	3.09%	0.11%	0.12%	1.00%	5.41%	44.28%

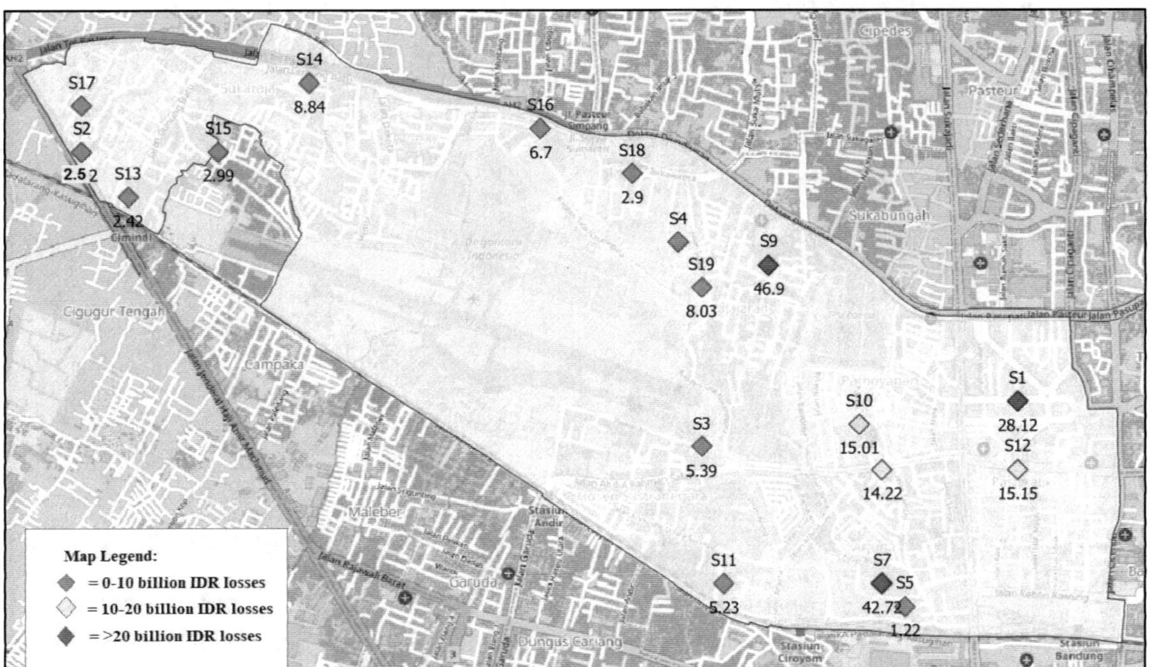

Figure 5. Total estimation map of earthquake losses of school buildings in Cicendo District, Bandung City (in billion IDR)

Table 4. Identifying potential development of the of HAZUS model for school buildings in Indonesia

Stage Analysis	Potential Development
Hazard	Various return period of seismic parameter with empirical approach for annual losses estimation
Structure	Modification of building capacity parameter (A_y, D_y, A_u, D_u) with local typical school building structure in Indonesia
Damage	Adjustment of fragility curve parameter ($\bar{S}_{d,ds}$ and β_{ds}) for each damage states with local structure type
Losses	- Development of losses parameter construction related ($RCS_{ds,I}$, $RCA_{ds,i}$, $RCD_{ds,i}$, RT_{ds}, LOF_{ds}) from earthquake reconstruction and rehabilitation of historical data - Modification of school data in MEC database with building taxonomy data

Figure 6 shows a comparison between FEMA structure criteria and typical of school building structure in this study. The C3 type of structure, typical school building in Indonesia, consist of 1-3 stories, and has non-rigid diaphragms on the roof level, although another level is a concrete rigid floor as shown in Figure 6c. The different rigid floors in the roof level would give different building capacity curves compared to expected HAZUS for C3 building capacity curve shown in Figure 6a. Typical building structure that was closely identified as RM1 has a large number of one story school building that is vulnerable to earthquake as shown in Figure 6d [19,20].

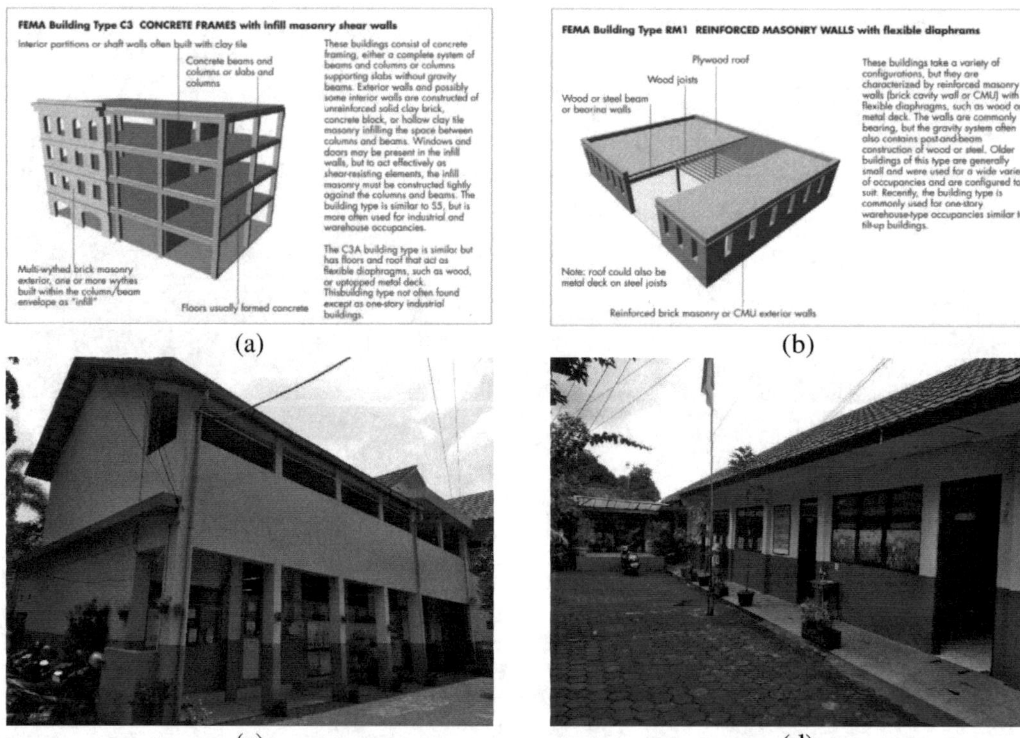

Figure 6. Building structure comparison (a) C3 FEMA structure criteria [13] (b) RM1 FEMA structure criteria [13] (c) Typical Building structure identified as C3 (d) Typical Building structure identified as RM1

In typical school building, non-rigid diaphragms consist of a wood truss/frame structure with or without concrete beam under the truss structure as the common non-engineered building in housing.

FEMA defines RM1's non-rigid diaphragms as a plywood roof with wood joist, which would give a different building performance from RM1 school building structure. The development of a structure type definition and typical capacity curve parameter would provide more accurate earthquake loss estimation. The specific structure definition would also give different fragility curve parameter. This development is developed in further studies for a typical school building structure in Indonesia.

The modification in building repair cost ratio is indicated by the deviation of construction costs and productivity between Indonesia and USA [21]. The development of construction-related losses parameter was conducted using historical school rehabilitation data on several impacted earthquake, which would give more accurate estimation in Indonesia. Another development in losses analysis is to update MEC school administration data base with building taxonomy data, such as (building area, story) to strengthen still analysis and spatially estimated losses.

5. Conclusion

The HAZUS model can be applied as a tool for estimating earthquake losses for school buildings in Indonesia, which needs several adjustments. Hazard data can be applied according to the 2017 Indonesia earthquake map, by adjusting to empirical parameters for several return periods to obtain annual losses. The structural characteristics of typical school buildings in Indonesia also needs modification to obtain the building capacity and the characteristics of the fragility curve, for each damage state, which are specific to local conditions in Indonesia. Meanwhile, adjustment of construction parameters (cost ratio and duration) in the calculation of losses is needed considering that there are significant differences between the USA as the HAZUS model database and conditions in Indonesia. It is also possible to develop the Indonesia's MEC school database with building taxonomy data such as (building area, story) to strengthen still and spatial analysis in estimating losses.

Acknowledgements

This study was funded by Center or Education Services (PUSLAPDIK), Ministry of Education, Culture, Research, and Technology of the Republic of Indonesia and Indonesia Endowment Fund for Education (LPDP), Ministry of Finance Republic of Indonesia. The author is grateful for the anonymous reviewer's constructive comments and suggestions as well as for the research team's feedback.

References

[1] UNDRR 2011 Indonesia Disaster Profile (1815-2019)
[2] BNPB 2021 Indonesia's Disaster Casualties and Damage by Time
[3] BPS 2020 *Indonesia's Education Statistics 2020* (Jakarta: Badan Pusat Statistik)
[4] The World Bank 2019 *Boosting Financial Resilience to Disaster Shocks* (Washington, DC)
[5] Wein A and Rose A 2011 Economic resilience lessons from the shakeout earthquake Scenario *Earthq. Spectra* **27** 559–73
[6] FEMA-NIBS 2020 HAZUS Earthquake Model Technical Manual *Fed. Emerg. Manag. Agency*
[7] Ramirez C M, Mitrani-Reiser J, Haselton C B, Spear A D, Steiner J, Deierlein G G . and Miranda E 2012 Expected earthquake damage and repair costs in reinforced concrete frame buildings *Earthq. Eng. Struct. Dyn.* **41**
[8] FEMA 2001 *Technical and User's Manual of Advanced Engineering Building Module (AEBM) "Hazus MH 2.1"* (Washington, DC: Federal Emergency Management Agency)
[9] Cao T and Petersen M D 2006 Uncertainty of earthquake losses due to model uncertainty of input ground motions in the Los Angeles area *Bull. Seismol. Soc. Am.* **96** 365–76
[10] PUSGEN 2017 *Indonesia Earthquake source and hazard map 2017* (Bandung: Ministry of Public Works and Housing of Indonesia)
[11] BSN 2019 *SNI 1726-2019 - Procedures for Planning Earthquake Resistance for Building and Non-Building Structures* (Jakarta: Badan Standardisasi Nasional)
[12] USGS 2007 USGS Vs30 Map Viewer

[13] FEMA 2006 *FEMA 454- Designing for Eartqhuakes*
[14] Muntafi Y, Nojima N and Jamal A U 2020 Damage Probability Assessment of Hospital Buildings in Yogyakarta, Indonesia as Essential Facility due to an Earthquake Scenario *J. Civ. Eng. Forum* **6** 225
[15] Mayor Bandung City 2021 *Bandung Mayor's Decision No.80-2021* (Indonesia)
[16] Ministry of Education and Culture of Indonesia School Data in Cicendo District, Bandung City
[17] Ministry of Education and Culture of Indonesia 2007 *PerMenDikNas no. 24 Tahun 2007* (Indonesia)
[18] Pribadi K and Yuliawati A K 2009 Earthquake Disaster Preparedness Education as an Effort to Improve Safety *J. ABMAS* **7** 1–8
[19] Boen T 2020 Rapid Screening Procedure for Buildings Earthquakes : The Need for Retrofitting by **01013**
[20] Idris Y, Cummins P, Rusydy I, Muksin U, Syamsidik, Habibie M Y and Meilianda E 2022 Post-Earthquake Damage Assessment after the 6.5 Mw Earthquake on December, 7th 2016 in Pidie Jaya, Indonesia *J. Earthq. Eng.* **26** 409–26
[21] Arcadis Indonesia 2018 Construction Cost Handbook In Indonesia 2018 1

Liquefaction Potential Analysis and Soil Improvement Method for Yogyakarta-Bawen Toll Road in Sleman, Yogyakarta

P Kevin[1], S P R Wardani[1,2], Muhrozi[1], I D Atmanto[1]

[1] Department of Civil Engineering, Universitas Diponegoro, Jl. Prof. Soedarto SH, Tembalang, Semarang, Indonesia

Corresponding author: wardani@live.undip.ac.id

Abstract. Based on soil investigation using SPT in the planned Yogyakarta-Bawen Toll Road in Sayengan District, Sleman Regency, the soil layers consist of sandy silt, fine-coarse sand, and gravel. Several studies that used the Idriss & Boulanger method showed the liquefaction potential in all observed locations is between high and very high. Therefore, the proposed soil improvement methods are stone columns and jet grouting. Further analysis is conducted to define the effective diameter of the column based on the excess pore pressure ratio (R_u) in both alternatives using the finite difference method (FDM). The result of R_u and CSR analysis in BH 78 showed that by using a 1.2 m diameter stone column in the triangular grid with 2.2 m spacing, $R_{u\ max}$ is reduced by 54.75%, and $SF_{liquefaction}$ is improved to 1.202. Alternatively, soil improvement by using a 1.0 m diameter jet grouting in the rectangular grid with 3.0 m spacing showed $R_{u\ max}$ is reduced by 11.41% and $SF_{liquefaction}$ is improved to 2.373. Therefore, improvement with the stone column simultaneously reduces $R_{u\ max}$ and cyclic stress ratio (CSR), on the other hand, the jet grouting only reduces the cyclic stress ratio.

Keywords: finite difference method, jet grouting, liquefaction, safety factor, stone column

1. Introduction

The previous studies [1,2] on the Southern and Eastern parts of Yogyakarta Special Province suggested that they are prone to liquefaction. However, the liquefaction potential and its mitigation method studies in the north-western part of Yogyakarta are quite limited thus they need to be conducted as the presence and importance of the Yogyakarta-Bawen Toll Road project in this area. Moreover, the analyzed locations in [3] are in a moderately susceptible liquefaction zone. This fact is supported by the presence of saturated sand and proximity to the earthquake source. Based on the soil investigation STA 0+280 – STA 1+330 of this project, it showed that most of the soil layers consist of sandy silt and fine–coarse sand with shallow groundwater (< 10 m from the soil surface). This finding is justified by the geological map [4] that the 1st section is on the Merapi-Merbabu undifferentiated volcanic rocks. Furthermore, the presence of the Opak Fault within a 15-25 km radius of the analysis location induces the liquefaction potential.

2. Analysis Methodology

Preliminary studies with slope stability analysis used the Fellenius equation where the safety factor of each embankment needs to satisfy the requirement by SNI 8460:2017 [5], which should not be less than 1.25. Also, the liquefaction potential analysis used the Idriss & Boulanger method [6] as seen in figure 1.

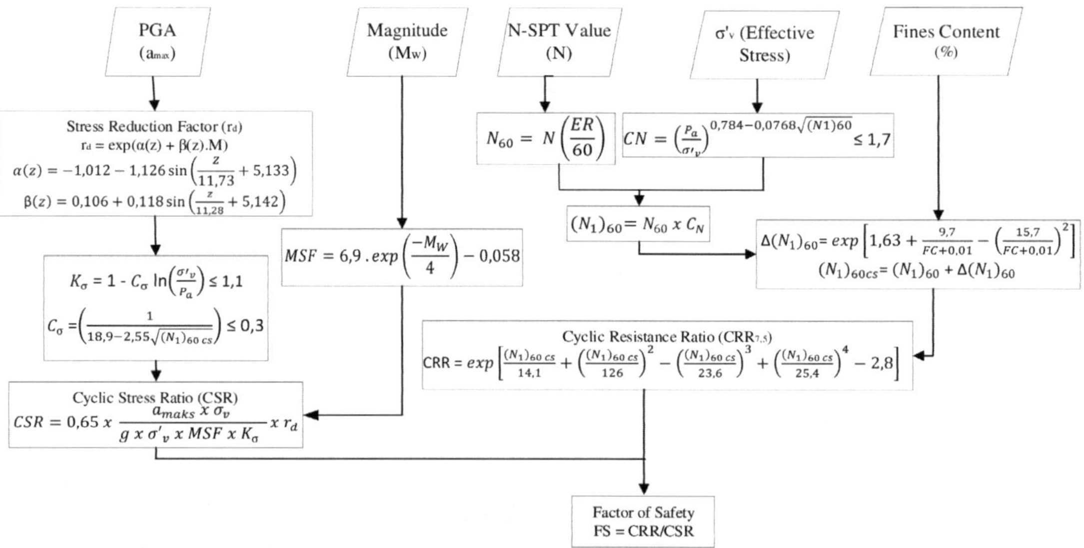

Figure 1. Flowchart Idriss & Boulanger Method.

Further analysis was conducted using the FDM model to identify excess pore water pressure ratio (R_u) in unimproved and improved soils. Also, the FDM analysis is only used to identify R_u reduction caused by soil improvement (stone column and jet grouting). The inputs of the FDM model are synthetic ground motion and soil properties. This synthetic ground motion is matched with a spectral response based on SNI 1726:2019 [7]. Furthermore, the reduction cyclic stress analysis in the stone column uses Baez & Martin [8] and Priebe [9] methods. For the jet grouting, the reduction cyclic stress analysis uses a method proposed by Özsoy & Durgunoğlu [10].

3. Geotechnical Data

BH 14 with 40 m depth boring is stratified into 4 layers. The first layer in 0-12 m depth is dominated by silt and fine-coarse sand with a medium-dense consistency. Meanwhile, the second layer in 12-24 m is mostly composed of sandy silt and gravel with a dense consistency, and the third in 24-30 m depth is majorly silty fine-coarse sand with a medium-dense consistency. Lastly, the fourth layer in 30-40 m has the same characteristic as the second. BH 16 has similar characteristics and thickness of layers as BH 14. Meanwhile, BH 78 with 30 m depth boring is stratified into 3 layers, the first layer in 0-8 m depth is dominated by loose-medium sand with a medium-dense consistency. The second layer in 8-16 m depth consisted of silty fine-medium sand with a medium-dense consistency, while the third in 14-30 m depth is dominated by silty fine-medium sand with dense consistency. Furthermore, BH 17 with 40 m depth boring is stratified into 3 layers, where the first in 0-14 m depth is majorly silty loose-medium sand with a medium-dense consistency. The second layer in 14-28 m depth consisted of fine-medium sand with a medium-dense consistency, while the third layer in 28-40 m depth is mostly composed of gravelly fine-coarse sand with dense consistency. All the investigated boreholes have groundwater levels of less than 10 m (0.8 – 4.75 m from the ground surface). In addition, the soil properties were obtained by using the SPT correlation table [11]. The results of soil layer stratification and simplification are shown in figure 2 and table 1.

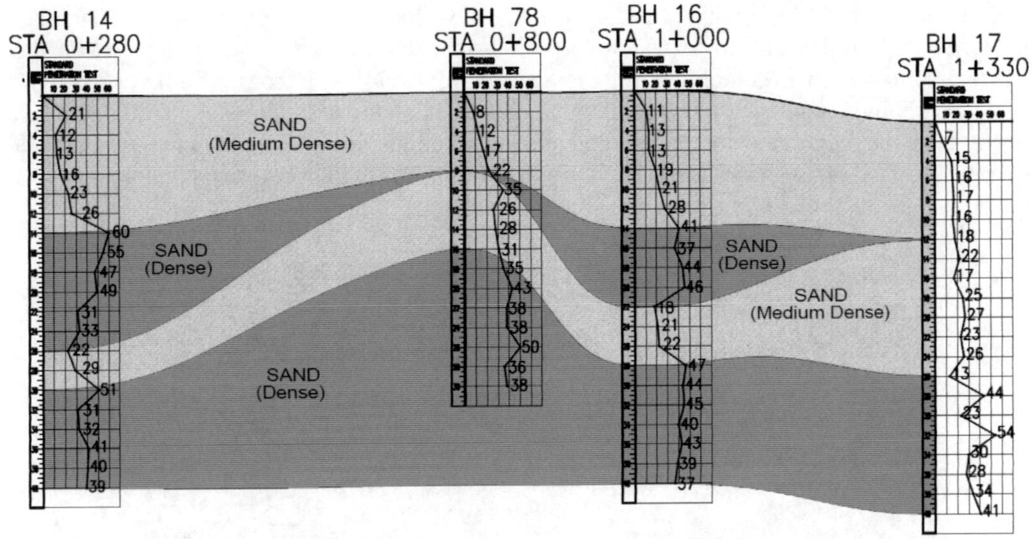

Figure 2. Soil Layer Stratification BH 14, BH 78, BH 16, and BH 17.

Table 1. Simplified Soil Properties for Static Analysis.

Borehole	Depth (m)	Soil Classification	N-SPT average	Density (γ)	The angle of Internal Friction(ϕ)	Poisson Ratio (ν)	Elastic Modulus (E)
	m			kN/m^3	°		MPa
BH 14	0-14	Sand (Medium Dense)	18.5	17	34.3	0.3	17.35
	14-26	Sand (Dense)	45.8	19	44.0	0.35	69.79
	26-30	Sand (Medium Dense)	25.5	17	37.8	0.3	25.05
	30-40	Sand (Dense)	39	19	42.3	0.35	61.25
BH 78	0-8	Sand (Medium Dense)	12.3	17	31.2	0.3	10.57
	8-16	Sand (Medium Dense)	27.8	17	38.9	0.3	27.53
	16-30	Sand (Dense)	38.6	19	42.2	0.35	60.78
BH 16	0-14	Sand (Medium Dense)	17.5	17	33.8	0.3	16.25
	14-22	Sand (Dense)	42	19	43	0.35	65
	22-28	Sand (Medium Dense)	20.3	17	35.2	0.3	19.37
	28-40	Sand (Dense)	42.1	19	43	0.35	65.18
BH 17	0-8	Sand (Medium Dense)	12.3	17	31.2	0.3	10.57
	8-16	Sand (Medium Dense)	27.8	17	38.9	0.3	27.53
	16-30	Sand (Dense)	38.6	19	42.2	0.35	60.78

4. Result and Discussion

After the soil stratification and simplification were conducted, the slope stability analysis before soil improvement used the Fellenius method. In the slope stability analysis location, the average height of the planned embankment is 8 m. The embankment safety factors of each borehole can be seen in table 2.

Table 2. The safety factor in analysis location.

Borehole	Height of Embankment	Safety Factor	Remarks
BH 14	7 m	2.166	Acceptable
BH 78	8.16 m	1.980	Acceptable
BH 16	8.11 m	2.052	Acceptable
BH 17	8.30 m	1.923	Acceptable

Liquefaction potential analysis with Idriss & Boulanger method can be seen in figure 1. The data used in liquefaction potential analysis, for example, PGA_M of each location were obtained from the RSA 2021 software and N-SPT data including depth groundwater level from soil investigation. Also, liquefaction potential analysis in a certain depth is indicated to be safe when the safety factor exceeds 1.1. Meanwhile, the magnitude (M_w) and epicentrum source distance are determined by the Indonesia Earthquake Hazard Deagregation Map [12] and Opak Fault Map [13]. The analyzed location is shown in figure 3.

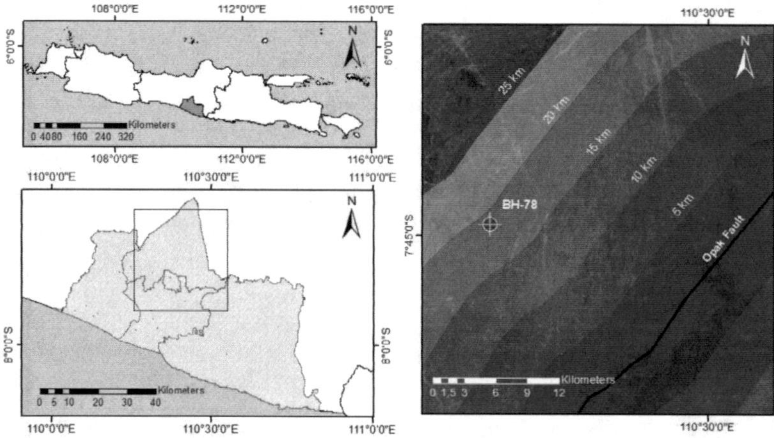

Figure 3. Opak Fault Map.

Based on the Map [12], the BH 78 has a deagregate magnitude (M_W) of 6.8-7.0 and epicentrum source distance (R) of 10-20 km. According to figure 4, BH 78 is located 15-20 km from Opak Fault with M_{max} 6.6 M. For conservative analysis results, the chosen design magnitude is 7.0 M with an epicentrum source distance (R) of 10-20 km. Also, the characteristics of earthquake sources are shallow crustal and the mechanism of fault is strike-slip. These characteristics and mechanisms are used as criteria for ground motion search in PEER NGA-West 2 database [14]. The ground motion synthetics are generated from spectral matching analysis with a wavelet algorithm. The selected original ground motions are shown in table 3.

Table 3. The selected original ground motion.

Event	Year	Station	Magnitude (M)	Epicentrum (km)	Duration (s)	Orientation
Imperial Valley, The USA	1940	El Centro Array #9	6.95	12.98	53.7	180°
Kobe, Japan	1995	Takatori	6.9	13.12	40.94	90°
Darfield, New Zealand	2010	DSLC	7.0	13.41	137.69	West (180°)
Darfield, New Zealand	2010	RKAC	7.0	15.51	65.08	East (0°)

The results of spectral matching are selected as the lowest average misfit which is El Centro Array #9 with a 1.9% average misfit. Furthermore, the matched ground motion is corrected to baseline with Seismosignal. The corrected ground motion can be seen in figure 4.

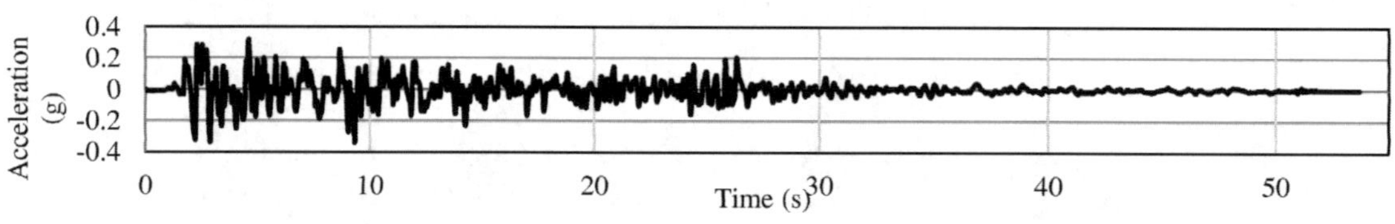

Figure 4. The Corrected Ground Motion.

Spectral matching is conducted with Seismomatch. The original and matched spectrum response can be seen in figures 5 & 6.

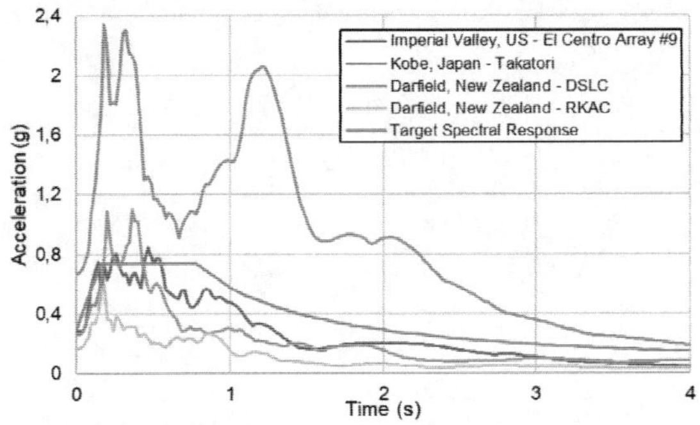

Figure 5. The original spectral response.

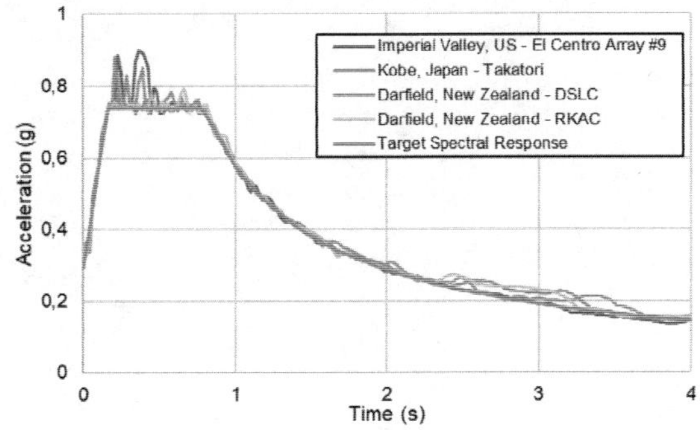

Figure 6. The matched spectral response.

After the earthquake design is determined, the liquefaction potential analysis and index are conducted. The liquefaction potential index [15] is obtained from equation (1).

$$\mathrm{LPI} = \int_0^{20} F(z)w(z) \quad (1)$$

where:
F = Liquefaction safety factor in z depth
w = Depth factor in z depth
The result of the liquefaction potential analysis is shown in table 4 and figure 7.

Table 4. Liquefaction Potential Analysis in Various Boreholes.

Depth (m)	BH 14			BH 78			BH 16			BH 17		
	N-SPT	FC (%)	L/NL	N-SPT	FC (%)	L/NL	N-SPT	FC (%)	L/NL	N-SPT	FC (%)	L/NL
2	21	25	L	8	30	L	11	25	L	21	25	L
4	12	37	L	12	25	L	13	25	L	12	37	L
6	13	43	L	17	25	L	13	25	L	13	43	L
8	16	25	L	22	25	L	19	25	L	16	25	L
10	23	25	L	35	20	NL	21	25	L	23	25	L
12	26	25	L	26	25	L	28	25	L	26	25	L
14	60	20	NL	28	25	L	41	20	NL	60	20	NL
16	55	20	NL	31	20	L	37	20	NL	55	20	NL
18	47	20	NL	35	20	NL	44	20	NL	47	20	NL
20	49	20	NL	43	20	NL	46	20	NL	49	20	NL
22	31	20	L	38	20	NL	18	25	L	31	20	L
24	33	20	L	38	20	NL	21	25	L	33	20	L
26	22	25	L	50	20	NL	22	25	L	22	25	L
28	29	25	L	36	20	L	47	20	NL	29	25	L
30	51	20	NL	38	20	NL	44	20	NL	51	20	NL
32	31	20	L				45	20	NL	31	20	L
34	32	20	L				40	20	NL	32	20	L
36	41	20	NL				43	20	NL	41	20	NL
38	40	20	L				39	20	L	40	20	L
40	39	20	L				37	20	L	39	20	L

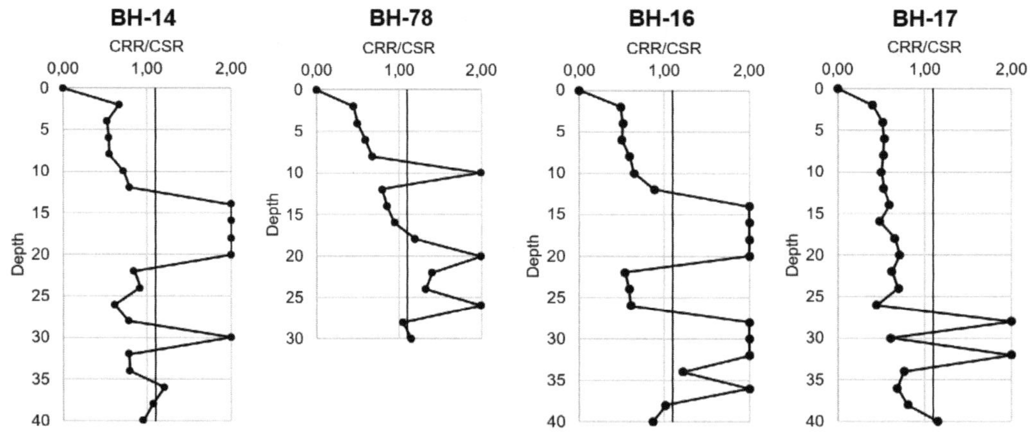

Figure 7. Liquefaction Potential Analysis in Various Boreholes.

In table 4, a liquefiable layer is indicated by L, and a non-liquefiable layer is indicated by NL. Also, the liquefaction potential index in BH 14 is high at 14.865, while BH 78 at 15.325, BH 16 at 16.665, and BH 17 at 22.335 are very high. The BH 78 FDM model is analyzed with a 30 m x 10 m dimension model and with a singular column. This FDM analysis only becomes a reference model in the effective column diameter determination. The dynamic soil properties in BH 78 and the singular column FDM model are presented in figure 8 and table 5.

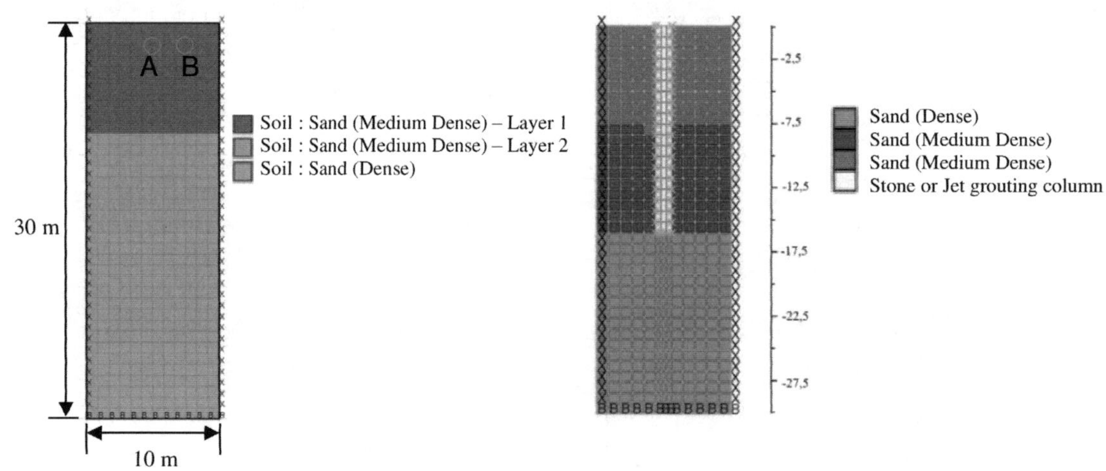

Figure 8. Singular column finite difference method model.

Table 5. Liquefaction Potential Analysis in Various Boreholes.

Depth (m)	Classification	ν	G_{max} (kPa)	K (kPa)	E (kPa)	$(N_1)_{60cs}$	φ (°)
0-8	Sand (Medium Dense)	0.3	64954.22	140734.1	168881	18.18	31.2
8-16	Sand (Medium Dense)	0.3	116781.5	253026.6	303631.9	29.97	38.9
16-30	Sand (Dense)	0.35	166216.2	498648.5	448783.7	33.27	42.2
-	Stone Column	0.3	627000	1358000	1630200	-	43
-	Jet Grouting	0.25	3277040.13	5461733.55	8192600	-	-

The point of analysis in the model without soil improvement is indicated as A and the column is shown as B. Based on figure 8, R_u decreases with increasing diameter of the stone column. The smallest $R_{u\ max}$

is in the 1.2 m diameter of a stone column with a total reduction of 54.75% from $R_{u\ max}$ without improvement (from 1.549 to 0.701).

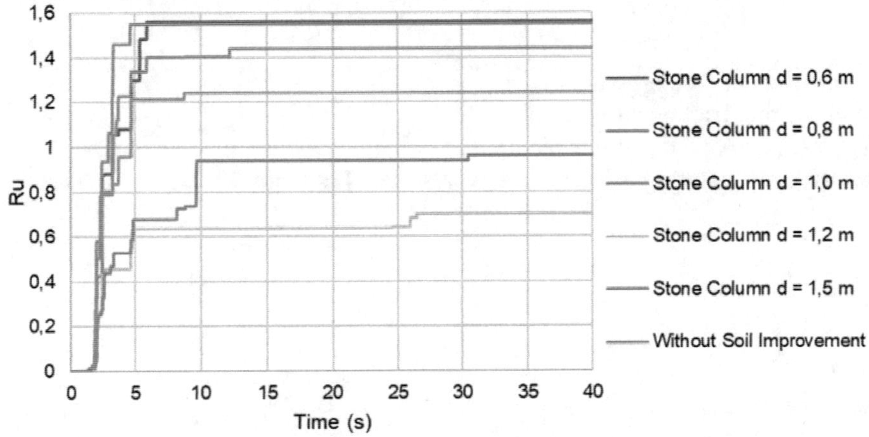

Figure 9. R_u against dynamic time with the stone column.

Based on figure 10, R_u is not significantly reduced by the use of jet grouting. Also, it is shown that the jet grouting does not contribute to R_u reduction because the function is not for drainage as that of the stone column [16]. The function of the jet grouting is to reduce the seismic load in soil because of its higher modulus than the original soil. In the next jet grouting analysis, the used diameter of the column is 1.0 m because of its minimum $R_{u\ max}$. The result of FDM analysis is excess pore water pressure ratio (R_u) in 40 seconds with column diameter variation as seen in figure 10.

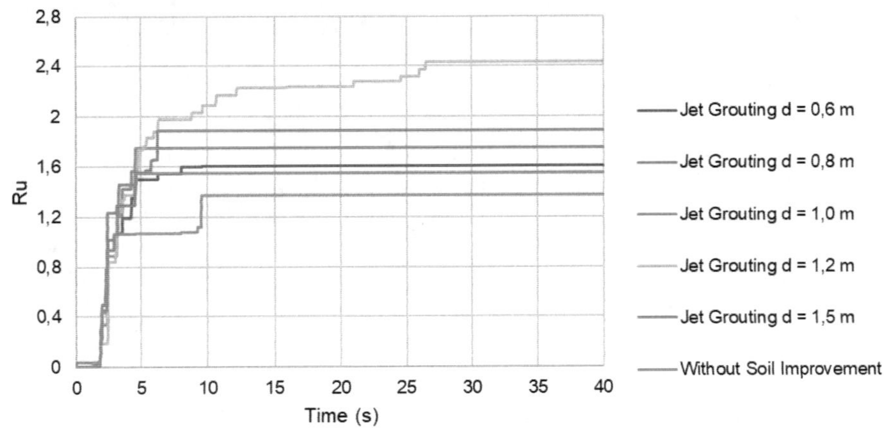

Figure 10. R_u against dynamic time with jet grouting.

After the effective diameter is obtained, the reduction cyclic stress analysis for the stone column uses the Priebe and Baez & Martin method. Also, for the Priebe method, the cyclic stress reduction analysis uses equations (2), (3), and (4).

$$\eta = \frac{A_{SC}}{A} \quad (2)$$

$$n_0 = 1 + \eta\left(\frac{1}{\tan^2\left(45^\circ - \frac{\varphi_c}{2}\right)(1-\eta)} - 1\right) \quad (3)$$

IOP Conf. Series: Earth and Environmental Science 1244 (2023) 012023 doi:10.1088/1755-1315/1244/1/012023

$$K_G = \frac{1}{n_0} \quad (4)$$

where:

n_0 = soil improvement factor

φ_c = angle of internal friction stone column

η = ratio of the stone column and tributary area

For Baez & Martin, the cyclic stress reduction analysis uses equation (5).

$$K_G = \frac{1}{1+\eta\left(\frac{G_{SC}}{G_S}-1\right)} \quad (5)$$

where:

G_{SC} = Shear modulus of stone column

G_S = Shear modulus of soil

Results of the cyclic stress reduction analysis for both methods can be seen in tables 6, 7, and 8. The safety factor is to not exceed 2.0 to avoid overestimation.

Table 6. Simplified Soil and Column Properties for Dynamic Analysis.

Depth	Space	Tributary Diameter	SC Area	Tributary Area	η	n_0	K_G	K_G
m	m	m	m²	m²			Priebe (1998)	Baez & Martin (1993)
0-8	1.8	1.89	1.131	2.806	0.403	4.169	0.240	0.147
	2.4	2.52	1.131	4.988	0.227	2.324	0.430	0.234
	3	3.15	1.131	7.793	0.145	1.753	0.571	0.323
	3.6	3.78	1.131	11.222	0.101	1.492	0.670	0.407
8-16	1.8	1.89	1.131	2.806	0.403	4.169	0.240	0.335
	2.4	2.52	1.131	4.988	0.227	2.324	0.430	0.473
	3	3.15	1.131	7.793	0.145	1.753	0.571	0.583
	3.6	3.78	1.131	11.222	0.101	1.492	0.670	0.668

Based on table 6, the space needed to achieve the target safety factor is less than 2 D (2.4 m). In the next analysis, the used space is 2.2 m with a triangle grid. The safety factors after improvement with Priebe and Baez & Martin method can be seen in table 7.

Table 7. The BH 78 CSR after stone column improvement.

Depth (m)	G (kN/m²)	CSR	$CSR_{required}$	Priebe (1998)				Baez & Martin (1993)			
				KG	CSR_{SC}	CRR	FS	KG	CSR_{SC}	CRR	FS
2	4064.103	0.302	0.122	0.372	0.112	0.135	1.202	0.204	0.062	0.135	2.000
4	4064.103	0.314	0.142	0.372	0.117	0.156	1.334	0.204	0.064	0.156	2.000
6	4064.103	0.337	0.179	0.372	0.125	0.197	1.573	0.204	0.069	0.197	2.000
8	4064.103	0.355	0.217	0.372	0.132	0.239	1.808	0.204	0.072	0.239	2.000
10	10586.54	0.396	0.720	0.372	0.147	0.792	2.000	0.429	0.170	0.792	2.000
12	10586.54	0.375	0.268	0.372	0.140	0.295	2.000	0.429	0.161	0.295	1.833
14	10586.54	0.380	0.293	0.372	0.141	0.322	2.000	0.429	0.163	0.322	1.976
16	10586.54	0.383	0.326	0.372	0.143	0.359	2.000	0.429	0.164	0.359	2.000

The minimum safety factor for cyclic reduction analysis in table 8 is 1.202. Further analysis of soil improvement with jet grouting is carried out using Özsoy & Durgunoğlu method and the cyclic reduction factor is calculated with equations (6), (7), and (8).

$$S_R = \frac{FS_L}{FS_{JG}} \tag{6}$$

$$a_r = \frac{\dfrac{1}{K_G(G_{JG}/G_S)} - \dfrac{1}{(G_{JG}/G_S)}}{1 - \dfrac{1}{(G_{JG}/G_S)}} = \frac{\pi}{4} x \left(\frac{D}{s}\right)^2 \tag{7}$$

where:
SR = Ratio of safety factor before and after soil improvement
Gjg = Shear modulus jet-grouting column
ar = Area replacement ratio
D = jet-grouting column diameter(m)
s = space jet-grouting column (m)

The jet grouting column diameter used in this analysis is 1.0 m with various spaces. The S_R and safety factor analysis with the Özsoy & Durgunoğlu method can be seen in tables 8 and 9.

Table 8. The BH 78 SR analysis using jet grouting (1.0 m column diameter) with Özsoy & Durgunoğlu Method.

Space (m)	Area Replacement (a_r)	0-8 m Depth (S_R)	8-16 m Depth (S_R)
1.5	0.349	0.055	0.096
2	0.196	0.093	0.158
2.5	0.126	0.139	0.227
3	0.087	0.188	0.297
3.5	0.064	0.240	0.366
4	0.049	0.292	0.429

Table 9. The Safety Factor after jet grouting improvement.

Depth (m)	G_{max} (kN/m²)	FS_L	$FS_{required}$	$S_{R\ Required}$	S_R	FS_{JG}
2	64954.22	0.446	2.000	0.223	0.188	2.373
4	64954.22	0.496	2.000	0.248	0.188	2.639
6	64954.22	0.585	2.000	0.293	0.188	3.000
8	64954.22	0.673	2.000	0.336	0.188	3.000
10	116781.49	2.000	2.000	1.000	0.297	3.000
12	116781.49	0.786	2.000	0.393	0.297	2.643
14	116781.49	0.848	2.000	0.424	0.297	2.850
16	116781.49	0.936	2.000	0.468	0.297	3.000

Based on table 8, the space needed to achieve the target safety factor (FS > 2.0) is less than 3.5 D (3.5 m). In the next analysis, the space used is 3.0 m with a rectangle grid. The safety factors after improvement with Özsoy & Durgunoğlu method can be seen in table 9. Based on table 9, jet grouting with a column diameter of 1.0 m and spacing of 3 m (rectangular grid) can comply with cyclic stress reduction requirements.

5. Conclusions

From the series of analyses, it can be concluded that slope stability in all borehole locations achieved the safety factor required by SNI 8460:2017. Furthermore, the liquefaction potential is high in BH 14 and very high in BH 78, BH 16, and BH 17, hence all the locations need soil improvement to prevent future liquefaction. Soil improvement using stone columns simultaneously reduces $R_{u\ max}$ and cyclic stress ratio (CSR). On the other hand, jet grouting only reduces the cyclic stress ratio. With the soil properties dominated by sandy silt and sand, stone columns become a more favorable option to be used rather than jet grouting, which is more expensive.

Acknowledgments
The authors are grateful to the Department of Civil Engineering, Faculty of Engineering, Diponegoro University.

References

[1] Mase L Z 2017 Shaking Table Test of Soil Liquefaction in Southern Yogyakarta *Int. J. Technol.* **8** 747–60
[2] Mase L Z, Fathani T F and Adi A D 2021 A Simple Shaking Table Test to Measure Liquefaction Potential of Prambanan Area, Yogyakarta, Indonesia *ASEAN Eng. J.* **11** 89–108
[3] Badan Geologi 2019 *Peta Zona Kerentanan Likuefaksi Indonesia* (Bandung)
[4] Pusat Survei Geologi 2019 *Peta Geologi Indonesia* (Bandung)
[5] Badan Standardisasi Nasional 2017 *SNI 8640:2017 Persyaratan Perancangan Geoteknik* vol 8460 (Jakarta)
[6] Idriss I M and Boulanger R W 2010 *SPT-Based Liquefaction Triggering Procedures (No. UCD/CGM-10/02)* (Davis)
[7] Badan Standardisasi Nasional 2019 *SNI 1726:2019 Tata Cara Perencanaan Ketahanan Gempa Untuk Struktur Bangunan Gedung dan Non Gedung* (Jakarta)
[8] Baez J I and Martin G R 1993 Advances in the Design of Vibro-systems for Improvement of Liquefaction Resistance. *Proceedings of the Symposium on Ground Improvement* (Vancouver: Canadian Geotechnical Society) pp 1–16
[9] Priebe H J 1998 Vibro Replacement to Prevent Earthquake Induced Liquefaction *Gr. Eng.* **31** 30–3
[10] Özsoy B and Durgunoğlu H T 2003 Sıvılaşma Etkilerinin Yüksek Kayma Modüllü Zemin – Çimento Karışımı Kolonlarla Azaltılması *Beşinci Ulus. Deprem Mühendisliği Konf.* **İstanbul**
[11] Look B G 2007 *Handbook of Geotechnical Investigation and Design Tables* (London: Taylor & Francis Group)
[12] Asrurifak M, Irsyam M, Hutapea B M, Mahesworo R P, Ridwan M and Aldiamar F 2012 Peta Deagregasi Hazard Gempa Indonesia untuk Periode Ulang Gempa 2475 Tahun *PIT-XVI HATTI : Geotechnical Challenges in Present on Coming Nationwide Construction Activities* (Jakarta: Himpunan Ahli Teknik Tanah Indonesia)
[13] PUSGEN (Pusat Studi Gempa Nasional) Pusat Litbang Perumahan dan Permukiman 2017 *Peta Sumber Dan Bahaya Gempa Indonesia Tahun 2017 (Map of Indonesia Earthquake Sources and Hazards in 2017)* (Jakarta)
[14] Ancheta T D, Darragh R B, Stewart J P, Seyhan E, Silva W J, Chiou B S, K.E. W, Graves R W, Kottke A R, Boore D M and Kishida T 2013 *PEER NGA-WEST2 Database*
[15] Iwasaki T, Arakawa T and Tokida K 1984 Simplified Procedures for Assessing Soil Liquefaction During Earthquakes *Int. J. Soil Dyn. Earthq. Eng.* **3** 49–58
[16] Ozener P, Dulger M and Berilgen M 2015 Numerical Study on Effectiveness of Jet-Grout Columns in Liquefaction Mitigation *6th Int. Conf. Earthq. Geotech. Eng.*

Evaluation of Liquefaction Potential Index in the Educational Infrastructure Reconstruction Project at the University of Tadulako, Palu City, Central Sulawesi

H R Pasin[1,2] and S Ismanti[1,*]

[1] Department of Civil and Enviromental Engineering, Faculty of Engineering, Universitas Gadjah Mada Yogyakarta, Indonesia
[2] Directorate General of Human Settlement, Ministry of Public Works and Housing, Republic of Indonesia.

[*]Corresponding Author's e-mail address: sito.ismanti@ugm.ac.id

Abstract. On September 28, 2018, The Palu-Koro fault triggered an earthquake of M_w 7.5. at a depth of 20 km and 72 km northeast of Palu City. As a result, the University of Tadulako suffered significant damage post this earthquake. The Ministry of Public Works and Housing was assigned to reconstruct this location, which was located near the epicentre, and there are alluvium deposits of the Holocene age with loose sand soil and shallow groundwater levels on-site. Based on the Indonesian Seismic Code SNI 1726:2019, it is necessary to propose liquefaction potential and mitigation for building planning. The study aims to calculate the Safety Factor (*S.F.*) against liquefaction using a simplified procedure by Seed-Idriss and developed methods by Idriss-Boulanger. The *S.F.* was used as an input to determine the Liquefaction Potential Index (*LPI*) by Iwasaki et al. The study showed different categories of *LPI* values. The highest *LPI* value was found at BH-05, with a very high liquefaction potential. The soil depths susceptible to liquefaction are up to 9 meters. Application of deep foundation penetrating liquefiable soil layer is recommended as the mitigation. It is necessary to conduct further studies on the deep foundation analysis to be applied to this location.

1. Introduction

On September 28, 2018, at 18.02 Time of Central Indonesian (WITA), UTC +8, The U.S. Geological Survey reported a magnitude M_w 7.5 earthquake in the northern city of Palu, Central Sulawesi, at depths 20 km and 72 km to the northeast of the Palu City. The USGS said the Palu-Koro fault was the quake's source [1,2]. This earthquake triggered liquefaction in Balaroa Village, Petobo Village, Lolu Village, Jono Oge Village, and Sibalaya Village [3]. This earthquake resulted in severe damage that destroyed 177 buildings, and around 2200 people died, with over 1000 still missing [4,5]. The recent event resulted in the worst natural disaster in recorded history, causing the loss of 85% of infrastructure and many public facilities in Palu, Sigi, and Donggala [4,6], including educational infrastructure facilities and the University of Tadulako, suffered significant damage post this earthquake.

The Ministry of Public Works and Housing has been assigned to reconstruct the University of Tadulako in Palu, Central Sulawesi, which has suffered massive damage and requires accelerated recovery to carry out post-disaster education activities [7]. The University of Tadulako was chosen because it is near the earthquake epicentre. Besides, there are alluvium deposits of the Holocene age [8], with loose sand soil and groundwater levels in shallow conditions [9]. This area has the Palu-Koro fault in the Tondo segment [10]. In addition, research on earthquake-induced liquefaction is still lacking in this area. According to Indonesia's Seismic Code, SNI 1726:2019, In earthquake-prone areas, buildings designated as seismic design category *C* to *F* must include details on how their structure could withstand liquefaction and suggestions on liquefaction mitigation measures. These suggestions must be evaluated at a 2% probability of exceedance in 50 years. [11]. Liquefaction occurs when loose, saturated soil loses its strength and appears to flow as fluid due to an earthquake [12].

This research was conducted at the University of Tadulako. The purpose of this research was to find out the effect of liquefaction around this zone. The analysis of liquefaction potential in this research is predicted to be helpful consideration in the preparation of the building reconstruction project plan in the University of Tadulako phase II, besides that to select appropriate liquefaction mitigation alternatives in the field and for spatial mapping of the potential of the liquefaction hazard map in the University of Tadulako.

2. Research Method

This research used a Soil Investigation Report [9] to evaluate the liquefaction potential, carried out for fourteen boreholes with Standard Penetration Test (*SPT*) data collection spread over the University of Tadulako area, as presented in the **Table 1**. The potential liquefaction analysis is calculated by the Safety Factor (*S.F.*) against liquefaction using a simplified procedure by Seed-Idriss [13] and developed using the *SPT*-based approaches method composed by Idriss and Boulanger, which the value was obtained by comparing the ratio of cyclic resistance (*CRR*) and the ratio of cyclic stress (*CSR*) [14]. For further analysis, the *S.F.* value was used for determining the potential level of liquefaction using the Liquefaction Potential Index (*LPI*) method composed by Iwasaki et al. [15].

Table 1. Soil investigation data results of the standard penetration test (*N-SPT*) value [9]

Depth (m)	***N-SPT* value**													
	BH -01	**BH -02**	**BH -03**	**BH -04**	**BH -05**	**BH -06**	**BH -07**	**BH -08**	**BH -09**	**BH -10**	**BH -11**	**BH -12**	**BH -13**	**BH -14**
1	8	6	7	4	1	4	3	5	5	7	7	6	3	15
3	16	11	14	8	3	7	9	10	10	13	13	11	6	29
5	17	15	18	10	15	11	27	15	15	16	16	21	13	33
7	36	20	48	21	14	20	31	46	46	32	27	35	45	42
9	40	26	60	51	13	21	47	55	55	35	46	35	40	53
11	31	26	60	60	35	60	60	60	60	60	60	48	60	60
13	32	35	60	60	52	60	60	60	60	60	60	60	60	60
15	31	41	60	60	55	60	60	60	60	60	60	60	60	60
17	41	47	60	60	60	60	60	60	60	60	60	60	60	60
19	50	60	60	60	60	60	60	60	60	60	60	60	60	60
20	50	60	60	60	60	60	60	60	60	60	60	60	60	60

2.1. Geological Conditions

A study of geological conditions is essential to determining the characteristics of soil deposits to liquefaction susceptibility at the research site. The potential liquefaction hazard can occur in soil types dominated by non-plastic loose sandy soil and cohesionless soil. This type of soil is often found in alluvial deposits, which have not undergone cementation [16]. Alluvial deposits of the Holocene age are very susceptible to liquefaction. The groundwater level also greatly affects the liquefaction potential. The shallower the groundwater level, the higher the potential for liquefaction hazards on-site [17]. In an earthquake, the shallow groundwater table coupled with loose soil grains will increase pore water pressure, resulting in a liquefaction phenomenon. This is related to the amplification that occurs during an earthquake.

There are at least 7 (seven) objects of geological condition study in this research to determine whether the area under review has the liquefaction potential. The first part is geological formation. The study of the geological formation is an important part of determining the characteristics of soil deposits to liquefaction susceptibility at the research site. **Figure 1.(a).** shows the geological map in the research location, based on a geological reconnaissance map of the Palu quadrangle, Sulawesi, composed by

Sukamto [8] in this, which is formed from Celebes Molasses of Sarasin and Sarasin Material (*QTms*), including conglomerate, sandstone, mudstone, coral limestone, and marl, some weakly hardened (mainly limestone) as well as Alluvium and Coastal Sediments (*Qap*) which include gravel, sand, mud, and coral limestone [8]. The second part is the stratigraphy condition. Several boreholes are divided into several sectional profiles. It is helpful to describe the characteristics of each dominant soil layer at the location to be reviewed, as an initial determination of whether the location under review has a soil type that is vulnerable to liquefaction disasters. It was intended to facilitate the identification of subsurface soils. There are at least 14 (fourteen) boreholes divided into 5 (five) sectional profiles, which consist of cross sections A to A' into E to E' as shown in **Figure 1.(b)**.

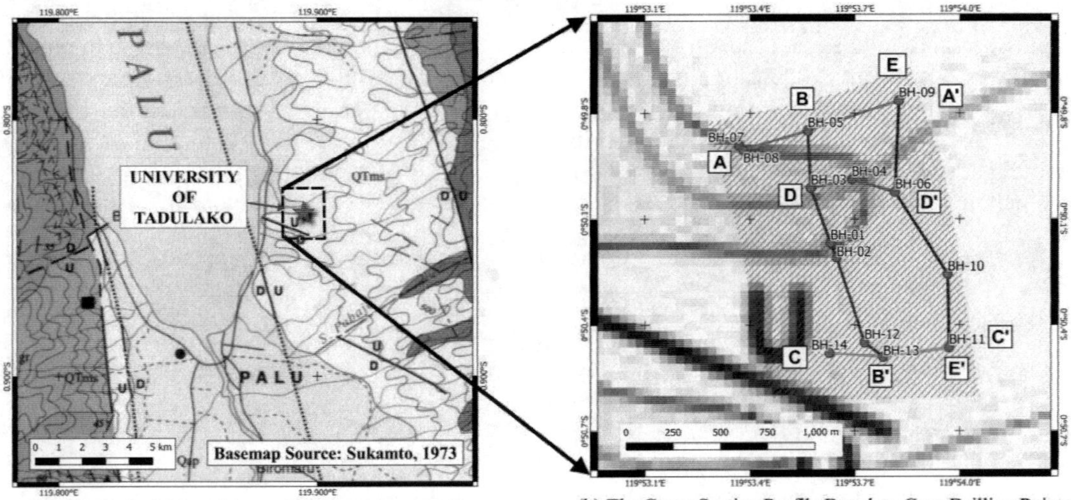

(a). Geological formation conditions at the research location

(b) The Cross-Section Profile Based on Core Drilling Points in the University of Tadulako Reconstruction Project Phase II

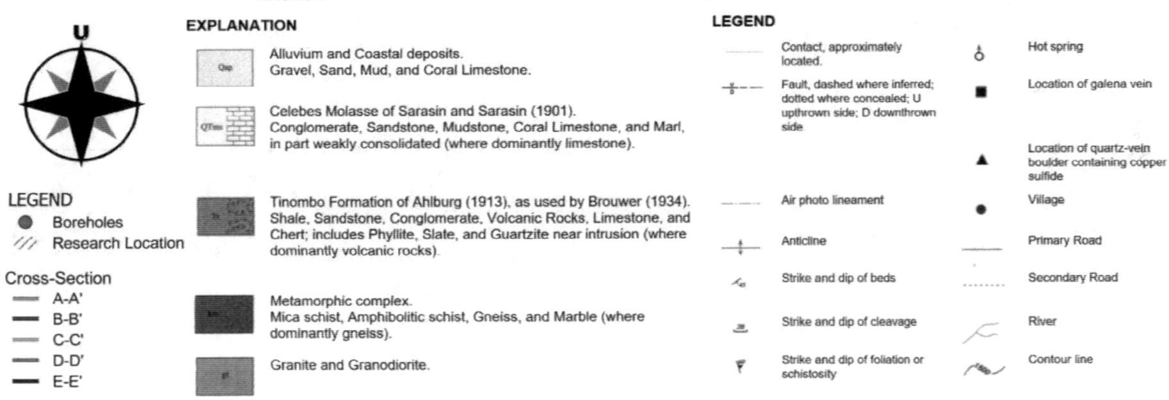

Figure 1. The Geological Map and The Distribution of Sectional Profile based on *SPT* Drilled Point in University of Tadulako, Palu City, Central Sulawesi (modified from [8]).

Figure 2 to **Figure 6** shows the result of the sectional identification of the stratigraphy in the study area. As seen in **Figure 2**, sectionals A-A' consisted of sandy silt with loose, medium, and dense condition and silty sand with loose, medium, and dense condition. The groundwater depth is around 3 meters. The soil type in this section is dominated by loose sand silt on the top layer of the soil surface, and the deeper the layer, the soil type is dominated by dense sandy silt. Sectional B-B' consisted of silty sand with loose, medium, and dense condition, clayey sand with medium and dense condition, sandy silt with loose and medium condition, and clayey silt with loose and medium condition. The groundwater depths are between 3 to 6 meters. This section is dominated by silty sand with loose conditions above the soil surface and the deeper the soil layer, the harder and denser the soil conditions. Sectional B-B' can be seen in **Figure 3**. **Figure 4** shows that sectional C-C' is fully dominated by silty sand with medium and dense soil condition. The groundwater depths are between 3 to 6 meters. The sectional D-D' consists of silty sand with medium and dense condition, clayey sand with dense condition, sandy silt with loose and medium condition, and a little dense clayey silt. The groundwater depth is around 3 meters. This section is dominated by sandy silt, clayey sand and silty sand, as shown in **Figure 5** And the last part of sectional stratigraphy, as shown in **Figure 6**, The sectional E-E' consists of silty sand with loose, medium, and dense condition, sandy silt with loose, medium, and dense condition, and a thin layer of medium gravel sand. The groundwater depth is around 3 meters. This section is dominated by silty sand with loose conditions above the soil surface and the deeper the soil layer, the harder and denser the soil conditions.

According to Thevanayagam et al. (2002) that indicates that soil types prone to liquefaction hazards are loose sand, silty sand, sandy silt containing non-plastic silt, and gravel. Liquefaction rarely occurs in clay soils (except for sensitive clays) [18]. Based on the results of identification and stratigraphic analysis at the research location, it was shown that the potential area for the building reconstruction project area of the University of Tadulako has liquefaction potential based on the dominant soil type, namely silty sand soil with loose to dense conditions, and shallow groundwater depths ranging from 3 – 6 meters from ground level.

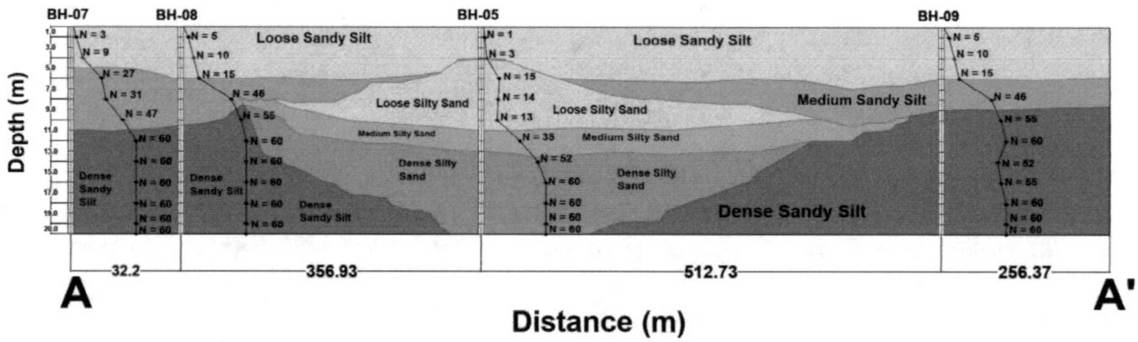

Figure 2. The Sectional of Soil Type for Profile A to A' of the Research Location.

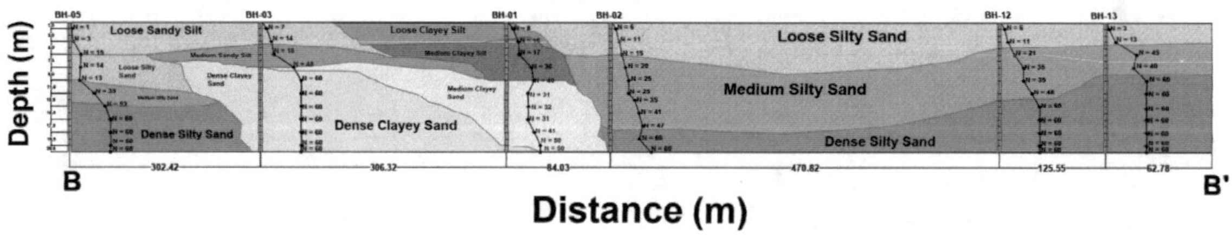

Figure 3. The Sectional of Soil Type for Profile B to B' of the Research Location.

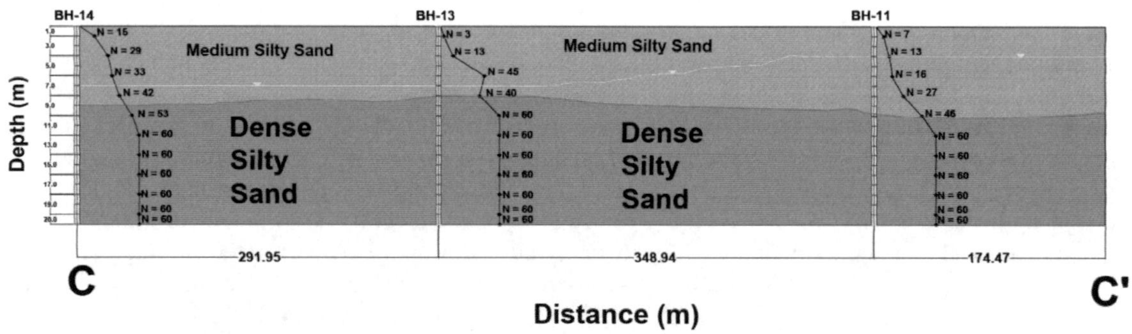

Figure 4. The Sectional of Soil Type for Profile C to C' of the Research Location.

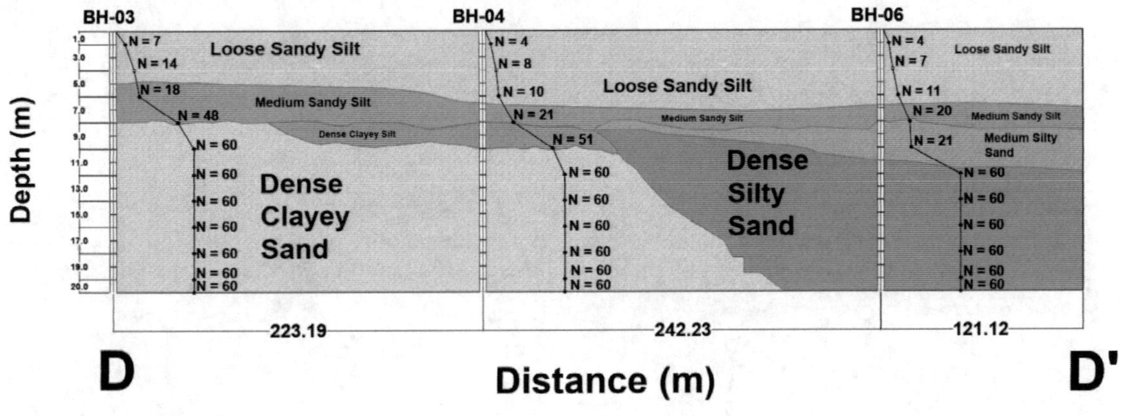

Figure 5. The Sectional of Soil Type for Profile D to D' of the Research Location.

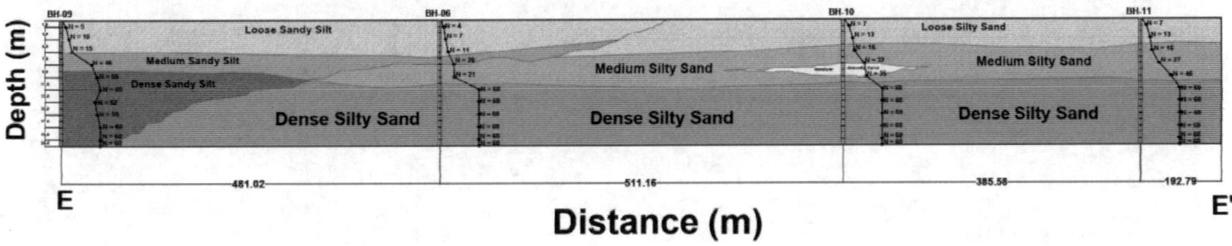

Figure 6. The Sectional of Soil Type for Profile E to E' of the Research Location

The third part is the physiographical condition. The University of Tadulako is located in the east Palu valley, especially in the Tondo Village, Mantikulore District, Palu City, Central Sulawesi. Physiographic conditions are related to geological conditions, geomorphology, topography, and geological structures that are formed as a result of developments and changes in natural conditions in an area due to the collision and subduction of tectonic plates as well as hydrological and climatological processes [19,20,21]. According to the Geological Agency (2018), the city of Palu is an area consisting of mountains, oceans, rivers, bays and valleys so that this city bears a five-dimensional city, as shown in **Figure 7** and **Figure 8**, which is a digital model of the shape of the earth's surface around the research location represented in two and three dimensional views [10,24,25,26].

The fourth part is the geomorphological condition, based on the Geomophological Map of Indonesia at a scale of 1:5000000, the Holocene alluvial valleys are alluvial plains bounded by slopes and hills that have the possibility of soil movement in areas with high topography with non-compact rocks. This area identified the presence of alluvial fans. The tectonic conditions in this area are very vulnerable to geological disasters [10,21]. The geographical location of the research location is in the eastern valley of Palu and is bordered by waters (Palu Bay) to the west and hills to the east of the research location. On the north side and the research location is bordered by 2 watersheds (*DAS*). Based on the soil investigation report belonging to the Geotechnical Consultant [9], the University of Tadulako is included in the denudation geomorphological unit and hills with steep slopes. Denudation is a phenomenon of lowering the relief in a plain that slopes down to the sea due to the very slow process of the strength of rainwater and wind [20]. Denudational geomorphology occurs due to degradation processes and soil movements such as erosion and landslides which then experience deposition [25]. When an earthquake occurs, geomorphological conditions like this can cause flow failure problems if they occur on loose, saturated, and sloped soils which are often found near water sources, such as lakes, rivers and beaches [26].

According to Day (2012), soil deposits that are located near the water table tend to be more susceptible to liquefaction because the structure of the soil layer is loose and is more easily separated by water [16]. **Figure 7** shows that the research location is located in an area close to water, which is the coast of Palu Bay and is flanked by 2 (two) watersheds, namely Watutela and Watutailo. This is related to the estimation of liquefaction vulnerability to environmental conditions (soil deposits) in the floodplain area around the research location [16], which with water flow makes it possible to increase the accumulation of groundwater in the area [10]. Elevation based on contour lines and geographical location of the research location can be seen in **Figure 7**.

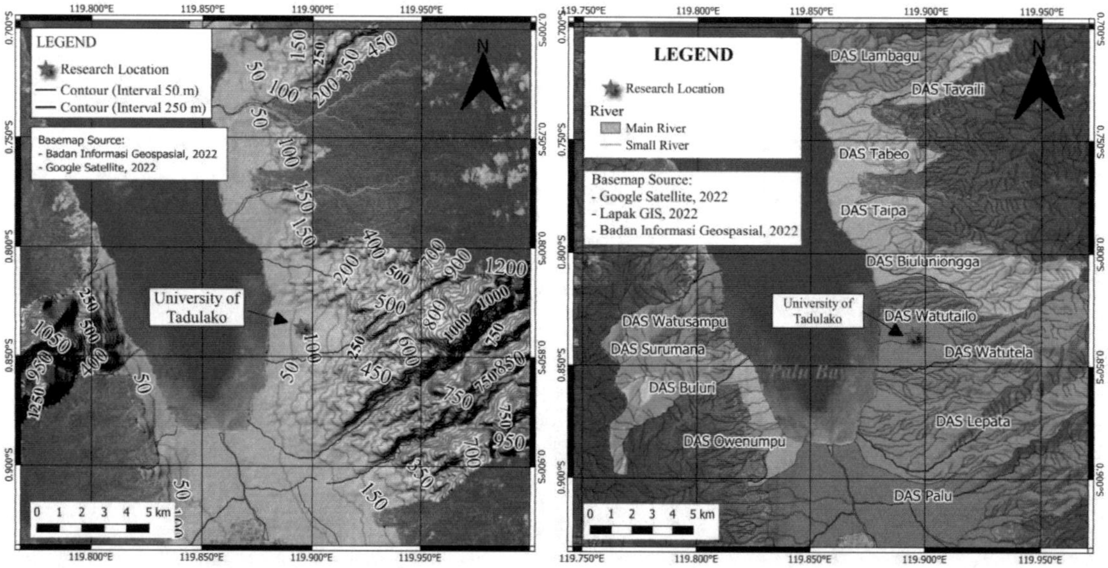

(a). Contour Line Condition in Palu City (b) Research Location Close to the Water Table

Figure 7. Elevation based on contour lines and geographical location of the research location (modified from [24,25,26]).

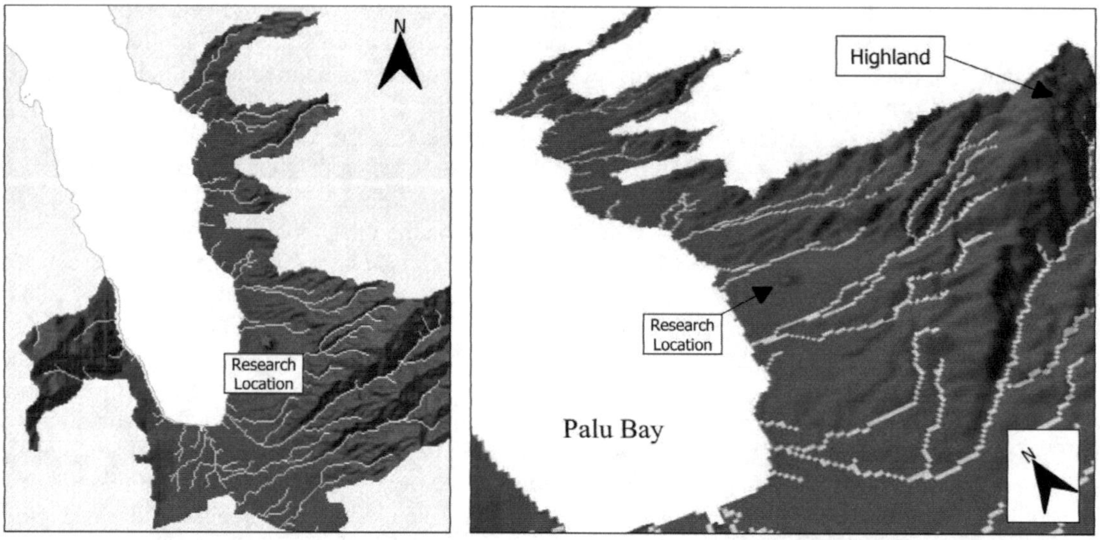

(a). Topography in a 2-dimensional (2D) view around the research location (b). Topography in a 3-dimensional (3D) view around the research location

Figure 8. Topography Condition in 2D and 3D View Resulting from Digital Elevation Model (DEM) Around in Research Location, the University of Tadulako (Basemap source: modified from [23]).

The fifth part is the topographical condition. **Figure 8** shows the research location located in a sloping hilly area not far from the seacoast, namely, coastal Palu bay. The sixth part, in hydrogeological condition, there is a Tawaeli groundwater basin in this area, which has moderate to high groundwater potential [27], with a depth of the free groundwater level between 6 to 15 meters, with depressed aquifer lithology between 8 to 50 meters [10]. **Figure 9** shows that the BH-03, BH-05, BH-07, and BH-08 in the research location, the University of Tadulako, intersect with the Tawaeli Groundwater Basin (GWB) [22,27].

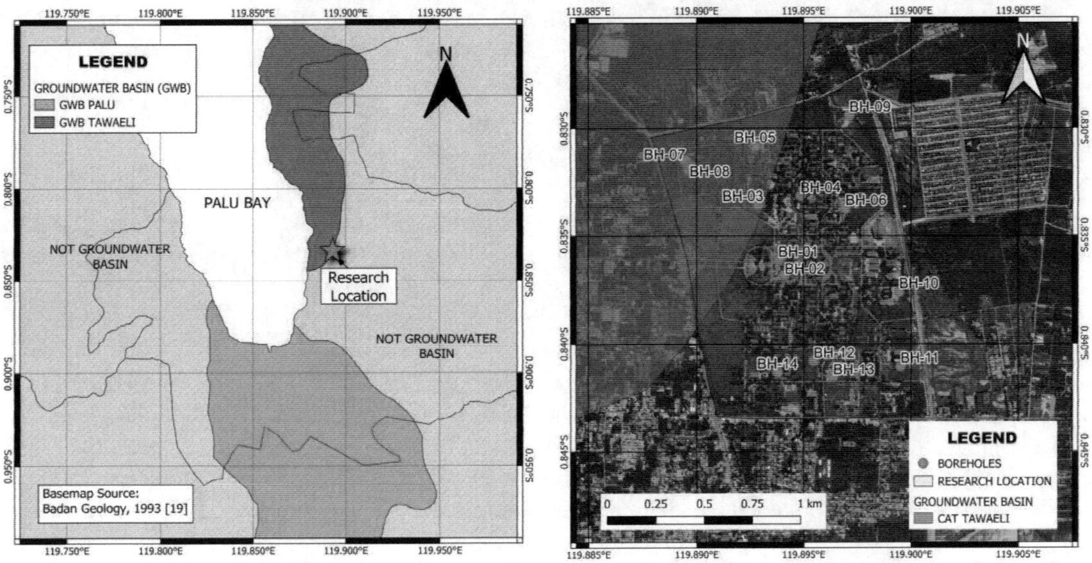

(a). Hydrogeological map in the form of the Groundwater Basins of Surrounding the Palu City

(b) Boreholes in Research Location Intersect with Tawaeli Groundwater Basin

Figure 9. Groundwater Basin Condition in Research Location [19,22].

Figure 10. Documentation of groundwater level rise when drilling the borehole BH-02 Rectorate Building, University of Tadulako [9].

Figure 11. Wet core box soil samples from a depth of 3 to 4 meters at the borehole BH-05 Dean of Teacher Training and Education Building [9].

This is also supported by the documentation attached Soil Investigation Report belonging to the geotechnical consultant project [9], in which when drilling was carried out at one-point BH-02 of the Rectorate Building, the groundwater level rose to the surface, as shown in **Figure 10**. and accompanied by documentation of a wet core box soil sample from a depth of 3 to 4 meters at the borehole BH-05 Dean of Teacher Training and Education as shown in **Figure 11.** Hence in this research, the Groundwater level is determined through the visualization of core box samples on wet soil for each borehole ranging from 3 to 6 meters, as presented in **Table 6.**

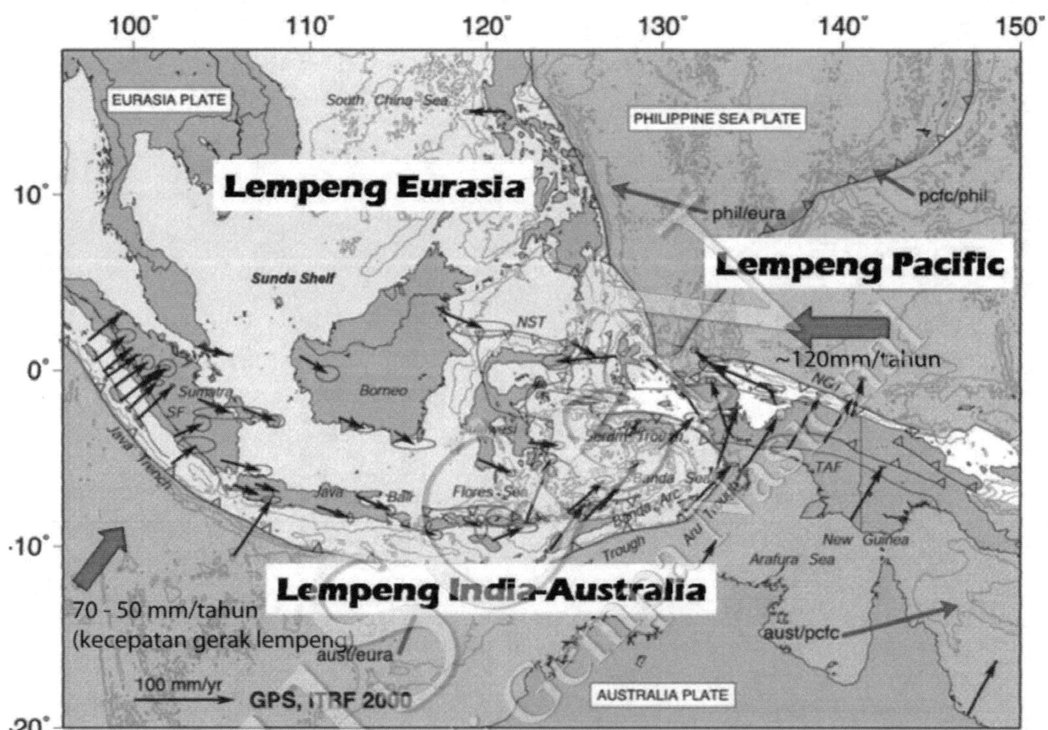

Figure 12. Active plates in the Indonesian region (Bock et al. (2003) in [28]).

The seventh part is the geological structure condition. Central Sulawesi is located between 3 of the Earth's tectonic plates, such as the Indo-Australian Plate, the Pacific Plate, and the Eurasian Plate, which meet each other as shown in **Figure 12** [6,10,28]. Therefore, this area is very vulnerable to natural disasters, especially those caused by tectonic movements that cause shifts in the Palu-Koro Fault and other faults identified in this region [10]. The Palu-Koro Fault is the main structure of Sulawesi Island which extends from Central Sulawesi to the Karimata Strait [28] with a length of more than 300 km [6]. These faults are always moving and colliding with each other, which results in such a large fault, and is followed by movement on the earth's surface [20]. Recent research estimates that the Palu-Koro fault is actively moving at a speed of 41-45 mm/year [29].

The University of Tadulako is located near to the Palu-Koro Fault and several other identified faults around it, so this area is very vulnerable to natural geological disasters such as earthquakes and liquefaction. Based on the map of the Palu disaster-prone space zone [30]. The University of Tadulako is located in a zone prone to high earthquakes and medium liquefaction disasters. When the 2018 earthquake occurred, the University of Tadulako suffered very significant damage. Post-earthquake rebuilding planning requires a more in-depth analysis of the seismic hazard at the study site, bearing in mind that this area is built on loose sandy soil with a shallow groundwater table, resulting in the possibility of liquefaction hazard that will occur in this area. Thus, mitigation efforts are needed to reduce the impact of the liquefaction that occurs.

2.2. Probabilistic analysis using Indonesian Seismic Code SNI 1726:2019

SNI 1726:2019 is the minimum standard for earthquake resistance planning for building and non-building structures. Referring to Indonesian Geotechnical Design Requirements Code SNI 8460:2017, stipulates references for the seismic design of buildings and non-buildings using Indonesia Seismic Code SNI 1726:2012, which has been updated to SNI 1726:2019 [31]. In addition, this probability analysis was chosen because the University of Tadulako is an educational facility that, according to SNI 1726:2019, is categorized as an important facility that is included in risk category IV (the highest category in this code), which is very vital for life safety so that the planned age used is exceeded 2% in 50 years, with a return period of 2500 years.

Determination of peak ground acceleration (*PGA*) using *N-SPT* value data with depth for each drill point and the assistance of the Design Spectrum Response application belonging to the Directorate General of Human Settlements, Ministry of Public Works and Public Housing, and the National Earthquake Study Center Republic of Indonesia [32]. Soil type greatly affects *PGA* value, the worse the soil type, the higher the *PGA* value on-site. Hence *PGA* value needs to be corrected with the coefficient of site class as shown in Table 2. [11] proposed the probabilistic analysis using Indonesian Seismic Code SNI 1726:2019 as:

$$PGA_M = F_{PGA} \; x \; PGA \qquad (1)$$

where PGA_M is peak ground acceleration at the surface, F_{PGA} is the coefficient of site class determined by **Table 2,** and *PGA* is peak ground acceleration (*g*) using [32].

Table 2. Coefficient of site class [11].

Site Class	***PGA (g)***					
	0.1	**0.2**	**0.3**	**0.4**	**0.5**	**0.6**
SA	0.8	0.8	0.8	0.5	0.8	0.8
SB	0.9	0.9	0.9	0.9	0.9	0.9
SC	1.3	1.2	1.2	1.2	1.2	1.2
SD	1.6	1.4	1.4	1.2	1.1	1.1
SE	2.4	1.9	1.9	1.4	1.2	1.1
SF	A special evaluation is required at each location					

2.3. Liquefaction Potential Analysis

One method of analyzing the liquefaction potential is by using the Simplified Procedure method by Seed and Idriss [13] and developed using *SPT*-based liquefaction triggering analysis for a single boring procedure to evaluate the liquefaction potential during earthquakes by Idriss and Boulanger [14] which obtained the Safety Factor (*S.F.*) value by comparing the ratio of cyclic resistance (*CRR*) and the ratio of cyclic stress (*CSR*), as seen in **Equation (2)** below:

$$SF=\frac{CRR_{M,\sigma'_{vc}}}{CSR} \tag{2}$$

The simplified procedure uses the equation Seed, and Idriss [13] developed to estimate the shear stress caused by cyclic loading. *CSR* is calculated as **Equation (3)** below:

$$CSR = 0.65\left(\frac{\sigma_{vc}}{\sigma'_{vc}}\right)\left(\frac{a_{max}}{g}\right)r_d \tag{3}$$

where σ_{vc} is the total of vertical stress (kN/m^2), $\sigma'_{vc}$is the effective vertical stress (kN/m^2) and r_d is the coefficient of shear stress reduction.

CRR has described as soil strength to resist liquefaction hazard. The analysis for *CRR* based on in situ soil investigation with *N-SPT* data which has been corrected for clean sand with M_w = 7.5 and σ'_{vc}=1 atm, that can be seen in the following **Equation (4)** to **(8)**:

$$CRR_{M=7.5,\ \sigma'_{vc}=1\ atm}=exp\left(\frac{(N_1)_{60cs}}{14,1}+\left(\frac{(N_1)_{60cs}}{126}\right)^2-\left(\frac{(N_1)_{60cs}}{23,6}\right)^3+\left(\frac{(N_1)_{60cs}}{25,4}\right)^4-2,8\right) \tag{4}$$

where $(N_1)_{60cs}$ is the *SPT* value of clean sand that has been corrected for overburden is 1 atm.

CRR value which has been corrected for clean sand with M_w = 7.5 and σ'_{vc}=1 atm, then corrected again with correction factor by Overburden Effective Stress and Magnitude Scaling Factor (*MSF*) hence *CRR* value equal to earthquake M_w 7.5.

$$CRR_{M,\sigma'_{vc}} = CRR_{M=7.5,\ \sigma'_{vc}=1\ atm} \cdot MSF \cdot K_\sigma \tag{5}$$

$$MSF = 6.9\ exp\left(\frac{-M}{4}\right)-0.058 \leq 1.8 \tag{6}$$

$$K_\sigma = 1 - C_\sigma\ ln\left(\frac{\sigma'_{vc}}{Pa}\right) \leq 1.1 \tag{7}$$

$$C_\sigma = \frac{1}{18.9-2.55\sqrt{(N_1)_{60cs}}} \leq 0.3 \tag{8}$$

where *MSF* is a correction factor for the magnitude scale equal M_w 7.5, K_σ is overburden correction factor and C_σ is overburden resistance correction factor.

2.4. Liquefaction Potential Index (LPI)

The liquefaction potential index is a method developed by Iwasaki *et al.* [15] to identify the level of vulnerability to potential liquefaction hazards. The *LPI* is an advanced analysis to define the thickness of the liquefied soil using data *S.F.* value from *SPT*-based procedure analysis by Idriss and Boulanger [16]. The *LPI* formula can be seen in the following **Equation (9)** until **(13)**:

$$LPI= \int_0^{20} F(z).w(z).dz \tag{9}$$

$$F(z) =1-FS \text{ for } FS < 1 \tag{10}$$

$$F(z) = 0 \text{ for } FS \geq 1 \tag{11}$$

$$w(z) =10 - 0.5z \text{ for } z < 20 \text{ meter} \tag{12}$$

$$w(z) = 0 \text{ for } z \geq 20 \text{ meter} \tag{13}$$

where F is the level of damage due to liquefaction in the soil layer. w is depth factor and Z is depth in meter (m). The liquefaction potential category is based on *LPI* value, as shown in **Table 3.**

Table 3. Liquefaction potential category based on *LPI* Value [15].

LPI Value	Liquefaction Potential Category
$LPI = 0$	Very Low
$0 < LPI \leq 5$	Low
$5 < LPI \leq 15$	High
$LPI > 15$	Very High

3. Results and Discussion

Based on the probabilistic analysis using the Indonesia's Seismic Code SNI 1726:2019 result and discussion, on determining the peak ground acceleration (*PGA*) obtained for seismic site classification in this study area is dominated by stiff soil or medium soil (*SD*) and one borehole BH-14 has very dense soil or hard soil (*SC*) [11]. The PGA_M values range from 0.543 g to 0.587 g, as shown in **Table 4**. Based on Fathani *et al.* [33], the *PGA* value is more than equal to 0.40 g ($PGA \geq 0.40$ g), and it has a very high level of risk of damage caused. Hence the result of the analysis is used for further analysis of the potential for liquefaction hazards.

Before estimating liquefaction potential, we must find the safety factor value first using *SPT*-based procedure analysis by Idriss and Boulanger [14]. Based on the results presented in **Table 5**, there are 6 (six) boreholes that liquefaction occurred in the boreholes number BH-01, BH-02, BH-05, BH-06, BH-10, and BH-11. The liquefied soil layer is marked with a red label sign. The depths susceptible to liquefaction ranging from 5 meters to 9 meters. Liquefaction occurs in vulnerable soil layers when the safety factor value less than 1 (*S.F.* < 1). The lower the safety factor value, the more vulnerable soil to liquefy. The analysis starts from a depth of 3 meters due to the availability of data obtained from the Soil Investigation Report by the Geotechnical Consultant [9]. This analysis cannot be applied or not applicable (n.a.) to unsaturated soils and clay, hence this is only applicable for saturated loose sand.

Table 4. Probabilistic Seismic Hazard Analysis Results.

Boreholes	Coordinates		Peak Ground Acceleration SNI 1726:2019				Building Reconstruction Plan
	X	Y	Site Class	F_{PGA}	*PGA*	PGA_M	
BH-01	-0.836	119.894	*SD*	1.111	0.489	0.543	Rectorate Building
BH-02	-0.837	119.894	*SD*	1.111	0.489	0.543	
BH-03	-0.834	119.893	*SD*	1.111	0.489	0.543	Dean of Social and Political Science Building
BH-04	-0.833	119.895	*SD*	1.111	0.489	0.543	College of Teacher Training and Education Building
BH-05	-0.831	119.893	*SD*	1.111	0.489	0.543	Dean of Teacher Training and Education Building
BH-06	-0.834	119.897	*SD*	1.111	0.489	0.543	Dean & College of Law Building
BH-07	-0.832	119.889	*SD*	1.111	0.489	0.543	Physical Education, Health and Recreation Study Building
BH-08	-0.832	119.890	*SD*	1.111	0.489	0.543	
BH-09	-0.829	119.897	*SD*	1.111	0.489	0.543	Dormitory Building
BH-10	-0.838	119.899	*SD*	1.111	0.489	0.543	Forestry Building
BH-11	-0.841	119.899	*SD*	1.111	0.489	0.543	Animal Husbandry and Fisheries Building
BH-12	-0.841	119.895	*SD*	1.111	0.489	0.543	Public Health Building
BH-13	-0.841	119.896	*SD*	1.111	0.489	0.543	Medical Building
BH-14	-0.841	119.894	*SC*	1.2	0.489	0.587	Architectural Design and Theory Laboratory Building

Table 5. The Safety Factor (*S.F.*) vs. Depth Analysis Result.

Depth (m)	Safety Factor (*S.F.*) in Boreholes Number													
	01	02	03	04	05	06	07	08	09	10	11	12	13	14
0	n.a.	n.a.	n.a.	n.a.	n.a.	n.a.	n.a.	n.a.	n.a.	n.a.	n.a.	n.a.	n.a.	n.a.
1	n.a.	n.a.	n.a.	n.a.	n.a.	n.a.	n.a.	n.a.	n.a.	n.a.	n.a.	n.a.	n.a.	n.a.
3	n.a.	n.a.	n.a.	n.a.	n.a.	n.a.	n.a.	n.a.	n.a.	n.a.	n.a.	n.a.	n.a.	n.a.
5	0.98	0.76	n.a.	n.a.	0.72	n.a.	n.a.	n.a.	n.a.	0.82	0.82	n.a.	n.a.	n.a.
7	4.35	1.03	4.33	1.12	0.50	0.88	n.a.	n.a.	n.a.	4.34	4.34	4.93	5.27	4.81
9	3.97	3.01	3.95	3.96	0.42	0.87	n.a.	n.a.	n.a.	3.96	3.96	4.36	4.59	4.19
11	3.73	1.91	3.71	3.72	3.73	3.73	n.a.	n.a.	n.a.	3.72	3.72	4.02	4.19	3.83
13	3.57	3.57	3.55	3.57	3.57	3.57	n.a.	n.a.	n.a.	3.56	3.56	3.80	3.93	3.60
15	3.26	3.46	3.44	3.46	3.46	3.46	n.a.	n.a.	n.a.	3.45	3.45	3.65	3.76	3.44
17	3.39	3.39			3.38							3.54		
19	3.33	3.34			3.33									
20	3.31	3.31			3.31									

From the analysis of safety factor, the liquefaction potential index (*LPI*) method by Iwasaki *et al.* [15] can be calculated for liquefaction potential and defining the thickness of the liquefied soil. The dominant soil in study area consist of 3 (three) boreholes is clayey sand, 8 (eight) boreholes is silty sand, and 3 (three) boreholes is sandy silt. Based on the results and discussion of liquefaction potential category, there are 8 (eight) boreholes that are very low, 5 (five) boreholes are low and only 1 (one) borehole is very high. The highest *LPI* value is 16.996 in borehole BH-05 with a very high category. The *LPI* value can be categorized as very high, if the value is more than 15 (fifteen) (*LPI* > 15). The dominant soil in this borehole (BH-05) is silty sand. **Table 6** shows the results of the liquefaction potential index in study area.

Table 6. The Liquefaction Potential Index (*LPI*) Result.

Boreholes	Groundwater Table	Dominant Soil	Depths Vulnerable to Liquefaction	*LPI* Value	*LPI* Categories
BH-01	3 meters	Clayey Sand	5 meters	0.233	Low
BH-02	3 meters	Silty Sand	5 meters	3.526	Low
BH-03	3 meters	Clayey Sand	n.a.	0.000	Very Low
BH-04	3 meters	Clayey Sand	n.a.	0.000	Very Low
BH-05	3 meters	Silty Sand	5 to 9 meters	16.996	Very High
BH-06	3 meters	Silty Sand	7 to 9 meters	3.003	Low
BH-07	3 meters	Sandy Silt	n.a.	0.000	Very Low
BH-08	3 meters	Sandy Silt	n.a.	0.000	Very Low
BH-09	3 meters	Sandy Silt	n.a.	0.000	Very Low
BH-10	3 meters	Silty Sand	5 meters	2.755	Low
BH-11	3 meters	Silty Sand	5 meters	2.628	Low
BH-12	5 meters	Silty Sand	n.a.	0.000	Very Low
BH-13	6 meters	Silty Sand	n.a.	0.000	Very Low
BH-14	6 meters	Silty Sand	n.a.	0.000	Very Low

The spatial evaluation of the liquefaction potential index is a spatial analysis based on the calculated LPI value and then mapped using *Quantum GIS Desktop 3.16.15.* The analysis was carried out using the Inverse Distance Weighting (IDW) Interpolation method. IDW Interpolation is a mathematical (deterministic) method by estimating the unknown value by assuming the closest value associated with the activity. **Figure 13** shows the spatial liquefaction potential index at the University of Tadulako based on the *LPI* value, which is categorized from very low to very high. In **Figure 13** can be seen that the borehole BH-05 has a brighter colour than the other boreholes, which indicates that the borehole BH-05 has a very high level of liquefaction potential. This liquefaction potential hazard map is useful for spatial planning for the reconstruction project of the University of Tadulako phase II and other future development in the university or around this region, especially for preventive measures against liquefaction hazards [9].

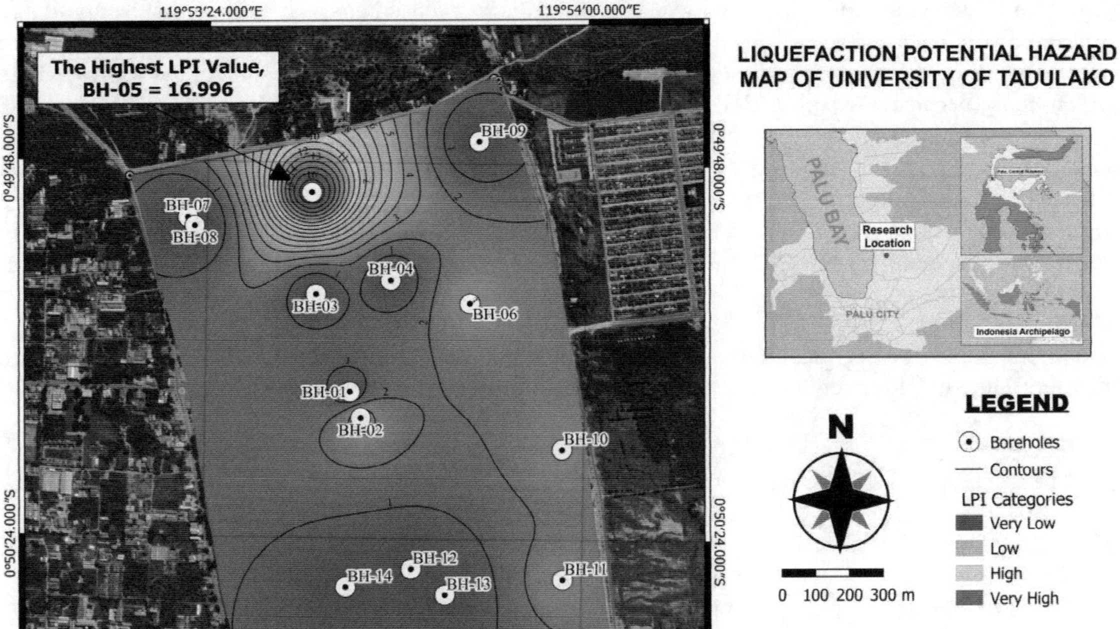

LPI Value	Liquefaction Potential Category
LPI = 0	Very Low
0 < LPI ≤ 5	Low
5 < LPI ≤ 10	High
LPI > 15	Very High

Figure 13. Liquefaction Potential Hazard Map of the University of Tadulako [22].

Based on the results of the analysis that can be seen in **Table 5**, **Table 6**, and **Figure 13**,Why does the largest *LPI* value occur at the borehole BH-05? Why does it anomaly only occur at that borehole while at other boreholes it doesn't? This anomaly can happen because the liquefied soil layer at the borehole number BH-05 was quite deep compared to other boreholes, as we can see from the SPT-based liquefaction triggering analysis for a single boring by Idriss and Boulanger method [14] in **Table 5** and liquefaction potential index by Iwasaki et al. [15] in **Table 6** , where the depth was prone to soil liquefaction at a depth of 5 to 9 meters, or in other words, the thickness of the liquefied soil was 4 meters deep, while others are thinner, only around 1 until 2 meters deep.

In addition, if seen in **Table 6**, the borehole number BH-05 is dominated by silty sand soil types with *N-SPT* values ranging from 1 to 15 to a depth of 9 m, as shown in **Table 1**. This value affects the value of the cyclic resistance ratio (*CRR*) of the soil so that the smaller the value of the *N-SPT*, the lower the value of the *CRR*, with a PGA_M value of 0.543 (g) and the magnitude of M_w 7.5, a safety factor of 0.42 to 0.72 (*S.F.* < 1) is obtained. The liquefied soil is at a depth of 5 to 9 meters. The value of the factor of safety at that depth produces a value of liquefaction potential index (*LPI*) of 16.996 (*LPI* > 15) so it is included in the very high category.

4. Conclusion

Based on geological condition studies results, the University of Tadulako has a potential liquefaction hazard. The results of the potential liquefaction analysis at the University of Tadulako show several liquefaction potential categories ranging from very low to very high, with various depths vulnerable to liquefaction. The highest *LPI* value was found at point BH-05, in which the *LPI* value is 16.996. For

LPI value greater than 15 (*LPI* > 15) is very high in the liquefaction potential category. It means the potential for liquefaction hazards occur in the soil at the borehole is very high, and the liquefied soil layer is thick. Besides that, some boreholes were found in clay, such as BH-07, BH-08, and BH-09. Hence the analysis cannot apply to the clay soil (n.a.) and is considered not to be liquefied with a category of liquefaction potential very low (*LPI* = 0). The *SPT*-based analysis also cannot be applied to unsaturated soil, such as the soil layer above the groundwater table. The soil depths are vulnerable to potential liquefaction ranging from a depth of 5 meters to 9 meters. From the results of the analysis there is an anomaly because the soil in BH-05 has very bad conditions with an *N-SPT* value of less than 15 (N < 15). Recommendations related to structural mitigation of liquefaction hazards are to place deep foundations under soil layers vulnerable to potential liquefaction hazards. It is necessary to conduct further studies on the appropriate mitigation plan alternatives to be applied for this location, given the various categories and depths that are prone to liquefaction hazards. In addition, the use of maps in evaluating the potential for liquefaction disasters is an aspect of disaster mitigation for regional planning.

5. Acknowledgments
The authors would like to thank the Balai Prasarana Permukiman Wilayah Sulawesi Tengah, Directorate General of Human Settlement, Ministry of Public Works and Housing, Republic of Indonesia.

6. References
[1] Mason B Hutabarat D and Prakoso W, 2019 Geotechnical Reconnaissance: The 28 September 2018 M7.5 Palu-Donggala, Indonesia Earthquake *Rep. Geotech. Extrem. Events Reconnaissance, Geotech. Extrem. Events Reconnaiss. Assoc.* April p. 1–86.
[2] The United States Geological Survey, 2018, M7.5 Palu, Indonesia Earthquake of September 28, 2018 Poster. [Online]. Available: https://earthquake.usgs.gov/product/poster/20180928/us/1540320089835/poster.pdf.
[3] Montgomery J Wartman J Reed A N Gallant A P Hutabarat D and Mason H B, 2021 Field reconnaissance data from GEER investigation of the 2018 MW 7.5 Palu-Donggala earthquake *Data Br.* **34** p. 106742.
[4] Kijewski-Correa T and Robertson I, 2018 StEER: Structural Extreme Event Reconnaissance Network PALU EARTHQUAKE AND TSUNAMI, SULAWESI, INDONESIA FIELD ASSESSMENT TEAM 1 (FAT-1) EARLY ACCESS RECONNAISSANCE REPORT (EARR) **1** p. 1–71.
[5] Asian Development Bank and Ministry of Public Works and Housing Republic of Indonesia, 2019 THE INITIAL ENVIRONMENTAL EXAMINATION (IEE) RECONSTRUCTION AND REHABILITATION PALU COASTAL PROTECTION November.
[6] Irsyam, 2018 *Kajian Gempa Palu Provinsi Sulawesi Tengah 28 September 2018 (M7.4)* **1** .
[7] Republik Indonesia, 2018, Instruksi Presiden Republik Indonesia Nomor 10 Tahun 2018 Tentang Percepatan rehabilitasi dan Rekonstruksi Pascabencana Gempa Bumi dan Tsunami di Provinsi Sulawesi Tengah dan Wilayah Terdampak Lainnya. p. 1–25.
[8] Sukamto, 1973, Peta Geologi Tinjau, Indonesia Lembar Palu Skala 1 : 250.000. Pusat Penelitian dan Pengembangan Geologi, p. 2015 & 2115.
[9] PT. Geocipta Bangun Optima, 2021, Laporan Penyelidikan Tanah Universitas Tadulako Zona 1, Bandung.
[10] Badan Geologi Kementerian ESDM, 2018 *Di Balik Pesona Palu Bencana Melanda Geologi Menata* Bandung: Badan Geologi, Kementerian ESDM.
[11] BSN, 2019 *SNI 1726: 2019 - Tata cara perencanaan ketahanan gempa untuk struktur bangunan gedung dan non gedung* .
[12] Kramer S L, 1996 *Geotechnical Earthquake Engineering* Prentice-Hall, Inc.
[13] Seed H B and Idriss I M, 1970 *A Simplified Procedure for Evaluating Soil Liquefaction Potential* November .
[14] Idriss I M and Boulanger R W, 2008 *Soil Liquefaction During Earthquakes* **13**, 2 .
[15] Iwasaki T Tokida K and Tatsuoka F, 1981 Soil Liquefaction Potential Evaluation with Use of the Simplified Procedure *Int. Conf. Recent Adv. Geotech. Earthq. Eng. Soil Dyn.* **12** p. 209–

214.
[16] Day R W, 2012 *Geotechnical Earthquake Engineering Handbook* 2nd Ed. Mc Graw Hill.
[17] Youd T L Tinsley J C Perkins D M King E J and Preston R F, 1979 Liquefaction Potential Map of San Fernando Valley, California. *Geol. Surv. Circ. (United States)* 807 p. 37–48.
[18] Thevanayagam S and Martin G R, 2002 Liquefaction in silty soils - screening and remediation issues *Soil Dyn. Earthq. Eng.* **22**, 9–12 p. 1035–1042.
[19] Suharyono, 2020 *Dasar - Dasar Kajian Geografi Regional* Yogyakarta: Penerbit Ombak.
[20] Sukandarrumidi Purnamawati D I Miftahussalam Rakhman A N and Maulana F ., 2021 *Geologi Umum Bagian Kedua* Edisi Kedu Yogyakarta: Gadjah Mada University Press.
[21] Verstappen H Sutikno and Suratman, 2022 *Garis Besar Geomorfologi Indonesia* Yogyakarta: Gadjah Mada University Press.
[22] Google Satellite, 2022, Peta Universitas Tadulako. [Online]. Available: https://www.google.com/maps/place/Tadulako+University/@-0.838742,119.8899821,4323m/data=!3m1!1e3!4m5!3m4!1s0x2d8be942bc0a6703:0xf8a9048c767a7489!8m2!3d-0.8364322!4d119.8936937. [Accessed: 06-Nov-2022].
[23] Badan Informasi Geospasial, 2022, Peta Kota Palu. [Online]. Available: https://tanahair.indonesia.go.id/portal-web/download/perwilayah. [Accessed: 01-Nov-2022].
[24] Lapak GIS, 2022, Peta Daerah Aliran Sungai. [Online]. Available: https://www.lapakgis.com/2019/07/shapefile-peta-daerah-aliran-sungai-indonesia-gis.html. [Accessed: 10-Nov-2022].
[25] Indriyani N K S, 2014, Geomorfologi Indonesia Klasifikasi Bentuk Lahan. [Online]. Available: https://mahasiswa.ung.ac.id/451412046/home/2014/10/27/geomorfologi-indonesia-bentuklahan.html. [Accessed: 10-Nov-2022].
[26] Hardiyatmo H ., 2022 *Rekayasa Gempa untuk Analisis Struktur dan Geoteknik* Yogyakarta: Gadjah Mada University Press.
[27] Arief and Hidayat, 1993, Peta Hidrogeologi Indonesia, Lembar Palu dan Parigi, Sulawesi Skala 1 : 250.000. Direktorat Geologi Tata Lingkungan, Kementerian ESDM.
[28] Pusat Studi Gempa Nasional and Pustlitbang PUPR, 2017 *Peta Sumber dan Bahaya Gempa Indonesia Tahun 2017* Bandung: Pusat Penelitian dan Pengembangan Perumahan dan Permukiman, Badan Penelitian dan Pengembamgan, Kementerian PUPR.
[29] Socquet A *et al.*, 2006 Microblock rotations and fault coupling in SE Asia triple junction (Sulawesi, Indonesia) from GPS and earthquake slip vector data *J. Geophys. Res. Solid Earth* **111**, 8 p. 1–15.
[30] Pemprov. Sulawesi Tengah *et al.*, 2019, Peta Zona Rawan Bencana Palu dan Sekitarnya.
[31] SNI-8460-2017, 2017 Persyaratan Perancangan Geoteknik *Badan Standarisasi Nas.* p. 2017.
[32] Ditjen Cipta Karya dan Pusgen, 2021, Spektrum Respon Desain. [Online]. Available: http://rsa.ciptakarya.pu.go.id/2021/index.php?pga=0.6000&ss=1.5000&s1=0.6000&tl=12&kelas=2#grafik. [Accessed: 08-Nov-2022].
[33] Fathani T F Adi A D Pramumijoyo S and Karnawati D, 2008 The Determination of Peak Ground Acceleration at Bantul Regency , Yogyakarta Province , Indonesia *Yogyakarta Earthq. May 27, 2006* p. 1–15.

Analysis of Liquefaction Potential in Solo-Yogyakarta-NYIA Kulon Progo Sta. 16+700 – 22+500

Yulianisa[1,2], Hardiyatmo HC[1], F Faris[1,2,3]

[1] Department of Civil Engineering and Environment Engineering, Gadjah Mada University, Yogyakarta, Indonesia
[2] Directorate General of Highways, The Ministry of Public Works and Housing, South Jakarta, Indonesia
[3] Centre of Excellence of Technological Innovation for Disaster Mitigation (GAMA-InaTEK) Gadjah Mada University

*Corresponding author: yulianisa@mail.ugm.ac.id

Abstract. The Solo-Yogyakarta-NYIA Kulon Progo is a National Strategic Project to accelerate the accessibility to New Yogyakarta International Airport (NYIA). This toll road will pass through Central Java Province and the Special Region of Yogyakarta Province. Previous studies have investigated that during the 2006 Yogyakarta Earthquake with a moment magnitude, M_w of 6.3, liquefaction was found in some areas in Klaten Regency of Central Java Province. This study aims to determine liquefaction potential in the Solo-Yogyakarta-NYIA Kulon Progo Toll Road Section 1, focusing on stations 16+700 to 22+500, which were located in Klaten Regency. This study first analyzes soil conditions and groundwater depth based on soil investigation data. Analysis of liquefaction potential was found out using Idriss-Boulanger based on Standard Penetration. Parameter of peak ground acceleration using two approaches; probabilistic and deterministic. For probabilistic, the peak ground acceleration using probability of 7% in 75 years and 2% in 50 years. For deterministic using the attenuation function of Campbell and Bozorgnia (2014). Then, the Liquefaction Potential Index was also used to determine the level of liquefaction vulnerability in the study area. The results of liquefaction potential analysis using each approach of peak ground acceleration show the differences in safety factor of liquefaction values. It can be concluded that several sites along the study area have the potential to liquefy at a very low to a very high level, with PGA value at least 0,43 for probability of 2% in 50 years, 0,32 for probability of 7% in 75 years, and 0,36 for deterministic approaches sourced by Opak Fault. Then, the location with the highest liquefaction potential is at stationing 20+550, with LPI values of each approach of 42.91, 33.36, and 34.70.

1. Introduction

Indonesia is a country surrounded by Eurasian, Indo-Australian, Philippine Sea, and Pacific Plates [1]. One of the areas which located to the north of the meeting point of the Indo-Australian and Eurasian Plates is Java Island. According to USGS Earthquake Summary Map, on May 27th, 2006, a strong earthquake of Mw 6.3 occurred on Java Island, about 25 km from Yogyakarta City, with a hypocentre at a depth of 10 km and the coordinate of 7.955°S and 110.430°E [2]. This earthquake occurred at a shallow depth in the overriding Sunda Plate well above the dipping Australian Plate. Some previous studies showed that earthquakes occurred due to the activity of the Opak Fault [3]. In addition, this earthquake resulted in widespread damage to buildings in Bantul, Prambanan, and Klaten areas, including Yogyakarta City. Also, it triggered liquefaction of soils, such as lateral spreads and sand boils [4] [5] [6]. During an earthquake, liquefaction occurs in saturated cohesionless soil that loses its strength in response to strong ground shaking. Seed (1970) showed that some significant factors that influence liquefaction potential depend on the soil type and the intensity of ground shaking [7].

Based on the liquefaction hazard, it is necessary to analyze the liquefaction potential in the area near Opak Fault to reduce the impact of an earthquake that triggers liquefaction. SNI 1726:2019 explained that one of the categories of near-fault is 10 km from a known active fault surface projection and is capable of producing an earthquake of Mw 6 or greater [8]. The Solo-Yogyakarta-NYIA Kulon Progo Toll Road is one of sites located near the Opak Fault. This toll, known as the National Strategic Project (PSN), is part of the Toll Road Development Program to support Borobudur as a National Tourism Strategic Area and accelerate the accessibility to New Yogyakarta International Airport (NYIA). Based on the geological map, this toll road in Klaten Regency is located in quaternary volcanic deposits. In Youd and Perkins, 1978, Youd, 1991, it is explained that soil layers susceptible to liquefy are generally formed in quaternary geological environments [9], [10]. The rock formation that makes up this area are the Merapi (Qvm) volcanic rocks that are thought to be Holocene-aged and consist of volcanic brection, lava, and tuff [11].

This paper aims to evaluate liquefaction potential in the Solo-Yogyakarta-NYIA Kulon Progo Toll Road Section 1, focusing on stations 16+700 to 22+500 as shown in Figure 1, which were located in Klaten Regency using deterministic and probabilistic approach for determining peak ground acceleration value. The liquefaction hazard was analyzed using the Idriss-Boulanger method based on SPT data. Then, the liquefaction vulnerability in the study area was determined using Liquefaction Potential Index categories.

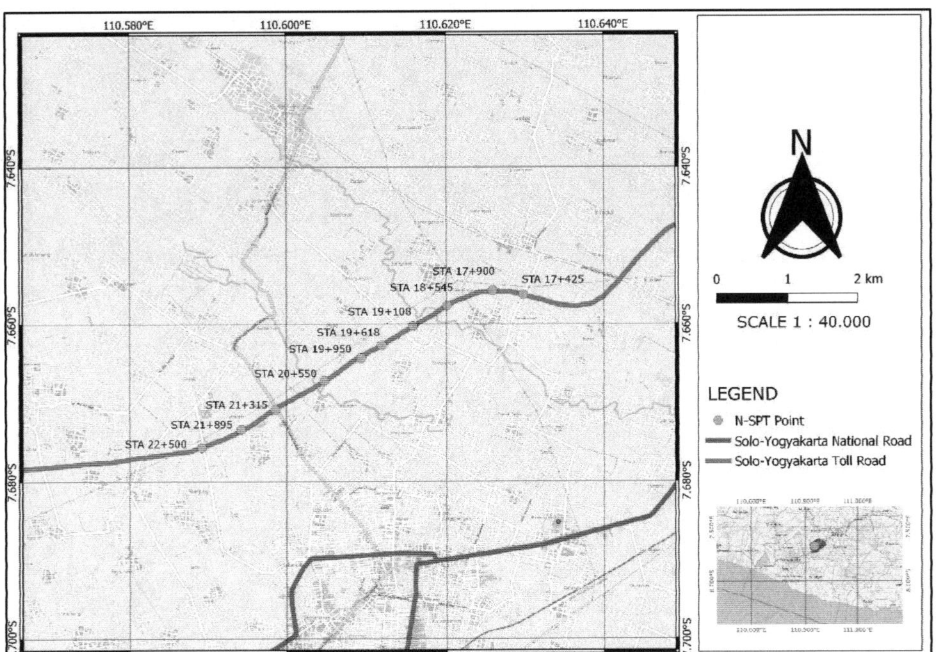

Figure 1. Location of Study Area (modified from [12])

IOP Conf. Series: Earth and Environmental Science 1244 (2023) 012025 doi:10.1088/1755-1315/1244/1/012025

2. Research Methods

The stages of this study consisted of seismic and liquefaction potential analysis. Seismic analysis was found by determining sites near-fault and site class based on the category in SNI 1726:2019 and identifying the history of earthquake events sourced by the Opak Fault. Then, peak ground acceleration (PGA) is analyzed based on deterministic and probabilistic approaches. The liquefaction potential analysis was found using a semi-empirical method by analyzing the value of cyclic stress and resistance ratios.

2.1. Seismic Analysis

Seismic analysis to determine the PGA value was evaluated by two approaches, namely probabilistic and deterministic. The probabilistic approach was determined based on The Seismic Source and Hazard Map of Indonesia (2017) using the Design Spectra Indonesia 2019 website for a probability of 2% in 50 years and the LINI Bina Marga website for a probability of 7% in 75 years. For the deterministic approach, it is determined by the attenuation formula by Campbell and Bozorgnia (2014). This formula was developed based on the earthquake source due to strike-slip mechanism of moment magnitude of $3.3 \leq M_w \leq 8.5$ [13] and similar to an earthquake produced by the Opak Fault. Seismic analysis for the worst scenario sourced by the Opak Fault using a moment magnitude of 6.3 and depth of 10 km, as shown in equation (1) to equation (10). This PGA calculates using the Pacific Earthquake Engineering Research (PEER) Centre website. In addition, this formula provides the minimum Root Mean Square (RMS) Error value for predicting PGA [14].
Meanwhile, the determination of PGA value based on Equation (1).

$$\ln Y = \ln PGA = f_{mag} + f_{dis} + f_{flt} + f_{hng} + f_{site} + f_{sed} + f_{hyp} + f_{dip} + f_{atn} \tag{1}$$

where, ln Y = ln PGA = peak ground acceleration, f_{mag} is a term of magnitude, f_{dis} is a term of geometric attenuation, f_{flt} is a term of fault style, f_{hng} is a term of hanging wall, f_{site} is a term of shallow site response, f_{sed} is a term of basin response, f_{hyp} is a term of hypocentral depth, f_{dip} is a term of fault dip, and f_{atn} is a term of anelastic attenuation. These terms were calculated using equation (2) to equation (8).

$$f_{mag} = c_0 + c_1 M_w + c_2(M_w - 4.5) + c_3(M_w - 5.5), for 5.5 < M_w \leq 6.5 \tag{2}$$

$$f_{dis} = (c_5 + c_6 M_w)\ln(\sqrt{R^2_{RUP} + c_7^2}) \tag{3}$$

where f_{flt} and $f_{hng} = 0$ for strike-slip mechanism, c-term is coefficient based on Table 1 and Table 2, M_w for moment magnitude of 6.3, and R_{RUP} is rupture distance to each borehole location.

$$f_{site} = \left\{ \begin{array}{l} c_{11}\ln\left(\frac{v_{s30}}{k_1}\right) + k_2\left\{\ln\left[A_{1100} + c\left(\frac{v_{s30}}{k_1}\right)^n\right] - \ln[A_{1100} + c]\right\} for v_{s30} \leq k_1 \\ (c_{11} + k_2 n)\ln\left(\frac{v_{s30}}{k_1}\right) for v_{s30} > k_1 \end{array} \right\} \tag{4}$$

$$f_{sed} = (c_{14})(z_{2.5} - 1); z_{2.5} \leq 1 \tag{5}$$

$$f_{hyp} = f_{hypH} + f_{hypM} \tag{6}$$

$$f_{dip} = c_{19}(5.5 - M_w)\delta; M_w > 5.5 \tag{7}$$

$$f_{atn} = 0; for R_{RUP} \leq 80 \tag{8}$$

where v_{s30} is the average shear wave velocity in the top of 30 m of the site, $z_{2.5}$ is the depth to the 2.5 km/s shear wave velocity, and δ is average dip angle of rupture.

Table 1: Coefficients of Campbell&Bozorgnia Method For Estimating PGA (1)

T (s)	c_0	c_1	c_2	c_3	c_5	c_6	c_7	c_{11}
PGA	-4.416	0.984	0.537	-1.499	-2.773	0.248	6.768	1.090

Table 2: Coefficients of Campbell&Bozorgnia Method For Estimating PGA (2)

T (s)	c_{14}	c_{16}	c_{17}	c_{18}	c_{19}	k_1	k_2
PGA	-0.0064	0.393	0.0977	0.00333	0.00757	865	-1.186

2.2. *Analysis of Liquefaction Potential*

Analysis of liquefaction potential was found using a simplified procedure based on SPT data by Idriss and Boulanger (2008) method. This method compares CRR with CSR [15], where CRR was calculated using equation (9).

$$CSR = 0.65 \times \frac{a_{\max}}{g} \times \frac{\sigma_v}{\sigma'_v} \times r_d \quad (9)$$

where, CSR is cyclic stress ratio, $a_{\max}$ is maximum horizontal acceleration (g), g is the acceleration of gravity (9,81 m/s^2), σ_v , σ'_v is total stress and effective stress (kN/m^2), respectively, whereas r_d is coefficient of reduction that was calculated as a function of moment magnitude (M) and the depth (z) as shown in equation (10).

$$r_d = \exp\left(-1.012 - 1.126\sin\left(\frac{z}{11.73} + 5.133\right) + 0.106 + 118\sin\left(\frac{z}{11.28} + 5.142\right)M\right) \quad (10)$$

The CRR value is the ratio of soil resistance to cyclic loads that cause liquefaction. Several factors that affect that value are corrected SPT value for 60% efficiency, overburden pressure, and fines content. The corrected SPT value was calculated using equation (11) to equation (14).

$$(N)_{60} = N_m \times C_E \times C_B \times C_R \times C_S \quad (11)$$

$$(N_1)_{60} = N_{60} \times C_N \quad (12)$$

$$C_N = \left(\frac{Pa}{\sigma'_{vc}}\right)^{0.784 - 0.0768\sqrt{(N_1)_{60-cs}}} \quad (13)$$

$$(N_1)_{60-cs} = (N_1)_{60} + \exp\left(1.63 + \frac{9.7}{FC + 0.01} - \left(\frac{15.7}{FC + 0.01}\right)^2\right) \quad (14)$$

where $(N_1)_{60}$ is corrected SPT value for 60% efficiency and effective stress of 1 atm, $(N_1)_{60-cs}$ is corrected SPT value of fines content (FC), N_m is SPT value of field test, whereas C_E, C_B, C_R, C_S , C_N is a factor of correction for energy ratio, a diameter of borehole, length of the rod, sampler liners, and effective stress, respectively.

Then, it is necessary to calculate the value of the Magnitude Scaling Factor (MSF) to adopt the CSR and/or CRR to the general value of M_w 7.5. MSF by Seed et al. (1975b, in Idriss & Boulanger, 2008) is defined as shown in equation (15).

$$MSF = 6.9\exp\left(\frac{-M}{4}\right) - 0.058; MSF \le 1.8 \tag{15}$$

where MSF is the factor of scaling magnitude and M is moment magnitude. Then, the correction factor of overburden was calculated using equation developed by Boulanger and Idriss (2004a, in Idriss and Boulanger, 2008) as shown in equation (16).

$$K_\sigma = 1 - \left(\frac{1}{18.9 - 2.55\sqrt{(N_1)_{60}}}\right)\ln\left(\frac{\sigma'_{vc}}{Pa}\right) \le 1.1 \tag{16}$$

where, K_σ is overburden correction factor, σ'_{vc} is effective stress, and Pa is atmospheric pressure. Then, the CRR value was calculated by SPT correlation for triggering liquefaction in clean sand as shown in equation (17) to equation (18).

$$CRR_{M=7.5,\sigma'_{vc}=1} = \exp\left(\frac{(N_1)_{60-cs}}{14.1} + \left(\frac{(N_1)_{60-cs}}{126}\right)^2 - \left(\frac{(N_1)_{60-cs}}{23.6}\right)^3 + \left(\frac{(N_1)_{60-cs}}{25.4}\right)^4 - 2.8\right) \tag{17}$$

$$CRR_{M,\sigma'_{vc}} = CRR_{M=7.5,\sigma'_{vc}=1} \times MSF \times K_\sigma \tag{18}$$

$CRR_{M=7.5,\sigma'_{vc}=1}$ is cyclic resistance ratio for M_w 7.5 and effective stress 1 atm, $(N_1)_{60-cs}$ is corrected SPT value of fines content, and $(N_1)_{60}$ is corrected SPT value for 60% efficiency. Meanwhile, the factor of safety against liquefaction can be evaluated according to equation (19).

$$FS_{liq} = \frac{CRR}{CSR} \tag{19}$$

Then, the factor of safety value was used to determine the level of potential liquefaction using Iwasaki et al. (1981) method, as shown in equation (20) to equation (23).

$$LPI = \int_0^{20} F(z).w(z).dz \tag{20}$$

$$F(z) = 0 \text{ for } FS_{liq} < 1.0 \tag{21}$$

$$w(z) = 10 - 0.5z \text{ if } z < 20 \tag{22}$$

$$w(z) = 0 \text{ if } z \ge 20 \tag{23}$$

where LPI is liquefaction potential index, $F(z)$ is level of damage for soil layer, $w(z)$ is a weighting factor of depth, and z is depth 0-20 m. The category of potential liquefaction based on LPI is shown in Table 3.

Table 3. LPI Category by Iwasaki, et al (1981) [16]

LPI value	Level
LPI = 0	Very Low
0< LPI ≤ 5	Low
5< LPI ≤ 15	High
LPI > 15	Very High

3. Result and Analysis

Based on analysis of 10 boreholes along The Solo-Yogyakarta-NYIA Kulon Progo focusing on Sta. 17+425 to Sta. 22+500, the results of a standard penetration test shows variations in the depth of groundwater levels at the study site ranging from 0.5 – 9.0 meter with an average SPT value from 14.44 to 27.16. From the results of the same soil test, it is found that the dominant soil type is sand with loose to medium consistency. Table 4 shows the location, average SPT values, depth of groundwater, and site distance to the closest coordinate of Opak Fault for each borehole.

Table 4. Soil Investigation Result for Sta. 16+700 to 22+500

Borehole (STA-)	Coordinate		Average SPT	Groundwater Level (m)	Site Distance to The Closest Coordinate of Opak Fault (km)
	Longitude	Latitude			
17+425	-7.65621	110.62975	19.59	-1.5	10.22
17+900	-7.65569	110.62583	36.33	-7.5	9.91
18+545	-7.65759	110.62010	14.89	-9.0	9.29
19+108	-7.66024	110.61583	17.63	-0.5	8.75
19+618	-7.66267	110.61193	27.16	-7.0	8.25
19+950	-7.66430	110.60936	32.83	-3.0	7.92
20+550	-7.66719	110.60475	14.44	-1.0	7.33
21+315	-7.67082	110.59861	14.93	-0.5	6.56
21+895	-7.67336	110.59425	23.05	-3.6	6.02
22+500	-7.67565	110.58927	25.10	-1.2	5.47

3.1. Seismic Analysis

The near-fault site analysis results showed that 90% of boreholes included in sites near-fault categories (<10 km apart). The results show that PGA for probabilistic of 2% in 50 years (2500 years) ranges from 0.43g-0.48g, while for probabilistic of 7% in 75 years (1000years) shows the exact value of 0.32g. It is indicated that the LINI website uses a large scale, which caused the map resolution not to be too detailed, as shown by the Design Spectra Indonesia 2019 in determining the PGA value for the probability of 2% in 50 years. For the deterministic approach based on equation (1), the earthquake scenario used M_w of 6.3, sourced from the Opak Fault, with a depth of 10 km. The results show peak ground acceleration based on Campbell and Bozorgnia (2014) method ranging from 0.36g-0.50g. The peak ground acceleration value at the surface for each approach can be seen in Table 5. The PGA results from each method show that the higher PGA value is shown at the site location with the closest distance to The Opak Fault.

Table 5. Peak Ground Acceleration (PGA_M) Value for Sta. 16+700 to 22+500

Borehole (STA-)	Site Class	The Seismic Source and Hazard Map of Indonesia 2017						Campbell and Bozorgnia (2014)		
		Prob. 2% in 50 years			Prob. 7% in 75 years					
		PGA (g)	F_{PGA}	*PGA_M (g)*	*PGA (g)*	F_{PGA}	*PGA_M (g)*	*PGA (g)*	F_{PGA}	*PGA_M (g)*
17+425	SD	0.43	1.27	0.55	0.32	1.18	0.38	0.36	1.24	0.45
17+900	SD	0.44	1.26	0.55	0.32	1.18	0.38	0.39	1.21	0.47
18+545	SE	0.44	1.52	0.67	0.32	1.14	0.36	0.38	1.45	0.54
19+108	SD	0.44	1.26	0.56	0.32	1.18	0.38	0.39	1.21	0.47
19+618	SD	0.45	1.25	0.56	0.32	1.18	0. 38	0.44	1.17	0.51
19+950	SD	0.45	1.25	0.56	0.32	1.18	0. 38	0.45	1.15	0.52
20+550	SE	0.46	1.49	0.68	0.32	1.14	0.36	0.41	1.39	0.56
21+315	SE	0.46	1.47	0.68	0.32	1.14	0.36	0.42	1.36	0.57
21+895	SD	0.47	1.23	0.58	0.32	1.28	0.38	0.47	1.14	0.53
22+500	SD	0.48	1.23	0.58	0.32	1.28	0.38	0.50	1.10	0.54

3.2. Analysis of Liquefaction Potential

The results of liquefaction potential analysis using Idriss and Boulanger (2018) method show that the factor of safety for each method resulted in similar values. The probability of 2% in 75 years shows a significant difference value. It is indicated the different parameters of moment magnitude that are used in the analysis. Analysis of liquefaction potential for probability approach used a moment magnitude, M_w of 7.5. Meanwhile, Campbell and Bozorgnia (2014) used a moment magnitude, M_w of 6.3, which was based on the highest moment magnitude produced by the Opak Fault. In addition, the minimum factor of safety is shown at Sta. 20+550, which has a shallow groundwater level and consists of sand as a dominant soil type. Liquefaction at this station occurred at 2-4 m and 4-12 m. Then, very low potential of liquefaction occurs at Sta. 19+950, which has dense sand as a dominant soil type. The factor safety value for each method is shown in Table 6.

Table 6. The Factor of Safety Value for Sta. 16+700 to 22+500

Borehole (STA-)	**The Seismic Source and Hazard Map of Indonesia (2017)**				**Campbell and Bozorgnia (2014)**	
	Prob. 2% in 50 years		**Prob. 7% in 75 years**			
	FS Minimum	**FS Maximum**	**FS Minimum**	**FS Maximum**	**FS Minimum**	**FS Maximum**
17+425	0.23	3.32	0.34	4.76	0.45	5.70
17+900	0.36	**5.35**	0.53	**7.80**	0.66	**7.29**
18+545	0.24	3.26	0.43	5.97	0.45	6.38
19+108	0.19	3.01	0.28	4.45	0.31	4.63
19+618	0.48	3.94	0.71	5.86	0.78	6.47
19+950	2.25	4.75	3.36	7.10	4.06	5.46
20+550	**0.16**	2.63	**0.29**	4.89	**0.27**	4.42
21+315	0.15	2.52	0.29	5.21	0.30	4.15
21+895	0.25	3.96	0.39	6.04	0.46	4.70
22+500	0.33	3.21	0.51	4.95	0.50	4.69

The results of liquefaction potential level in this study used the *FS* value to evaluate the category of LPI of each borehole. As shown in Table 7, the probability of 2% in 50 years shows 5 boreholes with a very high level, 3 boreholes with a high level, and 1 borehole with a low and very low level. In comparison, the probability of 7% in 75 years results in 4 boreholes with a very high level, 3 boreholes with a high level, and 3 boreholes with a low and very low level. The different result is shown by Campbell and Bozorgnia (2014), which resulted in 3 boreholes with a very high level, 3 boreholes with a high level, and 4 boreholes with a low and very low level. The different results of the LPI value are due to the different factors of safety values in each approach. The distribution of LPI value for each peak ground parameter approach can be seen in Figure 2 for the probability of 2% in 50 years, Figure 3 for the probability of 7% in 75 years, and Figure 4 for the worst earthquake scenario using attenuation formula.

Table 7. LPI Value for Sta. 16+700 to 22+500

Borehole (STA-)	The Seismic Source and Hazard Map of Indonesia (2017)				Campbell and Bozorgnia (2014)	
	Prob. of 2% in 50 years		Prob. of 7% in 75 years			
	LPI	Category	LPI	Category	LPI	Category
17+425	22.00	**Very High**	13.81	High	11.22	High
17+900	8.17	High	5.50	High	3.43	Low
18+545	17.44	**Very High**	9.84	High	9.45	High
19+108	26.07	**Very High**	21.32	**Very High**	19.37	**Very High**
19+618	6.24	High	3.43	Low	2.64	Low
19+950	0.00	Very Low	0.00	Very Low	0.00	Very Low
20+550	42.91	**Very High**	33.36	**Very High**	34.70	**Very High**
21+315	40.70	**Very High**	25.60	**Very High**	27.35	**Very High**
21+895	4.67	Low	2.91	Low	2.21	Low
22+500	13.30	High	19.68	High	9.89	High

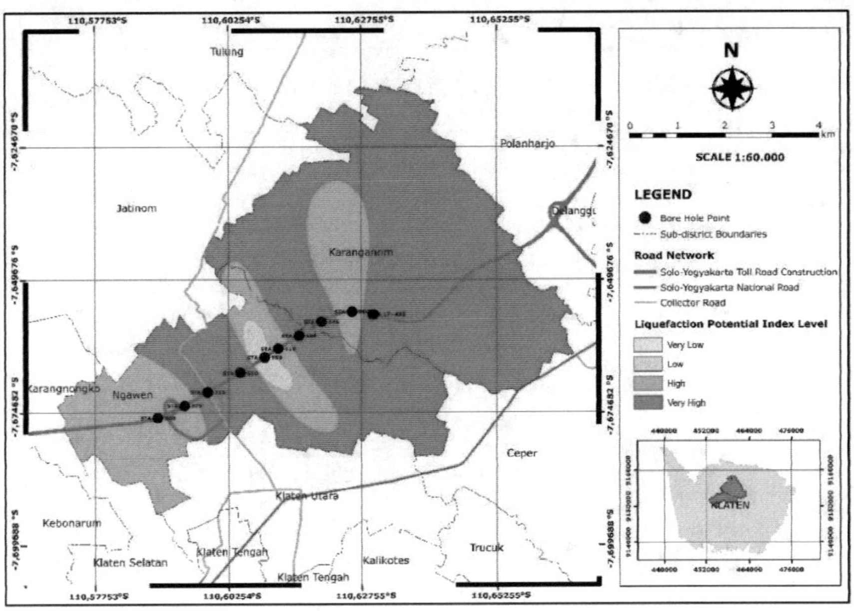

Figure 2. Liquefaction Potential Index for Probability of 2% in 50 years

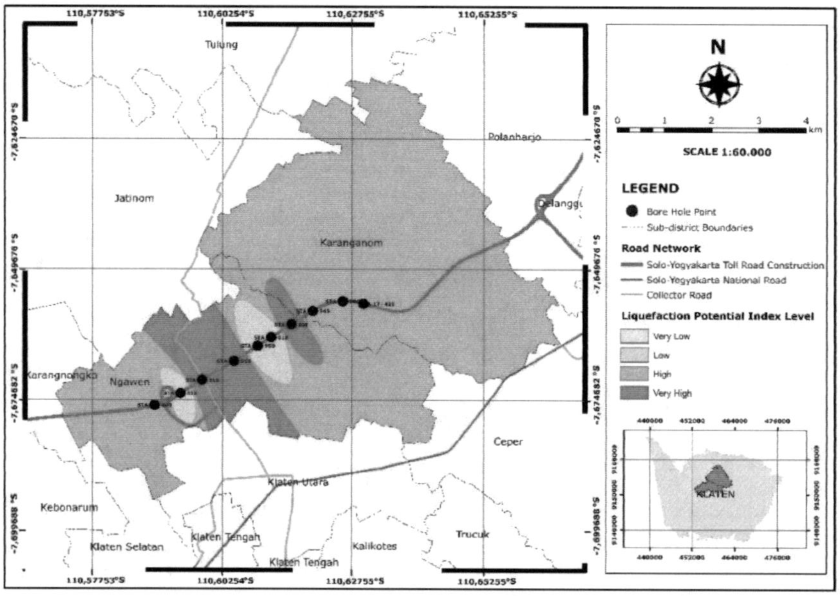

Figure 3. Liquefaction Potential Index for Probability of 7% in 75 years

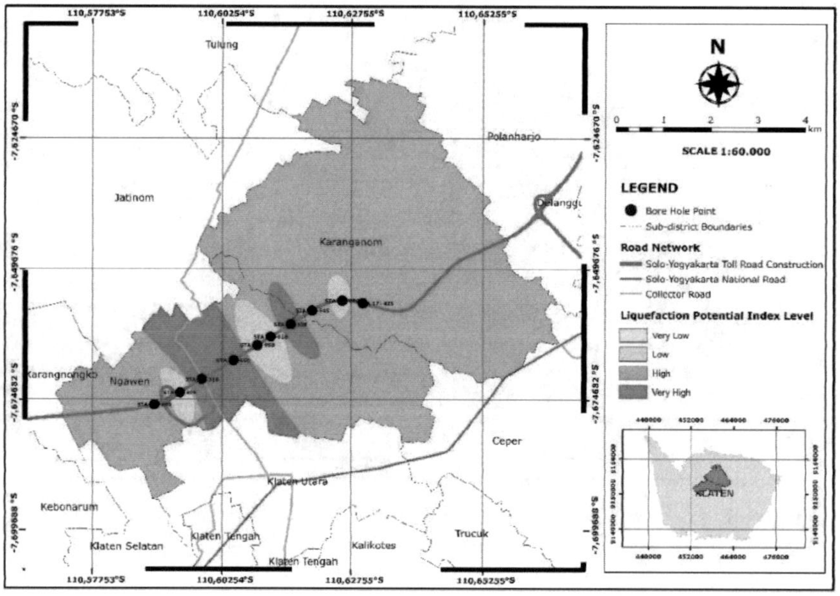

Figure 4. Liquefaction Potential Index Using Campbell and Bozorgnia (2014) method for Moment Magnitude, M_w of 6.3 and PGA of 0.36-0.50g

4. Conclusions

The Solo-Yogyakarta-NYIA Kulon Progo toll road as one of the toll road constructions located in a vulnerable area of earthquake-induced liquefaction, a detailed analysis of the potential for liquefaction is necessary. The existence of Opak Faults is one of the causes of seismic hazards that have the potential to occur in areas that are included in the category of sites near-fault. Liquefaction potential analysis has been found using three approaches; probability of 2% in 50 years, probability of 7% in 75 years, and Campbell and Bozorgnia (2014) method to determine a peak ground acceleration value. The probability of 2% in 50 years resulted in the highest value of peak ground acceleration and LPI value for each borehole. However, the use of this approach was not in accordance with the structure above toll roads, such as embankments and bridges. The results of liquefaction potential analysis show the differences in safety factor values due to the difference in the use of the moment magnitude. These results affect the CRR and CSR values in each layer. Based on the analysis of 10 boreholes along Sta. 16+700 to Sta. 22+500 shows that liquefaction potentially occurred with a very low to a very high level. Several methods for the analysis of liquefaction hazards at some stations with high and very high LPI values that potentially affect the stability of the structure are needed for further studies.

5. Acknowledgements

The author would like to express our appreciation to the Directorate General of Highways, The Ministry of Public Works and Housing, and PT Jogjasolo Marga Makmur as a Business Entity managing the Solo-Yogyakarta-NYIA Kulon Progo Toll Road for the support.

6. References

[1] Irsyam M, 2017 *Peta Sumber dan Bahaya Gempa Indonesia Tahun 2017 (Pusat Studi Gempa Nasional)* Pusat Penelitian dan Pengembangan Perumahan dan Permukiman, Badan Penelitian dan Pengembangan, Kementerian Pekerjaan Umum dan Perumahan Rakyat.
[2] USGS, M6.3 Java Earthquake of 26 May 2006. [Online]. Available: https://earthquake.usgs.gov/product/poster/20060526/us/1461770047809/poster.pdf. [Accessed: 12-Mar-2022].
[3] Sulaeman C, 2008 Karakterisasi sumber gempa Yogyakarta 2006 berdasarkan data GPS *Indones. J. Geosci.* **3**, 1 p. 49–56.
[4] Sutiono A Prastistho B Prasetyadi C and Supartoyo, 2018 Opak fault: A comparative review *IOP Conf. Ser. Earth Environ. Sci.* **212**, 1.
[5] Sarah D and Soebowo E, 2013 Liquefaction due to the 2006 Yogyakarta Earthquake: Field Occurrence and Geotechnical Analysis *Procedia Earth Planet. Sci.* **6** p. 383–389.
[6] Mase L Z, 2017 Experimental Liquefaction Study of Southern Yogyakarta Using Shaking Table *J. Tek. Sipil ITB* **24**, 1 p. 11–18.
[7] Bolton Seed H and Idriss I M, 1970 *National Technical Information Service Asimplified Procedure for Evaluating Soil Liquefaction Potential* November .
[8] Badan Standarisasi Nasional, 2019 Standar Nasional Indonesia 1726:2019 Tata cara perencanaan ketahanan gempa untuk struktur bangunan gedung dan nongedung 8.
[9] Youd T L, 1978 Mapping Liquefaction-Induced Ground Failure Potential April 1978.
[10] Youd T L, 1991 Mapping of earthquake-induced liquefaction for seismic zonation November.
[11] Surono B T I S, 1992, Peta Geologi Lembar Surakarta-Giritontro, Jawa Tengah. .
[12] PT Jogjasolo Marga Makmur; PT Adhi Karya (Persero) Tbk., 2020, Laporan Pendahuluan Rencana Teknik Akhir (RTA) Jalan Tol Solo - Yogyakarta - NYIA Kulon Progo Seksi 1.1 (STA 0+000 - 22+500).
[13] Campbell K W and Bozorgnia Y, 2014 NGA-West2 ground motion model for the average horizontal components of PGA, PGV, and 5% damped linear acceleration response spectra *Earthq. Spectra* **30**, 3 p. 1087–1114.
[14] Mase L Z, 2019 Performance of NGA Models in Predicting Ground Motion Parameters of The Strong Earthquake *J. Civ. Eng. Forum* **5**, 3 p. 227.
[15] Idriss, I. M.; Boulanger R W, 2008 *SOIL LIQUEFACTION DURING EARTHQUAKES* **160**, 4057 California.
[16] Iwasaki T Tokida K and Tatsuoka F, 1981 Soil Liquefaction Potential Evaluation with Use of the Simplified Procedure *Int. Conf. Recent Adv. Geotech. Earthq. Eng. Soil Dyn.* **12** p. 209–214.

Study of soil liquefaction potential at Anutapura General Hospital, Palu City, Central Sulawesi Province

D R Septiadi[1,2,*], **H C Hardiyatmo**[1] **and F Faris**[1]

[1]Department of Civil and Environmental Engineering, Faculty of Engineering, Universitas Gadjah Mada, Yogyakarta 55281, Indonesia
[2]Directorate General of Human Settlement, Ministry of Public Works and Housing, Republic of Indonesia
*Corresponding author: daniarrizkyseptiadi@mail.ugm.ac.id or daniar.rizky95@gmail.com

Abstract. On September 28, 2018, at 18:02 local time, a strong tectonic earthquake with a moment magnitude of 7.5 struck Central Sulawesi Province, Indonesia, triggering liquefaction disasters in Palu city and Sigi district. The project of the Anutapura Medical Centre (AMC) building at the Anutapura General Hospital in Palu city is one of the Indonesian government programs to support the rehabilitation and reconstruction of infrastructure after the earthquake. This hospital is located close to the liquefaction-affected area in Balaroa. Geologically, the hospital area is also located in the formation of Holocene deposits that have not been fully lithified. This study aims to analyze the liquefaction potential, which focuses on the construction area of the AMC building. The soil grain size distribution was used for the preliminary analysis of potentially liquified soils. The SPT-based liquefaction triggering procedure was used to calculate the safety factor (SF) against liquefaction. Subsequently, the SF value was further analyzed using the liquefaction severity index (LSI) to determine the severity level. The results show that soils in the study area have the potential to liquefy with a low to moderate severity level. The results are expected to be used as consideration to mitigate the possible liquefaction.

Keywords: liquefaction, hospital building, Balaroa, safety factor, liquefaction severity index.

1. Introduction

Sulawesi island, Indonesia, is located in a confluence zone of 3 plates, commonly called a triple junction, which causes complex deformations to form active faults that often cause several tectonic earthquakes [1,2]. Several earthquakes struck Central Sulawesi province with a 7.5 magnitude mainshock on September 28, 2018 [3]. This earthquake is thought to have occurred due to the Palu-Koro shear fault activity. **Figure 1** shows the Spread of the Central Sulawesi earthquake epicenters in 2018. The mainshock was located at geographical coordinates at 0.178 north latitude, 119.84 east longitude, with an epicenter depth of 20 km. The earthquake triggered a tsunami along the coast of Palu Bay and liquefaction in several locations, namely Balaroa, Petobo, Jono Oge, Lolu, and Southern Sibalaya villages. **Figure 2** shows Several locations affected by liquefaction in Palu city and Sigi district. It was reported that 3720 houses and their infrastructure were affected by liquefaction, with a total affected area of approximately 510 hectares [4].

Liquefaction can be interpreted as soil change from a solid to a liquid consistency [8]. It is caused by cyclic loading (earthquake), which causes excess pore water pressure and decreases effective stress to zero [9]. Non-cohesive soil (sand) is the type of soil that is most vulnerable to liquefaction [10]. Sand has a low cohesion, so it tends to be unstable and easily released when the pore water pressure increases.

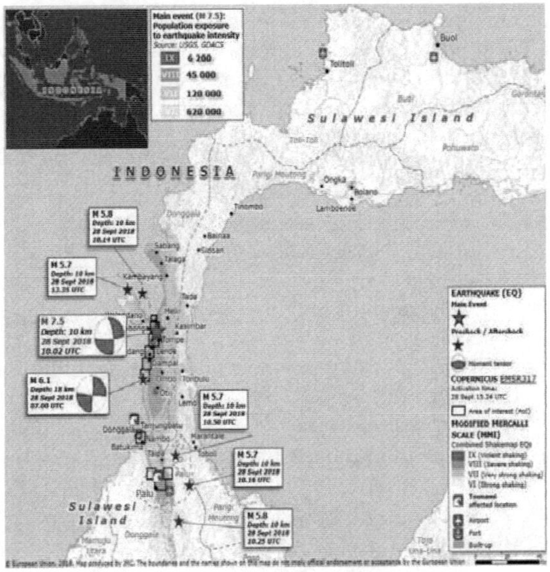

Figure 1. Spread of the Central Sulawesi earthquake epicenters in 2018 [5]

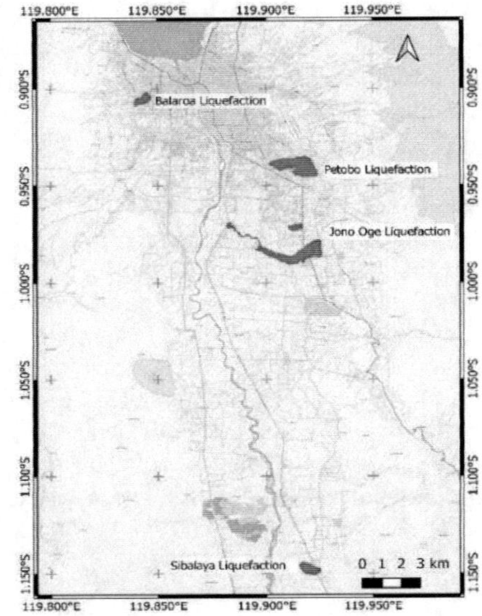

Figure 2. Several locations affected by liquefaction in Palu city and Sigi district [6] (base map source: Open Street Map [7])

Figure 3. Study location (base map source: Google Earth [11])

The Anutapura Medical Center (AMC) building project at the Anutapura general hospital in Palu is one of the Indonesian government programs to support rehabilitation and reconstruction activities after the 2018 Central Sulawesi earthquake. This hospital is located near the affected-liquefaction area in Balaroa village (**Figure 3**). Previous research has shown that areas affected by liquefaction in Balaroa still have the potential to liquefy if an earthquake occurs [13]. In order to ensure AMC building construction safety and usefulness, it is necessary to conduct a study regarding liquefaction potential.

This study aims to analyze the liquefaction potential, which focuses on the construction area of the AMC building. The soil grain size distribution was used for the preliminary analysis of potentialy liquified soils. The standard penetration test (SPT) based liquefaction triggering procedure was adopted to calculate the safety factor (SF) against liquefaction. Subsequently, the safety factor (SF) value was further analyzed using the liquefaction severity index (LSI) to determine the severity level.

2. Site Conditions

Geographically, the city of Palu is located in Palu Bay, which is the estuary area of the Palu river. The landscape of Palu City and its surroundings is an alluvial land that stretches from north to south. This land is flanked by 2 hills on the east and west sides.

Based on regional geological maps [14], the geological formations in Palu City and its surroundings consist of 5 formations, namely alluvium and coastal deposits, Celebes Molasses, Tinombo formation, metamorphic complex, and granite. **Figure 4** shows regional geological maps of Palu City and its surroundings. The Anutapura General Hospital is located in the western part of Palu City, which is geologically alluvium and coastal deposits (Qap). Qap formation material consists of gravel, sand, mud, and coral limestone. This formation is estimated to be of the Holocene age, which has not been completely lithified [15].

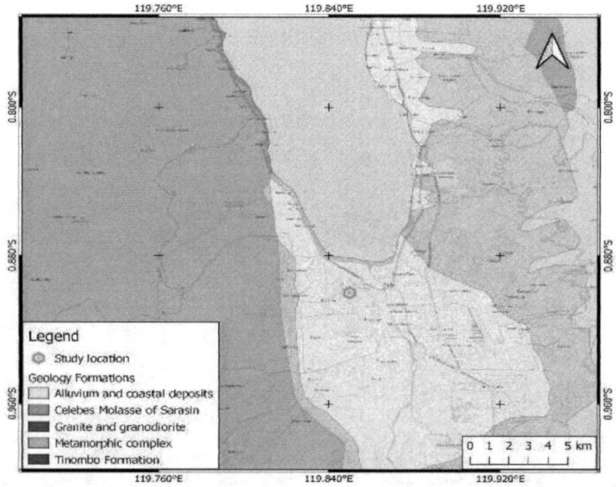

Figure 4. Regional geological maps of Palu City and its surroundings (modified from [14])

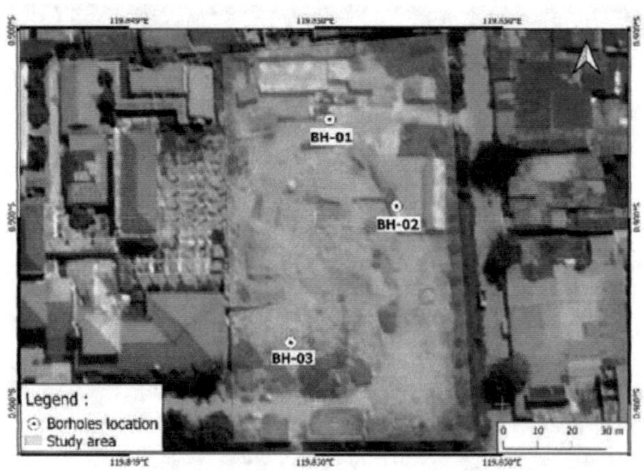

Figure 5. Borehole points location (base map source: Google Earth [12])

A soil investigation in the AMC building construction area consists of 3 borehole points equipped with standard penetration test (SPT) and laboratory test data. **Figure 5** shows the location of the 3 borehole points used for analysis at the AMC building construction area. **Figure 6** shows a section of the soil stratigraphic layer equipped with a standard penetration test (SPT) graph for each borehole from north to south direction. The soil in this area is dominated by sand with several layers of clay and silt. Several layers of saturated loose sand were found at depths of 5-19 meters. The hard soil layers (SPT > 50) were found at 43 meters below the ground surface. The groundwater levels were found at a depth of 2.3 - 4 meters.

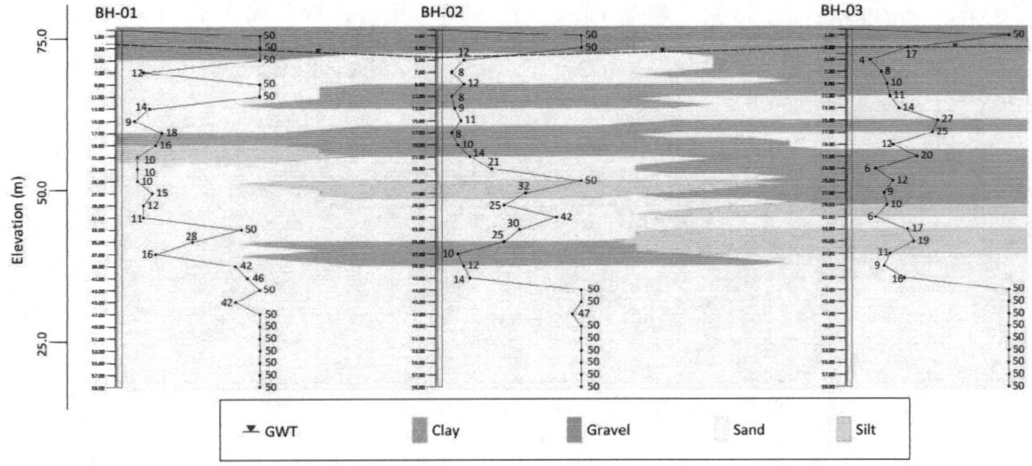

Figure 6. Section of the soil stratigraphic layer and SPT value in the AMC building construction area

3. Methods

3.1. Liquefaction potential based on soil grain size

The soil grain size distribution can be used in the initial liquefaction potential analysis. This study used the Tsuchida boundary graph [16] to categorize soil grain size distribution with the most and potentially liquefied soils, as shown in **Figure 7**. The data analyzed were non-cohesive soil grain size distributions of several soil samples taken from each borehole point.

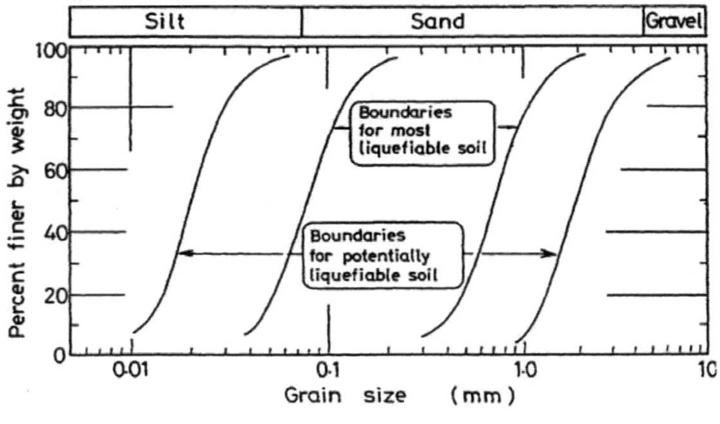

Figure 7. The boundary graph of potentially liquefied soil based on soil grain size distribution [16]

3.2. Standard penetration test (SPT) based liquefaction triggering procedure
Idriss and Boulanger [17] developed the SPT-based liquefaction triggering procedure to analyze the liquefaction potential quantitatively. This procedure is a development and improvement of the previous method, namely the simplified method by Seed-Idriss [18] and the Youd-Idriss method (NCEER) [19]. In this procedure, the safety factor (SF) empirically expresses the liquefaction potential, which can be calculated using equation (1).

$$SF = \frac{CRR_{M,\sigma'_{vc}}}{CSR} \tag{1}$$

SF value is a comparison of the cyclic resistance ratio (CRR) and the cyclic stress ratio (CSR) in each soil layer. Liquefaction had the potential to occur if the SF value was < 1.

CSR can be interpreted as the shear stress ratio in the soil caused by an earthquake. CSR was determined using the equation developed by Seed-Idriss [18] in equations (2) to (5) below.

$$CSR = 0.65\left(\frac{\sigma_{vc}}{\sigma'_{vc}}\right)\left(\frac{a_{max}}{g}\right)r_d \tag{2}$$

$$r_d = exp[\,\alpha(z) + \beta(z).M\,] \tag{3}$$

$$\alpha(z) = -1.012 - 1.126\,sin\left(\frac{z}{11.73} + 5.133\right) \tag{4}$$

$$\beta(z) = 0.106 + 0.118\,sin\left(\frac{z}{11.28} + 5.142\right) \tag{5}$$

Where, σ_{vc} is the total soil stress, σ'_{vc} is the effective overburden soil stress, a_{max} is the peak ground acceleration (PGA), r_d is the stress reduction factor, and z is the soil layer depth.

CRR can be interpreted as the ratio of soil resistance to liquefaction based on soil test data (SPT). SPT value must be corrected by several factors. The SPT correction value based on the test procedure was expressed by the N_{60} value which was calculated by equation (6).

$$N_{60} = C_E.C_B.C_R.C_S.N_m \tag{6}$$

The N_{60} value must then be corrected based on the overburden stress factor in each soil layer which is expressed by the $(N_1)_{60}$ value. The $(N_1)_{60}$ was obtained using equations (7) and (8) below

$$(N_1)_{60} = C_N.N_{60} \tag{7}$$

$$C_N = \left(\frac{P_a}{\sigma'_{vc}}\right)^{0.784-0.0768\sqrt{(N_1)_{60cs}}} \leq 1.7 \tag{8}$$

Where C_E is the correction factor based on the energy ratio, C_B is the drill diameter correction factor, C_R is the correction factor based on rod length, C_S is the samplers type correction factor, N_m is the value of the SPT, C_N is the correction factor based on the effective overburden stress, and P_a is the atmospheric pressure (100 kPa). The CRR value at the 7.5 magnitude value and the 1 atm effective overburden soil stress can be calculated using equations (9) to (11) below.

$$CRR_{M=7.5,\sigma'_{vc}=1\,atm} = exp\left(\frac{(N_1)_{60cs}}{14.1} + \left(\frac{(N_1)_{60cs}}{126}\right)^2 - \left(\frac{(N_1)_{60cs}}{23.6}\right)^3 + \left(\frac{(N_1)_{60cs}}{25.4}\right)^4 - 2.8\right) \tag{9}$$

$$(N_1)_{60cs} = (N_1)_{60} + \Delta(N_1)_{60} \tag{10}$$

$$\Delta(N_1)_{60} = exp\left(1.63\,\frac{9.7}{FC + 0.01} - \left(\frac{15.7}{FC + 0.01}\right)^2\right) \tag{11}$$

Where, $(N_1)_{60cs}$ is the corrected SPT value that has been adjusted for the fine content, and $\Delta(N_1)_{60}$ is the additional correction value of the fine content.

For certain moment magnitude and soil effective stress conditions, the CRR value must be adjusted to the magnitude scale factor (MSF) and the effective soil stress correction factor (K_σ) using the equations (12) to (15) below.

$$CRR_{M,\sigma'_{vc}} = CRR_{M=7.5,\sigma'_{vc}=1\,atm} \, .MSF\,.K_\sigma \tag{12}$$

$$MSF = 6.9\,exp\left(\frac{-M}{4}\right) - 0.058 \leq 1.8 \tag{13}$$

$$K_\sigma = 1 - C_\sigma\,ln\left(\frac{\sigma'_{vc}}{Pa}\right) \leq 1.1 \tag{14}$$

$$C_\sigma = \frac{1}{18.9 - 2.55\sqrt{(N_1)_{60cs}}} \leq 0.3 \tag{15}$$

3.3. Liquefaction severity index (LSI)

The liquefaction severity index (LSI) is an analysis that can be used to estimate the damage or severity to the soil caused by liquefaction. LSI was determined using the Sonmez and Gokceoglu method [20], which was developed from Iwasaki et al. method [21]. LSI was calculated based on the liquefaction SF value from the surface to a depth of 20 meters with equations (16) to (17).

$$LSI = \int_0^{20} P_L(z).w(z).dz \tag{16}$$

$$P_L(z) = \frac{1}{1 + \left(\frac{SF}{0.96}\right)^{0.45}} \; ; SF < 1.411 \tag{17}$$

$$P_L(z) = 0\,;\, SF \geq 1.411 \tag{18}$$

$$w(z) = 10 - 0.5z \tag{19}$$

Where $P_L(z)$ is the liquefaction probability function, *SF* is the liquefaction safety factor value, and z is the soil layer depth in meters. LSI values were classified into 6 levels with categories from not liquefied to high severity level, as listed in **Table 1.**

Table 1 Liquefaction severity index category [20]

LSI	Category
LSI = 0	Not liquefied
0 < LSI < 15	Very low
15 ≤ LSI < 35	Low
35 ≤ LSI < 65	Moderate
65 ≤ LSI < 85	High
LSI ≥ 85	Very high

4. Result and discussion

The soil grain size distribution represents the weight of the soil grains retained at a specific sieve diameter, producing a graph. In the case of liquefaction, non-cohesive soils with uniform gradation have higher liquefaction susceptibility than soils with good gradation [10]. The steeper the shape of the soil grain size graph, the more potential for liquefaction.

This study plotted the soil grain size distribution of non-cohesive soil samples from each borehole on the Tsuchida [16] boundary graph. The plotting results show that the non-cohesive soil samples in each borehole had potentially liquefied. The soil sample in BH-02 was the most potentially liquefied compared to the soil in BH-01 and BH-03, as shown in **Figure 8**.

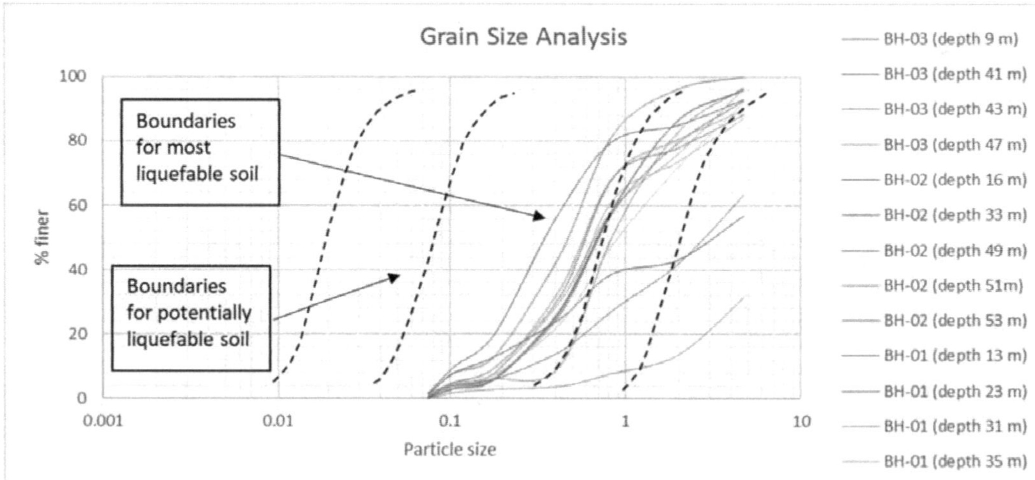

Figure 8. Plotted soil grain size distribution on the Tsuchida [16] graph

Analysis of liquefaction potential was carried out at each borehole up to a depth of 20 meters from the ground surface elevation. A magnitude of 7.5 was used in the analysis, which was the largest magnitude value in the history of the earthquake triggered by the Palu-Koro fault [22]. Based on the preliminary site-specific response analysis (SSRA) report at the project location conducted by the planning consultant team (PT. Adhi Karya), the predicted value of peak ground acceleration (PGA) at the Anutapura general hospital location was 0.63 g [23]. This PGA value was used in the liquefaction potential analysis.

For the analysis, soils containing clay and silt with a fine content of more than 50% were not analyzed (n/a) and considered not to have the potential for liquefaction and had no SF value. **Table 2** shows an example of an analysis table for the liquefaction potential analysis at borehole point BH-02. The SF value and the liquefied layer for each borehole point can be seen in **Figure 9**. Liquefaction potential occurs when SF < 1.

Table 2 Calculation of liquefaction potential analysis at BH-02

Depth (m)	**N**	**Soil Type**	***FC* (%)**	**σ_{vc} (kPa)**	**σ'_{vc} (kPa)**	**a_{max} (g)**	**$(N_1)_{60}$**	**$(N_1)_{60cs}$**	***CSR***	***CRR***	***SF***
1	50	Gravell	1.33	18.60	18.60	0.63	74.92	74.92	0.41	2.00	4.86
3	50	Gravell	1.33	55.80	55.80	0.63	63.59	63.59	0.40	2.00	4.95
5	12	Sand	0.13	90.93	71.33	0.63	17.16	17.16	0.50	0.18	0.36
7	8	Sand	0.13	130.73	91.53	0.63	10.26	10.26	0.55	0.12	0.22
9	12	Sand	0.13	165.86	107.06	0.63	14.93	14.93	0.58	0.15	0.27
11	8	Clay	90.56	203.59	125.19	0.63	n/a	n/a	0.59	n/a	n/a
13	9	Clay	90.56	241.62	143.62	0.63	n/a	n/a	0.59	n/a	n/a
15	11	Sand	0.13	276.31	158.71	0.63	11.09	11.09	0.59	0.12	0.20
17	8	Clay	90.56	314.31	177.11	0.63	n/a	n/a	0.58	n/a	n/a
19	10	Clay	90.56	352.61	195.81	0.63	n/a	n/a	0.57	n/a	n/a
21	14	Clay	90.56	392.06	215.66	0.63	n/a	n/a	0.55	n/a	n/a

The results of the liquefaction potential analysis show that 3 borehole points at the AMC building construction area had the potential to liquefy at a certain depth. At borehole point BH-01, the groundwater level depth was 2.3 meters. The potentially liquefied soil layer depths were between 5-7 meters and 11-17 meters. At borehole point BH-02, the groundwater level depth was 4.5 meters. The potentially liquefied soil layer depths were between 4.5-9 meters and 13-15 meters. At borehole point BH-03, the groundwater level depth was 3 meters. The potentially liquefied soil layer depths were between 11-15 meters and 17-19 meters. The analysis results show that the potentially liquefied soils were saturated sandy soils with loose to medium density with an SPT value of 8-27.

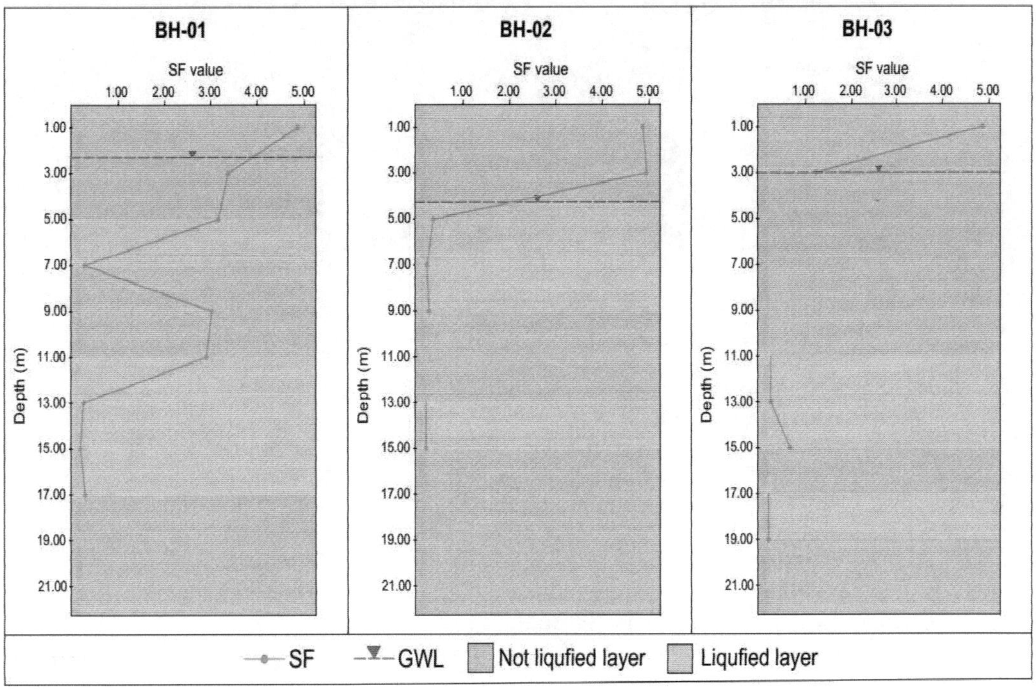

Figure 9. Safety factor (SF) chart with liquefied and not liquefied layers at borehole points BH-01, BH-02 and BH-03

Table 3 Recapitulation of LSI analysis results

Location	Coordinate	LSI	Category
BH-01	-0.8997 , 119.8495	27.94	Low
BH-02	-0.8999 , 119.8497	43.76	Moderate
BH-03	-0.9003 , 119.8494	15.79	Low

The analysis was continued by calculating the liquefaction severity index (LSI) based on the SF value in each soil layer that had been previously analyzed. **Table 3** shows the LSI analysis results for each borehole. BH-01 and BH-03 were categorized as low severity levels. Meanwhile, BH-02 was categorized as a moderate severity level. This low to medium level value occurs because the depth of the potentially liquefied soil layers are located in the deep layer so that the level of damage that may be caused is not so visible up to the ground surface. The LSI value was then used to create a liquefaction severity map at the project site, as shown in **Figure 10**.

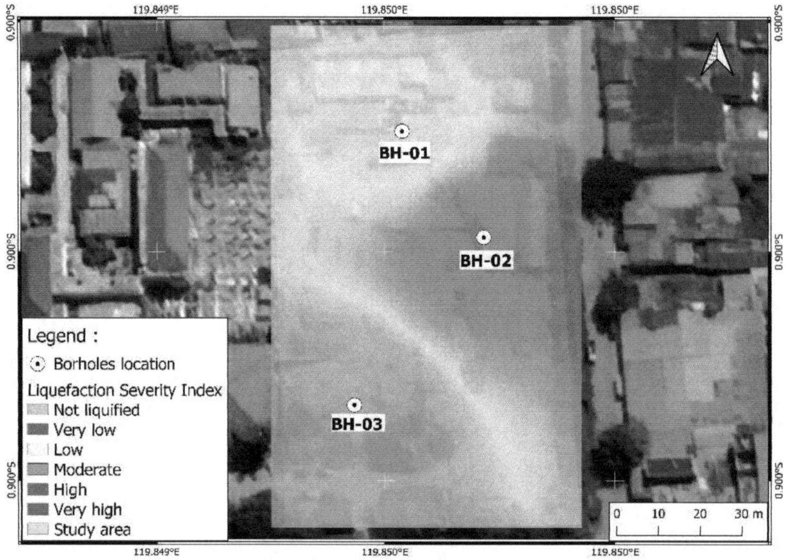

Figure 10. Liquefaction severity map of AMC building construction area (base map source: Google Earth [12])

Based on the analysis results, mitigation is needed to mitigate the possibility of liquefaction in the AMC building construction. Structural measurements must be carried out on the lower structure of the building, taking into account the effects of the liquefied soil layer. As known, liquefaction can cause loss of soil bearing capacity [24]. The foundation structure placed in the liquefied layer can cause the building to sink and fail. Deep pile foundations placed in hard soil layers through the liquefied layer can be one of the mitigations against the liquefaction hazard [25,26]. However, the pile foundation must consider the effect of liquefaction during the design process. According to the Indonesian national standard on geotechnical design requirements [27], the frictional resistance in the liquefied soil layer should be ignored when calculating the pile bearing capacity. The shear strength of the liquefied soil layer must also be reduced to obtain an additional value of bending moment and deflection in the foundation structure.

5. Conclusion

A study on the liquefaction potential has been carried out in the construction area of the AMC Building at the Anutapura general hospital. The study location is close to the Balaroa liquefaction-affected area with Holocene alluvial rock formations and low groundwater conditions. According to the soil investigations, the soil was dominated by sand with several clays and silt layers. Based on the soil grain size distribution, the non-cohesive soil at the location tended to have a uniform gradation, so it was categorized as potentially liquefied. The liquefaction potential analysis results show that the soils in the construction area of the AMC building had the potential to liquefy with low to moderate severity levels. The study results are expected to be taken into consideration in mitigation efforts. The potentially liquefied layer must be considered when designing the structure, especially in the lower structure of the building. The use of deep foundations, such as pile foundations, can be an option for mitigating liquefaction.

Acknowledgments

The authors express their gratitude to the Settlement Infrastructure Agency for Central Sulawesi Region, Directorate General of Human Settlement, Indonesian Ministry of Public Works and Housing for supporting the completion of the study.

References

[1] Walpersdorf A, Vigny C, Manurung P, Subarya C and Sutisna S 1998 Determining the Sula block kinematics in the triple junction area in Indonesia by GPS *Geophys J Int* **135** 351–61

[2] PuSGeN 2017 *Peta Sumber dan Bahaya Gempa Indonesia Tahun 2017* ed M Irsyam et. al. (Kabupaten Bandung: Puslitbang Perumahan dan Permukiman, Balitbang, Kementerian PUPR)

[3] United States Geological Survey (USGS) 2018 *M7.5 Palu, Indonesia Earthquake of September 28, 2018*

[4] Pemerintah Provinsi Sulawesi Tengah 2018 *Rencana Induk Pemulihan dan Pembangunan Kembali Wilayah Pascabencana Provinsi Sulawesi Tengah* ed D Lenggogeni et. al. (Jakarta: Kedeputian Bidang Pengembangan Regional, Kementerian PPN/ Bappenas)

[5] European Commission Joint Research Centre 2018 Mw 7.5 Earthquake in Indonesia 28, Sep 2018

[6] Mason B, Hutabarat D and Prakoso W 2019 Geotechnical Reconnaissance: The 28 September 2018 M7.5 Palu-Donggala, Indonesia Earthquake *Report of Geotechnical Extreme Events Reconnaissance, The Geotechnical Extreme Events Reconnaissance (GEER) Association* 1–86

[7] Open Street Map 2022 Palu and Sigi Map *[online] available at: https://www.openstreetmap.org/#map=11/-1.0189/119.8581&layers=H*

[8] Marcuson, William F. I 1978 Definition of Terms Related to Liquefaction *Journal of the Geotechnical Engineering Division* **104** 1197–200

[9] Das B M and Ramana G v 2010 *Principles of Soil Dynamics* (Cengage Learning)

[10] Day R W 2002 *Geotechnical Earthquake Engineering Handbook* (McGraw-Hill)

[11] Google Earth 2022 Balaroa Map *[online] available at: https://earth.google.com/web/@-0.90456502,119.84301127,26.44263545a,2886.82488481d,35y,0h,0t,0r*

[12] Google Earth 2022 Anutapura General Hospital Location Map *[online] available at: https://earth.google.com/web/@-0.90009395,119.84937922,15.04030731a,251.07850694d,35y,0h,0t,0r*

[13] Muhanifah H, Adi A D and Faris F 2021 Liquefaction investigation of Balaroa, Central Sulawesi on liquefied and non-liquefied areas *IOP Conf Ser Earth Environ Sci* **861** 052039

[14] Sukamto R 1973 *Peta Geologi Regional Tinjau Lembar Palu* (Bandung: Pusat Penelitian dan Pengembangan Geologi)

[15] Widyaningrum R 2012 Penyelidikan geologi teknik potensi liquifaksi daerah Palu , Provinsi Sulawesi Tengah *Kementrian Energi dan Sumber Daya Mineral* 43

[16] Tsuchida H 1970 Prediction and countermeasure against the liquefaction in sand deposits *Abstract of the Seminar in the Port and Harbor Research Institute* 31–333

[17] Idriss I M and Boulanger R W 2008 *Soil Liquefaction During Earthquakes* (Unites States of America: EERI Publication)

[18] Seed H B and Idriss I M 1971 Simplified Procedure for Evaluating Soil Liquefaction Potential *Journal of the Soil Mechanics and Foundations Division* **97** 1249–73

[19] Youd T L and Idriss I M 2001 Liquefaction Resistance of Soils: Summary Report from the 1996 NCEER and 1998 NCEER/NSF Workshops on Evaluation of Liquefaction Resistance of Soils *Journal of Geotechnical and Geoenvironmental Engineering* **127** 297–313

[20] Sonmez H and Gokceoglu C 2005 A liquefaction severity index suggested for engineering practice *Environmental Geology* **48** 81–91

[21] Iwasaki T, Tokida K and Tatsuoka F 1981 Soil Liquefaction Potential Evaluation with Use of the Simplified Procedure *International Conferences on Recent Advances in Geotechnical Earthquake Engineering and Soil Dynamics* **12** 209–14

[22] BMKG 2019 *Katalog Gempa Bumi Signifikan dan Merusak 1821-2018* ed T Prasetya and A Sudrajat (Jakarta: Pusat Gempabumi dan Tsunami BMKG)

[23] PT Adhi Karya (Persero) 2022 *Laporan Analisis Kegempaan Rumah Sakit Umum Anutapura (RSAP), Kota Palu, Sulawesi Tengah*

[24] Youd T L 1984 Geologic effects-liquefaction and associated ground failure *Proceedings of the Geologic and Hydrologic Hazards Training Program, US Geological Survey Open-File Report, USGS* 210–32

[25] Seed H B 1987 Design Problems in Soil Liquefaction *Journal of Geotechnical Engineering* **113** 827–45

[26] Hardiyatmo H C 2022 *Rekayasa Gempa untuk Analisis Struktur & Geoteknik* (Yogyakarta: Gadjah Mada University Press)

[27] Badan Standarisasi Nasional 2017 *SNI 8460:2017 Persyaratan perancangan geoteknik* (Jakarta: BSN)

The Stability of Sidoarjo Mud Volcano Embankment Induced By The Excess Pore Water And Mud Pressure Based On Numerical Simulation

K Ruus[1,2], A Rifa'i[1*] and A D Adi[1]

[1] Department of Civil and Environmental Engineering, Faculty of Engineering, Universitas Gadjah Mada, Yogyakarta, 55281, Indonesia
[2] Directorate General of Water Resources, Ministry of Public Works and Housing, South Jakarta, 12110, Indonesia

*Corresponding author: ahmad.rifai@ugm.ac.id

Abstract. Sidoarjo mud volcano is a geology phenomenon which is the worst in terms of disaster to the people and environment. This is partly due to the overpressure at depth and as well as the cracks and fracture existence. Furthermore, the overpressurization impacts the increasing of pore water pressure and causing the reduction of effective stress. The complexity of the subsurface's physical condition causes the instability of the embankment. This paper presents the evaluation of the stability of the embankment under the ideal and the presence of subsurface pressure conditions. The numerical simulation using Plaxis 2D has been conducted to determine the change of pore water pressure due to subsurface pressure and the safety factor of an embankment. The results show the subsurface pressure impact the increasing of pore water pressure around 208 kPa consequently. It reduces the safety factor around 4-30% from the ideal condition. The safety factor of embankment in the existing condition is still less than the minimum requirement which is 1,5 in both modeling conditions. The embankment collapse was caused by the landslide that leads upstream of the embankment as the effect of the tendency of upstream side landslides that is greater than the downstream side.

1. Introduction

Sidoarjo mud volcano, known as LUSI in terms of Indonesian is Lumpur Sidoarjo, is a geology phenomenon that occurred on May 29, 2006 located around the Watukosek fault zone [1]. LUSI mud volcano is the worst in terms of disaster to the people and environment, it is still erupting with the mudflow condition showing large, medium, and small fluctuating bursts. This is partly due to the high overpressure that is produced below the Upper Kalibeng Formation and nearly reached the lithostatic level so that the large energy from the deep formation can reach the surface [2][3]. Furthermore, it has an impact on the contribution of increasing pore water pressure and causing the reduction of effective stress in soil [3].The embankment has been constructed around the eruption to prevent the overflow of hot mud Sidoarjo into the surrounding area with a total length and height of the embankment are 14430 m and 11 m respectively [4]. The LUSI embankment is an earth-fill dam sitting on poor weak ground. Since the embankment was built in 2007, it has experienced many failures due to subsidence and mudflow activity. The embankment failures occurred such as hydraulic, seepage, and structural failures [5]. The LUSI material mostly consists of fine-grained soil which is dominated by clay particles about 54.47% of the total dry weight. According to USCS, it is classified as a high plasticity silt (MH) [6].

The complexity of the subsurface physical condition causes the instability of the embankment, it can be seen from the land subsidence estimated using GPS (Global Positioning System) and remote sensing [7][8]

The previous studies used FEM (Finite Element Model) and VLF-EM (Very-Low Frequency Electromagnetic) to determine the stability and potential collapse of the Sidoarjo mud volcano embankment [4] [5][9][10]. FEM results indicate that the collapse of the embankment that occurred was caused by landslides and differential settlements. The embankment failure was caused by a fault, which was the origin of a weak point in the embankment. The weak point was identified by applying VLF-EM data. These studies haven't analyze and measure the increase of pore water pressure which caused by overpressure from the subsurface that can reduce the strength of the soil. Thus, this study aims to evaluate the stability of the LUSI embankment, particularly on the north side of the embankment which is STA 250, since it is being considered as a critical side. Furthermore, this study will undertake the analysis by considering poor weak ground, mud eruption pressure, and changes in pore water pressure based on the numerical simulation using Plaxis software under plane strain conditions.

2. Analytical Approach

2.1. Settlement Analysis

The time rate and magnitude of settlement are need to be estimated for evaluation of embankment stability. Soil profile is generally separated into different layers depending on their properties. Thus thick layers with similar properties are often subdivided into subsoil layers for refinement of the analysis. The total settlement obtain by summing the settlement from each of the compressible layers [11]. The rate of primary consolidation can be calculated using one-dimensional consolidation theory proposed by Terzaghi [12]. The equation is defined:

$$T = \frac{c_v t}{H_{dr}^2} \tag{1}$$

Where T is the time rate of consolidation and serves as a measure of dimensionless time, H_{dr} is the length of the longest pore water drainage path, and c_v is the coefficient of consolidation. Total one-dimensional consolidation settlement, S_c, results from a change in void ratio, Δe, over the depth of the consolidation layer. The ultimate consolidation settlement for a normally consolidated soil is defined as

$$S_c = \sum_{i=1}^{n} \frac{C_{ci} H_{0i}}{1+e_{0i}} \log\left(\frac{\sigma'_{vfi}}{\sigma'_{v0i}}\right) \tag{2}$$

Where the summation is performed over n sublayer. If $\sigma'_{vfi} > \sigma'_p$, total consolidation settlement clays of an overconsolidated soil is,

$$S_c = \sum_{i=1}^{n} \frac{C_{ri} H_{0i}}{1+e_{0i}} \log\left(\frac{\sigma'_{vfi}}{\sigma'_{v0i}}\right) \tag{3}$$

and if $\sigma'_{vfi} < \sigma'_p$ is

$$S_c = \sum_{i=1}^{n} \frac{C_{ri} H_{0i}}{1+e_{0i}} \log\left(\frac{\sigma'_{pi}}{\sigma'_{v0i}}\right) + \frac{C_{ci} H_{0i}}{1+e_{0i}} \log\left(\frac{\sigma'_{vfi}}{\sigma'_{v0i}}\right) \tag{4}$$

Where C_c and C_r are compression index, H_{0i} is the initial thickness, e_{0i} is the initial void ratio of *i th* sublayer, σ'_{v0} is initial vertical stress, and σ'_{vfi} is the final vertical effective stress.

2.2. Phi-C Reduction Method

One of methods used to calculate the safety factor of embankment is numerical simulation using FEM (Finite Element Model) [13]. FEM provides an accurate and precise result to predict the behaviour of soil stress [14]. Strength reduction method is commonly used to evaluate the safety factor [15]. The process is termed *Phi-c reduction* in Plaxis software. In this approach the soil strength parameters and are reduced until failure of the structure occurs. The safety factor is defined:

$$SF = \frac{available\,strength}{strength\,at\;failure} = value\;of\;\Sigma Msf\;at\,failure \qquad (5)$$

3. Result And Discussion

3.1. Soil Investigation

The soil material of the embankment and ground was classified by using the SPT (Standard Penetration Test). There were BH-1 to BH-6 with depths up to 50 meterss to determine the soil layer of the north side of the embankment (Figure 1). The SPT data were correlated to obtain stratigraphy models of the ground (Figure 2). It shows that embankment materials are composed of silty sand with a consistency of very loose to very dense. The conditions of the ground are composed of silty clay with a soft to a very stiff consistency, interspersions of coarse-grained soils of silt and sand with a very loose to medium consistency, and clay soils with very soft to stiff consistency from top to the bottom layer respectively.

The CPTu (Cone Penetration Test with Pore Water Pressure Measurement) test was carried out to determine the condition of the pore water pressure. The highest pore water pressure occurs at CPTu 4 where there is an increase of ± 380 kPa from the normal pore water pressure of 200 kPa with the extreme pore water pressure around 580 kPa (Figure 3). Based on the evaluation result that there is an excess pore water pressure caused by residual pore water pressure that has not been dissipating due to the consolidation process and the effect of the subsurface pressure. Hence, the soil strength will be decreased due to the increasing of pore water pressure.

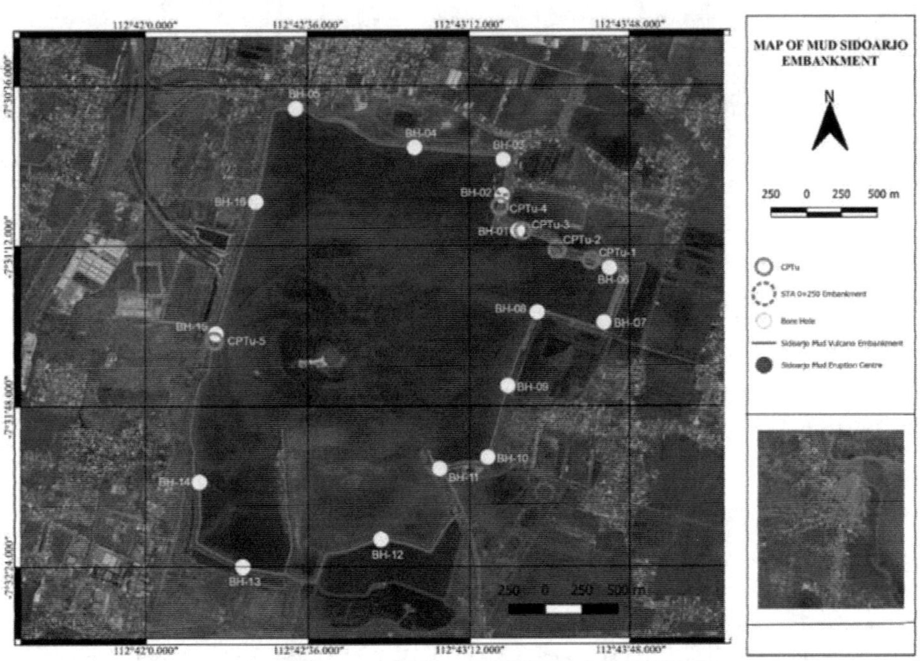

Figure 1 Location of Soil Investigation and STA-250 Embankment (Modification of Google Earth by Authors)

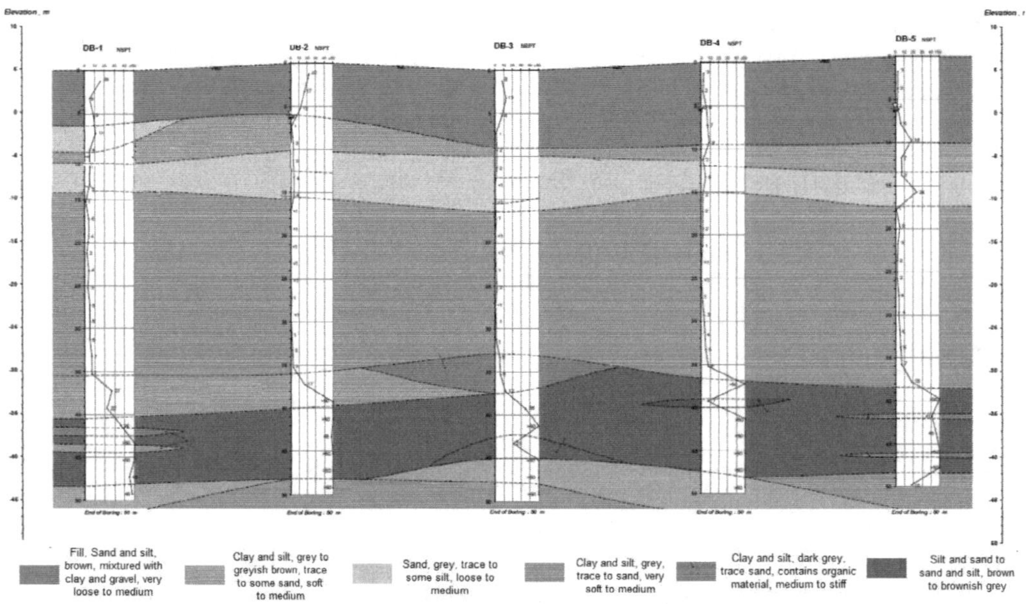

Figure 2 Soil Stratigraphy of North Side Embankment [16]

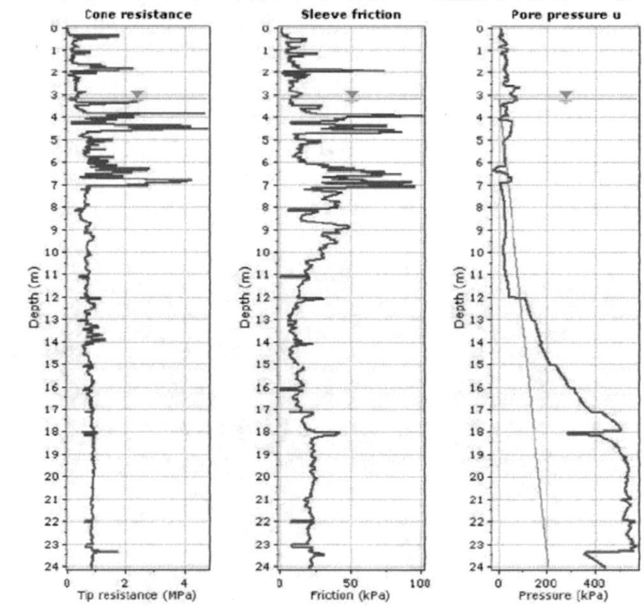

Figure 3 CPTu-4 Investigation Result [17]

3.2. STA 250 Embankment

The STA 250 embankment is the main mudflow retaining embankment, which was built on the outermost boundary of the northern affected area map. This embankment is located in Kendungbendo Village of Sidoarjo Regency, East Java, Indonesia. The location of STA 250 embankment is very close to the residential areas, so it has a very large potential danger if the embankment collapse (Figure 1). Based on the result of field measurement, it can be concluded that the STA 250 is the embankment with the highest rate of deformation compared to other embankments around - 1,092/year [17]. The initial design of the embankment was built at an elevation of + 11 meters from the ground level. The embankment constructed with berms on the downstream side at an elevation 3 m and 6 m, which is effectively devide a single slope to three slopes with ratio of 1,5:1 (H:V). The existing of STA 250 embankment cross section presented in Figure 4. The parameter soil properties of ground and STA 250 embankment used in stability modeling can be seen in Table 1 [16].

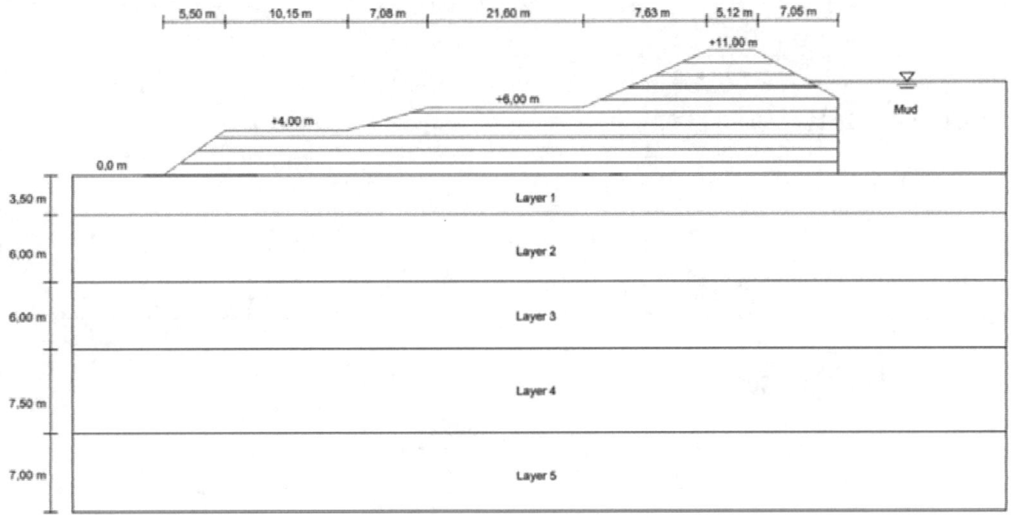

Figure 4 STA 250 Embankmkment Geometry [16]

Table 1 Input Ground and Embankment Materials in Plaxis Software

Layer	Ground and Embankment Materials	Water Condition	Unit Weight γ_{unsat} (kN/m³)	Unit Weight γ_{sat} (kN/m³)	Coef of Permeability $k_x = k_y$ 4. (m/day)	Elasticity Modulus E (kN/m²)	Effective Cohesion c' (kN/m²)	Effective Friction ϕ (deg)
1	Silt and Clay	Undrained	13.24	18.14	5.46 x 10^{-3}	3848	49.33	13
2	Sand and Silt	Drained	13.73	18.44	8.64 x 10^{-2}	10162	9.87	35
3	Silt and Sand	Undrained	11.38	16.67	1.47 x 10^{-2}	11247	17.76	29
4	Clay and Silt	Undrained	7.75	14.61	4.84 x 10^{-4}	4045	22.69	5
5	Clay and Silt	Undrained	7.45	14.42	4.57x 10^{-4}	3354	8.88	5
6	Embankment	Undrained	18.63	19.40	2.42x 10^{-2}	5750	10.06	26.97
7	Mud	Undrained	14	15	8.64x 10^{-4}	1000	3	5.46

3.3 Verification of Using Numerical Simulation

The process of validating slope stability calculations requires a lot of trial and error in order to representing the field condition [18]. A settlement analysis has been conducted, comparing the results from the numerical simulation with Plaxis software and with analytical calculation results. By means of conventional consolidation theory, which derived in equation 1-4. The time rate consolidation and ultimate settlement analysis has been obtained. The subsoil has been divided into 5 layers to increase the accuracy of the calculation. The settlement has then been calculated in each layer, resulting in a cumulative settlement after 3,41 years reaching 1,25 m (Table 2) while the numerical simulation results show the ultimate settlement about 1,27 m and took 3,79 years (Figure 5). By the comparing the two method results, it can be seen that the numerical analysis reasonably agreed with the analytical calculation

Table 2. The Analytical Calculation of Consolidation

Layer	e_0	Cc	Hc	$Cc.\ Hc$	σ'_0	$\Delta\sigma'$	$\frac{\sigma'_0 + \Delta\sigma'}{\sigma'_0}$	$Log\ \frac{\sigma'_0 + \Delta\sigma'}{\sigma'_0}$	Sc
			m	m	kN/m^2	kN/m^2			m
1	1,03	0,25	3,5	0,875	38,98	202,98	6,208	0,793	0,342
2	0,93	0,11	6	0,66	90,76	187,30	3,064	0,486	0,166
3	1,26	0,13	6	0,78	131,92	173,86	2,318	0,365	0,126
4	2,29	0,49	7,5	3,675	167,92	159,56	1,950	0,290	0,324
5	2,47	0,6	7	4,2	200,19	148,18	1,740	0,241	0,291
Ultimate Consolidation Settlement (m)									1,25

H_{dr}	t_v	c_v	t_{90}
m	U= 90%	m^2/sec	year
24	0,848	4,54E-06	3,41
Time Rate of Consolidation (Years)			3,41

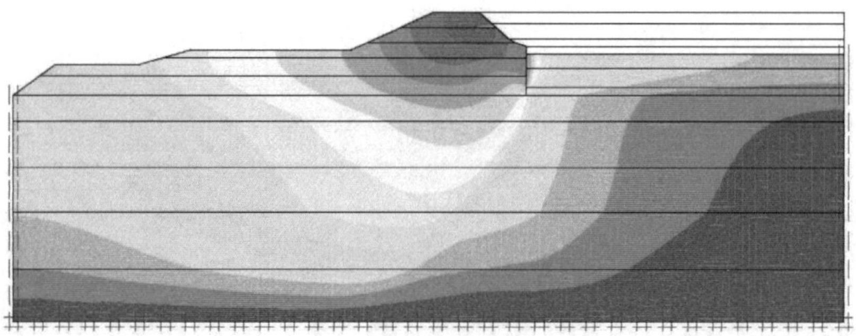

(a)

Plaxis - Finite Element Code for Soil and Rock Analyses

Project description : LUSI EMBANKMENT Date : 01/05/2022
User name : Koshiyuki Kabuto, Japan
Project name : Validasi Calculation.PLX
Output : Calculations list Page : 1

Identification	Phase no.	Start from	Calculation	Loading input	Time	Water	First	Last	Log info
Initial phase	0	N/A	N/A	N/A	0,00	0			N/A
Embankment 11 m	1	0	Plastic analysis	Staged construction	0,00	1	1	12	No errors.
Consolidation 365 days	2	1	Consolidation analysis	Staged construction	365,00	1	13	42	No errors.
Consolidation 730 days	3	1	Consolidation analysis	Staged construction	730,00	1	43	66	No errors.
Consolidation Min PWP	4	1	Consolidation analysis	Minimum pore pressure	1384,197	1	67	109	No errors.
SF Embankment Final Construction	5	1	Phi/c reduction	Incremental multipliers	0,00	1	110	209	No errors.
SF Embankment 365 days	6	2	Phi/c reduction	Incremental multipliers	0,00	1	210	309	No errors.
SF akhir	7	4	Phi/c reduction	Incremental multipliers	0,00	1	310	609	No errors.

Figure 5 Consolidation Analysis Using Plaxis: (a) Ultimate Consolidation Settlement = -1,27 m ; (b) Time of Rate Consolidation = 3,79 Years

The consolidation calculation in numerical simulation using Plaxis software has been obtained to determine the change in pore water pressure. The analysis shows that the pore water pressure due to the consolidation process only around 372,57 kPa, however the value a much lower when compare to the CPTu measurement of 580 kPa (Figure 3). Therefore, it is assumed that there is a contribution from subsurface pressure to an increase in pore water pressure. By comparing the numerical simulation result with the CPTu measurement, it can bee seen there is a difference in pore water pressure of 208 kPa (Table 3). The difference value indicates the contribution of subsurface pressure to the increasing of pore pressure.

Table 3 Pore Water Pressure Analysis Using CPTu and Plaxis Software

Pore Water Pressure	
CPTu	580 kPa
Plaxis (Consolidation Calculation)	372 kPa
Difference	208 kPa

3.4 Subsurface Pressure Modeling

Subsurface pressure modeling is carried out with 4 scenarios in Plaxis software, i.e. reducing the soil strength parameters assuming that an increase in pore water pressure can reduce the effective soil stress, provide an uplift force on the layer 4 and layer 5 of soil, and provide increment to the pore water pressure distribution using UDPP (User Define Pore Pressure) [19]. The analysis result of pore water pressure in 4 models can be seen in Table 4. It can be seen that scenarios 1-3 have no significant effect on changes in pore water pressure while scenario 4 using UDPP model gives the closest results to the CPTu value is 578,27 kPa. This is due to the increment to the pore water pressure distribution on layer 5 of the subsoil. In conditions where there is subsurface pressure, the pore water pressure will not be hydrostatic, so it cannot be defined by means of a phreatic level.

Table 4 Subsurface Modeling Results

Output	Reduce Soil Strength Parameters	Uplift Force Layer 5	Uplift Force Layer 4	User Defined PWP	CPTu
Extreme Pore Pressure (kPa)	372,81	372,49	371,79	578,27	580

3.5 Distribution of Vertical Displacement

The impact of subsurface pressure contributing to the increasing of pore water pressure is a reduction in the effective stress of the soil. With these conditions, it affects to the amount of soil deformation of embankment.The analysis result of the vertical distribution under ideal conditions showed that the embankment experienced a vertical deformation of -1.37 m (Figure 6a). In contrast to the presence of subsurface pressure cause embankment experienced uplift on the downstream side up to +0,6 m while the upstream subside for about -1,0 m (Figure 6b). Field GPS (Global Position System) surveys were conducted during the period of January 2021- February 2022 on top tanggul in elevation 11 meters [20]. The results show the ground has subside for about -1,077 meterss on February 2022, while the numerical simulation result of vertical simulation under the presence of subsurface condition is around -1,0 meters. In comparison with the numerical simulation, it seems that the numerical simulation reasonably agreed with the field measurements. The comparison of GPS surveys measurement of STA 250 embankment presented in Figure 7.

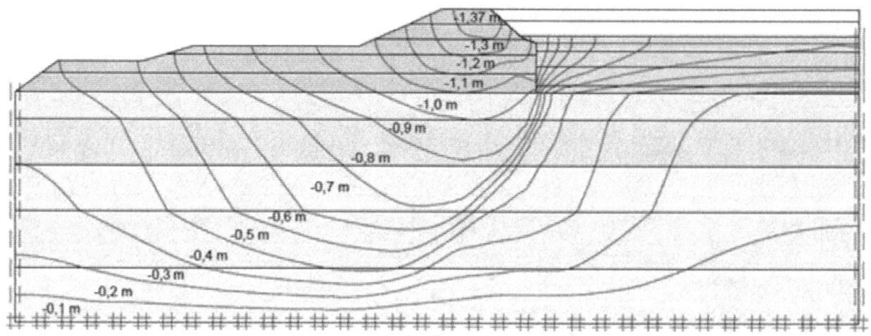

(a)

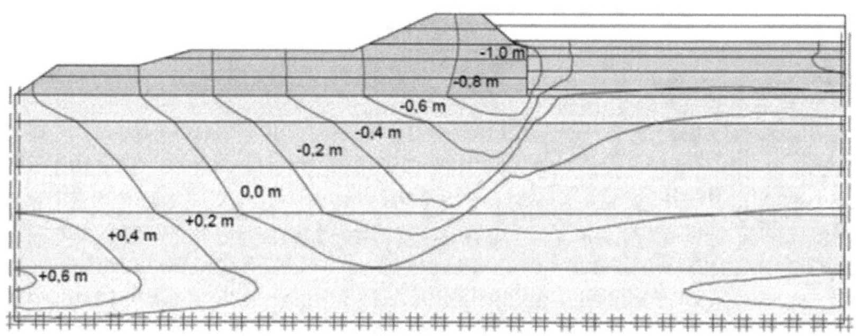

(b)

Figure 6 Distribution of Vertical Displacement: a) Ideal Conditions = -1,37 m; b) With Subsurface Pressure Conditions = -1,0 m (Upstream), +0,6 m (Downstream)

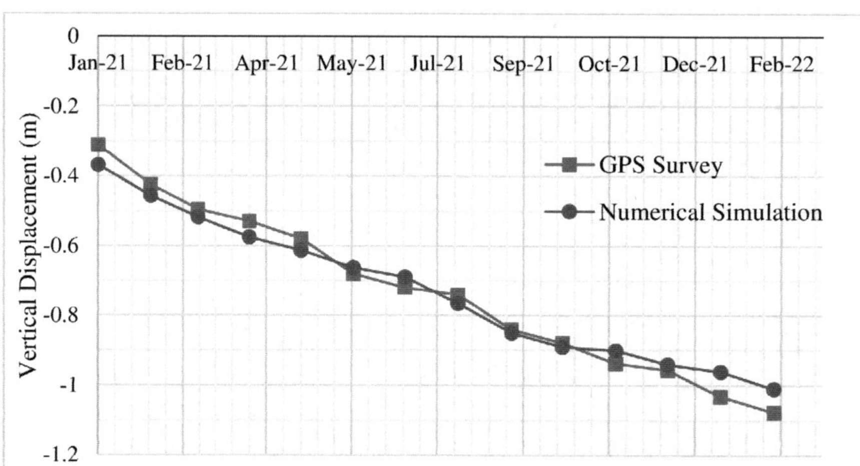

Figure 7 The comparison of GPS Surveys and Numerical Simulation Result

3.6 Safety Factor of Embankment Analysis

Analysis of the safety factor of the Sidoarjo mud volcano embankment was carried out under ideal conditions where there was no contribution of subsurface pressure and with the presence of subsurface pressure with the height of embankment is 11 meters in each condition.

The analysis result of the stability of the embankment at various mud elevations in the mud reservoir are presented in Table 5. It can be concluded that an increase in pore water pressure due to subsurface pressure can reduce the safety factor of the embankment by 4-30% from ideal conditions.The mud elevation in the mud reservoir is an elevation of + 6.5 meterss from the ground level in the existing conditions and the results show that the safety factor still less than the minimum requirement of SNI 8460:2017 and SNI 8064:2016, which is 1,5 in both modeling conditions (Figure 8).

Based on the numerical simulation of failure mode (Figure 9), it can be seen that the collapse of the Sidoarjo mud volcano embankment in both conditions was caused by a landslide leading to upstream of the embankment. With that said, the tendency of upstream side landslides is greater than the downstream side due to the steeper slope and mud deposits at the toe of the inner side embankment in a very soft slurry form. However, an assessment of the overall stability of the embankment is needed to be observed because the embankment at the field experienced a number of failures such a seepage and landslide along the north side (Figure 10).

Table 5 Safety Factor of Embankment

Mud Elevation	Safety Factor	
	Ideal Condition	With Subsurface Pressure Condition
+ 0,0 m	Collapse	Collapse
+ 1,0 m	Collapse	Collapse
+ 3,5 m	Collapse	Collapse
+ 5,5 m	1,0676	1,0262
+6,5 m	1,2334	1,1686
+ 7,5 m	1,4352	1,3213
+ 9,5 m	2,0091	1,4046

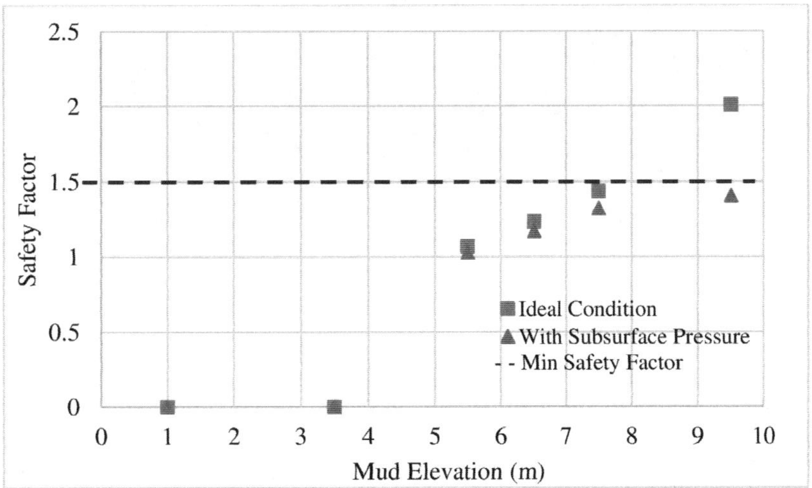

Figure 8 Safety Factor of Embankment With Minimum Requirement

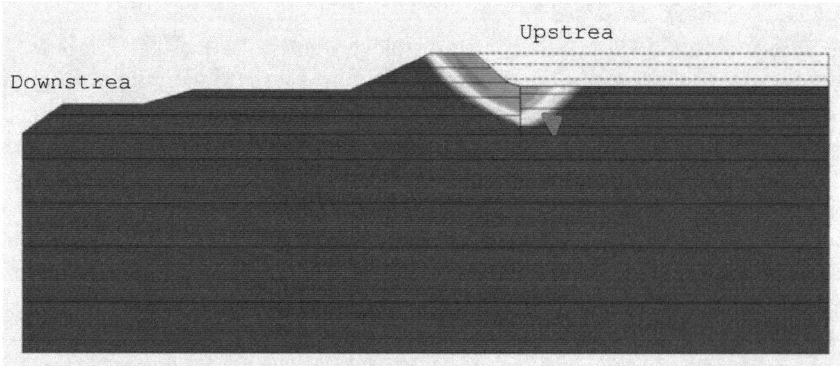

Figure 9 Failure Mode of Sidoarjo Mud Volcano Embankment

(a)

(b)

(c)

Figure 10 Site Investigation; (a) Seepage Along The North Side Embankment; (b) Landslide In The Outer Side of Embankment; (c) Landslide Potential Initiated by The Cracks

4 Conclusion

The stability of Sidoarjo mud volcano embankment has been analyzed based on the numerical simulation using Plaxis software by considering mud eruption and the change of pore water pressure due to subsurface pressure. It showed that there was an increase in pore water pressure of ± 380 kPa with the extreme pore water pressure is 580 kPa. This is due to the consolidation process of 372 kPa and a subsurface contribution of 208 kPa. Unfortunately, the increase of pore water pressure due to subsurface pressure can reduce the safety factor of the embankment by around 4-30% from ideal conditions.

The analysis result show that the collapse of the Sidoarjo mud volcano embankment was caused by a landslide leading to upstream of the embankment. However, an assessment of the overall stability of the embankment is needed to be observed for further study. Because at the field, it was found that the embankment experienced a number of failures such a seepage and landslide along the north side of the embankment. In consideration of the results, STA 250 has a stability problem consequently. So, the remedial measure is necessary to increase the stability of the embankment.

ACKNOWLEDGEMENT

The author would like to express gratitude to Center For Sidoarjo Mudflow Containtment and the Indonesian Ministry of Public Works and Housing for all of support so this research can be conducted well.

REFERENCES

[1] Mazzini A *et al.*, 2007 Triggering and dynamic evolution of the LUSI mud volcano, Indonesia *Earth Planet. Sci. Lett.* **261**, 3–4 p. 375–388.
[2] Tingay M, 2015 Abnormal pore pressure and associated environmental and geohazards Initial pore pressures under the Lusi mud volcano , Indonesia *Interpret. Vol. 3, No. 1 (February 2015); p. SE33–SE49, 5* **3**, 1.
[3] Tanikawa W Sakaguchi M Wibowo H T Shimamoto T and Tadai O, 2010 Fluid transport properties and estimation of overpressure at the Lusi mud volcano, East Java Basin *Eng. Geol.* **116**, 1–2 p. 73–85.
[4] Hakim, Abdul., Gunawan A, 2020 Evaluation of Sidoarjo mud volcano embankment *AIP Conf. Proc.* **2251**, August.
[5] Agustawijaya D and Sukandi, 2012 The Stability Analysis of the Lusi Mud Volcano Embankment Dams using FEM with a Special Reference to the Dam Point P10.D *Civ. Eng. Dimens.* **14**, 2 p. 100–109.
[6] Handoko L Rifa'i A Yasufuku N and Ishikura R, 2015 Physical properties and mineral content of Sidoarjo mud volcano *Procedia Eng.* **125**, December p. 324–330.
[7] Chaussard E Amelung F Abidin H and Hong S H, 2013 Sinking cities in Indonesia: ALOS PALSAR detects rapid subsidence due to groundwater and gas extraction *Remote Sens. Environ.* **128** p. 150–161.
[8] Abidin H Z Davies R J Kusuma M A Andreas H and Deguchi T, 2009 Subsidence and uplift of Sidoarjo (East Java) due to the eruption of the Lusi mud volcano (2006-present) *Environ. Geol.* **57**, 4 p. 833–844.
[9] Agustawijaya D and Sukandi, 2017 The Displacement Models of The Lusi Mud Volcano Embankment November 2013.
[10] Sungkono Husein A Prasetyo H Bahri A S Monteiro Santos F A and Santosa B J, 2014 The VLF-EM imaging of potential collapse on the LUSI embankment *J. Appl. Geophys.* **109** p. 218–232.
[11] Washington State Department of Transportation, 2022 *Geotechnical Design Manual M46-03.08* October .
[12] Hardiyatmo H C, 2019 *Mekanika Tanah II* Sixth Yogyakarta: Gadjah Mada University Press.
[13] Sijabat D M F, 2010, Analisa Perkuatan Daya Dukung Tanah Lunak Dengan Geosintetik Menggunakan Pemodelan Perangkat Lunak (Program Plaxis), Universitas Indonesia.
[14] Duda A and Siwowski T W, 2021 Stability and Settlement Analysis of a Lightweight Embankment Filled with Waste Tyre Bales over Soft Ground *Transp. Infrastruct. Geotechnol.*
[15] Luo N Bathurst R J and Javankhoshdel S, 2016 Probabilistic stability analysis of simple reinforced slopes by finite element method *Comput. Geotech.* **77** p. 45–55.
[16] PPLS, 2019, Laporan Review Design Supervisi Peningkatan Tanggul, Embung, Sistem

Drainase dan Embung di Kawasan Lumpur Sidoarjo (PT Tri exnas Consultant & Management Engineering KSO PT Atlantik Bina Persada Konsultan Teknik dan Supervisi), Surabaya.

[17] PPLS, 2021, Laporan Akhir Paket Pekerjaan Penilaian Kinerja Sarana dan Prasarana Pengendalian Lumpur (PT Aditya Engineering Consultant), Surabaya.

[18] Dayanun S Rifa 'i A and Suryolelono K B, 2014 Evaluation of Soil Embankment Mass Movement By Using Subdrain System and Reinforcement Based on the Numerical Simulation of Semarang-Solo Toll Road *Civ. Eng. Forum* **XXIII**, January p. 1475–1482.

[19] Brinkgreve R B J, 2007 *Plaxis 2D.V8* Delft University of Technology&PLAXIS b.v. Yhe Netherlands.

[20] PPLS, 2022, Laporan Unit Pengelola Bendungan, Surabaya.

A Study of Liquefaction Potential on Retaining Wall and Embankment in Parangtritis Village

C T S A Galag[1, 2], F Faris[1,*] and A D Adi[1]

[1] Department of Civil and Environment Engineering, Faculty of Engineering, Universitas Gadjah Mada, Yogyakarta 55281, Indonesia
[2] Directorate General of Highways, The Ministry of Public Works and Housing, South Jakarta 12110, Indonesia

*Corresponding author: fikri.faris@ugm.ac.id

Abstract. Parangtritis Village located in Bantul Regency, Special Region of Yogyakarta. Previous studies showed that the Opak Active Fault line passes right through the east side of the construction. This study aims to examine the potential for liquefaction in the retaining wall and embankment structures in Parangtritis Village in the event of an earthquake in the area. The subsurface evaluation was based on N-SPT drilling data as well as soil data at 4 drill point. The bedrock acceleration value was determined based on SNI 2833:2016. To obtain a more specific bedrock peak acceleration value, the Indonesian Spectra Design seismic data was used by the Directorate General of Highway, Ministry of Public Works and Housing, where the PGA bedrock acceleration value is 0.558 g. The stages of liquefaction potential analysis are divided into three stages, Cyclic Stress Ratio, Cyclic Resistance Ratio, Factor of Safety. Liquefaction Potential Index analysis as a determinant of potential damage. Data processing results at four drill points indicate that the soil layer at several depths has liquefaction potential with a safety factor value of <1. The results of these calculations show that the LPI value is in the LPI vulnerable > 15, indicating a high liquefaction potential.

1. Introduction

Parangtritis Village located in Kretek Kapanéwon, Bantul Regency, Yogyakarta Special Region is included in the development area of the southern route of Java island, especially the Kretek 2 Bridge Construction Project with the work location located at the mouth of the opak river of the Yogyakarta Special Region shown in **Figure 1**.

The Kretek 2 Bridge based on design drawings is planned to be a total of 598 m. The results of previous studies concluded that the Opak Active Fault line just passed through the east side of the bridge construction. The Opak fault can cause earthquakes that trigger earthquakes, such as in the history of earthquake events in 2006 [1]. Given the impacts and dangers caused by the liquefaction event, this study is an interesting topic and a special concern to find out the potential for liquefaction on the retaining wall and embankment in Parangtritis Village in the event of an earthquake in the area and minimize the damage caused. Thus, it is necessary to conduct research related to the analysis of potential liquefaction on the retaining wall and embankment in Parangtritis Village.

Figure 1. Kretek 2 Bridge Construction Location

2. Geological conditions and seismicity on site

The Java region is classified as one of the regions in the world that has a high level of seismicity due to the frequent occurrence of earthquakes. Plots of earthquakes recorded by the global seismicity network that have been localized in the EHDL catalog [2], shows clearly that this area is filled with epicenter points of earthquakes from the shallow hypocenter (below 60km) to the deep (above 60 km).

Earthquakes that occurred in the subduction boundary zone of the Indian-Australian Ocean plate to the bottom of Java Island which is the southern part of the Eurasian plate were seen dominating seismicity in this region. Shallow earthquakes off the southern coast of Java mostly originated from the brittle - seismogenic part of this subduction zone referred to as megathrusts (0 – 60 km).

Meanwhile, deep earthquakes are associated with the ductile part of the subduction zone which is often referred to as the Benioff Zone. In the Java region, this Benioff zone earthquake occurred to a depth of 600 km and even deeper. This is in stark contrast to Sumatra where the Benioff zone earthquake only occurred at a depth of no more than 300 km. This marked the difference and change from an oblique convergence system along southwest Sumatra at a relative speed of 60 to 45 mm/year to a frontal convergence system in Java at a speed of 70 mm/year [3]. The location of the bridge is approximately 130 m above the Benioff zone seen in **Figure 2.**

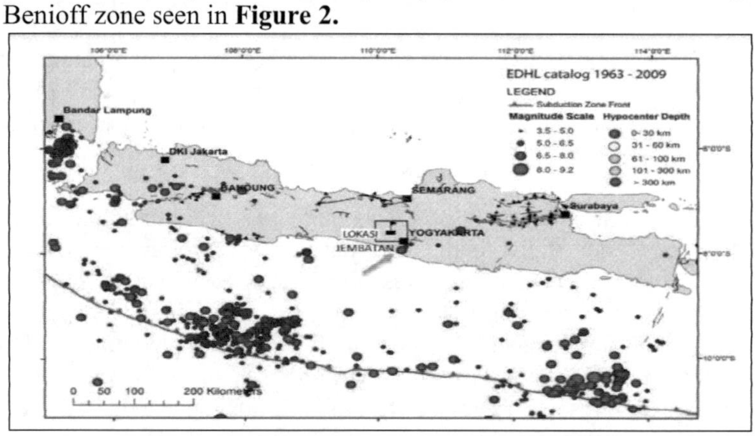

Figure 2. Shallow earthquake (0-30 km)

The plot of shallow earthquakes along with a depth of 30 km or less shows clearly that the potential for shallow earthquakes that can be destructive does not only occur in megathrusts but also the land area of active fault lines. It can be seen that there are records of earthquake events in the area around the bridge site.

3. Methods

The research location on the construction project of the Parangtritis Retaining Wall and Embankment of Kretek Bridge 2 is located in Depok, Kretek District, Bantul Regency, and D.I Yogyakarta Province. This project crosses the combined estuary of the Opak and Oyo Rivers which will later connect Parangtritis and Samas Beaches. The location of this study is just above the opak active fault with coordinates of 8° 0'36.35" LS, 110° 17'37.44" BT, so it is very vulnerable to earthquake events as can be seen in **Figure 3**.

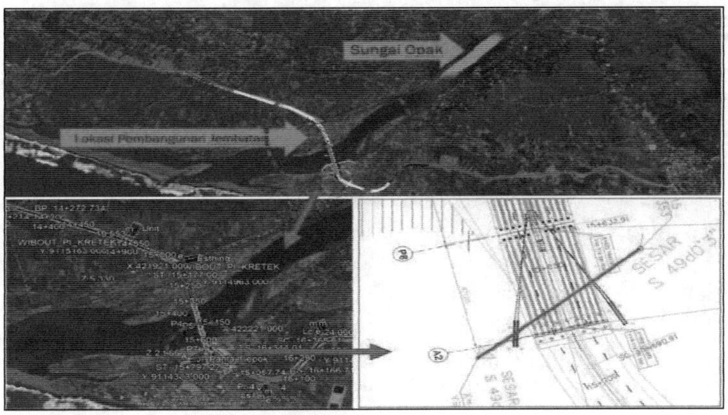

Figure 3. The location of the Retaining Wall is right above the Opak Fault [4]

This research was conducted using data obtained from PPK 1.4 as the owner of the Kretek Bridge 2 project. Drilling data obtained as many as 14 drill points. In the research conducted on the embankment work segment in the Parangtritis direction, the data used are drilling data for BH-8, BH-9, BH-13, and BH-14.

No	Lokasi	Koodinat	
		X	Y
1	BH-01	422085.747	9114831.457
2	BH-02	422081.383	9114787.827
3	BH-03	422104.084	9114708.224
4	BH-04	422100.243	9114703.647
5	BH-05	422123.627	9114703.725
6	BH-06	422118.421	9114619.369
7	BH-07	422142.729	9114580.693
8	BH-08	422138.053	9114534.978
9	BH-09	422161.306	9114498.234
10	BH-10	421997.024	9115069.673
11	BH-11	422035.577	9114991.696
12	BH-12	422067.227	9114913.310
13	BH-13	422290.491	9114397.059
14	BH-14	422429.463	9114330.459

Figure 4. Location of Bore Hole for Kretek Bridge 2

Subsurface soil conditions that are prone to liquefaction are uncrowded sand, there has been no cementation, and the depth of groundwater level near the soil surface. Because seismic shocks occur so quickly, the soil without cohesion will experience an increase in pore water pressure which causes water flow to the ground surface, which appears in the form of mudflows or sand boiling [5].

Fine sand soils have grain diameters between 0.08 mm to 0.2 mm which are loose, have no soil cohesion, and are easily subject to permeability. Propose a relationship between the relative density of D_r and the value of N-SPT for sandy soils [6], shown in **Table 1** below:

Table 1. Relative density and N-SPT

Relative Density (D_r)	Sand Density	**N-SPT**
< 0.2	Very Loose	< 4
0.2 – 0.4	Loose	4 -10
0.4 – 0.6	Medium	10 -30
0.6 – 0.8	Dense	30 – 50
0.8 – 1.0	Very Dense	> 50

Analysis of the gradation of the size of the granules is necessary to provide a close representation of the various compositions of the soil. This test is carried out to determine the size of the distribution of soil grains, which range from 4.75 mm to 0.0013 mm. Soil samples from boreholes are mechanically tested for every depth of 1-2 m to determine the distribution of soil grains in detail. The dispersion of soil granules can be utilized as an examination of liquefaction potential [7].

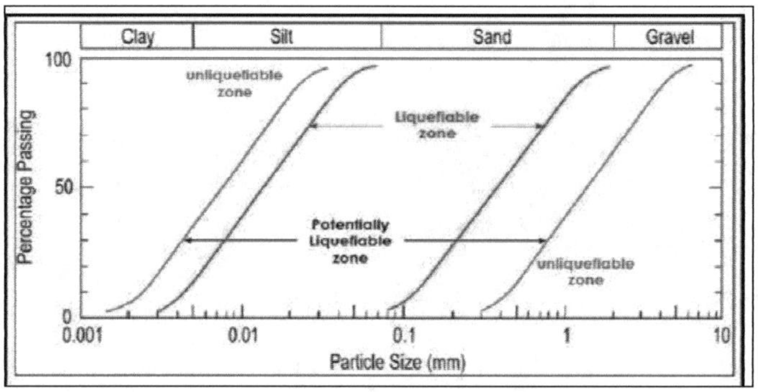

Figure 5. Gradient grading curves for soils with potential for liquefaction

In deciding the sort of soil area analyzed for spectral response design, the soil investigation data required within the frame of N-SPT values from drilling results at the soil site were reviewed and seismic downhole tests were carried out to obtain the values of the shear wave velocity (V_s). The information from these two comes about is utilized to classify the sort of soil or what is commonly known as the location classification. Each region features a diverse location or soil sort, so it is exceptionally imperative to classify the location into a few classes of sites. This study uses the reference of SNI 2833:2016 to determine the site class [8].

In the analysis, the maximum peak acceleration analysis for analysis used the maximum considered earthquake geometric mean (MCE_G) peak acceleration which has been adjusted to the influence of the site class, namely PGA_M. For the determination of the PGA_M value based on SNI 2833:2016 concerning Bridge Design against Earthquake Loads. The site coefficient is based on the site class in SNI 2833:2016 [8], as shown in the **Table 2**:

Table 2. Site Class and FPGA Value

Site Classification	V_{S30} (m/s)	N-SPT	PGA≤ 0,1	PGA= 0,2	PGA=0,3	PGA=0,4	PGA>0,5
SA	>1500	N/A	0.8	0.8	0.8	0.8	0.8
SB	750-1500	N/A	1.0	1.0	1.0	1.0	1.0
SC	350-750	≥ 50	1.2	1.2	1.1	1.0	1.0
SD	175-350	15-50	1.6	1.4	1.2	1.1	1.0
SE	<175	<15	2.5	1.7	1.2	0.9	0.9

IOP Conf. Series: Earth and Environmental Science 1244 (2023) 012028 doi:10.1088/1755-1315/1244/1/012028

The analysis of the liquefaction potential using a semi-empirical approach based on SPT data that had been developed by Idriss and Boulanger to evaluate the Cyclic Resistance Ratio (CRR) in the soil and also the Cyclic Stress Ratio (CSR) [9]. The ratio of CRR and CSR shows the safety factor of the soil.

Earthquakes can result in a Seismic Stress Ratio (SSR) or commonly called a Cyclic Stress Ratio (CSR) which can trigger liquefaction in saturated sand soils. Maximum earthquake acceleration, earth acceleration, total ground stress, effective ground stress, and also depth factor reduction can be factors that affect the magnitude of CSR. The value of CSR has been formulated by Seed and Idriss [10] which was later developed by Idriss and Boulanger [9] with the equation:

$$CSR = \frac{r_{av}}{\sigma'_v} = 0.65\frac{r_{max}}{\sigma'_v} = 0.65\,(\frac{\alpha_{max}}{g})(\frac{\sigma_v}{\sigma'_v})\,r_d \tag{1}$$

Whereas CSR is cyclic stress ratio, τ_{max} is cyclic shear stress, σ_v and σ'_v are total and effective vertical stress, a_{max} is maximum ground level acceleration in the horizontal direction, r_d is depth reduction coefficient, and g is gravitational acceleration 9.81 m/s2.

The soil's ability to withstand liquefaction or cyclic resistance ratio (CRR) can be determined based on test result data. Cetin in his journal [11] give corrections to get the value $(N_1)_{60}$, can be calculated by equation:

$$(N_1)_{60} = N_m C_N C_E C_B C_R C_S \tag{2}$$

In which N_m is N-SPT obtained from field tests, C_N is the normalization factor N_m against overburden stress, C_E is a correction of energy hammer ratio (ER), C_B is a correction for diameter borehole, C_R is the correction factor for the length of the bar, C_S is the correction for the sample.

The CRR value depends on the test result data such as the CPT or SPT test results because it is generally correlated with in situ parameters, such as the CPT resistance penetration value, the number of SPT strokes, or the shear wave velocity, V_s [9]. The CRR value can be calculated using the equation below:

$$CRR = \exp\left(\frac{(N_1)_{60cs}}{14.1} + \left(\frac{(N_1)_{60cs}}{126}\right)^2 - \left(\frac{(N_1)_{60cs}}{23.6}\right)^3 + \left(\frac{(N_1)_{60cs}}{25.4}\right)^4 - 2.8\right) \tag{3}$$

The factor of safety (FS) is a comparison between the Cyclic Resistance Ratio (CRR) and the Cyclic Stress Ratio (CRR). FS is used to indicate the danger of potential liquefaction. Idriss & Boulanger determines the Factor of Safety (FS) [9] in the following equation:

$$FS = \frac{CRR}{CSR} \tag{4}$$

If the FS value < 1 then the soil in the layer has the potential liquefaction while if FS ≥ 1 then the soil is safe from liquefaction [12].

The Liquefaction Potential Index is a method first introduced by Iwasaki [13] which aims to analyze the potential for liquefaction based on the thickness of the soil layer and the distance of the soil layer of potential liquefaction to the surface. Analysis of the potential for liquefaction using the LPI method is calculated to a depth of 20 meters from the ground level because liquefaction with a depth of more than 20 meters is very rare [14]. The LPI value is proportional to the thickness of the applied tenth layer, the depth of the applied soil layer, and the safety factor. For the calculation of LPI by Iwasaki [13] as in the following equation:

$$LPI = \int_{0}^{20} F.w(z)\,dz \tag{5}$$

Whereas F=1 – FS for FS<1.0, F=0 for FS>1.0, w(z)=10 – 0.5z for z < 20 with z in meters.

4. Result and Analysis

4.1. Grain Size Analysis

The results of this test are presented in a soil grain distribution chart illustrating the susceptibility criteria of soils susceptible to potential liquefaction. The level of potential liquefaction can be determined based on the distribution of the size of the material grains, which can be seen in the analysis chart below:

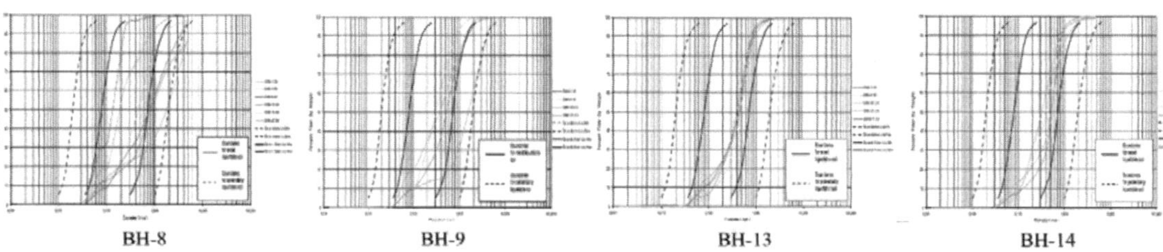

Figure 6. Grain size analysis charts of BH-8, BH-9, BH-13, and BH-14.

Based on the chart above, it is clear that in BH-8, BH-9, BH-13, and BH-14 the relationship between the distribution of granule size and the potential of liquefaction, so further analysis is needed regarding the potential for liquefaction in the study site.

4.2. Site Classifications

Based on the drilling data that has been carried out on BH-8, BH-9, BH-13, and BH-14, it shows that the four drilling locations are dominated by sandy soil and gravel sand with loose to dense densities.

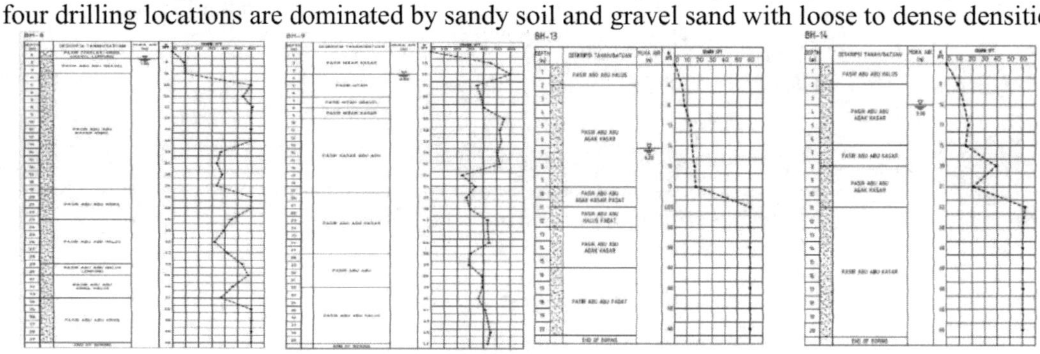

Figure 7. SPT data of BH-8,BH-9,BH-13 and BH-14

The groundwater table from drill test data was found at a depth of 1 m on BH-8, 3 m on BH-9, 9 m on BH-13, and 3 m on BH-14. Groundwater depth affects the susceptibility of the soil to liquefaction.

In this study, the dynamic parameters of the soil used were based on data from shear wave velocity measurements carried out at the BH-9 site. The data from the measurement is then re-evaluated for analysis purposes. In this analysis, the V_S value used is the result of the Seismic Down-hole Test (SDT) and the empirical correlation between the N-SPT and the V_S of the N-SPT test result [15]. Based on the results of analysis and evaluation according to SNI 2833: 2016, the average N-SPT value and the average V_S value of SDT results up to a depth of 30 m obtained the results of local soil types (site classes) around the bridge entered in the medium soil category (SD)[8]. The analysis used historical data on the earthquake in the Special Region of Yogyakarta in 2006 where the magnitude of the M_w earthquake that occurred was 6.3 [16].

In general, from **Figure 8**, it can be seen that the site of the work is dominated by the type of loose sandy soil. This suggests that the potential of liquefaction occurs because the site has an aspect of susceptibility to soil type and groundwater level depth.

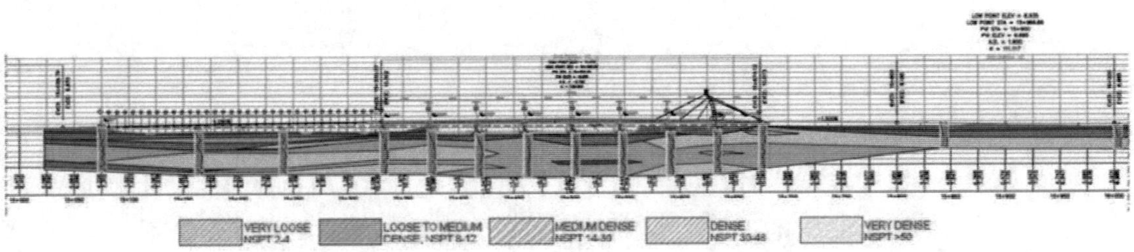

Figure 8. Soil profile at the research site

4.3. Determination of PGA

Based on data obtained from the Final Site Specific Response Spectra (SSRS) report of Kretek Bridge 2 Yogyakarta [2]. The results of the DSHA study at 50-percentile (median) obtained the value of the acceleration of the peak of the earthquake (PGA) in the bedrock due to the Main Opak fault of 0.538g with a distance of about 20 m. Probabilistic hazard analysis (PSHA) for the design of earthquake levels with a repeat period of 1000 has been carried out and obtained a peak acceleration value (PGA) in the bedrock is 0.477g.

Meanwhile, according to data from the Directorate General of Highway, Ministry of Public Works and Housing which can be accessed on the http:/lini.binamarga.pu.go.id page where the PGA base rock acceleration value of 0.558g for Bantul Regency was obtained. In this case, the author decided to use the data of 0.558g shown in Figure 9.

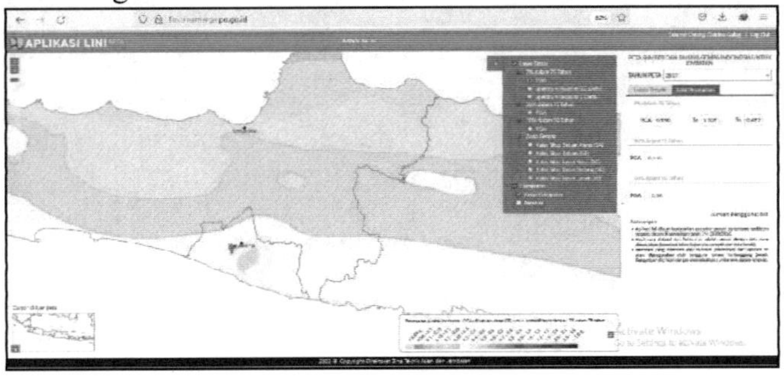

Figure 9. LINI Bina Marga Application (http://lini.binamarga.pu.go.id)

4.4. Liquefaction Potential Analysis

The analysis of liquefaction analysis was carried out at 4 drill test points at the study site using the simplified procedure method. The calculation of the esteem of the soil safety calculates against the cyclone utilizing the moment of the size of the seismic tremor, which is Mw 6.3 with the depth of the groundwater level according to the drill test results of each test point. The comes about of the calculation on each layer at the bore point utilizing using Idriss and Boulanger methods have an FS value of less than 1 as shown in **Table 3**.

Table 3. Factor of Safety Value

Boreholes	Depth (m)	*CSR*	*CRR*	FS_{Liq}
BH-8	0.0 – 1.5	0.434	0.231	0.531
	1.5 – 3.0	0.528	0.266	0.504
BH-9	15.0 – 16.5	0.413	0.333	0.805
BH-13	7.5 – 9.0	0.305	0.302	0.992
BH-14	1.5 – 3.0	0.350	0.315	0.901
	3.0 – 4.5	0.412	0.396	0.961
	4.5 – 6.0	0.443	0.335	0.756

Based on table 3 it can be seen that the soil layers that have potential liquefaction with low FS values (FS < 1) are at depth of 0 to 3 m at BH-8. 15 to 16.5 m at BH-9, 7.5 to 9 m at BH-13, and depth of 1.5 to 6 m at BH-14. The FS value at each review point has the potential for liquefaction in a particular subsoil can be depicted in graphic form as shown in **Figure 10**.

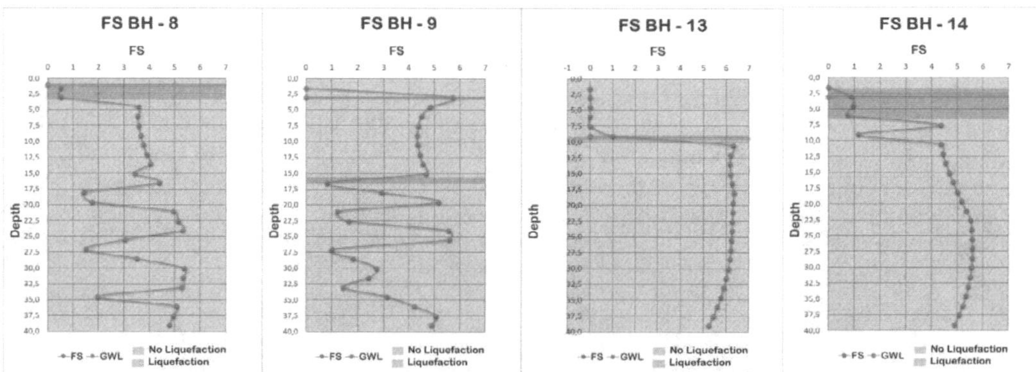

Figure 10. Graph of FS value on BH-8, BH-9, BH-13 and BH-14

4.5. Liquefaction Potential Index

The Liquefaction Potential Index esteem in this consideration was calculated based on the esteem of the soil layer safety figure against liquefaction (FS). The calculation of the LPI value at the research point is carried out by Equation (5) and the results of the calculation of the LPI value of each drill point can be seen in **Table 4**.

Table 4. LPI Result

Boreholes	LPI Value	LPI Classification
BH-8	12.831	High
BH-9	0.513	Low
BH-13	0.064	Low
BH-14	4.270	Low

The high susceptibility of liquefaction in BH-8 is due to the low value of the soil safety factor against liquefaction. The lower the esteem of the soil security factor against liquefaction at the survey point, the higher the LPI value will be so that the level of vulnerability to liquefaction will also become higher. According to Toprak and Holzer [14], the calculation only to a depth of 20 m because the effect of the occurrence of liquefaction on the ground surface rarely occurs after a depth of 20 m.

4.6. Liquefaction Potential Map

Based on the comes about of the investigation of LPI values at the research location. It can be depicted in the form of a map of the liquefaction zone for the study area as presented in **Figure 11**.

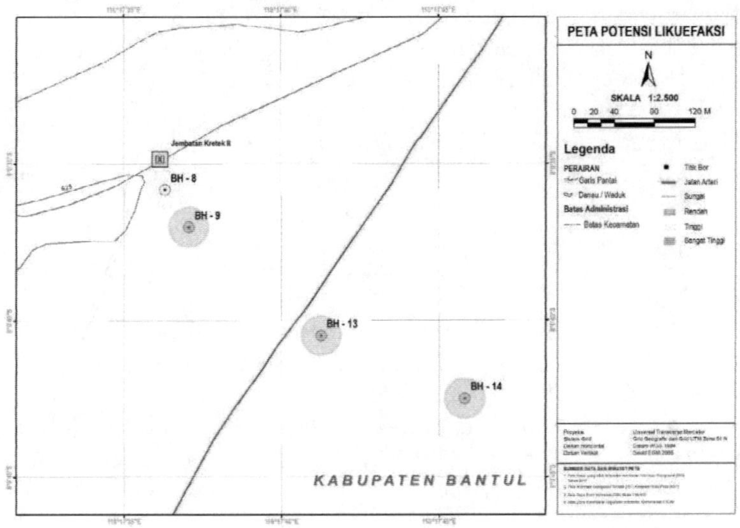

Figure 11. Liquefaction Potential Map

The colors, was obtained from the atlas of the liquefaction vulnerability zone issued by the Geological Agency, Ministry of Energy and Mineral Resources in 2019 [17]. The location of the research points BH-8, BH-9, BH-13 and BH-14 is depicted by point geometry, while the buffer from the research point location, based on assumptions still has the same Liquefaction Potential Index value, as the point described in polygon geometry, using geographic information system (GIS) applications.

5. Conclusions

Based on data on geological and geotechnical conditions at the location of the Retaining Wall and Embankment in Parangtritis Village, soil conditions are vulnerable to potential liquefaction.
Using the preliminary and simplified procedure methods, the same results were obtained for the research site. The research area was identified as vulnerable to potential liquefaction.
The results of the liquefaction potential analysis at 4 (four) drill points showed the value of the Safety Factor which has the potential to liquefaction at a depth of 0-16.5 m. BH-8 is the point with the greatest potential for liquefaction.
Based on the comes about of the investigation of the potential for liquefaction on the Retaining Wall and Embankment in Parangtritis Village, it is recommended for further research related to the mitigation plan.

Acknowledgment

The authors would like to specify their true appreciation to the Ministry of Public Works and Housing and Gadjah Mada University for funding this research.

References

[1] Buana T W and Agung M W, 2015 Liquefaction Characteristic Based on Ground Response Linier Equivalent Analysis and Cyclic Stress Concept on Young Merapi Volcanic Deposit in Bantul … *Jseg.or.Jp* p. 1–5.
[2] Asrurifak, 2019, Site Specific Respons Spectra Jembatan Kretek 2.

[3] McCaffrey R, 1988 Active tectonics of the eastern Sunda and Banda arcs *J. Geophys. Res.* 93, B12.
[4] Karya W, 2021, Laporan Evaluasi Data Teknis Acuan Pelaksanaan.
[5] Ishihara K and Yoshimine M, 1992 Evaluation of settlements in sand deposits following liquefaction during earthquakes *Soils Found.* 32, 1 p. 178–188.
[6] Youd T L Tinsley J C Perkins D M King E J and Preston R F, 1979 Liquefaction Potential Map of San Fernando Valley, California. *Geol. Surv. Circ. (United States)* 807 p. 37–48.
[7] Tsuchida H, 1970 Prediction and Countermeasure against Liquefaction in Sand Deposit *Abstr. Semin. Port Harb. Res. Institute, Minist. Transp. Yokosuka, Japan.*
[8] Badan Standardisasi Nasional Indonesia, 2016 Perencanaan Jembatan Terhadap Beban Gempa SNI 2833:2016 p. 1–70.
[9] Idriss&Boulanger, 2008 *Soil Liquefaction During Earthquakes* .
[10] Seed, H. B., & Idriss I M, 1971 Simplified Procedure for Evaluating Soil Liquefaction Potential *J. Soil Mech. Found. Div.* ASCE 97, SM9 p. 1249–1273.
[11] Cetin K O *et al.*, 2004 Standard Penetration Test-Based Probabilistic and Deterministic Assessment of Seismic Soil Liquefaction Potential December p. 1314–1340.
[12] Day R, 2002 *Earthquake engineering handbook* .
[13] Iwasaki T T K and T F, 1981 Soil Liquefaction Potential Evaluation with Use of the Simplified Procedure in *First Int. Conf. Recent Adv. Geotech. Earthq. Eng. Soil Dyn.* p. 209–14.
[14] Toprak S and Holzer T L, 2003 Liquefaction Potential Index: Field Assessment *J. Geotech. Geoenvironmental Eng.* 129, 4 p. 315–322.
[15] Wijaya K, Laporan Pekerjaan Lapangan Survey Downhole Seismik.
[16] Sulaeman C, 2008 Karakterisasi sumber gempa Yogyakarta 2006 berdasarkan data GPS *Indones. J. Geosci.* 3, 1 p. 49–56.
[17] Kementerian Energi Sumber Daya Mineral, 2019, ATLAS ZONA KERENTANAN LIKUEFAKSI INDONESIA EDISI PERTAMA 2019.pdf. .

IOP Conf. Series: Earth and Environmental Science 1244 (2023) 012029 doi:10.1088/1755-1315/1244/1/012029

Comparison Between Liquefaction Potential Index And Liquefaction Risk Index In Solo – Yogyakarta – YIA Kulon Progo Toll Road (STA. 07+500 – STA. 16+700)

F Hasiholan[1,2], S Ismanti[1*] and A Rifa'i[1]

[1] Department of Civil and Environmental Engineering, Faculty of Engineering, Universitas Gadjah Mada, Yogyakarta 55281, Indonesia

[2] Directorate General of Highways, The Ministry of Public Works and Housing, South Jakarta 12110, Indonesia

*Corresponding author: sito.ismanti@ugm.ac.id

Abstract. The construction of Solo – Yogyakarta – YIA Kulon Progo Toll Road aims to facilitate community economic connectivity from various sectors and increase the accessibility of tourism around southern Java. This research area has geological conditions consisting of tuff (volcanic ash sand) and shallow groundwater, which can trigger liquefaction. Historical earthquake data from The IRIS Earthquake Browser showed that an earthquake with Mw 5.2 occurred in Klaten Regency. This study investigated the index of potential liquefaction and the risk of ground failure potential due to liquefaction. The potential liquefaction analysis uses 10 SPT points at STA. 07+500 – STA. 16+700. The calculation safety factor of liquefaction analysis uses the method of Idriss-Boulanger. The index of liquefaction potential uses the LPI method, and the risk of ground failure potential due to liquefaction uses the LRI method. The result of this study shows the index of potential liquefaction and risk of ground failure potential due to liquefaction from low to very high. The correlation between the index of liquefaction potential and the risk of ground failure due to liquefaction is linear. Based on this study, further research is needed regarding liquefaction mitigation to be carried out in this study area.

1. Introduction

Solo – Yogyakarta – YIA Kulon Progo Toll Road construction is ± 96 Km. This toll road will connect Central Java Province with the Special Region of Yogyakarta Province. This toll road will be expected to facilitate community economic connectivity from the industrial, goods, and service sectors and improve tourism accessibility areas around southern Java, such as the Super Priority Tourist Destination Borobudur [1]. Solo – Yogyakarta – YIA Kulon Progo Toll Road will be built in 3 (three) sections. Section 1 is ± 42 Km from Kartasura Interchange to Purwomartani Interchange. Section 2 is ± 22 Km from Purwomartani Interchange to Gamping Interchange. Section 3 is ± 30 Km from Gamping Interchange to Kulon Progo Interchange. In section 1, the location of STA. 07+500 – STA. 16+700 are in Delanggu, Polanharjo, and Karanganom Districts, Klaten Regency [2].

Based on the Geological Map of Surakarta – Giritontro, Java, the construction of Solo – Yogyakarta – YIA Kulon Progo Toll Road (STA. 07+500 – STA. 16+700) located in the Merapi volcanic rock formation

(Qvm). This formation is composed of volcanic breccia, lava, and tuff [3]. Areas with geological conditions in the form of tuff (volcanic ash sand) have a more significant vibration hazard potential due to amplification and soil interaction [4].

Based on the history of earthquakes around the toll road construction site, the quake was felt several times with a strength greater than Mw 5.0. The earthquake generally originates in the sea area because of the zone of subduction activity system in consequence of the crash among the Indian Ocean Plate – Australia and the Eurasian Continental Plate. Even an earthquake in southern Yogyakarta in 2006 occurred due to active fault activity [5].

The Yogyakarta earthquake in 2006 caused geological damage observed on the ground surface, namely soil cracks, liquefaction, and landslides. The worst damage in Klaten Regency occurred in the Districts of Gantiwarno, Bayat, Prambanan, and parts of Wedi and Cawas. Even in Paseban Village, Bayat District, liquefaction phenomena occur with the appearance of fine sand along the ground cracks [5]. Due to this earthquake, many studies about liquefaction potential in the southern Klaten Regency. But there has been no study about liquefaction potential in the northern Klaten Regency, especially in the location of Solo – Yogyakarta – YIA Kulon Progo Toll Road (STA. 07+500 – STA. 16+700) [2].

Liquefaction is a geotechnical phenomenon that occurs when fine-grained soils are saturated with water and experience cyclic loading due to earthquakes for a certain period. The liquefaction phenomenon can increase the pore water pressure and reduce the effective soil stress so that the soil loses its bearing capacity [6]. Previous research suggests that high liquefaction susceptibility occurs due to several factors, such as the height of the water table, soil density, and distribution of soil particles [7]. The liquefaction potential also occurs because of the earthquake's magnitude at that location. The bigger the quake, the greater the liquefaction potential [8]. In the southern part of Java, the sandy soil layer has the potential to liquefy at a depth of 1-6 m from the ground surface [9].

This study investigated the index of potential liquefaction and the risk of ground failure on this toll road. The potential liquefaction analysis in this study uses 10 (ten) Standard Penetration Test (SPT) points on the construction of this toll road (Figure 1). The calculation safety factor of liquefaction analysis uses the method of Idriss-Boulanger (2008). The index of liquefaction potential uses the method of Liquefaction Potential Index (LPI), and the risk of ground failure potential due to liquefaction uses the method of Liquefaction Risk Index (LRI). The results of this study serve as material for consideration in mitigating the potential for liquefaction in this study area.

2. Research Method

2.1. Condition of geological and geotechnical

Based on the map of geological Surakarta - Giritontro, Java, Solo – Yogyakarta – YIA Kulon Progo Toll Road construction (STA. 07+500 - STA. 16+700) is above Merapi volcanic rock formation (Qvm), which belongs to the Holocene quaternary. This rock formation consists of volcanic breccia, lava, and tuff (Figure 1) [3]. The establishment of the Holocene Quaternary started 10,000 years ago and continues into the present [10]. An area with geological conditions in the form of tuff, generally in the state of volcanic ash sand, typically has a more significant potential hazard to vibration effects caused by amplification and interaction of soil vibrations [4]. Merapi volcanic rock formations, which are sediment resulting from the eruption of Mount Merapi, located in Klaten Regency that composed of sand, gravel, boulders, and some igneous rocks [11].

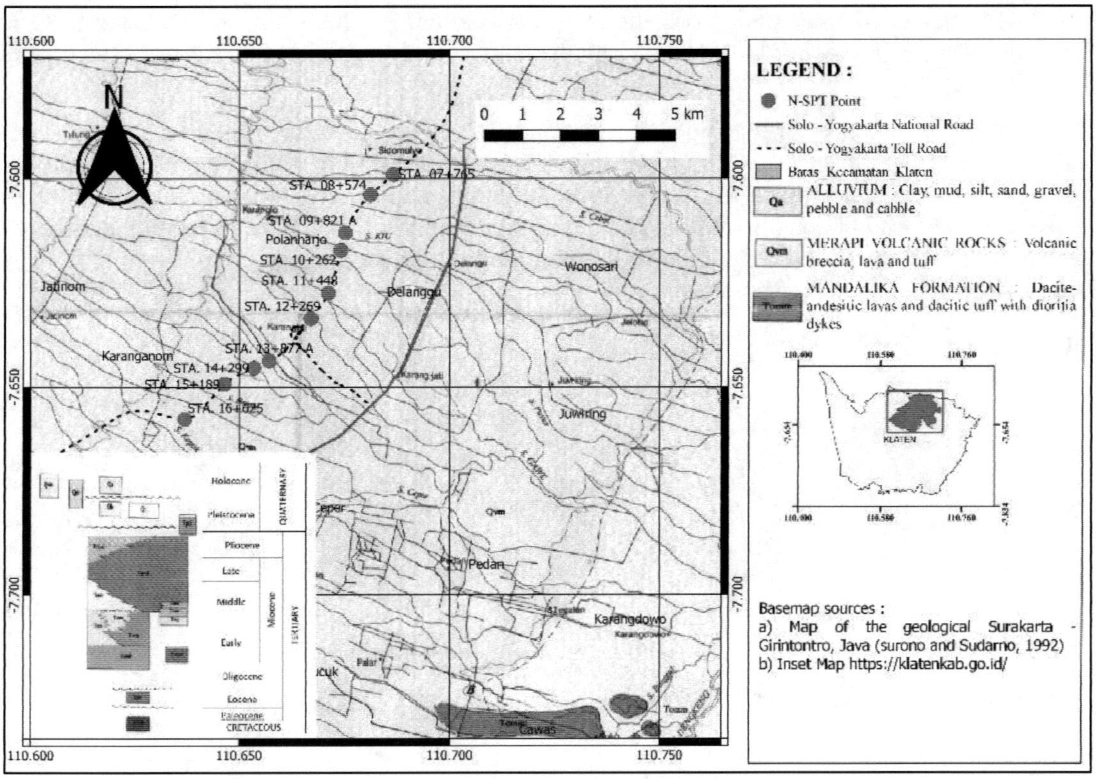

Figure 1. Map of geological of SPT points in the study area ([3]with modification)

This study uses geotechnical examination data from SPT testing on Solo – Yogyakarta – YIA Kulon Progo Toll Road construction (STA. 07+500 - STA. 16+700) in September – December 2020. The SPT data used is 10 points spread throughout the toll road construction (Table 1). In this study area, the geotechnical examination showed that the soil layer was sandy with loose to dense [12].

Table 1. The outcome of geotechnical test

SPT Points	Latitude	Longitude	Groundwater level (m)	$\overline{\text{N-SPT}}$	Dominant soil
STA. 07+765	- 7.599	110.687	- 2.20	17.74	silty SAND
STA. 08+574	- 7.604	110.681	- 1.50	19.80	silty SAND
STA. 09+821 A	- 7.613	110.675	- 1.70	31.72	sandy SILT
STA. 10+262	- 7.617	110.674	- 0.40	29.93	sandy SILT
STA. 11+448	- 7.627	110.671	- 1.20	18.36	sandy CLAY
STA. 12+269	- 7.634	110.667	- 3.00	22.87	silty SAND
STA. 13+877 A	- 7.644	110.657	- 4.00	15.73	silty SAND
STA. 14+299	- 7.645	110.654	- 1.90	23.75	silty SAND
STA. 15+189	- 7.649	110.647	- 1.40	23.96	SAND
STA. 16+625	- 7.658	110.637	- 1.50	28.51	SAND

2.2. Seismic design analysis

There are two methods to determine the seismic hazard estimation approach at a certain point: Probabilistic Seismic Hazard Analysis (PSHA) and Deterministic Seismic Hazard Analysis (DSHA) [13]. his study uses the SNI 8460:2017 (regulations concerning Indonesian Geotechnical Design Requirements). In SNI 8460:2017, construction of toll roads uses a 7% probability in 75 years with a 1,000-year return period [14].

The PSHA method in this study uses the Line application developed by the Directorate General of Highways of the Ministry of PUPR by entering the latitude and longitude coordinates of the seismic point to be analyzed [15].

The DSHA method uses the attenuation equation to determine the peak ground acceleration (PGA) that causes the most significant risk. Several parameters become decisive in this method. These parameters are the magnitude of the earthquake, the distance from the earthquake source, the mechanism of the earthquake source, and the location conditions from the earthquake recording. For infrastructure planning, generally use the DSHA method for acceleration which represents the 84th percentile at a median of 150% [13]. Table 2 shows the attenuation parameter to determine Peak Ground Acceleration in this study area.

Table 2. The attenuation parameter to determine Peak Ground Acceleration (PGA) [13], [16]

Parameter	Unit	Shallow Crustal (Fault and Shallow Background)			Benioff (deep intraslab)	
		Boore-Atkinson NGA. (Boore and Atkinson, 2014)	Campbell-Bozorgnia NGA. (Campbell and Bozorgnia, 2014)	Chio-Youngs NGA. (Chiou and Youngs, 2014)	Geomatrix slab seismicity rock, 1997 (Youngs et al., 1997)	AB 2003 intraslab seismicity worldwide (Atkinson and Boore Wordwide, 2003)
Moment magnitude	M_w	$3 \leq M_w \leq 8.5$ for strike slip	$3 \leq M_w \leq 8.5$ for strike slip	$3.5 \leq M_w \leq 8.5$ for strike slip	$M_w \geq 5.0$	$5.0 \leq M_w \leq 8.3$
Distance to surface projection	km	$0 \leq R_{jb} \leq 400$	$0 \leq R_{jb} \leq 300$	-	$10 \leq R_{jb} \leq 500$	-
The average velocity of shear waves for a depth of 30 meters	m/s	$150 \leq V_S30 \leq 1{,}500$	$150 \leq V_S30 \leq 1{,}500$	$180 \leq V_S30 \leq 1{,}500$	-	-
Depth to the Vs of 1 km/sec	km	$0 \leq Z_{1.0} \leq 3$	-	$Z_{1.0} \leq 20$	-	-
Distance to the Rupture	km	-	$0 \leq R_{rup} \leq 300$	$0 \leq R_{rup} \leq 300$	-	-
Depth to the Vs of 2.5 km/sec	km	-	$0 \leq Z_{2.5} \leq 2.5$	-	-	-
The depth of the hypocentre from the earthquake	km	-	$0 \leq Z_{hyp} \leq 20$	-	-	-
Depth to top of the coseismal rupture	km	-	$0 \leq Z_{tor} \leq 20$	$0 \leq Z_{tor} \leq 20$	-	-
The average dip of the rupture plane	degree	-	$15 \leq dip \leq 90$	-	-	-
Attenuation Equation		By Application NGA-West2			Manual	Manual

The most significant PGA result from the combination of the PSHA and DSHA methods at each SPT point is the bedrock PGA at that SPT point. Determination of liquefaction potential at that point using modified bedrock PGA at that SPT point. In SNI 8460:2017, the modified bedrock PGA value (PGA_M) calculation is obtained by multiplying the bedrock PGA value by the appropriate site coefficient (F_{PGA}) for the site class obtained from the average N-SPT at each SPT point [14].

2.3. Analysis of potential liquefaction

This study's potential liquefaction safety factor analysis uses SPT data with a simplified procedure by Idriss-Boulanger (2008). The calculation of the safety factor of liquefaction uses Equation (1) [17].

$$FS_{Liq} = \frac{CRR_{M,\sigma'_{vc}}}{CSR} \quad (1)$$

If the value of $FS_{Liq} < 1$, the soil lining is potential liquefaction but $FS_{Liq} \geq 1$ means that the soil lining is not potential liquefaction. The calculation of $CRR_{M,\sigma'_{vc}}$ (Cyclic Resistance Ratio for the earthquakes with other M_w and the effective overburden stress) uses Equation (2) – Equation (5).

$$CRR_{M,\sigma'_{vc}} = CRR_{M=7.5,\sigma'_{vc}=1} \cdot MSF \,.\, K_\sigma \quad (2)$$

$$K_\sigma = 1\text{-}C_\sigma \ln\left(\frac{\sigma'_{vc}}{P_a}\right) \quad (3)$$

$$C_\sigma = \frac{1}{18.9\text{-}2.55\sqrt{(N_1)_{60}}} \quad (4)$$

$$MSF = 6.9 \exp\left(\frac{-M_w}{4}\right) \text{-}0.058 \quad (5)$$

where MSF is the factor of scale magnitude, K_σ is the factor of the overburden correction restricted to the maximum value of 1.1, and C_σ is the factor of the overburden correction, which corrected N-SPT and the tool with the maximum value of 0.3.

The calculation of $CRR_{M=7.5,\sigma'_{vc}=1}$ (Cyclic Resistance Ratio for the earthquakes with M_w 7.5 and the effective overburden stress = 1) uses Equation (6) – Equation (9).

$$CRR_{M=7.5,\sigma'_{vc}=1} = \exp\left(\frac{(N_1)_{60CS}}{14.1} + \left(\frac{(N_1)_{60CS}}{126}\right)^2 - \left(\frac{(N_1)_{60CS}}{23.6}\right)^3 + \left(\frac{(N_1)_{60CS}}{25.4}\right)^4 - 2.8\right) \quad (6)$$

$$(N_1)_{60CS} = (N_1)_{60} + \Delta(N_1)_{60} \quad (7)$$

$$\Delta(N_1)_{60} = \exp\left(1.63 + \frac{9.7}{FC+0.01} - \left(\frac{15.7}{FC+0.01}\right)^2\right) \quad (8)$$

$$(N_1)_{60} = C_N.N_{60} = C_N.C_E.\,C_B.C_R.C_S\,.\,N_M \quad (9)$$

where $(N_1)_{60CS}$ is the correction of value N-SPT for fines content, FC is fines content, C_N is the factor of overburden stress, C_E is the factor of energy ratio correction, C_B is the correction of rod length factor, C_S is the correction of sampler factor, and N_M is the value of N-SPT in field.

Value of Cyclic Stress Ratio (CSR) was calculated using Equation (10) – Equation (13) by method of Seed and Idriss [18].

$$CSR = 0.65 \text{ x } (PGA_M) \text{ x } \left(\frac{\sigma_{vc}}{\sigma'_{vc}}\right) \text{ x } r_d \quad (10)$$

$$r_d = \exp[\alpha(z) + \beta(z)\,.\,M_w] \quad (11)$$

$$\alpha(z) = \text{-}1.012\text{-}1.126 \sin\left(\frac{z}{11.73} + 5.133\right) \quad (12)$$

$$\beta(z) = 0.106 + 0.118 \sin\left(\frac{z}{11.28} + 5.142\right) \quad (13)$$

where PGA_M is site modified of PGA, σ_{vc} is total of vertical stress, σ'_{vc} is effective of vertical stress, r_d is coefficient of stress reduction, z is the depth of each borehole below ground level in meters, and M_w is moment of magnitude.

2.4. Liquefaction Potential Index (LPI)

Iwasaki et al. developed the Liquefaction Potential Index (LPI) method. This method aims to predict the liquefaction potential in the study area. The calculation of this method reaches a depth of 20 meters from the ground surface. Equation (14) is the equation in the LPI method [19].

$$\text{LPI}= \int_0^{20} F(z).w(z)dz \tag{14}$$

where F(z) is the degree of damage to a layer in analysis of liquefaction analysis. F(z) = 1 - FS for FS < 1 and F(z) = 0 for FS ≥ 1. w(z) is the depth factor. The deeper the layer of liquefaction, the smaller the surface damage. w(z) = 10 - 0.5z for z < 20 m and w(z) = 0 for z ≥ 20 m. Based on the classification of liquefaction potential, if the value of LPI = 0 is very low; 0 < LPI ≤ 5 is low; 5 < LPI ≤ 15 is high; and if LPI > 5 is very high.

2.5 Liquefaction Risk Index (LRI)

Lee et al. defined the Liquefaction Risk Index (LRI) method in 2003. Equation (15) and Equation (16) are the equation in the LRI method [20]. LRI is an analysis to assess the risk of ground failure potential that can occur due to the liquefaction [21].

$$\text{LRI}= \int_0^{20} P_L w(z)dz \tag{15}$$

$$P_L = \frac{1}{1+\left(\frac{F_s}{0.96}\right)^{4.5}} \tag{16}$$

where P_L is the value of the probability liquefaction based on depth function. *w(z)* is factor of depth. *w(z)* = 10 – 0.5 *z*, where *z* is the maximum depth, maximum up to 20 meters. Based on the classification of potential risk for ground failure due to liquefaction, if the value of LRI < 20 is low; 20<LRI<30 is high; and if LRI > 30 is very high.

3. Results and Discussion

3.1. Seismic design analysis

Figure 2 shows historical earthquake data around the toll road location based on The Incorporated Research Institutions for Seismology (IRIS) Earthquake Browser and the results of bedrock PGA calculations based on the PSHA and DSHA methods [22].

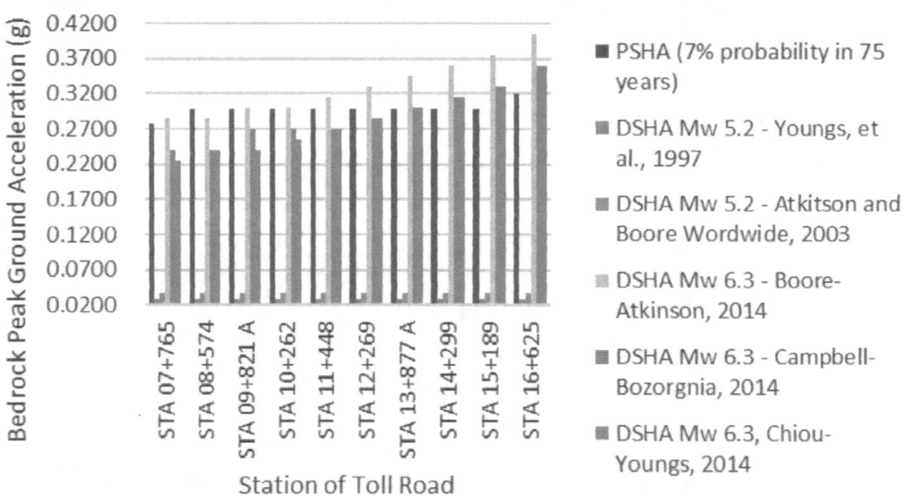

Figure 2. Bedrock Peak Ground Acceleration (PGA) at site

Based on Figure 2 and the site coefficient (F_{PGA}) from SNI 8460:2017, value of modified PGA at site (PGA_M) as shown in Table 3.

Table 3. Modified Peak Ground Acceleration at site (PGA_M)

SPT Points	Site Class	PGA Max (g)	F_{PGA}	PGA_M (g)
STA. 07+765	SD	0.2850	1.2300	0.3506
STA. 08+574	SD	0.2982	1.2035	0.3589
STA. 09+821 A	SD	0.3000	1.2000	0.3600
STA. 10+262	SD	0.3000	1.2000	0.3600
STA. 11+448	SD	0.3150	1.1850	0.3733
STA. 12+269	SD	0.3300	1.1700	0.3861
STA. 13+877 A	SD	0.3450	1.1550	0.3985
STA. 14+299	SD	0.3600	1.1400	0.4104
STA. 15+189	SD	0.3750	1.1250	0.4219
STA. 16+625	SD	0.4050	1.0950	0.4435

Based on Figure 2 and table 3, PGA will be more significant if the mechanism of the earthquake source comes from a fault (Mw 6.3), the earthquake's magnitude is powerful, and the site's distance is near the source of the earthquake.

3.2. Analysis of potential liquefaction

Based on Table 1, Table 3, and Equation (1) – Equation (13), the result of analysis of potential liquefaction as shown in Table 4.

Table 4. Summary of analysis potential liquefaction

SPT Points	Groundwater Level (m)	$\overline{\text{N-SPT}}$	PGA_M (g)	Dominant Soil	Potential Liquefaction at Depth (m)
STA. 07+765	- 2.20	17.74	0.3506	silty SAND	6-10 and 14-16
STA. 08+574	- 1.50	19.80	0.3589	silty SAND	0-2, 4-6, and 14-18
STA. 09+821 A	- 1.70	31.72	0.3600	sandy SILT	2-4
STA. 10+262	- 0.40	29.93	0.3600	sandy SILT	2-6
STA. 11+448	- 1.20	18.36	0.3733	sandy CLAY	2-4
STA. 12+269	- 3.00	22.87	0.3861	silty SAND	10-12 and 18-20
STA. 13+877 A	- 4.00	15.73	0.3985	silty SAND	4-6
STA. 14+299	- 1.90	23.75	0.4104	silty SAND	0-2, 4-6, and 16-18
STA. 15+189	- 1.40	23.96	0.4219	SAND	0-4, 8-10, and 12-14
STA. 16+625	- 1.50	28.51	0.4435	SAND	4-6 and 8-14

Table 4 describes the potential for liquefaction to occur at all points of the SPT. SPT Points that are the most susceptible to liquefaction in this study are in STA. 15+189 and STA. 16+625, which potential liquefaction reaches 8 meters. It can happen because the groundwater level in this location is shallow, and the average N-SPT in the site class is medium soil (SD), with sand as the dominant soil type.

3.3. Comparison Liquefaction Potential Index (LPI) and Liquefaction Risk Index (LRI)

Using Equation (14) – Equation (16), the result of comparing LPI and LRI in STA. 15+189 and STA. 16+625 (the most susceptible to liquefaction in this study), as shown in Figure 3.

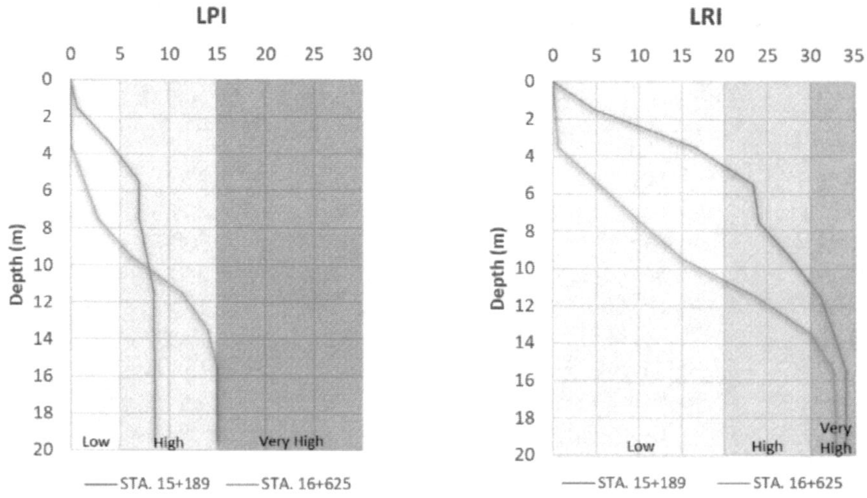

(a) (b)

Figure 3. Comparison LPI (a) and LRI (b) in STA. 15+189 and STA. 16+625

Figure 3 describes the results of the LPI and LRI linear. STA. 15+189 high liquefaction potential (LPI > 5) from a depth of 4 m, and the potential risk of ground failure due to liquefaction occurring is high (LRI > 20) from a depth of 4 m and very high (LRI > 30) from a depth of 10 m. At STA. 16+625 high liquefaction potential (LPI > 5) from a depth of 8 m and very high liquefaction potential (LPI > 15) from a depth of 14 m, and the potential risk of ground failure due to liquefaction occurring is high (LRI > 20) from a depth of 10 m and very high (LRI > 30) from a depth of 14 m. Figure 4 shows the results of the LPI and LRI plots on the map.

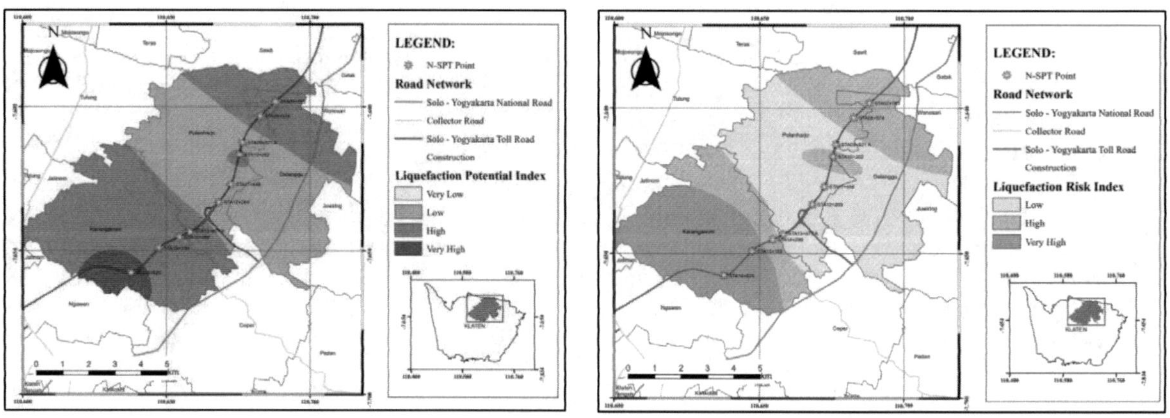

(a) (b)

Figure 4. Map of LPI (a) and LRI (b) in study area

Figure 4 shows that there are 3 (three) points of low liquefaction potential (STA. 09+821 A, STA. 11+448, and STA. 12+269), 6 (six) points of high liquefaction potential (STA. 07+765, STA. 08+574, STA. 10+262, STA. 13+877 A, STA. 14+299, and STA 15+189), and 1 (one) point with very high liquefaction potential (STA. 16+625). Based on the value of LRI, there are 4 (four) points of the potential risk of ground failure due to liquefaction occurring is low (STA. 09+821 A, STA. 11+448, STA. 12+269, and STA. 13+877 A), 4 (four) points of the potential risk of ground failure due to liquefaction occurring is high (STA. 07+765, STA. 08+574, STA. 10+262, and STA. 14+299), and 2 (two) points of the potential risk of ground failure due to liquefaction occurring is very high (STA. 15+189 and STA. 16+625). Figure 3 and Figure 4 in this paper illustrate that the relationship between potential liquefaction and the potential risk of ground failure due to liquefaction is linear. If the value of potential liquefaction (LPI) is high, it indicates that the potential risk of ground failure due to liquefaction (LRI) is also high. Moreover, vice versa, if the value of potential liquefaction (LPI) is low, the potential risk of ground failure due to liquefaction (LRI) is also low.

4. Conclusions

All SPT points indicated potential liquefaction with varying depth based on condition geological, geotechnical, and analysis of safety factor potential liquefaction. STA. 15+189 and STA. 16+625 is the most susceptible to potential liquefaction.

Based on the value of comparing LPI and LRI in STA. 15+189 is a high liquefaction potential, with the potential risk of ground failure due to liquefaction occurring being very high. In STA, 16+625 has a very high liquefaction potential and the potential risk of ground failure due to liquefaction occurring. The relationship between liquefaction potential and the risk of ground failure due to liquefaction is linear. Based on this study, further research to obtain liquefaction mitigation is recommended to be conducted in this study area.

5. Acknowledgment

The authors are grateful for the support of Ministry of Public Works and Housing and PT Jogjasolo Marga Makmur as operator of Solo – Yogyakarta – YIA Kulon Progo Toll Road Construction Project.

6. References

[1] BPJT 2021 Mulai Konstruksi, Tol Solo - Yogyakarta - Bandara YIA Kulon Progo Tingkatkan Konektivitas Destinasi Pariwisata Super Prioritas Borobudur

[2] PT Jogjasolo Marga Makmur, PT Adhi Karya (Persero) Tbk. and PT Perentjana Djaja 2021 *Rencana Teknik Akhir Jalan Tol Solo - Yogyakarta - NYIA Kulon Progo Seksi 1.1 (STA. 0+000 - 22+300) - Laporan Studi Geologi dan Sesar Aktif Rencana Centerline*

[3] Surono B T and Sudarno I 1992 Peta Geologi Lembar Surakarta dan Girintoro, Jawa

[4] Nakamura Y 2000 Clear identification of fundamental idea of Nakamura's technique and its applications *Proceedings of the XII World Conference Earthquake Engineering* **Paper no 2656**

[5] Supartoyo 2006 Gempabumi Yogyakarta Tanggal 26 mei 2006 *Buletin Berkala Merapi* **Volume 3, No. 2** 36–55

[6] Das B M and Rahmana G V 2011 *Principles of Soil Dynamics Second Edition* (Stamford CT: Cengage Learning)

[7] Sauri S, Rifa'i A and Hardiyatmo H C 2021 Liquefaction Vulnerability Analysis using N-SPT Value and Grain Size Analysis on Gumbasa Irrigation Canal in the Post-Disaster Petobo Area, Sulawesi *IOP Conf. Ser.: Earth Environ. Sci.* **930** 012081

[8] Fauzan, Rifa'i A and Ismanti S 2021 Study of Liquefaction Potential at Sabo dam Construction on Poi and Bangga River, Sigi Regency, Central Sulawesi *IOP Conf. Ser.: Earth Environ. Sci.* **930** 012083

[9] Rahman M, Fathani T, Rifa'i A and Hidayat M 2020 Analisis Tingkat Potensi Likuifaksi Di Kawasan Underpass Yogyakarta International Airpor *Jurnal Rekayasa Sipil (JRS-Unand)* **16** 91

[10] William M A J, Dunkerley D L, De Decker P, Kershaw A P and Stokes T J 1993 Quaternary Environments *Edward Arnold, A division of hodder & Stoughton, London New York Melbourne Auckland* 329pp

[11] Rahmahnia A 2017 *Mikrozonasi Kegempaan Berdasarkan Efek Lokal, Indeks Kerentanan Seismik, dan Percepatan Gerakan Tanah di Kabupaten Klaten, Jawa Tengah* (Surabaya: Institut Teknologi Sepuluh Nopember)

[12] PT. Jogjasolo Marga Makmur and PT. Adhi Karya (Persero) Tbk. 2021 *Laporan Pendahuluan Rencana Teknik Akhir (RTA) Jalan Tol Solo - Yogyakarta - NYIA Kulon Progo Seksi 1.1 (STA 0+000 - 22+500)* (Surakarta)

[13] Tim Pusat Studi Gempa Nasional 2017 *Peta Sumber dan Bahaya Gempa Indonesia Tahun 2017* (Bandung: Pusat Penelitian dan Pengembangan Perumahan dan Permukiman Badan Penelitian dan Pengembangan Kementerian Pekerjaan Umum dan Perumahan Rakyat)
[14] SNI 8460:2017 *Persyaratan Perancangan Geoteknik* (Jakarta: Badan Standardisasi Nasional)
[15] Bina Marga PUPR 2022 Aplikasi LINI versi Beta untuk Peta Sumber dan Bahaya Gempa Indonesia Untuk Jembatan
[16] Mase L Z 2019 Performance of NGA Models in Predicting Ground Motion Parameters of The Strong Earthquake *Journal of the Civil Engineering Forum* **5** 227–42
[17] Idriss I M and Boulanger R W 2008 *Soil Liquefaction During Earthquake* (Earthquake Engineering Research Institute)
[18] Seed H B and Idriss I M 1970 *A Simplified Procedure For Evaluating Soil Liquefaction Potential* (California: College of Engineering University of California)
[19] Iwasaki T, Tokida K and Tatsuoka F 1981 Soil Liquefaction Potential Evaluation with Use of the Simplified Procedure 209–14
[20] Lee D, Ku C and Yuan H 2003 A Study of Liquefaction Risk Potential at Yuanlin *Taiwan Eng Geol 71* pp: 97-117
[21] Mosavat N, Oh E and Chai G 2013 Liquefaction Risk Potential of Road Foundation in the Gold Coast Region, Australia **18**
[22] IRIS 2022 IRIS Earthquake Browser

Analysis of liquefaction potential in the Malalayang beach area, North Sulawesi

J Prayogo[1*,2] and F Faris[3]

[1] Master student in Natural Disaster Management Engineering, Department of Civil and Environmental Engineering Universitas Gadjah Mada, Yogyakarta, Indonesia

[2] Directorate General of Human Settlements, The Ministry of Public Works and Housing, South Jakarta, Indonesia

[3] Department of Civil and Environmental Engineering, Faculty of Engineering, Universitas Gadjah Mada, Yogyakarta, Indonesia

Corresponding author: jodieprayogo@mail.ugm.ac.id

Abstract. North Sulawesi Province was locating in a tectonically active area, affected by subduction of the Sulawesi Sea plate, Sangihe, and Halmahera. These plates are still active to provide an earthquake threat. Several large earthquakes that came from various plates around Manado City have been recorded. A large earthquake that shakes sandy soil can produce liquefaction events. The Malalayang beach area, located in the West of Manado City a part of Malalayang Beach Area Arrangement Work Package and the Bunaken Village Ecotourism Arrangement, was chosen as the research object to analyse the possibility of liquefaction. The research begin with collect boring testing data, lab test data, and earthquake data such as the moment of magnitude and Peak Ground Acceleration (PGA). This study also uses various methods or attenuation equations to determine the peak ground acceleration from the history of earthquakes obtained from multiple sources. Based on the data obtained from the field and laboratory tests, data processing is carried out to determine the liquefaction potential. From the results of the Liquefaction Potential Index (LPI) calculation, it was found that the Malalayang Beach Area has the potential for the occurrence of the liquefaction phenomenon.

1. Introduction

Liquefaction is a phenomenon that occurs in soils that have properties such as liquids. Liquefaction was discussed when an earthquake with a magnitude of 7.7 on Friday, September 28, 2018 in Palu triggered liquefaction. Liquefaction impacts the sinking of residential areas in Petobo village, Palu City.

Previously liquefaction has occurred in many similar events and various studies have been discussed in different places regarding the phenomenon. Multiple studies have confirmed that this incident needs several conditions to occur. All of these conditions must be met for an event to occur. So not all locations have the potential for liquefaction to happen and not all locations can escape the occurrence of liquefaction.

This research will focus on the location of the Malalayang Beach Area, which is part of the Malalayang Beach Area Arrangement Work and the Arrangement of Ecotourism Village Bunaken. The observations made were focused on the potential of liquefaction at the site. The analysis will be focused on the data obtained during the study.

The Malalayang Beach Area is part of the central government's strategic package that wants to elevate the Malalayang Beach Area and Bunaken Beach to become tourist attractions in North Sulawesi. The

IOP Conf. Series: Earth and Environmental Science 1244 (2023) 012030 doi:10.1088/1755-1315/1244/1/012030

internship location focuses on the area of Malalayang Beach in the west of Manado City, North Sulawesi as focus of the research location. The location of Malalayang Beach can be seen in Figure 1.

Figure 1. Research Location and Condition

2. Theoretical Studies

When the earthquake occurs, there is a reaction of loose sand due to an increase in pore water pressure. This increase in pore water pressure is caused by the presence of cyclic shear stress due to wave propagation. Water in the soil is trapped due to cyclic waves that occur very quickly. An increase in pore water pressure can result in a stream of water being pushed over the surface in the form of a torrent of mud or sand. The increase in pore water pressure that results in a flow up the water and transforms a mass of sand from a solid phase into a liquid phase is what liquefaction means [1]. Soil types with liquefaction potential are soils with grain sizes ranging from sand to silt [2]. As a result of the liquefaction event only occurs in water-saturated soils, so in general, liquefaction occurs around river areas, bays, and other water bodies [3].

The Central for Groundwater and Environmental Geology, Geological Agency, Ministry of Energy and Mineral Resources in 2019 has launched the Atlas of Indonesia's Liquefaction Vulnerability Zone. This atlas can be our initial guideline for determining which areas in Indonesia have the liquefaction potential. On the map of the vulnerability zone in North Sulawesi Province, there are several pink zones in the beach areas of North Sulawesi Province. One of them is around the coastline of Manado City, which is dominated by pink and yellow colors. Beach areas naturally have more potential for liquefaction because beach areas with sandier soil types and shallow groundwater elevations.

The most commonly used method of analysis to determine the liquefaction potential is to use a standard penetration test (SPT) as proposed [4]. The method was proposed by Seed and Idriss (1971) and is called a simplified procedure. The basis for analyzing potential liquefaction by calculating the soil factor of safety (FS) is a comparison between CRR and CSR [5]. On this basis, the Idriss-Boulanger Method (2008) [6] was born. This method bases its calculation theory on calculating PGA data (a_{max}), Earthquake Magnitude (M_w), N-SPT Value (N), σ_v (effective voltage), and the percentage of soil grain size (%). When the calculation results of an FS layer < 1, it is stated that the layer is subject to glycation but if FS > 1, it is considered that the layer does not experience liquefaction. We can also observe the liquefaction through modeling and laboratory testing in the form of shaking tables and centrifuge tests [7].

Then, a classification was carried out based on the division of the risk level of potential liquefaction to get the category of potential liquefaction [8]. This categorization has been confirmed in several other studies [9] and [10]. Then the category of potential liquefaction can be seen in Table 1.

Table 1. Categories of potential liquefaction based on LPI

LPI Value	Grade
LPI = 0	Very Low
$0 < LPI \leq 5$	Low
$5 < LPI \leq 15$	High
LPI > 15	Very High

Sulawesi island is composed of a complex tectonic order [11]. The structures identified in Sulawesi are still actively moving and often produce earthquakes. In the north of Sulawesi Island, there is north Sulawesi Subduction, also known as North Sulawesi Trench, with a geodetic shift speed of 42-50 mm / year [12]. North Sulawesi Trench, or Minahasa Trench, is the longest and largest thrust in Sulawesi. North Sulawesi Trench is a subduction zone between the Philippine Plate and the North Sulawesi Micro block. In addition, the position of North Sulawesi, which is adjacent to the Maluku Sea, which is the confluence of the world's three main plates, namely the Eurasian Plate, the Philippine Sea Plate, and the Australian Plate [13], caused this region also to get the influence of earthquakes from the three plates. Around this location, there are Halmahera Thrust and Sangihe Thrust.

3. Method

The research flow is a systematic stage that begins with collecting data to categorize potential liquefaction at the research site. The steps of the research activity can be seen in Figure 2.

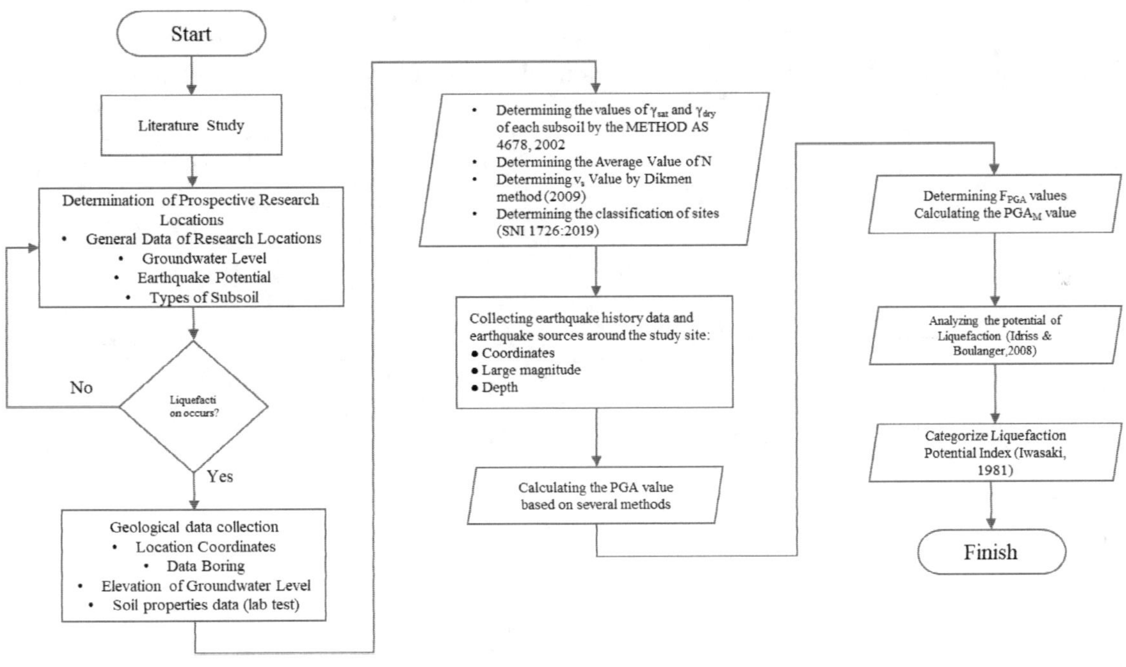

Figure 2. Research Flowchart

The research began with collecting cone penetration and boring testing data, lab testing data to obtain soil properties data and groundwater levels. The results of recording SPT test measurements are usually displayed as a boring log table. An example of data results of the MLY-BH-05 borehole test can be seen in Table 2.

Table 2. Boring Test Data MLY-BH-05 [14]

USCS	SPT				
	Depth (m)	*Blows*	*Penetration*	N_{60}	$(N_1)_{60}$
SP	1,15	8	30	7	11
SM	3,15	13	30	11	19
SM	5,15	18	30	17	28
SM	7,15	18	30	19	31
SM	9,15	27	30	28	38
SM	11,15	32	30	35	41
ANDESITE	13	50	4	>50	
ANDESITE	15	50	5	>50	
SM	17,15	37	30	40	34
SM	19,15	50	12	>50	>50
SM	21,15	50	13	>50	>50
SM	23,15	50	11	>50	>50
SM	25,15	50	9	>50	>50
SM	27,15	50	11	>50	>50
SM	29,15	50	9	>50	>50
SM	30,15	60	11	>50	>50

From the data results of each boring test point, an overview of the condition of subsurface soil/rock coating in the study area can be produced by plotting the type of layer of each boring test point from MLY-BH-01 to MLY-BH-05. An overview of the conditions of the plot of the five borehole points can be seen in Figure 3.

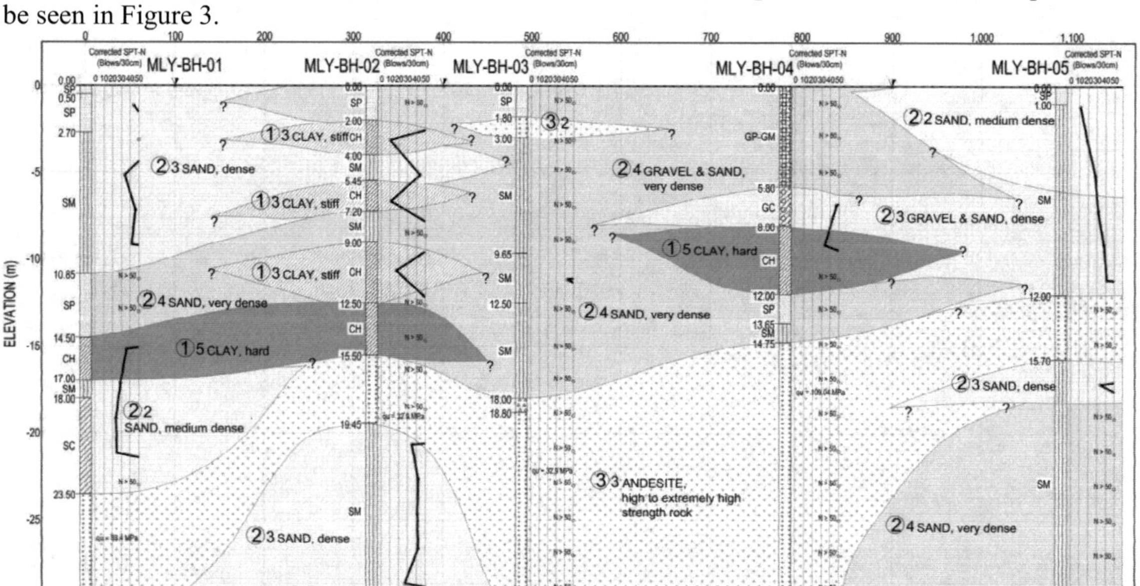

Figure 3. Condition of Soil/Subsurface Rock Layers in Areas Near Malalayang Beach based on Boring Hole MLY-BH-01 to MLY-BH-05 [14]

The boring test data takes some correlation of the N-SPT value to determine the shear wave velocity through the soil layer (v_s). It can be seen in the study by Ameratunga et al., 2016 [15], Dry Unit Weight (γ_{dry}), and the Weight of Saturated Unit Weight (γ_{sat}) in AS 4678-2002 [16].

IOP Conf. Series: Earth and Environmental Science 1244 (2023) 012030 doi:10.1088/1755-1315/1244/1/012030

The procedure for classifying a site provides seismic design criteria in the form of amplification factors in buildings. In formulating the seismic design criteria of a building at ground level or determining the amplification of the magnitude of the peak earthquake acceleration from bedrock to the ground surface for a site, the site must be classified first. The soil profile is classified on a site by looking at the soil profile in the top 30 m of the layer.

Table 3. Classification of sites based on SNI 1726:2019 [17]

Site Class	V_s (m/s)	N or N_{ch}	S_u (kPa)
SA (Heavy Rock)	>1500	N/A	N/A
SB (Rock)	750 - 1500	N/A	N/A
SC (Heavy Soil, very dense and soft rock	350-750	>50	≥100
SD (Medium Soil)	175-350	15-50	50-100
SE (Soft Soil)	<175	<15	<50
	Or any soil profile containing more than 3 m of soil with the following characteristics: 1. Plasticity index, PI > 20, 2. Water content, $w \geq 40\%$, 3. Zero shear strength $\bar{S}_u < 25$ kPA		
SF (special soil, which requires a specific geotechnical investigation and site-specific response analysis that follows)	Any subsoil profile that has one or more of the following characteristics: - Susceptibility and potentially fail or collapse due to earthquake loads such as light liquefaction, susceptible clay, weakly cemented soil - Highly organic clay and/or turf (thickness H > 3) - Very high plasticity clay (thickness H > 7.5 m with Plasticity Index PI > 75) Soft/semi-firm clay layer with thickness H > 35 m with $\bar{S}_u < 50$ kPA		

Notes: N/A = cannot be used

In addition, the collection of earthquake data from various other sources was also carried out, such as from the IRIS Earthquake Browser Website with a maximum filter of 2000 earthquakes, from January 1, 1970 to March 28, 2022, with earthquake magnitudes ranging from 6 to 10, and earthquake depths below 900 km (Figure 4). Coupled with some data from the Damaging Earthquake Catalog – BMKG ISSN 2477-0582. [18].

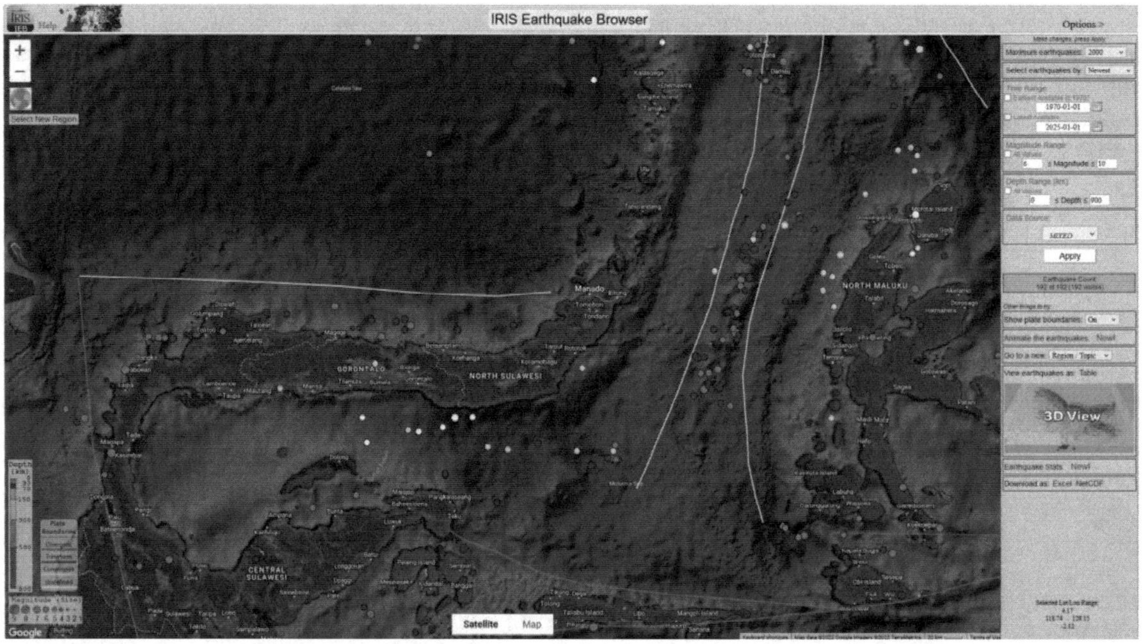

Figure 4. Collection of Earthquake Sources Around the Research Site from IRIS Earthquake Browser

Ground Motion Prediction Equation (GMPE) or soil acceleration attenuation equation is an empirical formulation used to estimate PGA values based on earthquake source parameters. GMPE is usually created empirically through regression from various soil acceleration measurement data [19].
The determination of the use of attenuation equations is based on several things such as the selection of the equations of Youngs (1997), Garcia (2005), Kanno (2006), Zhao (2006) and Hong (2009) based on the similarity of the geological and tectonic conditions of the regions where the equations are made. In addition, the selection of this equation is based on comparing the results of data processing by using the existing GMPE equation with the record data of the soil movement/accelerograph contained in Indonesia (Map of Sources and Dangers of earthquakes in Indonesia). The methods of Donovan (1973), Guirre (1978), Youngs (1988), and Lin & Lee (2008) are based on the large number of other studies that use this method for use in comparing PGA results from several attenuation methods. The effect of PGA on the incidence of liquefaction can also be seen in Mase, 2020 [20].

4. Results

4.1. Site Classification

Based on the calculation results of the values of $\bar{N}$ and $V_{s,30}$, it was found that each boring point was classified as belonging to the elementary school site class. Details of the results of the classification of such sites can be seen in Table 4.

Table 4. Classification of Malalayang Beach Sites

Boring Point	$\bar{N}$	$V_{s,30}$	Site Class
MLY-BH-01	34,98346	238,04	SD
MLY-BH-02	30,57778	283,13	SD
MLY-BH-03	47,99287	216,87	SD
MLY-BH-04	47,30598	198,46	SD
MLY-BH-05	28,76449	290,36	SD

4.2. Earthquake Records

From several sources such as Plate Plots from the 2017 Indonesian Earthquake Source and Hazard Map, the Catalogue of Damaging Earthquakes – BMKG ISSN 2477-0582 [15], the IRIS Earthquake Browser

Website (ds.iris.edu), and several earthquake sources from the website www.bmkg.go.id known some earthquake data that was successfully filtered, namely earthquakes with a large enough magnitude and with a distance not so far from the research location as follows (Table 5):

Table 5. List of Earthquake Events Around Research Location

No.	Date	Region	Coordinate		Depth	Magnitude	Epicentre	Hypocentre
			Lat.	*Long.*	*(km)*		*(km)*	*(km)*
History of Earthquakes Around Manado According to the Catalogue of Destructive Earthquakes - BMKG ISSN 2477-0582								
1	15-Nov-14	Ternate	1,95	126,49	10	7,3	197,2964	197,549634
2	21-Jan-07	Sulawesi Utara	1,18	126,42	63	6,7	184,5783	195,03374
Earthquake History (BMKG)								
1	14-Nov-19	Barat Laut Jailolo-Malut	1,67	126,39	73	7,1	180,0622	194,297215
2	11-Jun-21	Bolaanguki-Bolsel	-0,06	124,4	14	5,7	174,4351	174,996041
M_w Maximum Source and Hazard Map of the 2017 Indonesian Earthquake								
1		Northern Sulawesi Thrust 1	1,71372	123,9153	1	8,5	101,5907	101,595604
2		Northern Sulawesi Thrust 2	1,8494	122,9272	1	8,5	211,886	211,888382
3		Northern Sulawesi Thrust 3	2,006912	122,0153	1	8,5	314,687	314,688585
4		Northern Sulawesi Thrust 4	1,945827	120,8518	1	8,5	441,3948	441,395916
5		Northern Sulawesi Thrust 5	1,296776	119,8822	1	8,5	546,2358	546,236675
IRIS Earthquake Browser								
1	30/6/2004	Minahassa Peninsula; Sulawesi	0,7093	124,7697	96,7	6,2	83,61308	127,835975
2	13/12/1990	Minahassa Peninsula; Sulawesi	1,1511	124,0608	63,1	6,2	87,43237	107,824066
3	13/10/2001	Northern Molucca Sea	0,84	125,936	44,1	6	145,5463	152,080676
4	28/10/1998	Northern Molucca Sea	0,8201	125,9567	46	6,6	148,6248	155,580625
5	19/4/1990	Minahassa Peninsula; Sulawesi	1,1016	123,4704	45,3	6,5	151,358	157,991598
6	24/2/2001	Northern Molucca Sea	1,334	126,335	55,6	7,1	173,036	181,74934
7	21/1/2007	Northern Molucca Sea	1,0816	126,3618	23,5	7,5	180,4496	181,973373
8	14/11/2019	Northern Molucca Sea	1,6294	126,4144	33	7,1	182,2351	185,198891

4.3. Calculation of PGA Values with Empirical Formulas (GMPE)

After plotting each earthquake source consisting of the coordinate data of the earthquake location, its magnitude, and the earthquake's characteristics, the PGA value calculation was carried out based on each PGA value attenuation method. The calculation result can be seen as a graph in Figure 5 below.

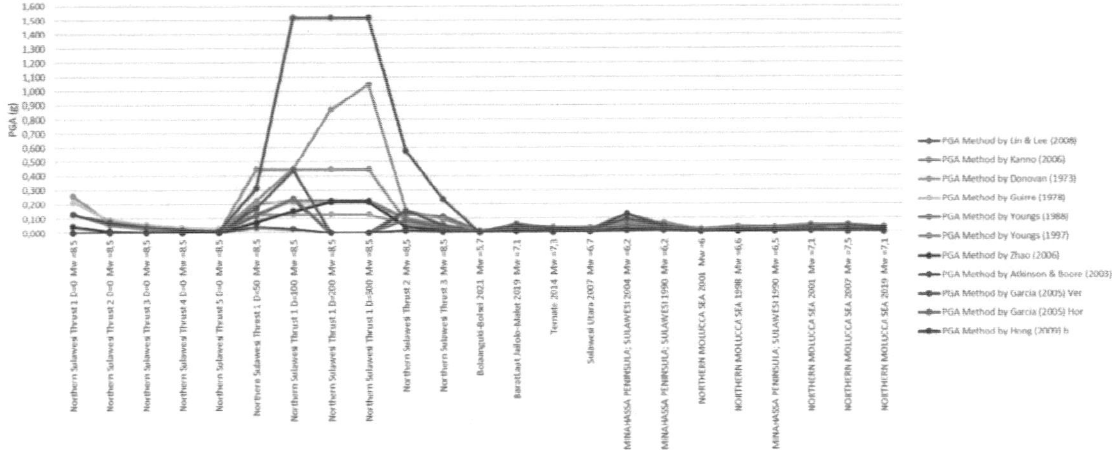

Figure 5: Graph of Comparison of PGA Values from Several GMPE Methods and Earthquake Sources

From Figure 5, Seen the largest PGA values in a row are 0.447 g (Kanno 2006), 0.440 g (Garcia 2005) and 0.398 g (Youngs 1997). In general, all GMPE methods will produce the same graph, and the difference is only noticeable if the graph crosses the depth limit for each method. The form of the graph produced by the Kanno, Garcia, and Youngs method has similarities. It happens even though these three methods take different data sources because they allow the use of a large enough earthquake source depth to produce a PGA value that tends to be large in line with the increasing depth of the earthquake source. Some of the PGA values taken can be seen in Table 6.

Table 6: Recap of Selected PGA Values for Liquefaction Potential Analysis

PGA	Earthquake Source	Method	Description
0,447	Northern Sulawesi Thrust 1 D=50	Kanno (2006)	Maximum PGA value from all earthquake sources and methods
0,259	Northern Sulawesi Thrust 1 D=10	Kanno (2006)	The maximum PGA value from the Northern Sulawesi Thrust earthquake source with a depth considered to be 10 km
0,097	MINAHASSA PENINSULA; SULAWESI	Garcia (2005) Ver	The maximum PGA value of all earthquakes that have ever happened
0,600	NORTHERN MOLUCCA SEA	RSA Cipta Karya	PGA Value from the Indonesian Spectra Design website

4.4. Liquefaction Analysis

The analysis of liquefaction uses the table attached in the appendix of the book Soil Liquefaction During Earthquakes [6]. It is an example of the results of the analysis of liquefaction using the Idriss & Boulanger (2008) method PGA 0.259 from Kanno (2006) attenuation method can be seen in the form of a graph in Figure 6. This graph shows the calculation results of the potential liquefaction at 5 borehole points. The results show only the MLY-BH-05 point indicating the potential for liquefaction because it has a Safety Factor < 1.

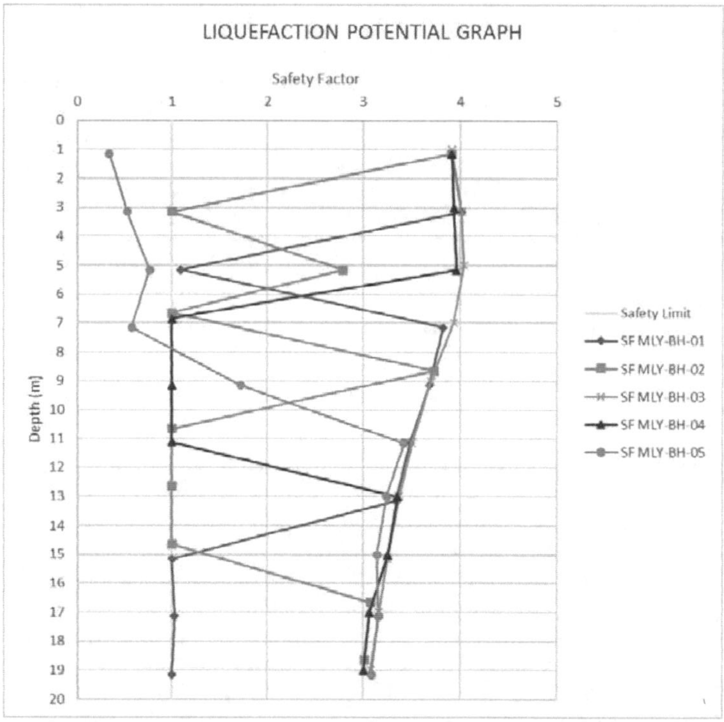

Figure 6. Graph of Relationship between Depth and Safety Factors from PGA Liquefaction Analysis Results in 0.259 g from Earthquake Source Northern Sulawesi Thrust 1 D=10

4.5. Potential Categories

Based on data processing, a recap of the liquefaction potential index (LPI) category using the Iwasaki Method, 1981 [7] for each PGA value can be seen in Table 9 as follows:

Table 7. Liquefaction Potential Index (LPI) Recap

Earthquake Source	Northern Sulawesi Thrust 1 D=50
PGA Method	Kanno (2006)
Peak Ground Acceleration (g)	0,447
Magnitude	8,5
Epicenter (km)	101,59

BH	Site Class	F_{PGA}	PGA_M	LPI	Description
01	SD	1,147	0,513	18,899	Very High
02	SD	1,147	0,513	0,000	Very Low
03	SD	1,147	0,513	0,000	Very Low
04	SD	1,147	0,513	0,000	Very Low
05	SD	1,147	0,513	31,268	Very High

Earthquake Source	Minahassa Peninsula; Sulawesi
PGA Method	Garcia (2005) Ver
Peak Ground Acceleration (g)	0,097
Magnitude	6,2
Epicenter (km)	83,61

BH	Site Class	F_{PGA}	PGA_M	LPI	Description
01	SD	1,600	0,155	0	Very Low
02	SD	1,600	0,155	0	Very Low
03	SD	1,600	0,155	0	Very Low
04	SD	1,600	0,155	0	Very Low
05	SD	1,600	0,155	0	Very Low

Earthquake Source	Northern Sulawesi Thrust 1 D=10
PGA Method	Kanno (2006)
Peak Ground Acceleration (g)	0,259
Magnitude	8,5
Epicenter (km)	101,59

BH	Site Class	F_{PGA}	PGA_M	LPI	Description
01	SD	1,359	0,352	0	Very Low
02	SD	1,359	0,352	0	Very Low
03	SD	1,359	0,352	0	Very Low
04	SD	1,359	0,352	0	Very Low
05	SD	1,359	0,352	29,39	Very High

Earthquake Source				Northern Molucca Sea	
PGA Method				RSA	
Peak Ground Acceleration (g)				0,600	
Magnitude				7,5	
Epicenter (km)				180,45	
BH	Site Class	F_{PGA}	PGA_M	LPI	Description
01	SD	1,100	0,660	14,05	High
02	SD	1,100	0,660	0	Very Low
03	SD	1,100	0,660	0	Very Low
04	SD	1,100	0,660	0	Very Low
05	SD	1,100	0,660	30,88	Very High

The LPI recap results show that the high PGA and Magnitude values will result in high potential for liquefaction at the location. In this case, the high PGA value used to analyze LPI values results in high potential for liquefaction at MLY-BH-01 and MLY-BH-05 points. Different things can be seen with earthquakes that have already happened, only producing the high PGA value of 0.097 g of all methods. This PGA value cannot create liquefaction potential at that location. The Kanno (2006) method with a PGA 0.259 g showed that the liquefaction event only occurred at the MLY-BH-05 point but in a very high category. Meanwhile, the PGA value from the spectra design application of Indonesia has made high category liquefaction potential at MLY-BH-01 and very high category at MLY-BH-05 point. From the different results of these calculations, MLY-BH-01 & MLY-BH-05 points need to be reviewed to ensure the liquefaction potential that can occur at these points and prepare for mitigation.

5. Conclusion

High PGA values increase the liquefaction potential at the study site. GMPE attenuation calculation methods resulted in a larger PGA value according to the increase in the depth of the earthquake source. Some methods have limited the depth of earthquakes that can be used in calculations, but some methods also do not provide clear limits regarding the maximum depth. The largest PGA value of all attenuation calculations (GMPE) and earthquake sources used in this study was 0,447 g (Kanno), 0,259 g (Kanno), 0,097 g (Garcia), and 0,600 g (RSA CK). The PGA value indicates that the MLY-BH-05 point has a considerable liquefaction potential, so efforts are needed to mitigate the potential liquefaction at a particular point.

6. Acknowledgments

The authors would like to express their gratitude for the support given by The Indonesian Ministry of Public Works and Housing.

7. Reference

[1] Day R W 2002 Geotechnical Earthquake Engineering Handbook, *Mc Graw-Hill, New York: USA. Foundations Division, ASCE*, Vol. **95**, No. SM5, pp. 1199-1218

[2] Aydan O 2008 Seismic and Tsunami Hazard Potentials in Indonesia with a special emphasis on Sumatra Island, *Journal of the School of Marine Science and Technology* – Tokai University (Japan), Vol. **6**, No. 3, pp. 19 – 38

[3] Kramer S L 1996 *Geotechnical Earthquake Engineering*, Prentice Hall, Englewood Cliffs, New Jersey, 653

[4] Seed H B, Tokimatsu K, Harder L F, and Chung R M 1985 Influence of SPT Procedures in Soil Liquefaction Resistance Evaluations *Journal of Geotechnical Engineering* 111(12)

[5] Seed H B and Idriss I M 1971 Simplified Procedure for Evaluation Soil Liquefaction Potential *Journal of Soil Mechanics and Foundation*, Division, ASCE, vol. **97**, No.9, pp 1249-1273

[6] Idriss I M and Boulanger R W 2008 Soil Liquefaction During Earthquakes *Earthquake Engineering Research Institute*, 136(6), 755

[7] Pramaditya A and Fathani T F 2020 Physical Modelling of Earthquake-induced Liquefaction on Uniform Soil Deposit and Settlement of Earth Structures *Journal of the Civil Engineering Forum*, 1000(1000)
[8] Iwasaki T, Tokida K, and Tatsuoka F 1981 Soil Liquefaction Potential Evaluation with Use of the Simplified Procedure *International Conferences on Recent Advances in Geotechnical Earthquake Engineering and Soil Dynamics.* 12
[9] Muhanifah H, Adi A D, and Faris F 2021 Liquefaction Investigation of Balaroa, Central Sulawesi on Liquefied and Non-Liquefied Areas *IOP Conference Series: Earth and Environmental Science*, 861(5)
[10] Andiny A N, Faris F, and Adi A D 2021 Re-liquefaction Hazard Evaluation in Flow-Slide Affected Area of Jono Oge, Central Sulawesi, Indonesia *IOP Conference Series: Earth and Environmental Science*, 861(5)
[11] Hall, R 2011 Australia–SE Asia Collision: Plate Tectonics and Crustal Flow *Geological Society*, London, Special Publications **355**, 75-109.
[12] Socquet A, Simons W, Vigny C, and Mccaffrey R 2006 Microblock Rotations and Fault Coupling in SE Asia Triple Junction (Sulawesi, Indonesia) From GPS and Earthquake Slip Vector Data *Journal of Geophysical Research Atmospheres* 111(B8)
[13] Widiwijayanti C, Tiberi C, Deplus C, Diament M, Mikhailov V, and Louat R 2004 Geodynamic Evolution of the Northern Molucca Sea Area (Eastern Indonesia) Constrained by 3-D Gravity Field Inversion *Tectonophysics* **386**, 203–222
[14] PT. SOILENS 2021 *Laporan Penyelidikan Tanah untuk Proyek Kawasan Pantai Malalayang dan Penataan Ecotourism Village Bunaken Sulawesi Utara Indonesia* PT. SOILENS
[15] Ameratunga J, Sivakungan N, and Das B M 2016 Correlation of Soil and Rock Properties in Geotechnical Engineering *Springer India* New Delhi
[16] AS 4678-2002 2002 *Earth Retaining Structures* Austalian Standard
[17] SNI 1726:2019 2019 *Tata Cara Perencanaan Ketahanan Gempa untuk Struktur Bangunan Gedung dan Nongedung* Jakarta: Badan Standarisasi Indonesia.
[18] Pusat Gempabumi dan Tsunami 2019 *Katalog Gempabumi Signifikan dan Merusak Tahun 1821 - 2018.* Cetakan Pertama. Pusat Gempabumi dan Tsunami Kedeputian Bidang Geofisika Badan Meteorologi Klimatologi dan Geofisika
[19] Douglas J 2011 Ground-Motion Prediction Equations 1964-2010 *PEER Report 2-11/102 Pacific Earthquake Engineering Research Centre College of Engineering University of California* Berkeley April 2011
[20] Mase L Z, Farid M, Sugianto N, and Agustina S 2020 The Implementation of Ground Response Analysis to Quantify Liquefaction Potential Index (LPI) in Bengkulu City, Indonesia *Journal of the Civil Engineering Forum* 6(3), 319

Liquefaction Potential Index Analysis in Solo – Yogyakarta - NYIA Kulon Progo Toll Road (Section Karanganyar to Klaten Regency) at Central Java

Askaviolita[1,2], **Sito Ismanti**[1,*] **and Teuku Faisal Fathani**[1]
[1] Department of Civil and Environmental Engineering, Faculty of Engineering, Universitas Gadjah Mada, Yogyakarta 55281, Indonesia
[2] Directorate of Freeway, Directorate General of Highway, Ministry of Public Works and Housing, Republic of Indonesia

*Corresponding author: sito.ismanti@ugm.ac.id

Abstract. Liquefaction had occurred in Yogyakarta due to an earthquake of moment magnitude, Mw 6.3 on May 27, 2006. This earthquake caused severe damage in several areas in Central Java. The study area in this construction was located mostly in some areas of Central Java Province that had a shallow groundwater table and often occurred earthquakes that might trigger liquefaction. The objective of this study was to analyze liquefaction potential in the construction of Solo – Yogyakarta – NYIA Kulon Progo toll road especially sta. 00+000-07+500 From Colomadu, Karanganyar Regency to Delanggu, Klaten Regency using two scenarios based on the existing earthquake history according to USGS data. That were earthquakes with Mw 6.3 and Mw 5.2 on a total of 47 boreholes. This study used data from Standard Penetration Tests (SPT) and laboratory tests. The liquefaction potential analysis used the simplified procedure and Liquefaction Potential Index method. Based on liquefaction potential analysis, an earthquake of Mw 5.2 has a very low liquefaction potential compared to an earthquake of Mw 6.3. According to analysis results and hazards caused, an earthquake with Mw 6.3 is recommended to be used as a basis for mitigation against liquefaction planning.

KEYWORDS: *earthquake, liquefaction potential index, safety factor, simplified procedure, toll road*

1. Introduction

Increasing population and economic needs, especially in big cities will increase the movement of people and goods from one place to another. In Indonesia, The construction of toll roads is very necessary for transportation because facilitates the smooth movement of people and goods [1]. The Solo – Yogyakarta – NYIA Kulon Progo toll road is a toll road that connects two big cities, namely Solo City in Central Java Province and Yogyakarta City in D.I Yogyakarta Province. In planning the infrastructure construction for public facilities, it is required to consider preventing damage to infrastructure due to natural disasters. Disasters that often occur in the construction of this toll road are earthquakes. Earthquakes are one the catastrophic phenomena that cause destruction and damage to the surrounding area and can trigger liquefaction [2].

Based on the history of earthquakes this construction of toll roads, the strong earthquake that ever occurred was on May 27th, 2006 at 05.54 WIB in Java Indonesia with a moment magnitude, *Mw* of 6.3.

This earthquake caused severe damage in several areas in Central Java, such as Klaten, Boyolali, Surakarta, and Salatiga. This earthquake was caused by the existence of the Opak Fault [3]. The Solo – Yogyakarta – NYIA Kulon Progo toll road will be built near the Opak Fault area which will affect the seismic conditions. Figure 1 (a) shows a planned route of that toll road located near the Opak Fault area.

According to USGS (United States Geological Survey), this strong earthquake was located at 7.955°S, 110.430°E about 25 km south-southwest (SSW) of Yogyakarta or about 440 km east-southeast (ESE) of Jakarta [4]. The earthquake also caused liquefaction that damaged buildings and infrastructures. The liquefaction phenomenon occurred near the Opak Fault area [5]. Liquefaction is a phenomenon when a mass of soil loses most of its shear resistance when subjected to a monotonous, cyclic, or shocking load so that the soil resembles a liquid [6].

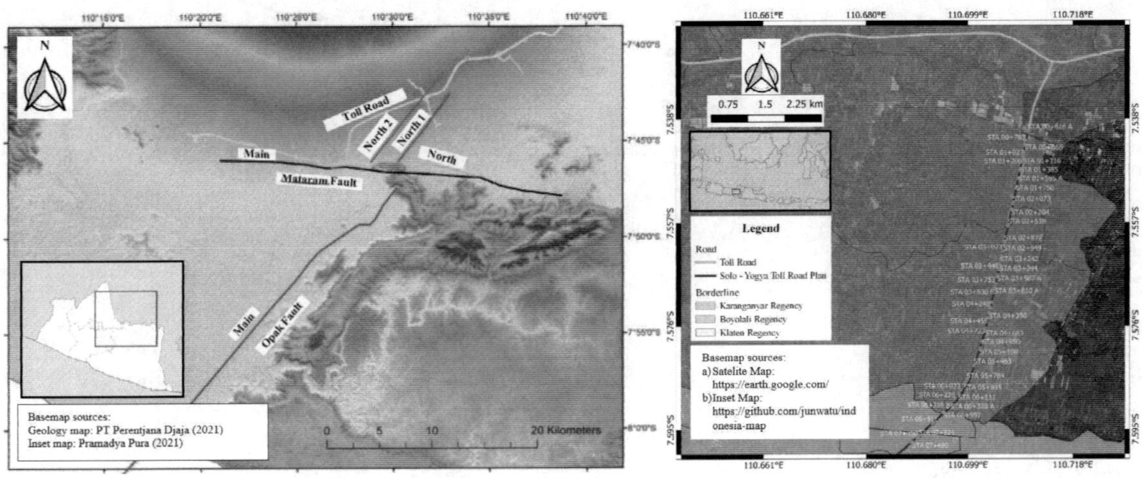

(a) (b)

Figure 1. (a) Location of Opak Fault and Solo – Yogyakarta – NYIA Kulon Progo toll road and (b) Location of Research Study

This research aims to determine the *liquefaction potential index* of Solo – Yogyakarta – NYIA Kulon Progo toll road especially stationing 00+000 - 07+500 in Central Java, shown in **Figure 1 (b)** based on the history of earthquakes that have occurred. Stationing 00+000–07+500 located in Colomadu, Karanganyar Regency until Delanggu, Klaten Regency. Liquefaction potential was analyzed with a simplified procedure by Idriss & Boulanger (2008) to calculate *the factor of safety* against liquefaction and the *liquefaction potential index* method by Iwasaki et al. (1984) to determine the classification of liquefaction potential [7],[8].

2. Geological Condition of Central Java Province

Structural control on Java Island is strongly influenced by tectonic activity. There are Eurasian Plate and the Indo-Australian Plate [9]. The Earthquake that happened in Yogyakarta Province and Central Java Province was due to tectonic activity and active faults. The study areas are in 3 regencies in Central Java Province, namely Karanganyar, Boyolali, and Klaten Regency. The research area had dynamic development of the geological condition. It means that tectonic and volcanic activity has happened in the research area multiple times [10]. The sediments that are most prone to liquefaction are newly deposited sediments, which become more resistant with age [7]. The geological condition of the research area was derived from the Merapi volcanic rocks that consist of volcanic breccia, lava, and tuff of the Holocene age, presented in **Figure 2** Merapi volcanic deposits are very vulnerable to liquefaction potential because the dominant consists of sand deposits that triggered liquefaction [11].

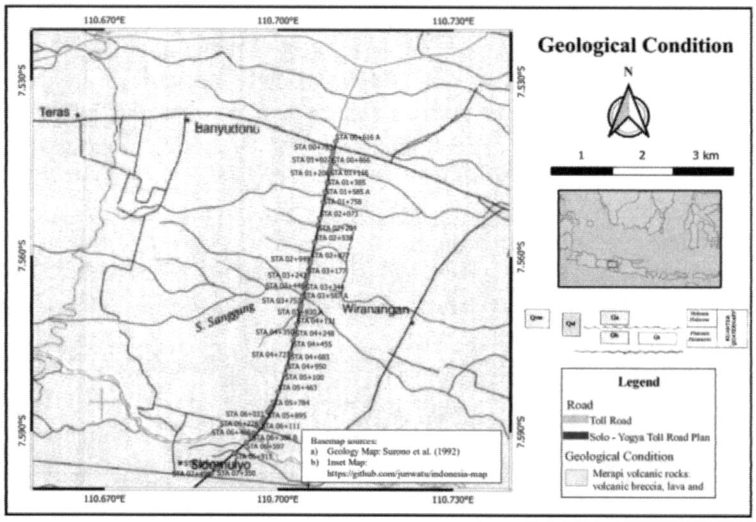

Figure 2. Geological Condition in Research Area

3. Research Methods

3.1 Research Data

In this research, the data used from soil investigation data (boring, SPT, and laboratory data) from the boring surveys. The borehole data was obtained from a boring survey in 2021 by PT Jogjasolo Marga Makmur. The analysis was carried out in 3 regencies in Central Java with a total of 47 boreholes data. The regencies were Karanganyar, Boyolali, and Klaten Regency. There are 1 borehole data in Karanganyar regency, 41 boreholes data in Boyolali Regency, and 5 boreholes data in Klaten Regency. **Figure 3** shows the location of the survey soil investigation.

(a) (b)

Figure 3. (a) The location of soil investigation survey at STA 02+204 at Boyolali Regency and (b) STA 07+499 at Klaten Regency

In the soil investigation, the SPT data is conducted at a depth of 0 to 30 meters from the ground surface with a depth interval of every 2 meters. However, liquefaction potential is analyzed up to a depth of 20 meters because at a depth of more than 20 m liquefaction is rarely reported [12]. The *N-SPT* values in the field that are more than 60 are assumed to have a maximum value of 60. Groundwater table values were obtained during a boring survey, with the values mostly shallow.

3.2 Seismic Analysis

Earthquakes that occurred in the study area are very common over time. In the event of an earthquake, design seismic provides a dynamic response to the structure, which causes large losses, so seismic analysis needs to be carried out in infrastructure construction planning. The soil conditions affect the determination of seismic analysis [13]. Seismic analysis was carried out to determine the value of maximum Peak Ground Acceleration. Firstly, the determination of moment magnitude for analysis is obtained from earthquake history around the study area that had a moment magnitude above *Mw* 5. An earthquake that had a moment magnitude below *Mw* 5 did not have the liquefaction potential [14]. According to USGS, there are 2 earthquakes that have moment magnitude above *Mw* 5, shown in **Table 1** [15].

Table 1. History of an earthquake in the study area [15]

Year	Month	Day	Lat	Lon	Depth (km)	*Mw*	Location
2006	5	26	7.961°S	110.446°E	12.5	6.3	Pundong, Yogyakarta
1979	10	7	7.672°S	110.755°E	180	5.2	Baki, Central Java

Of those earthquakes, an earthquake with *Mw* 6.3 is an earthquake with a source originating from an Opak fault [3], while an earthquake with *Mw* 5.2 is an earthquake with a source originating from the Benioff subduction. Because the earthquake with a source depth of more than 50 km is included in the Benioff subduction (intraslab) [16].

Based on the earthquake history, then an analysis is carried out to determine the peak ground acceleration (*PGA*). *PGA* is the acceleration of bedrock caused by an earthquake that occurs within a certain time. To calculate the *PGA* value using the attenuation method by Kanno et. al (2006) with earthquakes based on the earthquake history in **Table 1**, with the equation below [17].

$$\log pre = a_1 M_w + b_1 X - \log(X + d_1 \times 10^{e_1 M_w}) + c_1 + \varepsilon_1 \; (\mathrm{D} \leq 30 \text{ km}) \tag{1}$$

$$\log pre = a_2 M_w + b_2 X - \log(X) + c_2 + \varepsilon_2 \; (\mathrm{D} \geq 30 \text{ km}) \tag{2}$$

Which, pre is *PGA* (cm/sec^2), D is earthquake depth (km), $a_1, b_1, d_1, c_1, \varepsilon_1$ is regression coefficients for $\mathrm{D} \leq 30$ km and $a_2, b_2, c_2, \varepsilon_2$ for $\mathrm{D} \geq 30$ km. The value is shown in **Table 2.**

Table 2. The Regression Coefficient for the *PGA* period [17]

Event Model	*a*	*b*	*c*	*d*	*ε*
Shallow	0.56	-0.0031	0.26	0.0055	0.37
Deep	0.41	-0.0039	1.56	-	0.40

The *PGA* value is to determine the value of maximum *PGA* in the ground surface (A_S). Based on Indonesia National Standard (SNI) 2833:2016 [18], the value of A_S for a probability of 7% in 75 years to highway and bridge design was determined by multiplying the value of *PGA* and the site coefficient.

$$A_S = F_{PGA} \times PGA \tag{3}$$

Which, A_S is maximum *PGA* at the ground surface adapted from site coefficient, F_{PGA} is site coefficient for a probability of 7% in 75 years and *PGA* is the acceleration of bedrock. The soil site coefficient was classified based on Indonesia National Standard SNI 2833:2016 by the SPT value of each soil layer.

3.3 Liquefaction Potential Analysis

Liquefaction potential was analyzed by the simplified procedure method by Idriss and Boulanger (2008) [7]. The calculation used to analyze the liquefaction potential in each soil layer is to determine the value of the Factor of Safety. The value of *FS* is calculated as given in the equation below.

$$FS = \frac{CRR}{CSR} \quad (4)$$

Where, if the *FS* value < 1 then the soil layer has the potential to liquefaction while if $FS \geq 1$ then the soil layer hasn't the potential to liquefaction.

Cyclic Resistance Ratio (CRR) is the soil ratio value to withstand cyclic shear stresses during an earthquake. Idriss & Boulanger, 2008 determined the correlation between the *CRR* value and the SPT value with the following equation [7].

$$CRR_{M=7.5;\sigma'vc=1} = \exp\left(\frac{(N_1)_{60cs}}{14.1} + \left(\frac{(N_1)_{60cs}}{126}\right)^2 - \left(\frac{(N_1)_{60cs}}{23.6}\right)^3 + \left(\frac{(N_1)_{60cs}}{25.4}\right)^4 - 2.8\right) \quad (5)$$

$$CRR_{M;\sigma'vc} = CRR_{M=7.5;\sigma'vc=1} \times MSF \times K_\sigma \quad (6)$$

Where $(N_1)_{60cs}$ is correction factor of fines content, $CRR_{M;\sigma'_{vc}}$ is the *CRR* value for the other moment magnitude earthquake, *MSF* is Factor Scaling of earthquake magnitude, and K_σ is a factor correction of overburden.

Cyclic Stress Ratio (CSR) is the shear stress due to the earthquake. *CSR* occurs when the ground receives earthquake loads. The most important factor in determining *CSR* is the determination of the peak ground acceleration value. The calculation of *CSR* is analyzed using the Idriss & Boulanger, 2008 method based on the SPT values [7].

$$CSR = 0.65 \frac{\sigma_{vc}}{\sigma'_{vc}} \frac{a_{max}}{a} r_d \quad (7)$$

Where σ_{vc} is the total vertical stress, σ'_{vc} is the effective vertical stress r_d is the stress reduction coefficient, and a_{max} is maximum peak ground acceleration.

3.4 Liquefaction Potential Index

Liquefaction Potential Index (*LPI*) is a method that was first introduced by Iwasaki et. al which aims to analyze the liquefaction potential based on the thickness of the soil layer and the distance of the liquefaction potential soil layer to the surface. Analysis of liquefaction potential using the *LPI* method is calculated to a depth of 20 meters from the ground surface. The *LPI* value is calculated by the thickness, depth, and *safety factor* of the liquefied soil layer. For the calculation as in the following equation [8].

$$LPI = \int_0^{20} F.w(z)dz \quad (8)$$

With:

F=0 for FS>1.0 (9)

F=1 – FS for FS<1.0 (10)

w(z)=0 for z 20 m (11)

w(z)=10 – 0.5z for z < 20 m (z in meters) (12)

Where *F* is the damage level of a layer in liquefaction analysis and *w(z)* is depth weight factor, where *z* is the depth under consideration, maximum up to 20 meters. The value of *LPI* by Iwasaki was classified into several categories as shown in **Table 3.**

Table 3. *Liquefaction Potential Index* Classification [8]

LPI Value	Classification
$LPI = 0$	Very Low
$0 < LPI \leq 5$	Low
$5 < LPI \leq 15$	High
$LPI > 15$	Very High

4. Results and Discussion

The Solo – Yogyakarta – NYIA Kulon Progo Sta. 0+000 – 07+500 Toll Road (Colomadu, Karanganyar Regency until Delanggu, Klaten Regency) has soil condition dominated by silty sand soil and shallow ground water table with a dominated value of less than 2 m based on soil investigation survey, shown in **Figure 4.** The site classification according to SNI 2833:2016 for toll road planning in the research area was obtained from the average SPT (*N-SPT*) value with the result ranged between medium until soft soil. The result classified as medium soil (SD) if the average *N-SPT* valued between 15 to 50 and soft soil (SE) for the value of average *N-SPT* <15.

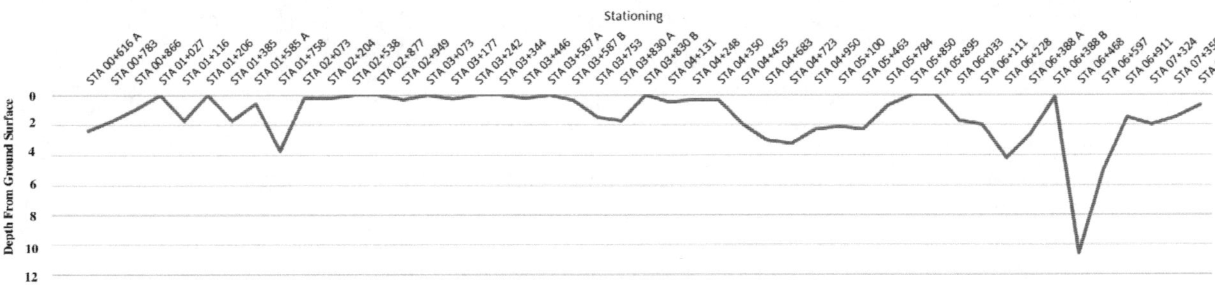

Figure 4. Ground Water Table in Research Area

The liquefaction potential was analyzed using two scenarios based on the existing earthquake history according to USGS, that is earthquakes with *Mw* 6.3 and *Mw* 5.2 on a total of 47 boreholes. Of a total of 47 boreholes, there is 1 borehole in Karanganyar Regency at STA. 00+616 A, 41 boreholes in Boyolali Regency at STA. 00+783 – STA. 06+468 and 5 boreholes in Klaten Regency at STA 06+597 – STA 07+499. The calculation result of peak ground acceleration and liquefaction potential for the earthquake *Mw* 6.3 and *Mw* 5.2 is presented in **Table 4.**

Table 4. The value of peak ground acceleration and liquefaction potential

No	Borehole	Sub-Disrict	An earthquake with *Mw* 6.3				An earthquake with *Mw* 5.2			
			PGA	*As*	*LPI*	Classification	*PGA*	*As*	*LPI*	Classification
1	STA 00+616 A	Colomadu, Karanganyar	0.15	0.23	0	Very Low	0.01	0.02	0	Very Low
2	STA 00+783	Banyudono, Boyolali	0.15	0.32	8.78	High	0.01	0.03	0	Very Low
3	STA 00+866	Banyudono, Boyolali	0.15	0.23	0.55	Low	0.01	0.02	0	Very Low
4	STA 01+027	Banyudono, Boyolali	0.15	0.23	0.72	Low	0.01	0.02	0	Very Low
5	STA 01+116	Banyudono, Boyolali	0.15	0.23	0	Very Low	0.01	0.02	0	Very Low
6	STA 01+206	Banyudono, Boyolali	0.15	0.23	4.68	Low	0.01	0.02	0	Very Low
7	STA 01+385	Banyudono, Boyolali	0.16	0.23	0	Very Low	0.01	0.02	0	Very Low
8	STA 01+585 A	Banyudono, Boyolali	0.16	0.23	2.29	Low	0.01	0.02	0	Very Low
9	STA 01+758	Banyudono, Boyolali	0.16	0.23	1.88	Low	0.01	0.02	0	Very Low
10	STA 02+073	Banyudono, Boyolali	0.16	0.32	26.38	Very High	0.01	0.03	0	Very Low
11	STA 02+204	Banyudono, Boyolali	0.16	0.23	5.2	High	0.01	0.02	0	Very Low

No	Borehole	Sub-Disrict	An earthquake with *Mw* 6.3				An earthquake with *Mw* 5.2			
			PGA	A_S	*LPI*	Classification	*PGA*	A_S	*LPI*	Classification
12	STA 02+538	Sawit, Boyolali	0.16	0.32	24.47	Very High	0.01	0.03	0	Very Low
13	STA 02+877	Sawit, Boyolali	0.16	0.24	6.08	High	0.01	0.02	0	Very Low
14	STA 02+949	Sawit, Boyolali	0.16	0.32	26.52	Very High	0.01	0.03	0	Very Low
15	STA 03+073	Sawit, Boyolali	0.16	0.24	15.38	Very High	0.01	0.02	0	Very Low
16	STA 03+177	Sawit, Boyolali	0.16	0.24	0.49	Low	0.01	0.02	0	Very Low
17	STA 03+242	Sawit, Boyolali	0.16	0.24	7.71	High	0.01	0.02	0	Very Low
18	STA 03+344	Sawit, Boyolali	0.16	0.24	12.96	High	0.01	0.02	0	Very Low
19	STA 03+446	Sawit, Boyolali	0.16	0.24	1.74	Low	0.01	0.02	0	Very Low
20	STA 03+587 A	Sawit, Boyolali	0.16	0.24	3.57	Low	0.01	0.02	0	Very Low
21	STA 03+587 B	Sawit, Boyolali	0.16	0.24	1.46	Low	0.01	0.02	0	Very Low
22	STA 03+753	Sawit, Boyolali	0.16	0.24	2.68	Low	0.01	0.02	0	Very Low
23	STA 03+830 A	Sawit, Boyolali	0.16	0.24	3.55	Low	0.01	0.02	0	Very Low
24	STA 03+830 B	Sawit, Boyolali	0.16	0.24	0	Very Low	0.01	0.02	0	Very Low
25	STA 04+131	Sawit, Boyolali	0.16	0.24	3.43	Low	0.01	0.02	0	Very Low
26	STA 04+248	Sawit, Boyolali	0.16	0.33	16.94	Very High	0.01	0.03	0	Very Low
27	STA 04+350	Sawit, Boyolali	0.17	0.24	8.22	High	0.01	0.02	0	Very Low
28	STA 04+455	Sawit, Boyolali	0.17	0.24	0	Very Low	0.01	0.02	0	Very Low
29	STA 04+683	Sawit, Boyolali	0.17	0.24	0.88	Low	0.01	0.02	0	Very Low
30	STA 04+723	Sawit, Boyolali	0.17	0.24	0.11	Low	0.01	0.02	0	Very Low
31	STA 04+950	Sawit, Boyolali	0.17	0.25	2.05	Low	0.01	0.02	0	Very Low
32	STA 05+100	Sawit, Boyolali	0.17	0.25	0	Very Low	0.01	0.02	0	Very Low
33	STA 05+463	Sawit, Boyolali	0.17	0.25	1.54	Low	0.01	0.02	0	Very Low
34	STA 05+784	Sawit, Boyolali	0.17	0.25	11.54	High	0.01	0.02	0	Very Low
35	STA 05+850	Sawit, Boyolali	0.17	0.33	8.53	High	0.01	0.03	0	Very Low
36	STA 05+895	Sawit, Boyolali	0.17	0.33	22.27	Very High	0.01	0.03	0	Very Low
37	STA 06+033	Sawit, Boyolali	0.17	0.25	4.26	Low	0.01	0.02	0	Very Low
38	STA 06+111	Sawit, Boyolali	0.17	0.25	2.12	Low	0.01	0.02	0	Very Low
39	STA 06+228	Sawit, Boyolali	0.17	0.33	0.99	Low	0.01	0.03	0	Very Low
40	STA 06+388 A	Sawit, Boyolali	0.17	0.33	5.6	High	0.01	0.03	0	Very Low
41	STA 06+388 B	Sawit, Boyolali	0.17	0.25	3.08	Low	0.01	0.02	0	Very Low
42	STA 06+468	Sawit, Boyolali	0.17	0.25	0	Very Low	0.01	0.02	0	Very Low
43	STA 06+597	Polanharjo, Klaten	0.17	0.25	0	Very Low	0.01	0.02	0	Very Low
44	STA 06+911	Polanharjo, Klaten	0.18	0.25	0	Very Low	0.01	0.02	0	Very Low
45	STA 07+324	Delanggu, Klaten	0.18	0.33	8.6	High	0.01	0.03	0	Very Low
46	STA 07+350	Delanggu, Klaten	0.18	0.26	0.68	Low	0.01	0.02	0	Very Low
47	STA 07+499	Delanggu, Klaten	0.18	0.26	0	Very Low	0.01	0.02	0	Very Low

Based on **Table 4**, an earthquake of *Mw* 5.2 has a very low liquefaction potential compared to an earthquake of *Mw* 6.3. There are several parameters besides the earthquake magnitude that cause the determination of liquefaction potential. Some of these parameters are summarized in **Table 5.** An earthquake with *Mw* 6.3 was affected by an Opak fault with a source originating from Yogyakarta Province, which has a source depth of 12.5 km from the ground surface. While the earthquake with *Mw* 5.2 was affected by Benioff subduction (intraslab) with a source originating from Central Java Province, which has a source depth of 180 km from the ground surface. Different earthquake sources can affect the depth of the earthquake. The source of the Benioff subduction (Intraslab) earthquake occurs at a depth of more than 50 km [16]. The depth of the earthquake is also a factor that affects the impact of damage caused by an earthquake. The impact of damage caused by an earthquake with a shallow depth has a higher risk of damage compared to a deep depth. This is because the shallower the distance from the hypocenter of the earthquake, the higher the value of *PGA*, and conversely the deeper the hypocenter of the earthquake, the smaller the value of *PGA* [19].

Table 5. Differences in earthquake parameters

Earthquake Magnitude	Depth (km)	Epicenter (km)	Hypocenter (km)	Earthquake source	*PGA average* (g)	A_S *average* (g)
6.3	12.5	48.46	50.05	Fault	0.16	0.26
5.2	180	11.07	180.34	Subduction	0.01	0.02

IOP Conf. Series: Earth and Environmental Science 1244 (2023) 012031 doi:10.1088/1755-1315/1244/1/012031

According to **Tabel 5**, The liquefaction potential of the research study is strongly influenced by fault conditions because the *PGA* and A_s value is greater than subduction conditions. The maximum peak ground acceleration at the ground surface (A_s) is one of the important roles that affect the liquefaction potential. The value of A_s is influenced by earthquake source, depth, epicenter, hypocenter, and site class. The average value of A_s with an earthquake magnitude of 6.3 is 0.26 g compared to the earthquake magnitude of 5.2 which is only 0.02 g. Based on Fathani et.al (2008), A_s values ranging from 0.20 to 0.30 have a medium risk of damage. Meanwhile, A_s value < 0.10 has a very low risk of damage [20]. Based on this comparison, the value of A_s generated by a magnitude 6.3 earthquake is more destructive than an earthquake of 5.2.

Based on the *LPI* value, The liquefaction potential with an earthquake of *Mw* 6.3 varied between very low to very high potential. There are 6 boreholes with a very high liquefaction potential and 10 boreholes with a high liquefaction potential, 21 boreholes with a low liquefaction potential, and 10 boreholes with a very low liquefaction potential. The boreholes that had a very high liquefaction potential is at STA 02+073; STA 02+538; STA 02+949; STA 03+073; STA 04+248; and STA 05+895. Meanwhile, the liquefaction potential with an earthquake of *Mw* 5.2 has very low liquefaction potential at all boreholes in the research study. Based on the results of the liquefaction potential, then hazard map is created using QGIS software as shown in **Figure 5.**

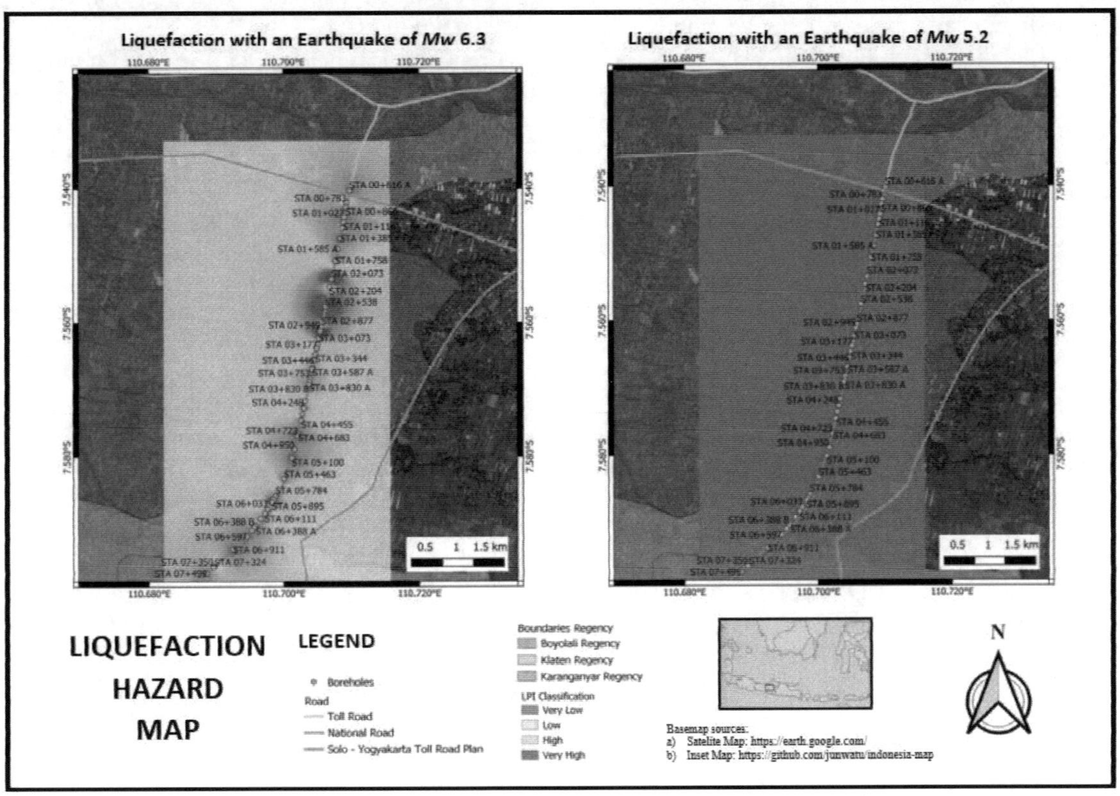

Figure 5. Liquefaction Hazard Map based on *LPI* method

Based on **Figure 5**, The liquefaction potential with an earthquake of *Mw* 6.3 in Karanganyar Regency in the research area has very low liquefaction potential. Meanwhile, Boyolali regency has liquefaction potential with a range of very low to very high potential. The areas with very high liquefaction potential especially from STA 02+073 until STA 03+073. However, Klaten Regency has a liquefaction potential varied from very low to high. While the liquefaction potential with an earthquake of *Mw* 5.2 appears to have the same color which indicates there is very low liquefaction potential in every regency. So, there is a need for further studies in liquefaction mitigation planning in this research area using earthquake parameters with a magnitude of 6.3.

5. Conclusion

Based on the soil investigation, the type of soil in the study area was dominated by silty sandy soil and had a shallow ground water level with a dominated value of less than 2 m below ground level. That caused the study area to have liquefaction potential.

According to earthquake history, there are 2 earthquakes with a moment magnitude above *Mw* 5 in the research study. First, an earthquake with *Mw* 6.3 was affected by an Opak fault with a source originating from Yogyakarta Province, which has a source depth of 12.5 km from the ground surface. Second, the earthquake with *Mw* 5.2 was affected by Benioff subduction (intraslab) with a source originating from Central Java Province, which has a source depth of 180 km from the ground surface. The depth of an earthquake with a magnitude of 6.3 is shallower than an earthquake with a magnitude of 5.2. The depth of the earthquake is also a factor that affects the impact of damage caused by the earthquake. The impact of damage caused by an earthquake with a shallow depth has a higher risk of damage compared to a deep depth.

Based on the value of A_s, the average value of A_s with an earthquake magnitude of 6.3 is 0.26 g compared to the earthquake magnitude of 5.2 which is only 0.02 g. The value of As generated by a magnitude 6.3 earthquake is more destructive than an earthquake of 5.2.

The liquefaction potential analysis was conducted at 47 boreholes that spread in several regions in Central Java Province using the simplified procedure and *LPI* method. Based on the analysis, an earthquake of *Mw* 5.2 has a very low liquefaction potential compared to an earthquake of *Mw* 6.3. The liquefaction potential with an earthquake of *Mw* 6.3 has a classification range from very low to very high potential. There are 6 boreholes with a very high liquefaction potential and 10 boreholes with a high liquefaction potential, 21 boreholes with a low liquefaction potential, and 10 boreholes with a very low liquefaction potential.

Based on the analysis result and hazard caused, an earthquake with *Mw* 6.3 is recommended to be used as a basis for mitigation against liquefaction planning. Further studies are needed regarding alternative choices of mitigation against liquefaction

Acknowledgement

The authors gratefully thank for the support given by the Directorate of Freeways, Directorate General of Highway, Ministry of Public Works and Housing, Republic of Indonesia, and PT Jogjasolo Marga Makmur.

References

[1] Astutik H P and Dewanti D, 2020 The Effect of Toll Gate Type on the Queue of Vehicles in Connecting Roads: A case study of Bawen – Yogyakarta Toll Road *J. Civ. Eng. Forum* **6**, 1 p. 1.

[2] Pramaditya A and Fathani T F, 2020 Physical Modelling of Earthquake-induced Liquefaction on Uniform Soil Deposit and Settlement of Earth Structures *J. Civ. Eng. Forum* **1000**, 1000 p. 85–96.

[3] Sutiono A Prastistho B Prasetyadi C and Supartoyo, 2018 Opak fault: A comparative review *IOP Conf. Ser. Earth Environ. Sci.* **212**, 1.

[4] United States Geological Survey, 2006, M6.3 Java Earthquake of 26 May 2006.

[5] Sarah D and Soebowo E, 2013 Liquefaction due to the 2006 Yogyakarta Earthquake: Field Occurrence and Geotechnical Analysis *Procedia Earth Planet. Sci.* **6** p. 383–389.

[6] Sasanti C A, 2015, Studi Potensi Soil Liquefaction Di Rencana Lokasi Genting Oil Kasuri Papua Barat.

[7] Idriss, I. M.; Boulanger R W, 2008 *Soil Liquefaction During Earthquakes* **160**, 4057 California.

[8] Iwasaki T Tadhasi A and Tokida K, 1984 Simplified procedures for assessing soil liquefaction

during earthquakes **3** p. 49–58.
[9] Bachri S, 2014 Pengaruh Tektonik Regional Terhadap Pola Struktur dan Tektonik Pulau Jawa *J. Geol. dan Sumberd. Miner.* **15**, 4 p. 215–221.
[10] Sutikno S, 2016 Earthquake Disaster of Yogyakarta and Central Java, and Disaster Reduction, Indonesia *Forum Geogr.* **21**, 1 p. 1–16.
[11] Buana T W Wafid M and Sadisun I A, 2016 Hubungan Potensi Likuifaksi Pada Endapan Gunungapi Merapi Muda Dengan Kerusakan Bangunan Di Kabupaten Bantul Pada Kasus Gempabumi 27 Mei 2006 *J. Environ. Geol. Hazards* **7**, No 2 p. 103–111.
[12] Toprak S and Holzer T L, 2003 Liquefaction Potential Index: Field Assessment *J. Geotech. Geoenvironmental Eng.* **129**, 4 p. 315–322.
[13] Akter S, 2021 Seismic Ground Response Analysis of Input Earthquake Motion and Site Amplification Factor at KUET *J. Civ. Eng. Forum* **8**, January p. 45–54.
[14] Green R A and Bommer J J, 2019 What is the smallest earthquake magnitude that needs to be considered in assessing liquefaction hazard? *Earthq. Spectra* **35**, 3 p. 1441–1464.
[15] United States Geological Survey, 2022, Latest Earthquake. [Online]. Available: https://earthquake.usgs.gov/earthquakes/map/. [Accessed: 22-Feb-2022].
[16] Erlangga W, 2020 Karakteristik Dan Parameter Subduksi Sumber Gempa Pulau Jawa *Teknisia* **XXV**, 2 p. 30–40.
[17] Kanno T Narita A Morikawa N Fujiwara H and Fukushima Y, 2006 A new attenuation relation for strong ground motion in Japan based on recorded data *Bull. Seismol. Soc. Am.* **96**, 3 p. 879–897.
[18] Badan Standardisasi Nasional Indonesia, 2016 SNI 2833:2016 Perencanaan Jembatan Terhadap Beban Gempa p. 1–70.
[19] Haris A and Irjan I, 2013 Analisis Percepatan Getaran Tanah Maksimum Wilayah Yogyakarta Dengan Metode Atenuasi Patwardhan *J. Neutrino* p. 66–72.
[20] Fathani T F Adi A D Pramumijoyo S and Karnawati D, 2008 The Determination of Peak Ground Acceleration at Bantul Regency , Yogyakarta Province , Indonesia *Yogyakarta Earthq. May 27, 2006* p. 1–15.

Ground Motion and Liquefaction Study at Opak River Estuary Bantul

A Zakariya[1,2], A Rifa'i[1*], S Ismanti[1]

[1]Department of Civil and Environmental Engineering, Faculty of Engineering, Universitas Gadjah Mada, Yogyakarta, Indonesia
[2]Directorate of Bridge Construction, Directorate General of Highways, Ministry of Public Works and Housing, Jakarta, Indonesia

*Corresponding email: ahmad.rifai@ugm.ac.id

Abstract. In Mei 2006, an earthquake occurred, and liquefaction was induced in Bantul Regency in the form of lateral spreading and sand boiling. Based on Indonesia's liquefaction susceptibility zone map 2019, the Bantul coast was identified as a high liquefaction hazard zone. There is a high risk of liquefaction in that location, and if liquefaction occurs, the effective stresses of the soil will be lost. This study aimed to determine ground motion-induced liquefaction potential in Opak river estuary in Bantul. The peak ground acceleration can be predicted by probabilistic seismic hazard analysis (PSHA) based on Indonesian spectra design application, deterministic seismic hazard analysis (DSHA) based on the mechanism of a recent earthquake in Bantul, and non-linear earthquake site response analysis (NERA) based on the 2004 Parkfield California earthquake, the 1986 Chalfant Valley California earthquake, and the 1987 Superstition Hill California earthquake. The value of acceleration from site-specific responses with DSHA and NERA is 0.558 g. Then, the maximum acceleration, seismic magnitude, N-SPT bore log soil, and other soil parameters are analyzed with a semi-empirical simplified method to estimate the safety factor of liquefaction potential using Settle3. The liquefaction safety factor was calculated by dividing the cyclic resistance ratio by the cyclic stress ratio; liquefaction occurs if the safety factor is less than 1. The result indicates that the Opak river estuary has a liquefaction potential beneath the surface from -1.5 to -16.5 m depth with various thick each boreholes. It implies that the bearing capacity design of the foundation system in this area and the liquefaction effect that will emerge when an earthquake occurs must be appropriately considered.

Keywords: DSHA, liquefaction potential, NERA, Settle3, site-specific response

1. Introduction

After the Palu earthquake in 2018, liquefaction has become a prominent topic for research on the impact of liquefaction on infrastructures such as buildings, roads, and bridges in Indonesia. Since 2019, an Indonesian liquefaction susceptibility zone map [1] has been provided, with areas classified as non-liquefied risk zone, low-liquefied risk zone, moderate-liquefied risk zone, and high-liquefied risk zone. The Directorate General of Highways has a road and bridge development plan for south Java Island, particularly the Java southern line (JLSS). This research focuses on the Opak river estuary area, Kretek District, Bantul Regency. The Opak Estuary was later classified as a high-liquefied risk zone. Due to the significant risk of liquefaction in that area, designing and constructing infrastructure must be considered for liquefaction potential analysis that would appear when a strong earthquake occurs.

2. Material and Method

2.1. Recent Earthquake and Liquefaction

The recent earthquake in Bantul has been recorded on a catalog of significant and catastrophic earthquakes 1821 – 2018 [2] since 1867 with a scale of VIII – IX MMI (see Table 1). On May 27, 2006, an earthquake with a moment magnitude of 6.2 afflicted Bantul Regency at a depth of 17.2 km, and a location of 8.01°S – 110.18°E originated from Opak fault (Figure 1.a). The trench test was conducted to prove that the Opak fault is located beneath the soil surface (Figure 1.b) [3]. The earthquake caused extensive damage and liquefaction, such as sand boiling and lateral spreading (Figure 2.a & 2.b). Liquefaction is a phenomenon that occurs when soil becomes liquefied due to an earthquake, with the condition of loose sandy soil and high groundwater level. Meanwhile, in 2018, an earthquake with a magnitude of 7.4 M_w hit Palu, causing large-scale liquefaction into a form of liquefied mass flow slide on Balaroa, Petobo, and Jono Oge [4].

Table 1. Recent earthquakes near Bantul Regency

Seismic Name	Date (dd/mm/yyyy)	Location	Depth (km)	Magnitude Scale	MMI Scale
Jawa Tengah	10/06/1867	-	-	-	VIII – IX
Jawa Tengah	27/09/1937	8.70°LS – 110.80°BT	-	-	VIII – IX
Yogyakarta	23/07/1943	8.60°LS – 109.90°BT	-	-	VIII – IX
Yogyakarta	13/03/1981	8.76°LS – 110.43°BT	51	5.6 SR	VII
Kulon Progo	25/05/2001	7.87°LS – 110.18°BT	143	6.3 SR	V
Bantul	27/05/2006	8.01°LS – 110.18°BT	17.2	6.2 M_w	VII
Jawa Barat	08/08/2007	6.13°LS – 107.68°BT	284	6.9 SR	IV
Yogyakarta	21/08/2010	8.03°LS – 110.30°BT	10	5.0 SR	V
Kebumen	25/01/2014	8.48°LS – 109.17°BT	48	6.5 SR	IV

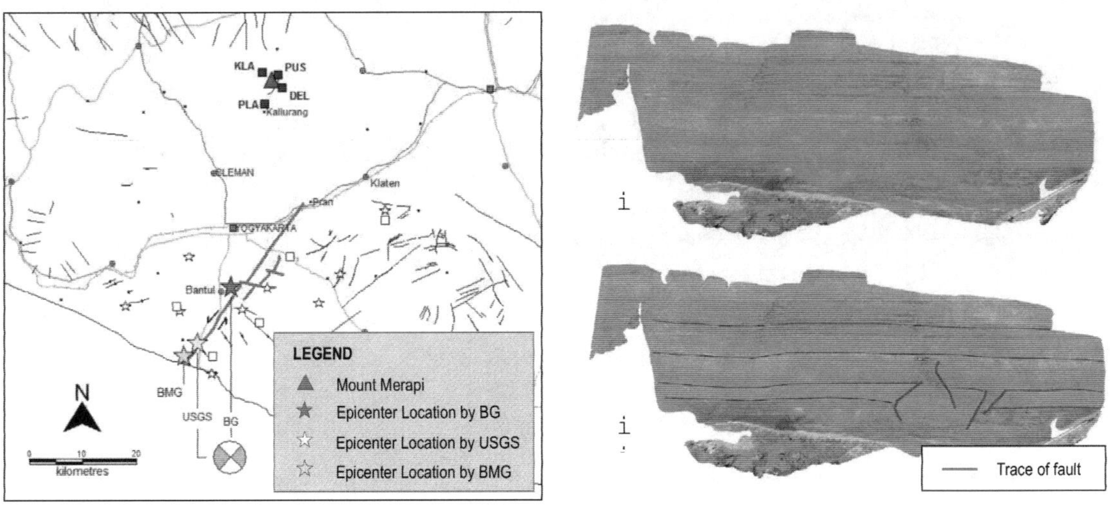

(a) (b)

Figure 1. The 2006 Yogyakarta Earthquake; (a) epicenter of earthquake and Opak fault [5], and (b) Opak fault trench test before marked and after marked, respectively

2.2. Ground Motion Prediction

Earthquake cause ground motion waves from deep earth to surface. Ground motion prediction equation (GMPE) or attenuation function is needed to estimate peak ground acceleration (PGA) and spectral acceleration (SA) at 0.2 sec and 1 sec. Ground motion recorded worldwide was used to develop the attenuation function. The main factors for the function are the type of earthquake, the magnitude scale, distance, and local site condition. The 2017 earthquake source and hazard map is the recent code in Indonesia which uses three seismic source types; shallow crustal, subduction (megathrust), and benioff (deep intraslab) [6]. Most of the attenuation in this code is adopted from New Generation Attenuation (NGA) West 1 and West 2. Several researchers, including Boore et al. (2014) [7], Campbell and Bozorgnia (2013) [8], and Chiou and Youngs (2014) [9], joined NGA West with the aim of predicting the ground motion value due to seismic from fault events in the United States for the western region. These attenuation models have been adopted not only in Indonesia but throughout the world. Mase (2019) [10] has examined NGA West 1 and West 2 with the result that both those models could be reliable in predicting the ground motion, but NGA West 2 has the prediction improved for PGA calculation. Seismic hazard analysis from attenuation function could be estimated in two methods of seismic hazard analysis: probabilistic seismic hazard analysis (PSHA) and deterministic seismic hazard analysis (DSHA).

(a) (b) (c) (d)

Figure 2. Liquefaction impact; (a) (b) sand boiling event at an open well and lateral spreading event on edge state road at Bantul [11] respectively, and (c) (d) lateral spreading event at Balaroa and Palu IV Bridge reportedly collapsed respectively [12]

2.3. Probabilistic Seismic Hazard Analysis

Based on earthquake incidence data in the past, this approach determines the probability of a future earthquake. To minimize deviation, PSHA considers the seismic source, frequency of occurrence, and data deaggregation. According to the 2017 earthquake source and hazard map, there are two user-friendly web applications: Indonesia spectra design, developed by Earthquake Study Center (PUSGEN), and LINI Binamarga, developed by Directorate General of Highways. The Indonesian spectra design application to get a PGA on bedrock with a 2% probability of exceedance in 50 years or equivalent to 2500 year return period. Considering SNI 1726-2019, the Indonesian spectra design for a building or non-building structures design with same characteristics [13]. Meanwhile, the LINI Binamarga application to get a PGA on bedrock with a 7% probability of exceedance in 75 years or equivalent to 1000 years period. Considering SNI 2833-2016, the LINI Binamarga for a bridge structures design [14].

In this study, LINI Binamarga is applied, and the PGA maximum result for 110.2931°E – 8.0084°S coordinate location is 0.447 g.

2.4. Deterministic Seismic Hazard Analysis

This method is helpful for designing special and important structures such as dams, power plants, and structures near faults. DSHA can be described in four steps; 1) identifying all earthquake sources near the site, 2) selecting a distance parameter for each zone based on source-to-site, 3) selecting the controlling earthquake dominantly, and 4) defining the hazard in terms of ground motions generated at the site by the controlling earthquake dominantly [15]. Another researcher, Muntafi et al. (2020) has concluded that the closer distance to the epicenter of earthquake, the higher damage to the building [16]. This study uses the recent attenuation function developed by NGA West researchers for the shallow crustal model. Boore et al. (2014) developed the equation from ground motion regression data with 0 – 400 km distance and magnitude scale 3 – 8.5 M_w in strike-slip type. The equation was expressed as follows:

$$ln\,Y = F_E(M, mech) + F_{P,B}(R_{JB}, M, region) + F_{S,B}(V_{s30}, R_{JB}, M, z_1) + \varepsilon_n\sigma(M, R_{JB}, V_{s30}) \quad (1)$$

where $\ln Y$ represents the ground motion natural logarithm, F_E for the source-dependent, $F_{P,B}$ for the path function, $F_{S,B}$ for the site amplification function. The predictor variables are M represent the moment magnitude, $mech$ represent the fault type, R_{JB} represent the Joyner-Boore distance, and V_{s30} represent the average shear-wave velocity at 30 m of topsoil. $\varepsilon_n\sigma$ is the fractional number of standard deviations.

Campbell and Bozorgnia (2013) model was intended for seismic sources from shallow crustal (strike-slip, reverse, and normal fault). It was developed by data regression from 322 earthquake occurrences around the world with a 500 km maximum distance, magnitude scale 3 – 7.9 M_w, and rupture length 0 – 80 km, and the equation was expressed in the following:

$$ln\,Y = \begin{cases} ln\,PGA \\ f_{mag} + f_{dis} + f_{flt} + f_{hng} + f_{site} + f_{sed} + f_{hyp} + f_{dip} + f_{atn} \end{cases} \quad (2)$$

where $\ln Y$ represents the natural logarithm of a RotD50 horizontal ground motion, the terms represent the earthquake magnitude, distance of geometric attenuation, faulting style, geometry of hanging-wall, shallow site response, basin response, depth of hypocentral, dip of rupture, and (apparent) anelastic attenuation respectively.

Chiou and Youngs (2014) model was developed by data regression from 243 earthquake occurrences in California (United States), Japan, New Zealand, Taiwan, and Turkiye with a magnitude scale of 3.1 – 7.6 M_w and rupture length of 0 – 80 km. The equation was expressed as follows:

$$ln(y_{ij}) = ln\left(y_{ref_{ij}}\right) + \emptyset_1 min\left(ln\left(V_{s30_j}/1130\right), 0\right) + \emptyset_2\left\{e^{\emptyset_3(min(V_{s30ij},1130)-360)} - e^{\emptyset_3(1130-360)}\right\} ln\left(\left(y_{ref_{ij}}e^{\eta_i} + \emptyset_4\right)/\emptyset_4\right) + \emptyset_5\left\{1 - e^{-\Delta Z_{1.0_j}/\emptyset_6}\right\} + \eta_i \quad (3)$$

where $\ln(y_{ij})$ represents the ground motion natural logarithm, $\ln\left(y_{ref_{ij}}\right)$ represents the natural logarithm of a ground motion reference (included for moment magnitude, closest distance to the rupture plane, Joyner-Boore distance, faulting style, and fault dip angle), $\emptyset$ represents the coefficients that are given by the model, V_{s30} represents the average shear-wave velocity at 30 m of topsoil, $Z_{1.0}$ represent the depth to shear wave velocity of 1.0 km/sec, $\Delta Z_{1.0_j}$ represent $Z_{1.0}$ (m) centered on the California-specific average $Z_{1.0}$ model, and η_i represents the inter-event residual. The reference site condition with V_{S30}=1130 m/sec.

Another researcher, Kanno et al. (2006), developed an attenuation model regression from strong ground motion records data on Japan between 1963 and 2003 [17]. The equation was expressed in the following:

$$log\,pre = a_1M_w + b_1X - log(X + d_1 10^{e_1M_w}) + c_1 + \varepsilon_1(D \leq 30km) \quad (4)$$

$$log\, pre = a_2 M_w + b_2 X - log(X) + c_2 + \varepsilon_2 (D > 30km) \quad (5)$$

where pre is the predicted PGA (g), M_w for the magnitude scale, X for the source distance, a_1, b_1, c_1, d_1, a_2, b_2, and c_2 are the regression coefficients, $e_1 = 0.5$ was selected for all periods, and ε_1 and ε_2 are errors between predicted and observed values. Equation (4) for focal depth (D) $\leq$ 30 km and Equation (5) for focal depth (D) > 30 km. Taruna et al. (2016) [18] have used Kanno models to predict the PGA value for the East Java region with a reliable result.

From the 4 attenuation functions discussed above, Boore et al. (2014), Campbell and Bozorgnia (2013), Chiou and Youngs (2014), and Kanno et al. (2006) are shown PGA maximum values about 0.503, 0.612, 0.521, and 0.592 g, respectively (see Figure 3).

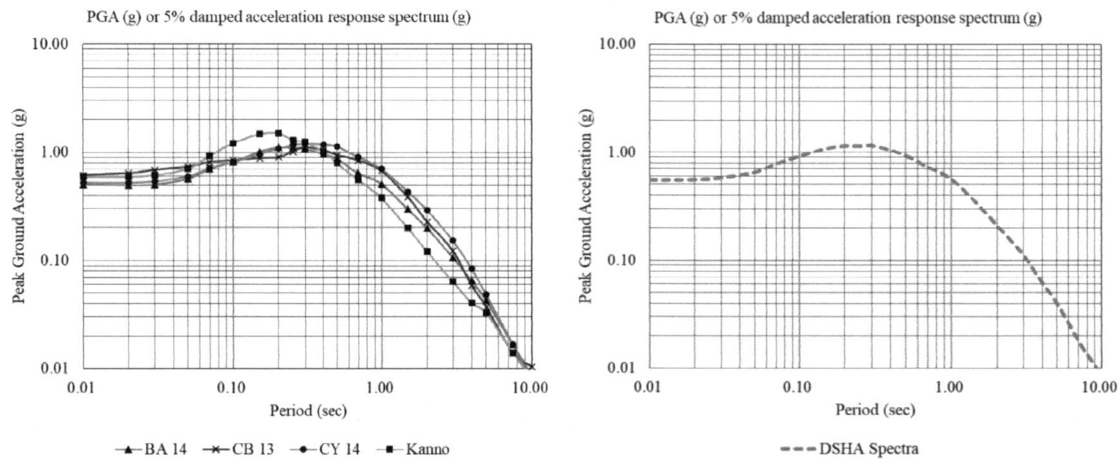

(a) (b)

Figure 3. Target Spectra from; (a) 4 attenuation function of GMPE, and (b) DSHA

2.5. Non-linear Earthquake Site-Response Analysis

Non-linear earthquake site response analyses (NERA) is an excel-based software developed by Bardet and Tobita (2001) to calculate one-dimensional site response in layered soil, considering shear wave velocity at each layer and acceleration response spectrum (including PGA) as output [19]. Fathani and Wilopo (2017) have used the use of NERA in determining the PGA value for Yogyakarta City, and the result PGA value from NERA is higher than other methods (Attenuation function and microtremor) [20]. The output of NERA is dependent on ground motion input selection. The source ground motion must be typically similar to the 2006 Bantul earthquake (EQ) for mechanism fault, magnitude scale, and rupture depth. Based on the ground motion database from Pacific Earthquake Engineering Research Center (PEER) [21], the ground motion used are Parkfield (Cholame 1E) 2004 EQ, Chalfant Valley-02 1986 EQ, and Superstition Hills (Poe Road) 1987 EQ (see Figure 4.a). Shear wave velocity for this study from the downhole seismic test carried out from field test [22] (see Figure 4.b). For the result, PGA value is 0.507, 0.679, and 0.450 g, respectively, as shown in Figure 5.

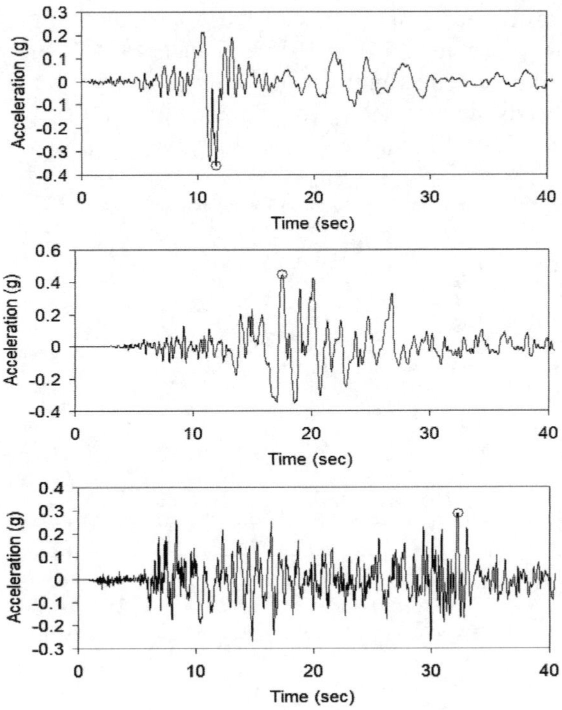

Depth (m)	N-SPT	Soil Type	V_{S30} (m/s)
1.5	45	SP	310.900
3.0	60	SP	327.590
4.5	35	SP	392.745
6.0	38	SP	360.750
7.5	60	SP	224.755
9.0	55	SP	234.440
10.5	52	SP	283.530
12.0	53	SP	386.740
13.5	50	SP	208.120
15.0	52	SP	236.840
16.5	24	SP	272.645
18.0	34	SP	394.170
19.5	41	SP	167.610
21.0	30	SP	355.180
22.5	33	SP	159.240
24.0	43	SP	368.660
25.5	44	SM	269.265
27.0	30	SM	373.820
28.5	36	SM	410.360
30.0	39	SM	374.430
		Average	305.590

(a) (b)

Figure 4. (a) Ground Motion Recorded; the 2004 Parkfield (Cholame 1E) EQ, the 1986 Chalfant Valley-02 EQ, and the 1987 Superstition Hills (Poe Road) EQ respectively, (b) Shear Wave Velocity from Downhole Seismic Test

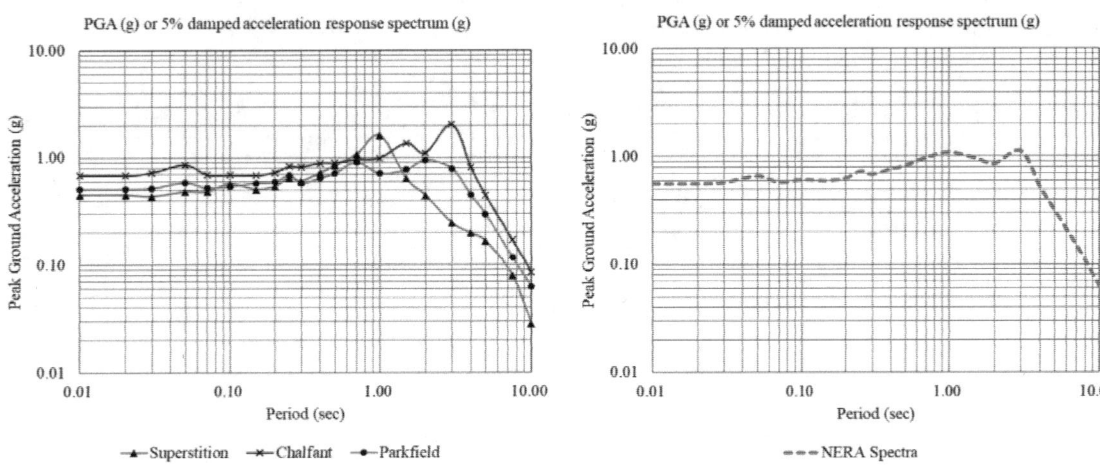

(a) (b)

Figure 5. Acceleration response spectrum; a) Superstition Hills (Poe Road), Chalfant Valley-02, Parkfield (Cholame 1E), respectively and b) Target Spectra

The logic tree framework is used in this study to allow uncertainties in the selection of models for recurrence model, maximum magnitude, distance from the seismic source, and attenuation function to be considered. The weighting of logic tree used in this analysis is shown in Figure 6.a. PGA value with logic tree framework in DSHA and NERA is shown in Figure 6.b with values 0.557 and 0.559 g, respectively. The logic tree was also used in considering DSHA and NERA for minimizing the uncertainties models. The PGA value result from those methods is 0.558 g, then PGA must be considered by site class amplification factor (F_{PGA}) to get the maximum peak ground acceleration at the surface (PGA_M). The site class amplification factor can be determined by the average of N-SPT or VS30 value (see Table 2). The ground motion prediction result of this study is shown in Table 3.

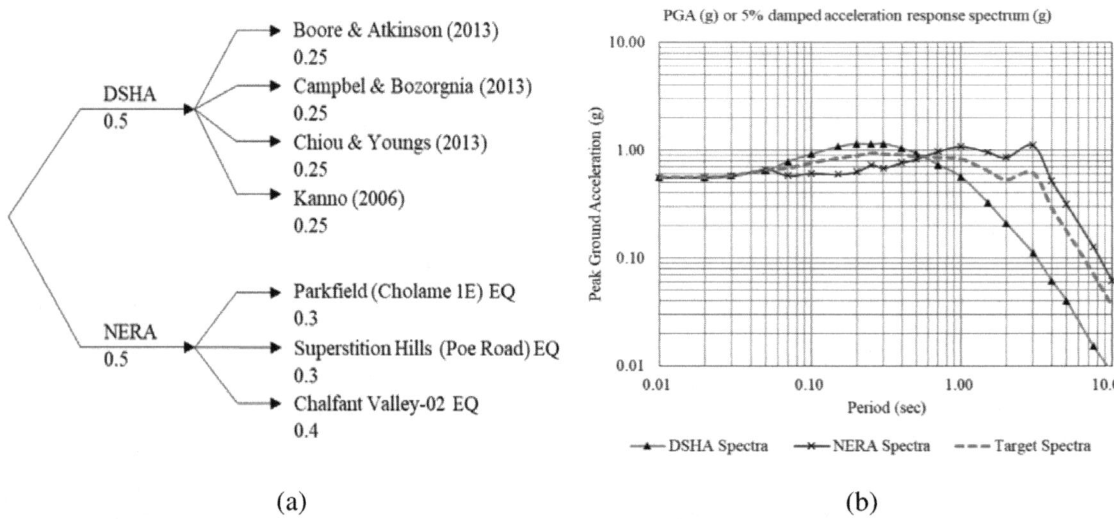

Figure 6. (a) Logic tree framework, and (b) Target Spectra from DSHA and NERA

Table 2. Site class amplification factor for bridge [14]

Site Class	N_{SPT}	V_{S30} (m/s)	PGA				
			≤0.1	0.2	0.3	0.4	≥0.5
SA	-	>1,500	0.8	0.8	0.8	0.8	0.8
SB	-	750 – 1,500	1.0	1.0	1.0	1.0	1.0
SC	>50	350 – 750	1.2	1.2	1.1	1.0	1.0
SD	15-50	175 – 350	2.5	1.7	1.2	1.1	1.0
SE	<15	<175	2.5	1.7	1.2	0.9	0.9

Table 3. PGA result from site-specific response

Method	Source	PGA	F_{PGA}	PGA_M
Spectra design of Indonesia 7% or 75 years with the 1000 years return period	Seismic source and hazard map of Indonesia	0.447	1.047	0.468
DSHA and NERA target spectra	4 GMPE and 3 ground motion	0.558	1.000	0.558

2.6. Liquefaction Potential Analysis

2.6.1. N-SPT Soil Data in Location

The study located in the Opak river estuary area, Kretek District, Bantul Regency, Indonesia (see Figure 7 and Table 4). Another researcher, Mase et al. (2013), investigated using an experimental study with a shaking table test that Opak river sand in Imogiri district has liquefaction potential [23]. This research used bore log data from the borehole test and soil engineering properties from the laboratory test. All bore log depths are 39 m below the surface, and soil characteristics are mostly in various ranges, with very loose sand to very dense sand. However, some borehole has a clay layer >-15 m depth (see Figure 8). Groundwater level varies from -0.5 to -1.5 m. Soil unit weight (bulk and submerged) can be estimated by N-SPT correlation shown in Table 5.

Figure 7. Soil investigation borehole location for liquefaction study [24]

Table 4. Boreholes coordinate

Borehole	Location	Groundwater Level
BH-1	110.2930°E - 8.0073°S	-1.2 m
BH-2	110.2929°E - 8.0077°S	-1.5 m
BH-3	110.2931°E - 8.0084°S	-1.0 m
BH-4	110.2931°E - 8.0084°S	-1.0 m
BH-5	110.2933°E - 8.0084°S	-0.5 m
BH-6	110.2933°E - 8.0092°S	-1.0 m
BH-7	110.2935°E - 8.0095°S	-1.0 m
BH-8	110.2934°E - 8.0100°S	-1.0 m
BH-9	110.2937°E - 8.0103°S	-1.5 m

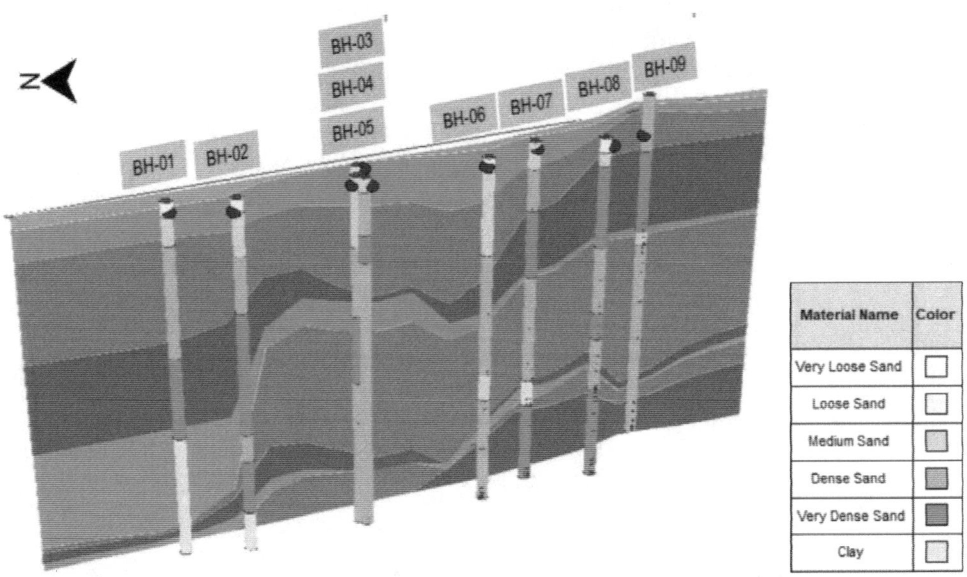

Figure 8. Soil Stratigraphy

Table 5. Correlation soil properties in cohesionless soil from Bowles 1997 [25]

Compactness	Very Loose	Loose	Medium	Dense	Very Dense
N-SPT value	0 – 4	4 – 10	10 – 30	30 – 50	>50
Relative Density, Dr (%)	0 – 15	15 – 35	35 – 65	65 – 85	85 – 100
Internal Friction, φ (°)	<28	28 – 30	30 – 36	36 – 41	>41
Bulk unit weight (kN/m)	<15.7	14.9 – 19.6	17.3 – 20.4	17.3 – 22.0	>20.4
Submerged unit weight (kN/m)	<9.4	8.6 – 10.2	9.4 – 11.0	10.5 – 13.4	>11.8

2.6.2. Simplified Procedure for Liquefaction Analysis with Settle3

The semi-empirical approach proposed by Idriss and Boulanger (2004) [26] was used for the liquefaction analysis in this study. This method is deliberated to analyze a potential liquefaction occurrence by determining the factor of safety against liquefaction using the following equation:

$$FS_{liq} = CRR/CSR \tag{6}$$

where FS_{liq} is the safety factor, CSR is the cyclic stress ratio, and CRR is the cyclic resistance ratio. CSR was developed by Seed and Idriss (1970) as known as The Seed-Idriss simplified procedure [27], and it was modified by Idriss and Boulanger (2004) using the following equation:

$$CSR = 0.65 \frac{a_{max}}{g} \frac{\sigma_{vc}}{\sigma'_{vc}} \frac{1}{MSF} \frac{1}{K_\sigma} r_d \tag{7}$$

where 0.65 coefficient represents a value equal to 65% of the peak cyclic stress, a_{max} is the maximum peak ground acceleration on the surface, σ_{vc} and σ'_{vc} are the total and effective vertical stress, *MSF* is the magnitude scaling factor of the moment magnitude 7.5, K_σ is the overburden correction factor, and r_d is the shear stress reduction coefficient that correlates the depth function and magnitude scale. Furthermore, CRR value can be calculated using the following equation:

$$CRR = exp\left(\frac{(N_1)_{60cs}}{14.1} + \left(\frac{(N_1)_{60cs}}{126}\right)^2 - \left(\frac{(N_1)_{60cs}}{23.6}\right)^3 + \left(\frac{(N_1)_{60cs}}{25.4}\right)^4 - 2.8\right) \tag{8}$$

where $(N_1)_{60cs}$ is N-SPT after correction with soil investigation boring equipment, finest content of soil, and effective overburden pressure. The calculation result can be used to predict whether a particular layer is liquefying. This study used the Settle3 application, part of Rocscience that helps in liquefaction and settlement estimates. For calculating liquefaction analysis in Settle3, this study is based on Idriss and Boulanger (2004) approach. Other researchers, Maulana et al. (2020), has used Settle3 to compare software and manual calculation with relatively similar result [28]. Beside N-SPT and unit weight soil data, Settle3 also requires groundwater level, fines content percentage, and D_{50} grain size of sand as data input. Moment Magnitude and PGA values determined from the previous step are 6.2 M_w and 0.558 g, respectively. Figure 9 – 11 shows the liquefaction potential analysis results for all boreholes.

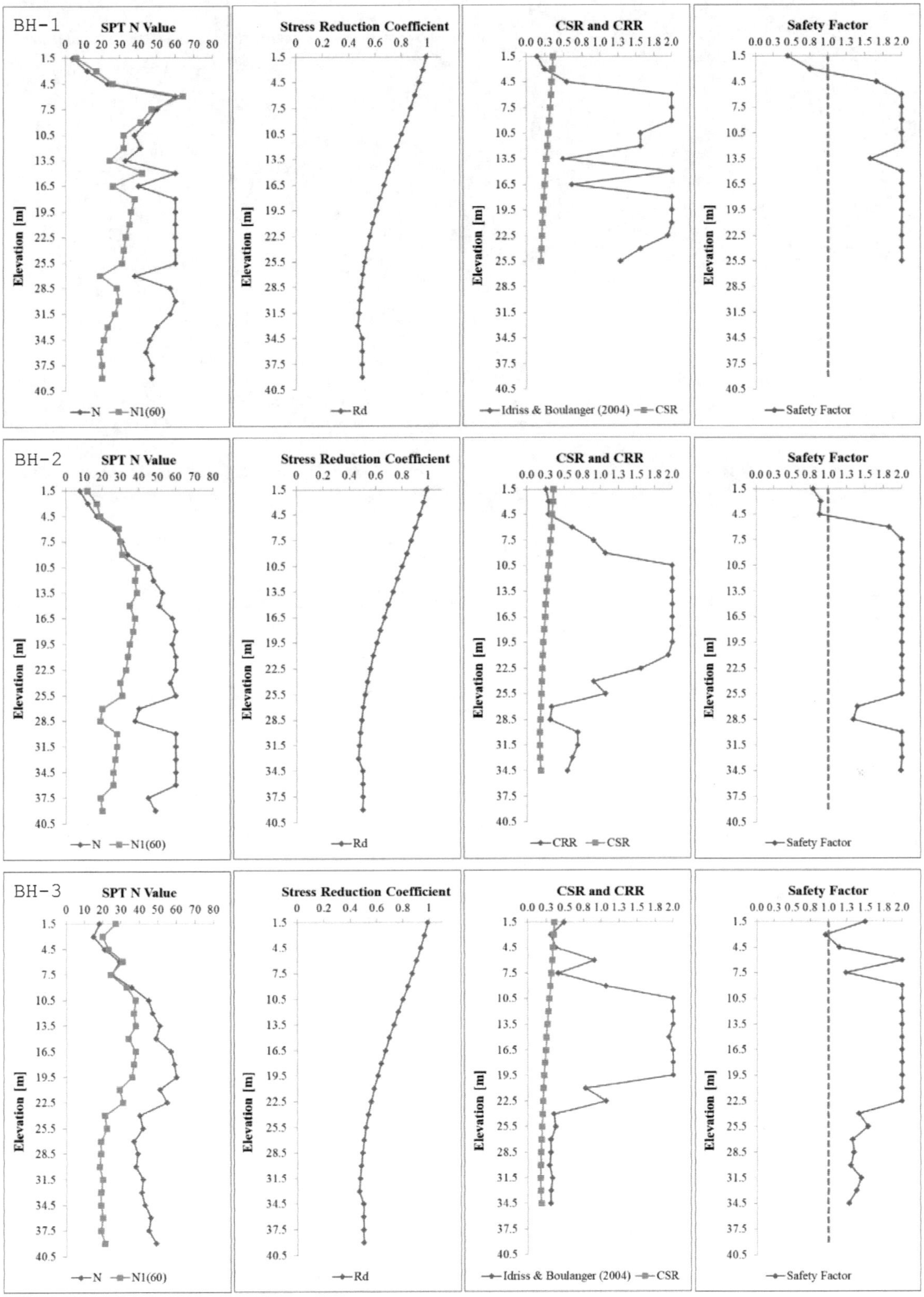

Figure 9. CSR, CRR, and safety factor at BH-1, BH-2, and BH-3 from Settle3 result

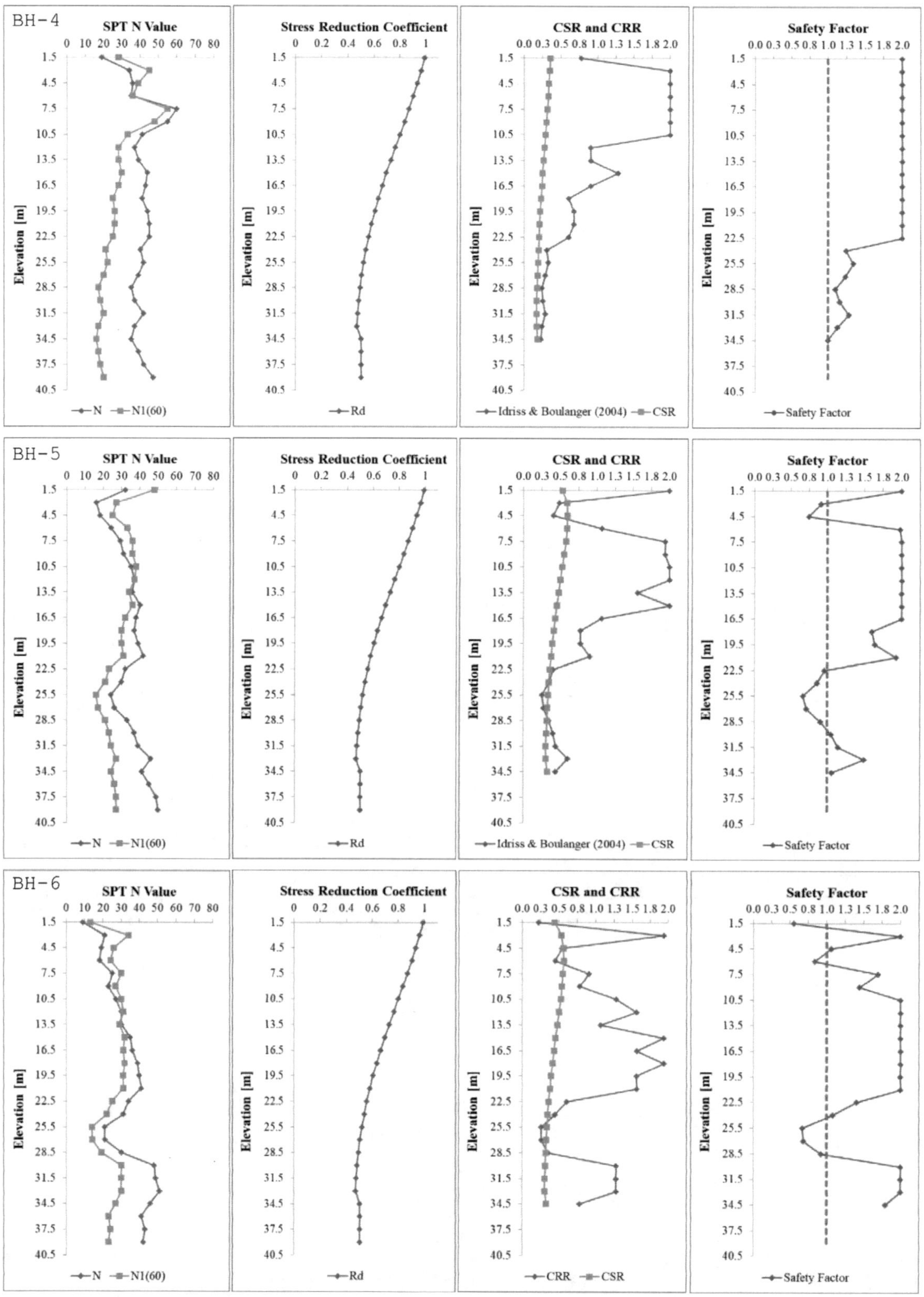

Figure 10. CSR, CRR, and safety factor at BH-4, BH-5, and BH-6 from Settle3 result

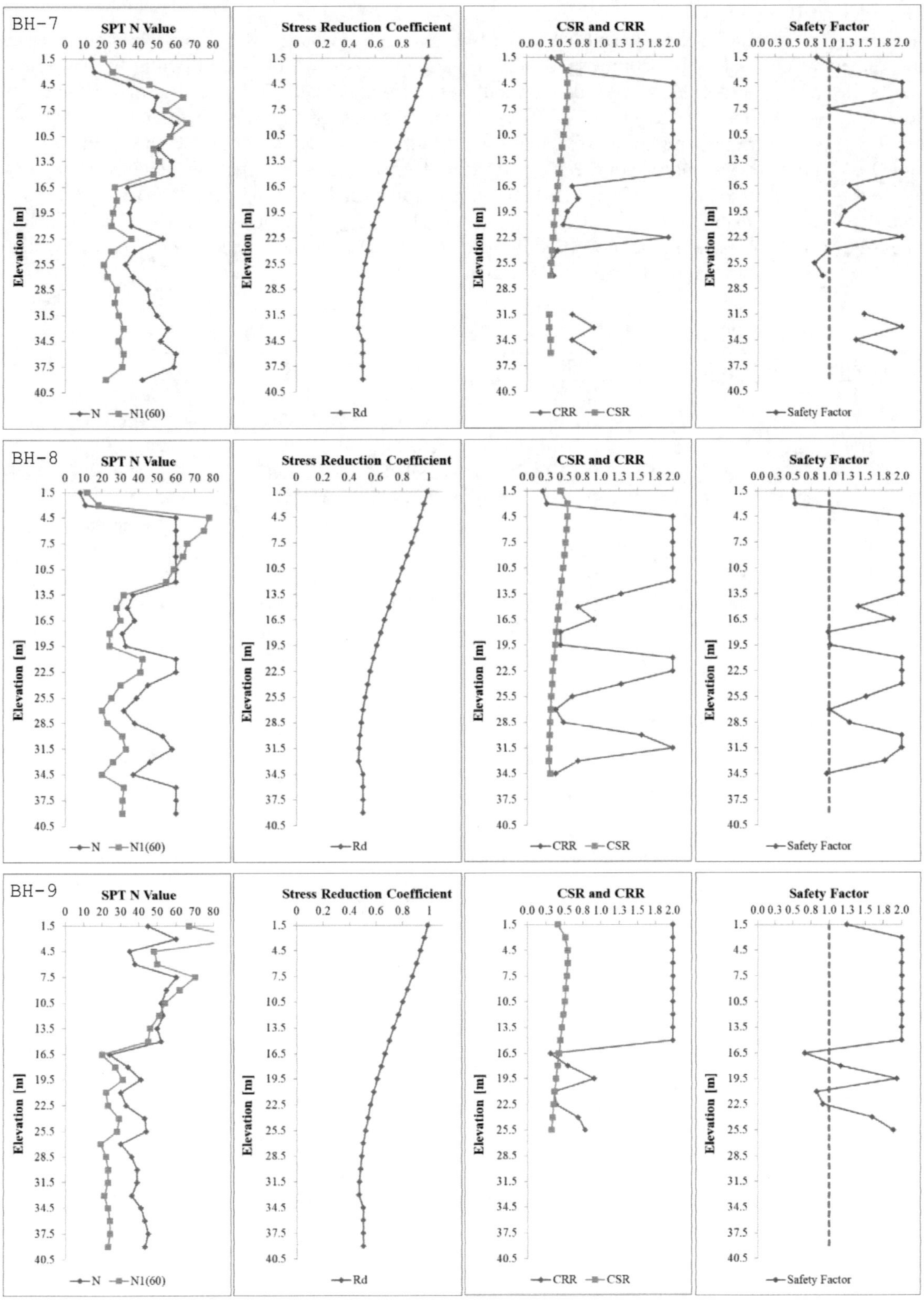

Figure 11. CSR, CRR, and safety factor at BH-7, BH-8, and BH-9 from Settle3 result

3. Result and Discussion

The Opak river estuary is generally composed of sandy soil with various densities, but mainly in a dense sand category. Very loose to medium sand only in the upper layer with 0 – 6 m depth. In the deeper layer, soil tends to be dense and very dense sand. Several boreholes in BH-1, BH-2, and BH-7 have clayey soil layer with various depth of 27 – 39 m. However, liquefied soil only occurs maximum depth of 20 m. This soil stratigraphy implies the analysis of liquefaction potential result, with seismic source from Opak fault with a magnitude of 6.2 M_w and PGA of 0.558 g, the layer that has potential to liquefy only in 0 – 6 m depth, except for BH-8 and BH-9, it has liquefied layer in 16.5 – 18.0 and 15 – 16.5 depth due to the inclusion of medium dense sand, and BH-4, it has no liquefied layer due the inclusion of dense to very dense sand in the most of soil layer. Settle3 can calculate precisely depending on data input sufficiently. The recapitulation result is shown in Table 6 below:

Table 6. Liquefaction potential analyzed result

Bore Hole	Depth (m)	CSR	CRR	SF	Potential Result
BH-1	0.0 – 1.5	0.358	0.147	0.452	Liquefied
	1.5 – 3.0	0.349	0.244	0.756	Liquefied
BH-2	0.0 – 1.5	0.358	0.258	0.794	Liquefied
	1.5 – 3.0	0.349	0.289	0.894	Liquefied
	3.0 – 4.5	0.339	0.289	0.881	Liquefied
BH-3	1.5 – 3.0	0.349	0.308	0.955	Liquefied
BH-4	0.0 – 1.5	0.358	0.781	>2.0	Non- liquefied
BH-5	1.5 – 3.0	0.590	0.487	0.909	Liquefied
	3.0 – 4.5	0.599	0.408	0.749	Liquefied
BH-6	0.0 – 1.5	0.439	0.220	0.550	Liquefied
	4.5 – 6.0	0.562	0.444	0.839	Liquefied
BH-7	0.0 – 1.5	0.434	0.328	0.831	Liquefied
BH-8	0.0 – 1.5	0.443	0.208	0.516	Liquefied
	1.5 – 3.0	0.535	0.258	0.530	Liquefied
	16.5 – 18.0	0.444	0.389	0.987	Liquefied
BH-9	15.0 – 16.5	0.419	0.308	0.666	Liquefied

4. Conclusion

The most crucial step before the liquefaction potential analysis is determining the value from the seismic source, moment magnitude, and PGA that implies the site. The seismic source and moment magnitude can be found in the authority's report. Still, the PGA value must be determined by the site condition, distance from the seismic source, mechanism fault, and infrastructure type that would develop (building, bridge, or dam). The result from the deterministic seismic hazard analysis method and non-linear earthquake site-response analysis method with logic tree framework is 0.558 g for maximum PGA value. This study used a simplified procedure by Idriss and Boulanger (2004) to estimate liquefaction potential. For practical purposes, this study used Settle3 as an application helping in liquefaction and settlement estimates. The Settle3 result showed that the Opak river estuary has liquefied soil potential indicated by BH-1 in a depth of 0 – 3 m, BH-2 in a depth of 0 – 4.5 m, BH-3 in a depth of 1.5 – 3 m, while BH-4 has no liquefied soil. Moreover, liquefied soil potentially found from BH-5 with depth between 1.5 – 4.5 m, BH-6 between 0 – 1.5 m and 4.5 – 6 m, BH-7 in between 1 – 1.5 m, BH-8 in between 0 – 3 m and 16.5 – 18 m, and also BH-9 in between 15 – 16.5 m. The layer containing liquefied soil must be adequately considered in designing the bearing capacity of the foundation system built in this area.

References

[1] Badan Geologi, 2019, Atlas Zona Kerentanan Likuefaksi Indonesia Bandung: Kementerian Energi Sumber Daya Mineral. In Bahasa.

[2] Pusat Gempabumi dan Tsunami, 2019, Katalog Gempabumi Signifikan dan Merusak 1821-2018

Jakarta: Badan Meteorologi Klimatologi dan Geofisika. In Bahasa.
[3] Satuan Kerja Pelaksanaan Jalan Nasional Wilayah Provinsi DIY, 2021, Uji Paritan Sesar Aktif Opak di Jembatan Kretek II , Yogyakarta, Sleman. In Bahasa.
[4] Jalil A Fathani T F Satyarno I and Wilopo W, 2021, Liquefaction in Palu: the cause of massive mudflows *Geoenvironmental Disasters* **8**, 1.
[5] Supartoyo, 2006, Gempabumi Yogyakarta Tanggal 27 Mei 2006 *Bul. Berk. Merapi* **3**, 2 p. 36–55. In Bahasa.
[6] Pusat Studi Gempa Nasional, 2017, Peta Sumber dan Bahaya Gempa Indonesia Tahun 2017 Jakarta: Kementerian Pekerjaan Umum dan Perumahan Rakyat. In Bahasa.
[7] Boore D M Stewart J P Seyhan E and Atkinson G M, 2014, NGA-West2 Equations for Predicting Response Spectral Accelerations for Shallow Crustal Earthquakes.
[8] Campbell K W and Bozorgnia Y, 2013, NGA-West2 Campbell-Bozorgnia Ground Motion Model for the Horizontal Components of PGA , PGV , Response Spectra for Periods Ranging from 0.01 to 10 sec, PEER Report 2013/06.
[9] Chiou B S J and Youngs R R, 2014, Update of the Chiou and Youngs NGA model for the average horizontal component of peak ground motion and response spectra.
[10] Mase L Z, 2019, Performance of NGA Models in Predicting Ground Motion Parameters of The Strong Earthquake *J. Civ. Eng. Forum* **5**, 3 p. 227.
[11] Mase L Z, 2017, Liquefaction potential analysis along coastal area of Bengkulu province due to the 2007 Mw 8.6 Bengkulu earthquake *J. Eng. Technol. Sci.* **49**, 6 p. 721–736.
[12] Nurdin S, 2021, Study Forensik terhadap Kerusakan Struktur Akibat Likuifaksi Pada Gempa Palu 2018 di Sulawesi Tengah, Palu. In Bahasa.
[13] Badan Standarisasi Nasional Indonesia, 2019, SNI 1726:2019 Tata cara perencanaan ketahanan gempa untuk struktur bangunan gedung dan nongedung, Jakarta. In Bahasa.
[14] Badan Standarisasi Nasional Indonesia, 2016, SNI 2833:2016 Perencanaan Jembatan Terhadap Beban Gempa, Jakarta. In Bahasa.
[15] Kramer S, 1996, Geotechnical Earthquake Engineering, New Jersey: Prentice-Hall, Inc.
[16] Muntafi Y Nojima N and Jamal A U, 2020, Damage Probability Assessment of Hospital Buildings in Yogyakarta, Indonesia as Essential Facility due to an Earthquake Scenario *J. Civ. Eng. Forum* **6**, 3 p. 225.
[17] Kanno T Narita A Morikawa N Fujiwara H and Fukushima Y, 2006, A new attenuation relation for strong ground motion in Japan based on recorded data *Bull. Seismol. Soc. Am.* **96**, 3 p. 879–897.
[18] Taruna R M Rohadi S Rudyanto A and Heryanto D T, 2018, Determining Ground Motion Prediction Equations (GMPEs) Using Euclidean and Likelihood Method for East Java *J. Meteorol. dan Geofis.* **17**, 3 p. 177–190.
[19] Bardet J and Tobita T, 2001, NERA : A Computer Program for Nonlinear Earthquake site Response Analyses of Department of Civil Engineering A Computer Program for N onlinear E arthquake site R esponse A nalyses of Layered Soil Deposits February.
[20] Fathani T F and Wilopo W, 2017, Seismic microzonation studies considering local site effects for Yogyakarta City, Indonesia *Int. J. GEOMATE* **12**, 32 p. 152–160.
[21] Pacific Earthquake Engineering Research Center, 2014, PEER Ground Motion Database. [Online]. Available: https://ngawest2.berkeley.edu/. [Accessed: 14-Feb-2022].
[22] Satuan Kerja Pelaksanaan Jalan Nasional Wilayah Provinsi DIY, 2021, Laporan Pekerjaan Lapangan Survey Downhole Seismik, Sleman. In Bahasa.
[23] Mase L Z Fathani T F and Adi A D, 2013, Studi Eksperimental Potensi Likuifaksi di Kali Opak Imogiri Provinsi Daerah Istimewa Yogyakarta *17th Annu. Sci. Meet.* April. In Bahasa.
[24] Google Earth, 2021, Google Earth Image Satellite - Kretek 2 Bridge Area. [Online]. Available: https://earth.google.com/web/. [Accessed: 05-Dec-2021].
[25] Bowles J E, 1997, Foundation Analysis and Design International Fifth Edition, Fifth Illnois: The McGraw-Hill Companies, Inc.
[26] Idriss I M and Boulanger R W, 2004, Evaluating The Potential for Liquefaction or Cyclic Failure of Silts and Clays *Neurosci. Lett.* **339**, December p. 123–126.

[27] Seed H B and Idriss I M, 1970, A Simplified Procedure for Evaluating Soil Liquefaction Potential November California: University of California.
[28] Maulana A Ahmad R and Fikri F, 2020, Liquefaction potential analysis on runway construction based on soil engineering properties *E3S Web Conf.* **156**, March.

Acknowledgments
The authors express their gratitude to Ministry of Public Work and Housing and Gadah Mada University for fully funded support to this research. The authors also would like to thank Unit Work of National Road Implementation (PJN) Special Region of Yogyakarta Province Territory, National Road Implementation Agency (BBPJN) of Central Java - Special Region of Yogyakarta Territory for site investigation data used in this study.

IOP Conf. Series: Earth and Environmental Science 1244 (2023) 012033 doi:10.1088/1755-1315/1244/1/012033

Dynamic Soil-Structure Interaction Through 2-D Site-Specific Response Analysis

W Sengara[1*, 3], R Sargawi[2], D Komerdevi[3] and A Sulaiman, [2]

[1]Professor, Civil and Environmental Engineering Faculty, Institut Teknologi Bandung, Bandung –Indonesia
[2]Researcher, Geotechnical Engineering Research Group, Civil and Environmental Engineering Faculty, Institut Teknologi Bandung, Bandung - Indonesia
[3]Senior Engineer, WSP Engineering Consultant, Bandung-Indonesia

*Corresponding email: wayansengara@itb.ac.id

Abstract. Seismic response of structures involved soil-structure-interaction (SSI) that need to consider both inertial and kinematic interaction. Many underground SSI has been developed such as modelling the structures and soils as continuum with seismic load simplified as pseudo-static or static equivalent. More recent and more realistic SSI model consider non-linear time-history analysis in the form of 2-D site-specific response analysis (SSRA), in which both inertial and kinematic interaction in combination with non-linear soil model, as well as time-history ground-motions are simulated. This paper presents simplified pseudo-static soil-structure interaction finite-element (FE) response of underground box-station structure under prescribed maximum strain displacement as resulted from 1-D site-specific response analysis under various ground-motions characteristics of subduction and shallow crustal earthquakes. More complex model of SSI system of basement is also presented. This basement SSI model incorporates non-linear time-history analysis with dynamic soil properties considered through 2-D SSRA simulating seismic wave propagation from reference base-rock to ground surface. Various base-rock seismic ground-motions are simulated from results of de-aggregation analysis from probabilistic seismic hazard analysis considering both subduction and shallow crustal earthquakes scaled at various ground-motion periods. The analysis adopts dynamic soil model with non-linear modulus and damping using FLAC computer program.

1. Introduction

Earthquake resistance design and construction of buildings and infrastructures includes enhancement of seismic design criteria through various process, from seismic hazard analysis, building-codes update, site-specific response analysis (SSRA), including selection of ground-motion and matching with the proposed site and structure spectrum. Analysis and design of structures require seismic design criteria that comply with seismic building codes and supporting guidelines for earthquake resistance of structures. Site-specific ground-motion procedures might be used to evaluate ground-motions for any structures (ASCE-7-2016;ASCE-7-2022,. Site-specific response analysis (SSRA) can be utilized through wave-propagation analysis from reference bedrock through soil profile to the ground surface. Appropriate computational techniques shall be adopted by treating the soil as a non-linear or equivalent-linear model. Having been analysed, the soil profile responses need to be evaluated and ground motion time histories at ground surface are calculated. As common practice the SSRA as also stated in the codes, the recommended ground-motions at ground surface resulted from wave propagation analysis need to represent deliberation from response sensitivity to uncertainty in assumed depth of soil model, soil non-linear properties, and variations in the input ground-motions.

SSRA analysis can be carried out by modelling the behaviour of the soil when the earthquake wave propagates from the bedrock to the ground surface, to get the respond then SSRA analyzed with model 1D. 1-D SSRA does not capture several features such as interaction between structure and buildings during earthquake. The effect of both inertial and kinematic interaction needs to be considered to provide more representative response to the system. 1-D waves propagation conducted to horizontally homogenous soil layers is subjected to free-field motions and do not consider non-horizonal profiles or geometry and do not consider presence of structures. If there are embankments, slopes, or structures at the ground surface, then 2-D SSRA is needed. If there are structures involved, then higher stiffness of the structure would affect the ground-motion response. To identify the influence of the structure response to the soil motions, the structure need to be modelled in 2-dimension analysis that include structural parameters in an integrated soil-structure-interaction analysis.

This paper presents simplified pseudo-static soil-structure interaction finite-element (FE) response of underground box-station structure under prescribed maximum strain displacement as resulted from 1-D site-specific response analysis under various ground-motions. More complex model of SSI system of basement is also presented. This basement SSI model incorporates non-linear time-history analysis with dynamic soil properties considered through 2-D SSRA simulating seismic wave propagation from reference base-rock to ground surface. Various base-rock seismic ground-motions are simulated from results of de-aggregation analysis from probabilistic seismic hazard analysis considering both subduction and shallow crustal earthquakes scaled at various ground-motion periods. The analysis adopts dynamic soil model with non-linear modulus and damping using FLAC computer program.

2. Analysis Methodology

2.1. One-Dimensional Site-Specific Response Analysis (SSRA)

1-D SSRA does not capture several features such as interaction between structure and buildings during earthquake. The effect of both inertial and kinematic interaction needs to be considered to provide more representative response to the system as identified by many previous studies (Sengara and Wijayanto, 2008). Modelling the soil-structural system in the SSRA through two-dimensional (2-D) wave propagation analysis is considered to better represent the soil-structural system compared to more common 1-D SSRA. This is with assumption that the 2-D SSRA need to be conducted with the same of similar soil constitutive modelling of 1-D currently available, where non-linearity in the soil layer moduli and damping ratios are represented in the modelling.

This paper presents dynamic soil-structure interaction through simplified and relatively new technique in 2-D consideration of practical SSRA. Methodology of SSRA and soil-structure interaction is described, including its nonlinear constitutive soil models. A finite-difference program is utilized with the purpose of assessing the importance of several aspects incorporating soil properties, geometry, and structural effects. Analysis comparing and matching 1-D and 2-D analysis in free-field conditions to calibrate several parameters, such as dynamic soil properties and influence of boundary conditions is presented. This paper then presents several case studies of 2-D SSRA with consider effect of interaction between soil structure. The analysis explores the variations in the ground-motions response through 1-D and 2-D analysis with various base ground-motions input.

$$\rho \frac{\partial^2 d}{\partial t^2} + \eta \frac{\partial d}{\partial t} = \frac{\partial \tau}{\partial z} \quad (1)$$

where $\boldsymbol{\rho}$ is the unit mass of soil; **d** is the displacement; $\boldsymbol{z}$ is the depth; $\boldsymbol{t}$ is the time; $\boldsymbol{\tau}$ is the shear stress; and $\boldsymbol{\eta}$ is a damping coefficient of mass proportional. Results of the analysis would be ground-motions time-history (acceleration, velocity, and displacement) at ground surface or at any level of interest.

Response spectra at ground surface representing each of the propagated base-rock ground motions could be developed as recommended spectra for design criteria of structures.

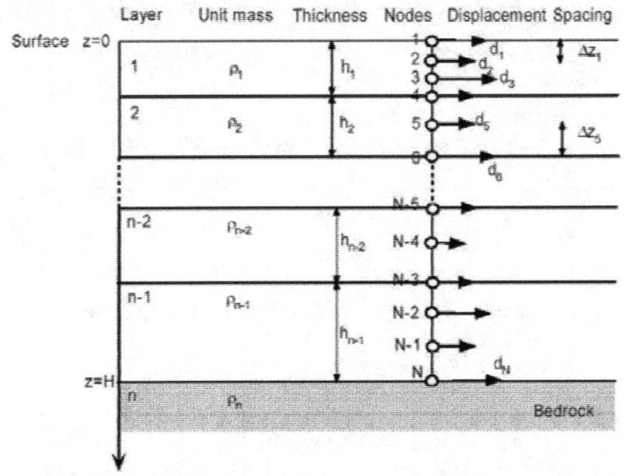

Figure 1. Illustration of 1-D Layered Soil Deposit System and its Spatial Discretization

2.2. Two-Dimensional Site-Specific Response Analysis

1-D waves propagation conducted to horizontally homogenous soil layers is subjected to free-field motions and do not consider non-horizonal profiles or geometry and do not consider presence of structures. If there are embankments, slopes, or structures at the ground surface, then 2-D SSRA is needed. If there are structures involved, then higher stiffness of the structure would affect the ground-motion response. To identify the influence of the structure response to the soil motions, the structure need to be modelled in 2-dimension analysis that include structural parameters in an integrated soil-structure-interaction analysis. Methodology in 2-D SSRA involves the main steps and procedures in a seismic effective stress analysis (Cubrinovski, 2011) shown in Figure 2.

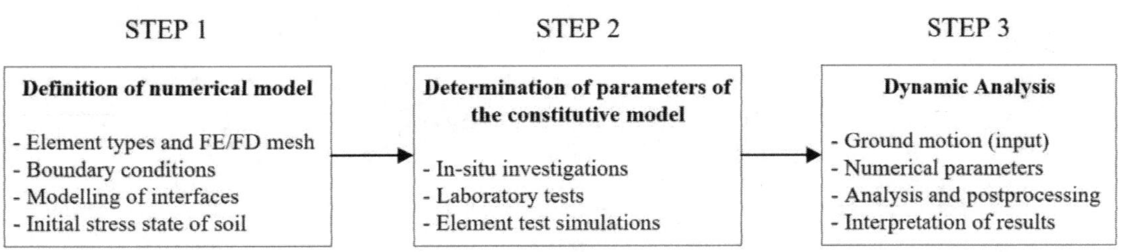

Figure 2. Main Steps and Procedures in Dynamic Analysis (Cubrinovski, 2011)

After both soil and structural geometry numerical model is discretized with the necessary boundary conditions definition are established, then initial or in-situ stress state need to be generated to determine the initial stresses in the soil due to gravity and calculate their effects on the stress-strain behaviour. Later, the model of constitutive parameters shall be identified utilizing data derived from primary sources such as field or soil laboratory test results. In the most cases, soil model parameters can be determined using formula correlation with the in-situ as well as laboratory tests. In some instances, calibration using an element simulation might be needed to evaluate the model parameters. Bes-fit values such as damping coefficients of the model parameters may be selected through simulation process.

As in the 1-D SSRA, an acceleration or velocity time-history should be defined and applied on the model base or reference base-rock. In the integration process, numerical parameters, and numerical damping should be used considering the geometry and soil behaviour. The 2-D dynamic analysis or analysis involving SSI then could be processed to obtain the dynamic response including soil-structure response. 2-D SSRA involving liquefaction potential could also be modelled adopting appropriate constitutive soil model that capable of modelling excess pore-water pressure to evaluate liquefaction potential and stability associated with the liquefactions.

3. Nonlinear Dynamic Soil Model

Two elasto-plastic dynamic models are described in this paper, namely Mohr-Coulomb and UBCHYST models. The Mohr-Coulomb model is relatively simpler with its model parameters already common in geotechnical engineering practice. The UBCHYST model on the other hand is considered relatively new and still rarely adopted in common practice since it is more complex in model formulation and with more numbers of model parameters. These two models are used in the 2-D SSRA involving SSI as case study in this paper.

3.1. Mohr Coulomb Model

The Mohr-Coulomb model is a simple model that evaluate the stress-strain of the soil as a linear elastic-perfectly plastic. Some engineers used this model in their analysis as a first soil behaviour approximation. The common two plastic parameters in the yield functions for the Mohr-Coulomb model are friction angle (φ) and the cohesion (c). The Mohr-Coulomb model supposes to input five parameters such as young's modulus, poisson's ratio, cohesion, friction angle, dilatancy angle. These parameters are generally common to geotechnical engineers and can be correlated from in-situ tests.

The Mohr-Coulomb models cannot capture either the dilation or the densification effect in the time of cycles. To modify this condition, the Mohr-Coulomb model is adopted using additional hysteretic damping. Therefore, the Mohr-Coulomb model was modified to model the changes of permanent volumetric strains in drained cyclic loading or pore pressures in undrained cyclic loading using Hysteretic Damping. The example of hysteretic damping is shown in the **Figure 3**.

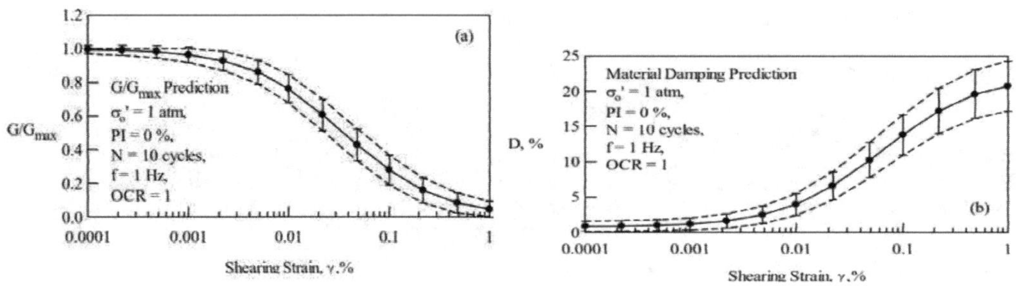

Figure 3. An Example of Normalized Shear Modulus and Material Damping Curves (Darendeli, 2001)

3.2. UBCHYST Model

UBCHYST was developed at the University of British Columbia for dynamic analyses of soil subjected to earthquake loading as the total stress model (Byrne and Naesgaard, 2010). The model can be used in soil that has low permeability such as clay or silt. This model may also be utilized in the granular soils that has a high permeability or in unsaturated granular soils. The model shall be evaluated using undrained shear strength. The model has been implemented in the 2D (two-dimensional) finite difference program FLAC (Itasca, 2011). The parameters associated with UBCHYST model are described in the literature (Byrne and Naesgaard, 2010). The model was derived using a calibration from cyclic simple shear laboratory test results to a published modulus reduction and damping curves.

However, the model typically calculates an overestimation of the damping responses at medium to large shear strains (>0.1%). The reason for this overestimation appeared because of the hysteresis loop width in the UBCHYST model. In this model, the shear modulus is calculated using a stress ratio function. The stress ratio magnitude is limited by a Mohr Coulomb failure criterion. The typical hysteretic stress-strain loop with the characteristic is illustrated in **4**.

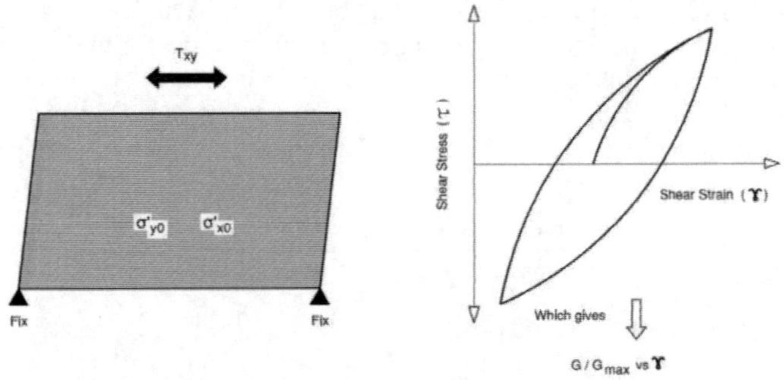

Figure 4. Typical hysteretic response of UBCHYST model and loop-size (damping) varying with shear strain.

4. Dynamic SSI and 2D SSRA Case Studies

4.1. Simplified SSI Through Racking Analysis of Underground Box Structures

Application of 1-D SSRA and simplified SSI through free-field racking deformation method and pseudo-static 2-D finite element analysis was carried out for the case of underground box station of MRT-Jakarta, as illustrated in **Figure 5.**

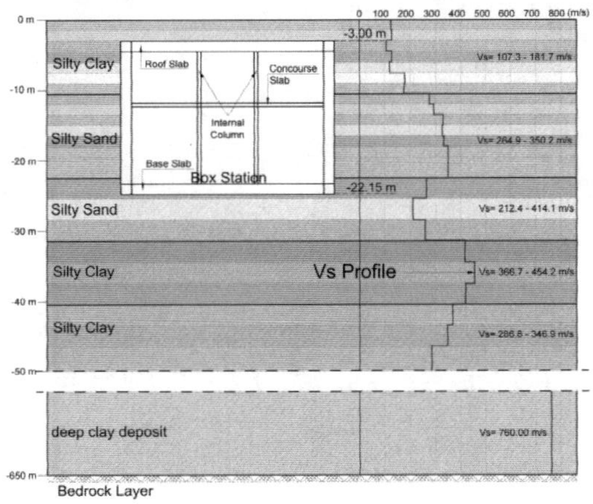

Figure 5. Soil Layer and Box-Station Profile of the Simplified SSI Case Study

The box station was designed and to be constructed on relatively soft soil layers underlain by relatively tuffaceous gravelly sand lenses on top of deep sedimentary stiff to hard clay of Jakarta formation to estimated bed-rock level of -650m from ground surface. The estimated bedrock of -650m from ground surface (Ridwan et al., 2019). The first analysis is to conduct typical 1-D SSRA through wave propagation analysis from bedrock to the ground surface. This analysis was conducted to provide maximum strain-displacement to be adopted in simplified SSI analysis.

The typical 1-D SSRA was conducted using NERA computer program. Eleven (11) earthquake ground-motions were generated at the reference base-rock, representing potential earthquake sources from both subduction and shallow crustal, as well as back-ground surround the site, as indicated from de-aggregation analysis within the probabilistic seismic hazard analysis of the specific site. The generated ground-motions were also scaled for several oscillatory periods of T=PGA, T=0.2 second and 1.0 second period to represent variation in the soil layers and structural periods of the box station. The 2500 years return period MCE is specified for the design of the box station. The Vs profile from ground surface to base-rock level of -650m was obtained from seismic downhole test for the upper 30m shallow layer in combination with the micro-tremor data for Mm layer to base-rock. The Vs-profile was also supported by several borehole N-SPT data correlated to Vs to complete continuous Vs profile data for the wave propagation analysis. Results of the analysis in the form of maximum strain displacement at the box-station elevation is presented in **Figure 6**.

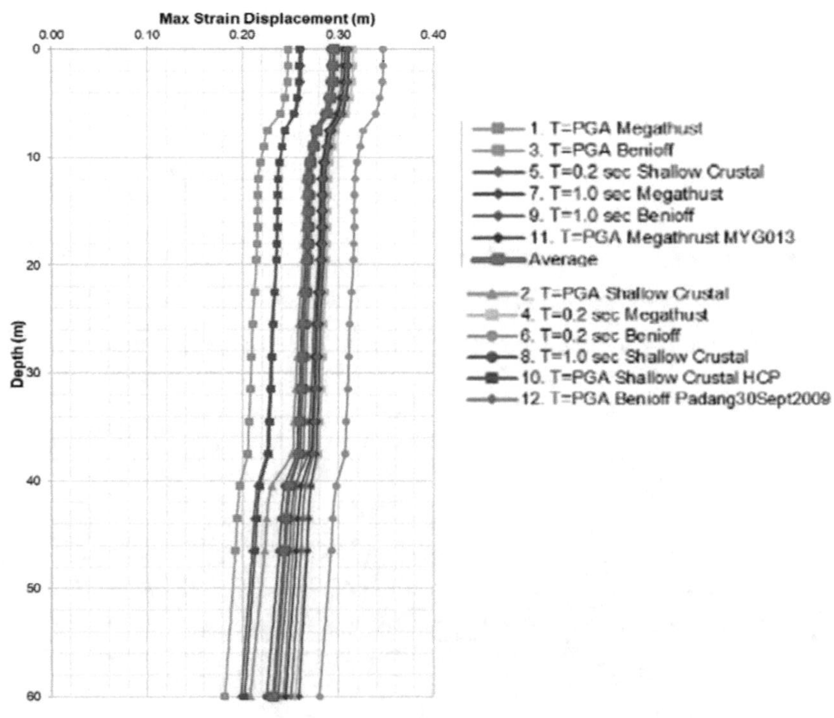

Figure 6. Maximum Strain Displacement Profile for 2% PE in 50 years

The maximum strain displacement response resulted from the 1-D SSRA case study shown in **Figure 6** is adopted to perform simplified SSI analysis of the box-station. In this case, pseudo-static analysis using Plaxis-2D finite element method has been carried out by prescribing 1-D SSRA free field line shear displacement responses at both side of the model boundary. This analysis incorporating cohesive soil layers adopting Hardening-Soil model under undrained condition. Geometry and the structural parameters of the box-station was assigned and the displacement response of the box-station is evaluated. 2-D geometrical model of the box-station under the 1-D prescribed displacement response is presented in **Figure 7.**

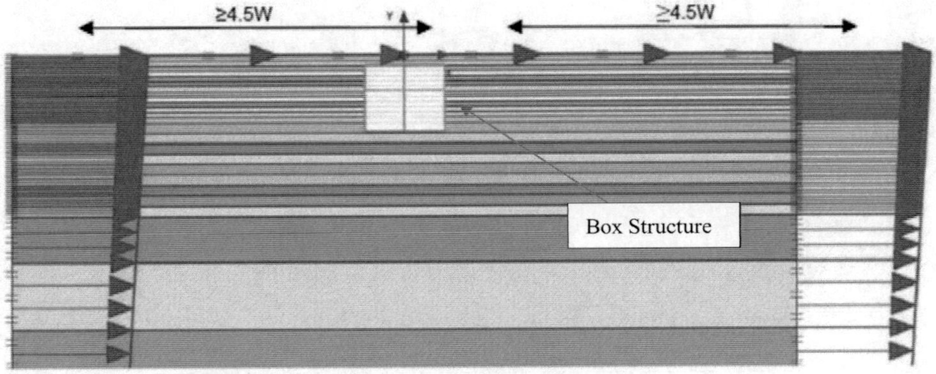

Figure 7. Geometrical Model and Deformation Response

Further analysis was conducted using free-field racking deformation method by calculating the racking ratio (R) to obtain structural deformation from the prescribed deformation input, in accordance with method proposed by Wang (Wang, 1993) and Hashash (Hashas, 2010), as illustrated in **Figure 8**. Five input motions representing different source mechanisms and periods within the prescribed shear displacements were conducted and variations on the box-structure deformation response, identifying various flexibility ratio R of < 1.0 showing that deformation of the underground box-station is relatively stiffer compared to that of the ground. In addition, F<1 shows that the surrounding soil is relatively soft, as also supported by results of the site class, namely soft soil (SE).

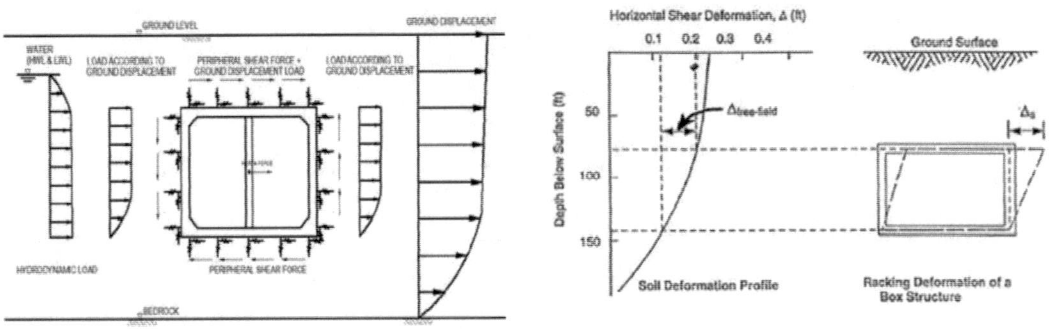

Figure 8. Illustration on Free-field Soil Shear Displacement Profile and Racking Deformation of a Box-Structure (Wang, 1993)

Comparisons on results of pseudo-static FEM model and racking deformation analysis have been made. The comparisons indicate that box-structure deformations using the racking free-field deformation method are more conservative compared to that of the pseudo-static deformation analysis. Both analyses conclude that the difference in the deformation results of does not exceed 20%. **Figure 9** shows comparison of structural deformation using free-field racking deformation method and those of the pseudo-static FEM analysis.

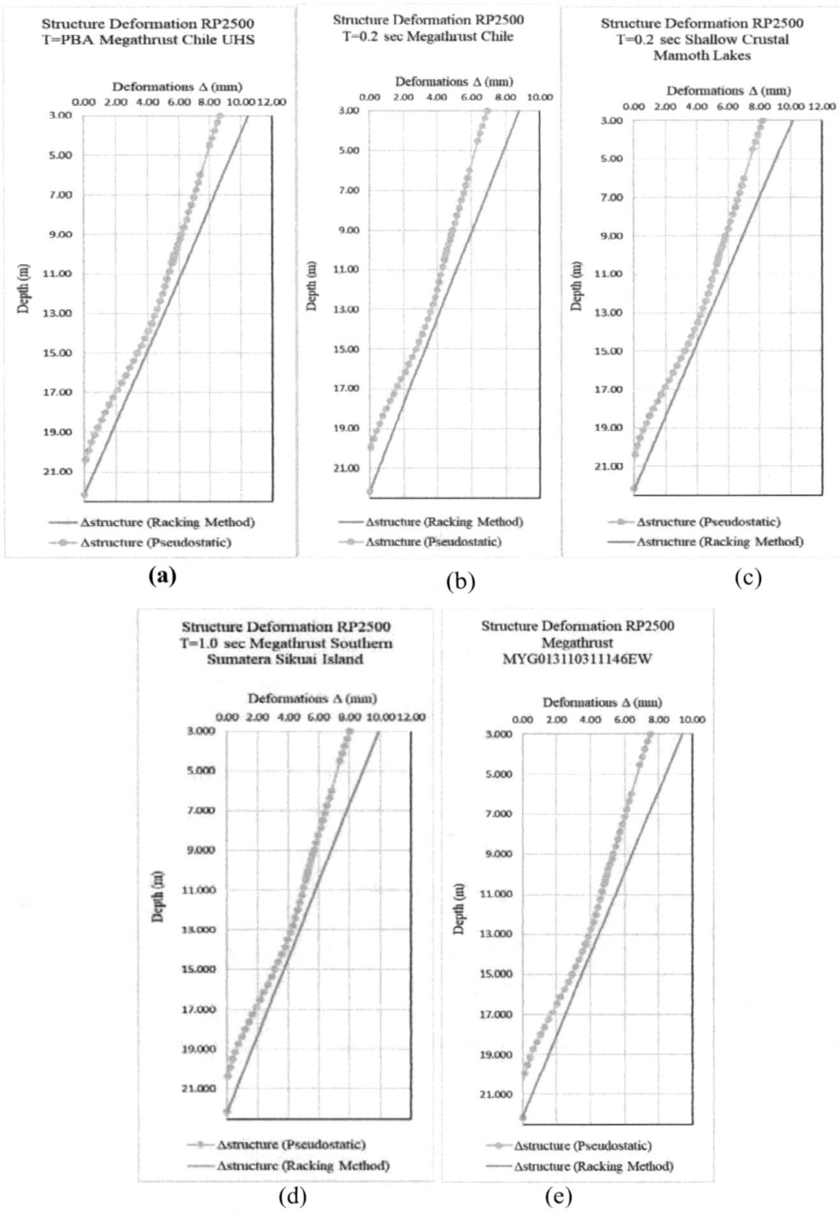

Figure 9. Comparison of Structural Deformation Using Free-Field Racking Deformation Method and FEM Pseudo-Static Analysis

4.2. 2-D SSI Non-Linear Dynamic Time-History Analysis

A case study on evaluating effect of basement structure of proposed high-rise building on spectral response at ground surface as well as at basement level surface, in comparison to that of 1-D free field typical SSRA is presented herein. **Figure 10** shows geometry of the basement, description of soil-layers and its shear wave velocity (Vs) profile to depth of reference base-rock. In this case study, two (2) earthquake probability levels are investigated, namely service level earthquake (SLE) representing 50% PE in 30 years, and risk-targeted maximum considered earthquake (MCE_R), as required in Indonesian seismic building codes SNI-1726-2019. Eleven (11) ground-motions have been selected at bedrock considering de-aggregation analysis of PSHA regarded to source contribution of seismic sources from Shallow Crustal, Megathrust and Benioff and evaluated the magnitude, soil condition, usable frequency of the ground motion, period/frequency sampling, spectral shape, and site-to-source distance that are close/similar to the de-aggregation result. The ground motions at bedrock also generated using maximum direction (RotD100) to provide maximum response based on structural period of building and scaling 0.2T to 2T.

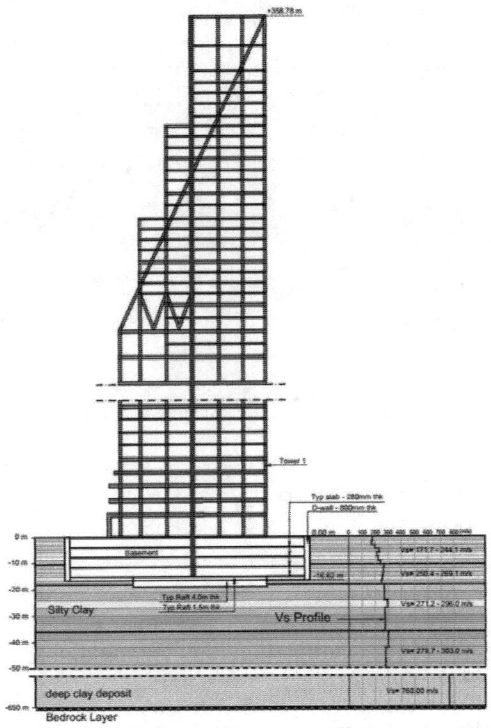

Figure 10. Shear Wave Velocity Data and Basement Geometry of 2-D SSI Case Study

Local site condition in the form of Vs profile is obtained from results of seismic downhole tests (SDT) for the top 30 meters. The overall Vs profile at the site for the whole considered depths is determined based on combination of the seismic downhole test (SDT) and also Vs correlation from N-SPT (Ohta & Goto 1978;Seed & Idriss 1981). secondary data of microtremor test at nearby site. Bedrock reference (S_{BC} with Vs=760 m/s) is assumed at depth of ±650 m (Ridwan et al., 2016). This Vs profile is used as input in the 1-D and 2-D wave propagation analysis from bedrock to the ground-surface using NERA and FLAC2D computer program, respectively.

Before conducting full-scale analysis involving inertial and kinematic basement effect, it is necessary to verify the spectral acceleration on the surface after earthquake excitation load from the bedrock. This analysis is referred to as free-field analysis, where the structure has not been installed and there is only the soil layer model. Free-field analysis is carried out to verify the vertical boundaries (quiet boundaries) and horizontal boundaries (free field boundaries). Additionally, the objective of this analysis is to calibrate between wave propagation results of the non-linear 1-D and the nonlinear 2-D analysis. In the free-field analysis, a horizontal soil column was modelled in 2-D plane-strain represented by a stack of rectangular soil element in vertical direction that has one unit length in horizontal direction. On the other hand, 1-D wave propagation analysis had previously been conducted using NERA computer program, as previously explained.

In this case study both 1-D and 2D SSRA are carried out. For 2-D SSRA, somehow a calibration model of 1-D and 2-D have been made so that similar response for free-field condition prior to 2-D SSI analysis has been achieved. For this calibration purposes, spectral accelerations at ground-surface resulted from 1-D NERA and 2-D FLAC2D are compared. The results have shown that the typical FLAC2D analysis provides quite similar pattern and very close agreement with that of non-linear 1-D NERA wave propagation results, as shown in **Figure 11 (a)** and **(b)** for both SLE and MCE_R earthquake, respectively. The calibration was conducted adopting scaled SLE Coastal Chile ground-motion and scaled MCE_R Takachi Oki ground-motion.

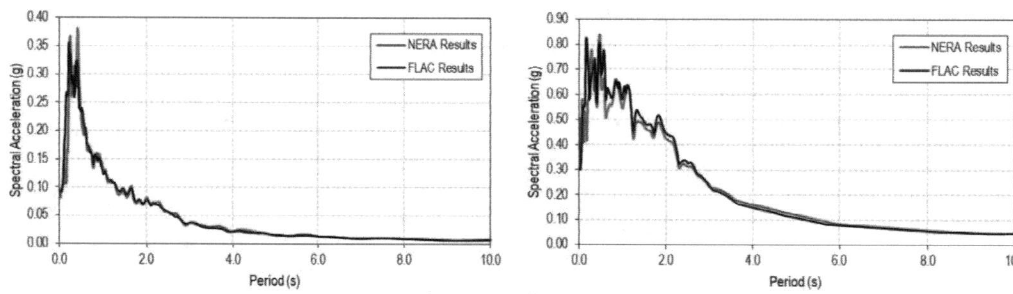

(a) **(b)**

Figure 11. Typical Comparison of Spectral Acceleration in Ground Surface between NERA and FLAC Results (a) for Scaled SLE and (b) for Scaled MCE_R

Further, SSI analysis is conducted with four-level basement structure was installed in the model after initial stresses were generated. Several structures such as the tower, typical slab, typical raft, shear wall, and D-Wall are modeled after the free-field analysis has been completed. The analysis was carried out in approximated plane-strain conditions. The parameters are already converted to the equivalent plane strain model from the 3D model according to the stiffness, spacing and dimension for each material. The approximate stiffnesses and the masses of diaphragm-wall and upper structural components have been included in the SSI model. In this analysis, the point locations of interests for spectral acceleration comparisons were selected to be at the bottom and the top right of basement. The detail of 2DFLAC mesh involving structural element are illustrated in **Figure 12**.

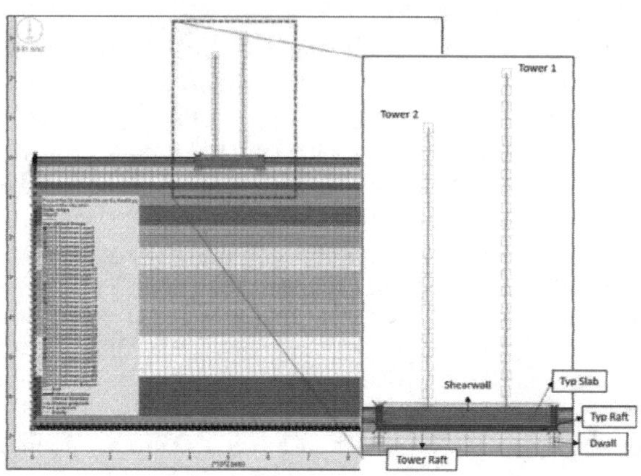

Figure 12. Detail Mesh Distribution and Model for 2 Dimensional SSRA including Kinematic Effect

Results of the analysis by identifying and evaluating the spectral response at the specified point locations of interest. Comparisons were made on the spectral response from free field and to those of SSI effect for each of the point locations. **Figure 13** shows the average spectra acceleration ratio between several interested points in 2D SSRA, i.e., free-field, top of basement, and bottom of basement and 1D SSRA response spectra for 11 ground motions.

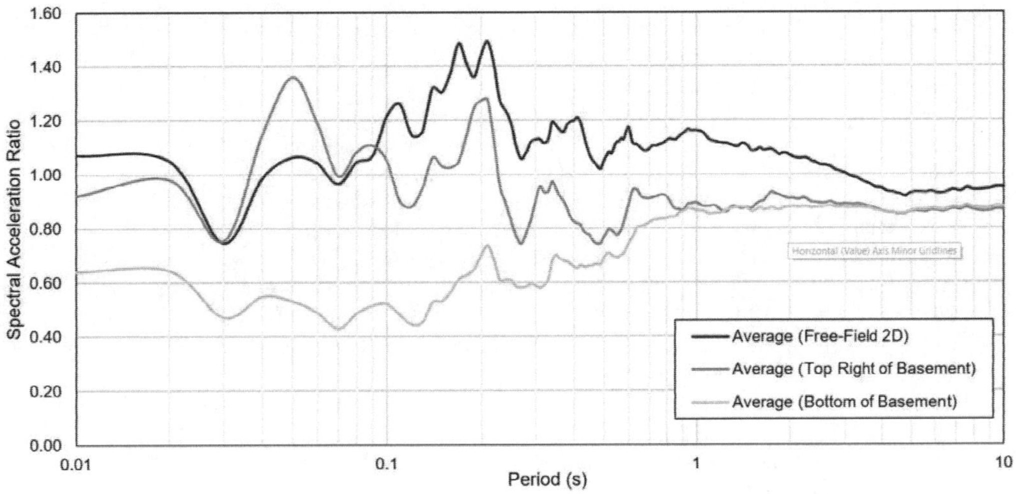

Figure 13. The Ratio of Average Spectral Acceleration for MCE_R Earthquake

5. Concluding Remarks

Dynamic soil-structure interaction and 2-D site-specific response analyses have been presented in this paper. The simplified SSI through pseudo-static and racking-deformation analysis of box-station structural model in combination with 1-D SSRA as free-field maximum shear strain-displacement prescribed in the structural model is considered to provide reasonable response of soil-structure response.

Two case studies of 2-D SSRA with SSI analysis provide comprehensive stress-deformation response of soil layers and structural system, implicitly incorporate inertial and kinematic effects of the soil-structural system. The first case study of SSRA through 2-D soil-basement-structure interaction model could be considered as an alternative method to better represent dynamic soil-structure response and better recommend design response spectra of structure. In addition, the effect of the soil-structure interaction system could provide optimization on its practical implications in non-linear time-history analysis of structural systems. Nevertheless, for practical purposes, some limitation requirements of the codes and guidelines may need to be considered and followed.

6. Acknowledgements

Assistance by Ms. Sarah Sibagariang her contribution in providing data analysis in this paper is highly appreciated. The authors also thankful to Dr. Yuamar Basarah during discussion on some analysis presented in this paper. Case geotechnical and project data provided by MRT Jakarta and Arup Consultant is appreciated. Other colleagues who have participated in the contribution in the supporting materials referred in this paper is also appreciated.

7. References

[1] American Society of Civil Engineer Standards, ASCE SEI 7-16. (2017). *Minimum Design Loads and Associated Criteria for Buildings and Other Structures*. Structural Engineering Institute ASCE, Virginia, USA.

[2] American Society of Civil Engineer Standards, ASCE SEI 7-22. (2022). *Minimum Design Loads and Associated Criteria for Buildings and Other Structures*. Structural Engineering Institute ASCE, Virginia, USA.

[3] Sengara, IW., Wijayanto, N. (2008). "Soil-Basement Structure Interaction Analysis On Dynamic Lateral Earth Pressure on Basement Wall." Proceeding International Conference on Earthquake Engineering and Disaster Mitigation-Bali.

[4] Sengara, IW., (2009), "Development of Earthquake Scenario for Three Different Sites within City of Jakarta." Earthquake Geotechnical Engineering Satellite Conference XVIIth International Conference on Soil Mechanics & Geotechnical Engineering, Alexandria, Egypt

[5] Komerdevi, D., and Sengara, IW., (2019). "Site-Specific Response Analysis and Pairs of Ground-Motions Time-History Generation of A Site in Jakarta." Proceeding 4th International Conference on Earthquake Engineering and Disaster Mitigation-Padang.

[6] Cubrinovski, M., Bray, J.D., Taylor, M.L., Giorgini, S., Bradley, B.A., Wotherspoon, L. and Zupan, J. (2011) "Soil Liquefaction Effects in the Central Business District During the February 2011 Christchurch Earthquake." Seismological Research Letters, Focused Issue on the 2011 Christchurch New Zealand Earthquake.

[7] Darendeli, M. B. (2001). *Development of a New Family of Normalized Modulus Reduction and Material Damping Curves*. Ph.D. Thesis, Univ.of Texas at Austin, Austin, TX.

[8] Byrne, P.M. and Naesgaard, E., (2010). Personal Communications.

[9] ITASCA. (2011). *Manual FLAC Version 7.0*, ITASCA Company.

[10] Ridwan, M., Widiyantoro, S., Irsyam, M. (2016). "Development of an Engineering Bedrock Map Beneath Jakarta Based on Microtremor Array Measurements." Geological Society, London, Special Publications, Vol. 441, pp. 153-165.

[11] Wang, J.N., (1993). *Seismic Design of Tunnels*: *A State-of-the-Art Approach, Monograph*, Monograph 7. Parsons, Brinckerhoff, Quade and Douglas Inc, New York.

[12] Hashash, Y. M. A., Hook, J. J., Schmidt, B. and Yao, J.C., (2001). "Seismic Design and Analysis of Underground Structures." Tunneling and Underground Space Technology Journal, Vol 16, 247-293.

[13] Badan Standar Nasional Indonesia. (2019). *Standar Nasional Indonesia 1726:2019*: Tata Cara Perencanaan Ketahanan Gempa untuk Struktur Bangunan Gedung dan Non-Gedung. Jakarta: Badan Standarisasi Nasional.

[14] Ohta Y., Goto N. (1978). "Empirical Shear Wave Velocity Equations in Terms of Characteristic Soil Indexes." Earthquake Engineering and Structural Dynamics. 6:167–187

[15] Seed H.B., Idriss I.M. (1981). "Evaluation of liquefaction potential of sand deposits based on observations and performance in previous earthquakes, Preprint No. 81–544, In situ testing to evaluate liquefaction susceptibility, Session No. 24, ASCE Annual Conference, St. Louis, Missouri

[16] Ellison KC, Masroor AM, Almufti I, Willford M, O'Riordan N. (2015). "Structure-Soil-Structure Interaction Analysis in a Highly Seismic, Dense Urban Regeneration Zone." 6th International Conference on Earthquake Geotechnical Engineering, Christchurch, New Zealand.

[17] Shiming and Gang, (1998). *Dynamic Soil-Structure Interaction Current Research in China and Switzerland*, pp. 203-216.

[18] Kramer, S., *Geotechnical Earthquake Engineering*, Upper Saddle River, New Jersey: Prentice Hall (1996).

[19] Pacific Earthquake Engineering Research Center (PEER). (2017). *Guidelines for Performance-Based Seismic Design of Tall Buildings Version 2.03*.

Seismic Load Design of Rukoh Dam Suppletion Tunnel in Pidie Regency, Aceh, Indonesia, Based on Deterministic and Probabilistic Seismic Hazard Analyses

M R Nuraulia[1, 2], I G B Indrawan[1], F Aldiamar[2]

[1]Department of Geological Engineering, Universitas Gadjah Mada, Yogyakarta, Indonesia
[2]Implementation Unit for Geotechnics, Tunnels, and Structures, Directorate General of Highways, Ministry of Public Works and Housing, Indonesia

Email: igbindrawan@ugm.ac.id

Abstract. The Rukoh Dam suppletion tunnel construction site in Pidie Regency, Aceh province, is located 5.6 kilometers from the Seulimeum-South segment of the active Sumatran fault. The active fault features a strike-slip fault mechanism with a slip rate of 7.0 millimeters per year and a maximum earthquake-recorded magnitude of 6.9 Mw. Considering that the construction site is located less than 10 kilometers from the active fault, which may cause earthquakes with a magnitude of at least 6 Mw, the SNI 1729:2019, which describes the procedures for planning earthquake resistance for buildings and non-building structures, requires the seismic load to be designed based on the site-specific responses analysis. Determination of the seismic load using a deterministic seismic hazard analysis was the objective of this study. The seismic load based on the probabilistic seismic hazard analysis by Pusat Studi Gempa Nasional (PusGen) in 2017 is also compared. The 2017 Indonesian Earthquake Source and Hazard Map were used to find out about the fault mechanism, slip rate, deterministic magnitude, orientation, and coordinates of traces. The deterministic seismic hazard analysis was conducted using EZ-FRISK 8.07 software and attenuation equations with shallow crustal earthquake sources for fault and shallow background earthquake sources, i.e., the Boore-Atkinson NGA 2014, the Campbell-Bozorgnia NGA 2014, and the Chiou-Youngs NGA 2014 equations. The deterministic analysis based on those attenuation equations predicted the active fault to result in a peak ground acceleration (PGA) value of 0.66 g at the construction site of the Rukoh Dam suppletion tunnel. In comparison, the probabilistic hazard analysis conducted by Pusat Studi Gempa Nasional (PusGen) in 2017 indicated a peak ground acceleration (PGA) of 0.537 g. Therefore, the seismic load is recommended for the structural designs of the embankment dam and the other infrastructures, particularly the suppletion tunnel.

Keywords: active fault, deterministic seismic hazard analysis, peak ground acceleration, Pidie regency, tunnel

1. Introduction

Since damage to such structures causes significant financial loss and environmental risks, monitoring seismic risks at key infrastructure locations, including dams, tunnels, and other infrastructure, is essential in earthquake engineering. Deterministic seismic hazard analysis (DSHA) and probabilistic seismic hazard analysis (PSHA) are two well-known methods for measuring seismic hazards. Both techniques follow a four-step procedure, with identical first two phases [1]. First, the deterministic technique evaluates the ground-shaking hazard based on the controlling source, which provides the most significant threat among all potential sources at each given position. In order to calculate the hazard with a given likelihood of exceeding a specific period, the probabilistic method accounts for the mistakes associated with estimating earthquake size, location, and ground motion [2].

Near-fault sites must fit one of the following criteria: they must be within 15 km of a known active fault surface projection capable of generating Mw 7 or more significant earthquake occurrences or within 10 km of a known active fault surface projection capable of producing Mw 6 or more significant earthquakes [3]. This study aims to determine the peak ground acceleration in the Rukoh dam suppletion tunnel using deterministic and probabilistic seismic hazard analysis.

2. Study Area: Geology and Seismicity

The research location is in the Rukoh Dam suppletion tunnel, which is administratively located in Panton Beunot Village, Tiro/Truseb District, Pidie Regency, Aceh Province, Indonesia. Geographically, it is situated in the northern part of Aceh mountainous block at coordinates 95°56'1.159" E 5°9'28.504" N to 95°55'51275" E 5°9'46,609" N as indicated by the green dot in Figure 1. The lithological group of the area is dominated by Middle to Late Miocene mudstone and calcareous siltstone, with some minor sandstone from the Kotabakti Formation.

Seismotectonically, the construction site of the suppletion tunnel area falls in the Sumatran fault zone with a length of 1900 km and is located across the backbone of the island of Sumatra, which grows in or near an active volcanic arc [4][5][6][7]. Therefore, the Nicobar, Seulimeum-North, Aceh-North, Pidie, and Lhokseumawe parts of the Sumatran Fault have been identified and are situated in the northernmost region of Sumatra Island. On the other hand, Seulimeum-South, Aceh Central, and Aceh-South are parts of the Aceh region's southern section (Figure 1).

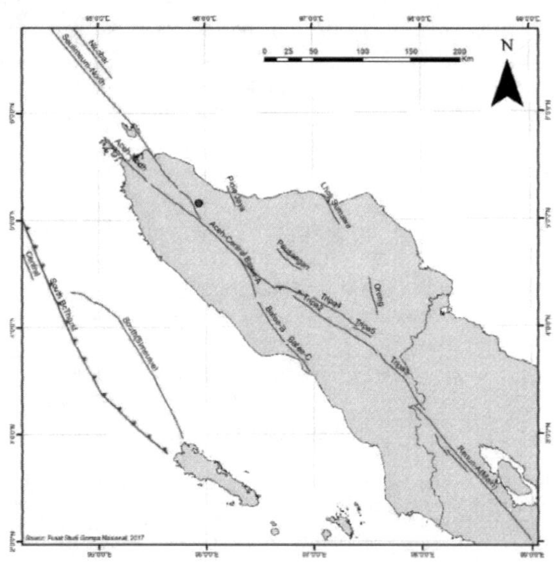

Figure 1. Study area map to active faults segment [8]

3. Methodology

When using the deterministic approach, a scenario assuming an earthquake of a specific magnitude will happen at a predefined place is considered. This method is generally used to calculate seismic acceleration in planning important strategic buildings, considering that it will be hazardous if damage occurs due to earthquake shocks. The deterministic method is a four-stage process, such as the following [1][2] (Figure 2).

1. Locating and identifying all earthquake sources in a place that might cause considerable ground motion is crucial. Each source geometry and potential are defined as part of the characterization of the earthquake source.
2. Establish the specifications for the distances between the earthquake source and the research location for each earthquake source zone. The deterministic approach is then used to determine the closest distance between the earthquake source zone and the study site. The distance used could be the distance from the epicenter of the hypocenter, depending on the empirical equation employed.
3. By analyzing the magnitude of the shock caused by the earthquake at a specific distance and place, one can choose the controlling earthquake that is anticipated to create the most shock. The magnitude and distance from the research location characterize a controlling earthquake.
4. Hazards that arise at a site owing to earthquake control in the form of ground motion (controlling earthquake). One or more ground motion parameters are derived from empirical equations to explain these properties. Peak acceleration, peak velocity, and response spectrum ordinates provide an estimation of the parameters of the earthquake hazard. Expert subjective assessments are always considered when applying the deterministic approach to assess the likelihood of seismic threats. This may occasionally make it difficult to agree with the possibility of an earthquake.

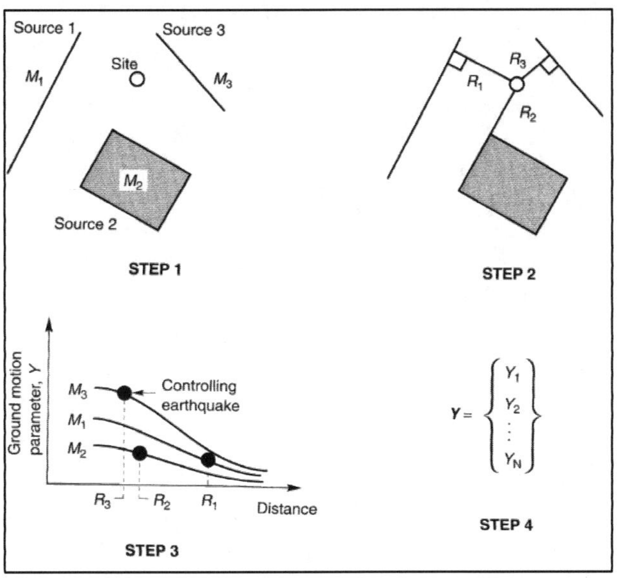

Figure 2. The stages of the deterministic seismic hazard method [1][2]

4. Earthquake Data

A comprehensive earthquake catalog is needed for deterministic methods to determine the largest earthquake sources that could occur. The information for this study was contributed by earlier researchers, particularly those in Sumatra. The mechanism or type of fault, dip, length, maximum magnitude, and slip rate are among the information provided. Table 1 summarizes these findings [8].

Table 1. Data and parameters for earthquake fault sources in Sumatra and surrounding areas [8]

No.	Structure Name		Type	Dip	Length (km)	Mmax (Mw)	Sliprate (mm/year)
	Main	Segment					
1	West Andaman Fault	North-West	SS	~90	478	7.9	5
2	West Andaman Fault	North-East	SS	~90	669	8.0	5
3	West Andaman Fault	Central	SS	~90	168	7.3	10
4	West Andaman Fault	South-Simelue	SS	~90	196	7.8	10
5	Sumatran Fault	Oreng Fault	SS	~90	36	6.9	1
6	Sumatran Fault	Lhok Sumawe	SS	~90	36	6.5	1
7	Sumatran Fault	Nicobar	SS	~90	120	7.1	18
8	Sumatran Fault	Seulimeum-North	SS	~90	143	7.7	18
9	Sumatran Fault	Seulimeum-South	SS	~90	38	6.9	8
10	Sumatran Fault	Aceh-North	SS	~90	65	7.2	2
11	Sumatran Fault	Aceh-Central	SS	~90	140	7.6	14
12	Sumatran Fault	Aceh-South	SS	~90	27	6.7	16
13	Sumatran Fault	Pidie Jaya	SS	~90	17	6.5	1
14	Takengon Fault	Lot Tawar	SS	~91	21	6.6	0.5
15	Takengon Fault	Peusangan	SS	~92	32	6.8	0.5
16	Sumatran Fault	Tripa1-Thrust	SS	45W	32	6.8	16
17	Sumatran Fault	Tripa 2	SS	~90	103	7.4	6
18	Sumatran Fault	Tripa 3	SS	~90	75	7.3	12
19	Sumatran Fault	Tripa 4	SS	~90	55	7.1	6
20	Sumatran Fault	Tripa 5	SS	~90	21	6.6	6
21	Batee Fault	Batee-A	SS	~90	44	7.0	6
22	Batee Fault	Batee-B	SS	~90	50	6.5	0.5
23	Batee Fault	Batee-C	SS	~90	37	6.9	0.1

5. Ground Motion Prediction Equation

The ground motion prediction function employed in Indonesia is separated into three categories depending on the earthquake source mechanism: subduction zone earthquake source, transform fault zone, and the equation produced based on the depth restriction of the earthquake focus. The following is the ground motion prediction function utilized in this research for each earthquake source. Shallow crustal earthquake sources, for fault and shallow background earthquake sources:

- Boore-Atkinson Ground Motion Prediction Equation NGA 2014 [9]
- Campbell-Bozorgnia Ground Motion Prediction Equation NGA 2014 [10]
- Chiou-Youngs Ground Motion Prediction Equation NGA 2014 [11]

6. Logic Tree Framework and Earthquake Scenarios

The earliest application of logic trees was in the probabilistic seismic hazard analysis [12]. They were first developed as a technique for capturing and quantifying the uncertainties related to the inputs required to carry out such an analysis, and they are now a typical component of the probabilistic seismic hazard analysis [1][13]. Despite the concept's widespread use, there are currently no established guidelines for building logic trees and assigning branch weights. Although the probabilistic method is where logic trees are most frequently employed, the deterministic method can also use logic trees [14].

The weighting of each parameter calculation function is determined before the logic tree is calculated. Each calculation function's output is assigned a particular weight. The weighted value represents the amount of confidence in the parameters in question. The final result, the total of the different weight values, may be calculated using the weight value. When all parameters are added together, a value of 1 is achieved. The final result is calculated by adding the weighted results of each parameter. The logic tree that was employed is shown in Table 2 below.

Table 2. Research logic tree based on the faults earthquake source [8]

Fault Models		Magnitude Uncertainity		Ground Motion Models	
Faults Trace	Characteristic	$M_{max} - 0.2$	0.2	Boore-Atkinson NGA 2014	0.33
	0.66	M_{max}	0.6	Campbell-Bozorgnia NGA 2014	0.33
		$M_{max} + 2$	0.2	Chiou-Youngs NGA 2014	0.33
	Gutenberg Richter	$M6.5 - (M_{max} - 0.2)$	0.2	Boore-Atkinson NGA 2014	0.33
	0.34	$M6.5 - M_{max}$	0.6	Campbell-Bozorgnia NGA 2014	0.33
		$M6.5 - (M_{max} + 2)$	0.2	Chiou-Youngs NGA 2014	0.33

The suppletion tunnel at the Rukoh Dam site has an interplate seismic control area (shallow crustal earthquake). According to Table 1, the Sumatran Fault Zone, which primarily controls the research's earthquake source, extends up to 500 km from the study area and features a strike-slip mechanism.

In this study, the fault modeling employed characteristic criteria that consider the slip rate and Gutenberg Richter criteria that consider the activity value. The earthquake scenario used a logic tree in line with Table 2. By adding weights to the logic tree, it is possible to measure how uncertain the size of the earthquake is.

7. Peak Ground Acceleration Calculations

Data of the peak ground acceleration (PGA), short-term spectral acceleration at period 0.2 s (S_S), and spectral acceleration at period 1 s (S_1) provided in the Source and Earthquake Hazard Map 2017 [8] were calculated by the probabilistic seismic hazard method. The outcomes of the calculations made in this study utilizing the deterministic seismic hazard analysis method are then compared to them. The deterministic seismic hazard analysis was conducted using EZ-FRISK 8.07 software. Table 3 compares calculating peak ground acceleration values using various ground motion prediction equations, whereas Table 4 compares calculating peak ground acceleration using probabilistic and deterministic seismic hazard analyses.

IOP Conf. Series: Earth and Environmental Science 1244 (2023) 012034 doi:10.1088/1755-1315/1244/1/012034

Table 3. Comparison of calculating peak ground acceleration values with three ground motion prediction equation

GMPE	PGA	S_S (g)	S_1
Boore-Atkinson NGA 2014 [9]	0.590	1.440	0.590
Campbell-Bozorgnia NGA 2014 [10]	0.783	1.707	0.610
Chiou-Youngs NGA 2014 [11]	0.600	1.550	0.520

PGA = peak ground acceleration / spectral acceleration at period 0.01 s
S_S = short-term spectral acceleration at period 0.2 s
S_1 = spectral acceleration at period 1 s

Table 4. Comparison of the peak ground acceleration calculation between probabilistic [8] and deterministic method

Analysis	PGA	S_S (g)	S_1
Probabilistic method	0.537	1.291	0.475
Deterministic method	0.660	1.566	0.574

PGA = peak ground acceleration / spectral acceleration at period 0.01 s
S_S = short-term spectral acceleration at period 0.2 s
S_1 = spectral acceleration at period 1 s

8. Response Spectra Study Area

The Rukoh Dam suppletion tunnel estimated bedrock-level 84th percentile deterministic response spectra have a 5% dampening. The three ground motion prediction equations above are used to calculate the PGA value. The calculation is based on research conducted by each ground motion prediction equation developer in various areas. The Campbell-Bozorgnia NGA 2014 yields the calculation of the most significant PGA value among the three ground motion prediction equations employed. The Boore-Atkinson NGA 2014, on the other hand, yields the lowest result. Therefore, the Chiou-Youngs NGA 2014 is quite near to the calculation's average outcome (Figure 3). The most significant variations are seen in situations where the data for these three attenuation functions is sparse, such as for close-proximity large earthquakes (Mw 8) and hanging-wall locations over rupture planes of shallow dipping earthquakes [15]. This is because the deterministic method always considers the worst-case scenario of an earthquake event, while the probabilistic method will yield a lower value because of the uncertainty surrounding the possibility of an earthquake [16][17]. The results of the deterministic method, which was used to figure out the PGA value, are compared with the chart in Figure 4 of the response spectra, which was used by the probabilistic method [8] to figure out the PGA.

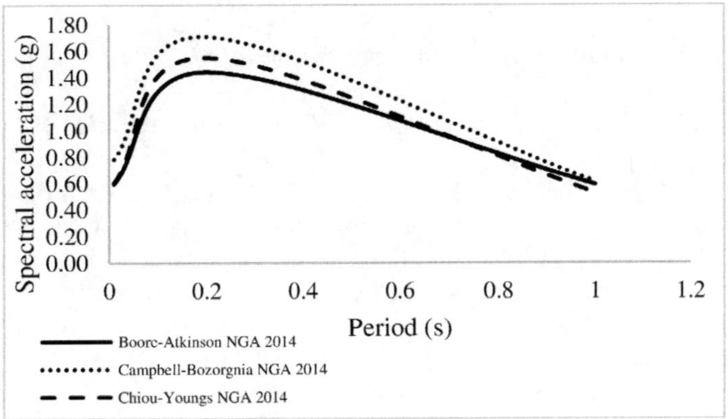

Figure 3. Comparison table of the response spectra for each of the ground motion prediction equations utilized in this study

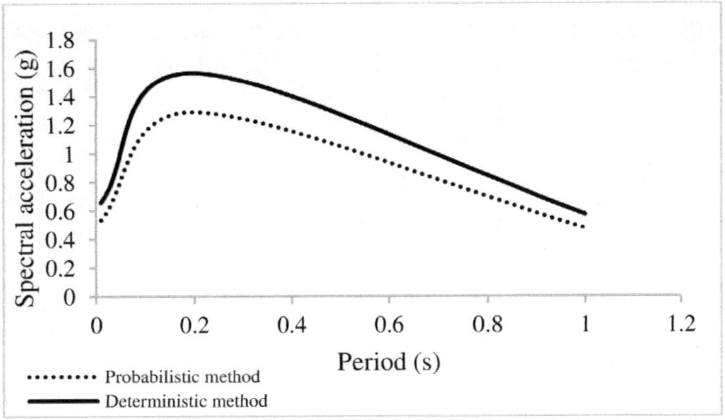

Figure 4. Comparison chart of the response spectra for probabilistic and deterministic method

9. Conclusions

The Rukoh Dam suppletion tunnel's seismic hazard is quantified using a deterministic approach that considers three scenarios for typical fault models and Gutenberg Richter models. The logic tree approach handles epistemic uncertainty by incorporating alternative theories. The results are given in terms of spectrum acceleration, which shows the variance of the maximum PGA. The PGA value is 0.66 g, according to the calculations from the deterministic method. Compared to the calculation with the probabilistic method that had been done earlier with a value of 0.537 g, the deterministic method applied in this study predicted higher values of PGA and spectral accelerations at 0.2 s and 1 s.

Acknowledgments

The use of the EZ-FRISK 8.07 software in this study was made possible by the Implementation Unit for Geotechnics, Tunnel, and Structures, Directorate General of Highways, Ministry of Public Works and Housing, Republic of Indonesia, for which the authors are grateful. The authors acknowledge the editor and anonymous reviewer's incisive critique and attentive reading of this paper, which significantly elevated the quality.

References

[1] Reiter, L. (1990). *Earthquake Hazard Analysis: Issues and Insights*. Columbia University Press, New York.

[2] Kramer, S. L. (1996). *Geotechnical Earthquake Engineering*. Prentice-Hall Inc., Upper Suddle River, New Jersey.

[3] Badan Standarisasi Nasional (2019). *SNI 1729:2019 tentang Tata cara perencanaan ketahanan gempa untuk struktur bangunan gedung dan nongedung*. Badan Standarisasi Nasional, Jakarta.

[4] Katili, J. A. and F. Hehuwat (1967). "On the occurrence of large transcurrent faults in Sumatra, Indonesia." *Journal of Geoscience, Osaka City University,* 10: p. 5-17.

[5] Bellier, O. et al. (1997). "Paleoseismicity and seismic hazard along the Great Sumatran fault (Indonesia)." *Journal of Geodynamics,* 24 (1-4): p. 169-183.

[6] Sieh, K. and Natawidjaja, D. (2000). "Neotectonics of the Sumatran fault, Indonesia." *Journal of Geophysical Research,* 105(B12): p. 28295-28326.

[7] Natawidjaja, D. (2001). "Large active faults along the Sumatran plate margin and their seismic threat to Indonesia, Malaysia, and Singapore." In *GEOSEA-Indonesian Association of Geologist GEOSEA 2001, Yogyakarta, Indonesia*.

[8] Pusat Studi Gempa Nasional (2017). *Peta Sumber dan Bahaya Gempa Nasional Tahun 2017.* Pusat Penelitian dan Pengembangan Perumahan dan Permukiman, Bandung.

[9] Boore, D. M., Stewart, J. P., Seyhan, E., and Atkinson, G. M. (2014). "NGA-West2 Equations for Predicting PGA, PGV, and 5% Damped PSA for Shallow Crustal Earthquakes." *Earthquake Spectra,* 30(3): p. 1057-1085.

[10] Campbell, K. W. and Bozorgnia, Y. (2014). "NGA-West2 Ground Motion Model for the Average Horizontal Components of PGA, PGV, and 5% Damped Linear Acceleration Response Spectra" *Earthquake Spectra,* 30(3): p. 1087-1115.

[11] Chiou, B.S.-J. and Youngs, R. R. (2014). "Update of the Chiou and Youngs NGA Model for the Average Horizontal Component of Peak Ground Motion and Response Spectra" *Earthquake Spectra,* 30(3): p. 1117-1153.

[12] Kukarni, R. B., R. R. Youngs, and K. J. Coppersmith (1984). "Assessment of confidence intervals for results of seismic hazard analysis" *Proceedings of Eighth World Conference on Earthquake Engineering*. 1: p. 263-270.

[13] Coppersmith, K. J., and R.R. Youngs (1986). Capturing uncertainty in probabilistic seismic hazard assessments within intraplate tectonic environments, in *Proceedings of the Third U.S. National Conference on Earthquake Engineering*. 1: p. 301–312.

[14] Bommer, J. J., Scherbaum, F., Bungum, H., Cotton, F., Sabetta, F., and Abrahamson, N. A. (2005). "On the Use of Logic Trees for Ground-Motion Prediction Equations in Seismic-Hazard Analysis." *Bulleting of the Seismological Society of America.* 95(2): p. 377-389.

[15] Gregor, N., Abrahamson, N.A., Atkinson, G.M., Boore, D.M., Bozorgnia, Y., Campbell, K.W., Chiou, B.S.J., Idriss, I.M., Kamai, R., Seyhan, E., Silva, W., Stewart, J.P. & Youngs, R. (2014). "Comparison of NGA-West2 GMPEs." *Earthquake Spectra*, 30(3): p. 1179–1197.

[16] Pailoplee, S., Sugiyama, Y. & Charusiri, P. (2009). "Deterministic and probabilistic seismic hazard analyses in Thailand and adjacent areas using active fault data." *Earth, Planets and Spaces*, 61: p. 1313–1325.

[17] Bakhshi, H., Alavi, S., Rezaie, Z., Hassani, M. (2022). "Deterministic and Probabilistic Seismic Hazard Analysis of Petroleum Storage of Sabzevar." *Journal of Rehabilitation in Civil Engineering*, 10(2): p. 17-34.

Analytical Study on Ground Behavior During the 2018 Sulawesi Earthquake, Indonesia

Ren Hori

Master's student, School of Engineering, University of Tokyo, Tokyo, Japan hori@iis.u-tokyo.ac.jp

Kazuma Inoue

Associate Professor, National Institute of Technology, Gunma College, Gunma, Japan, inoue@cvl.gunma-ct.ac.jp

Takaaki Ikeda

Professor, Nagaoka University of Technology, Niigata, Japan, ikeda@vos.nagaokaut.ac.jp

Takashi Kiyota

Associate Professor, Institute of Industrial Science, University of Tokyo, Tokyo, Japan, kiyota@iis.u-tokyo.ac.jp

Abstract. The 2018 Sulawesi earthquake, Indonesia, caused long-distance ground flow in several areas in Palu city, about 80 km south of the epicenter, with the largest area reaching about 3.5 km × 1.5 km. The common feature of the long-distance ground flow area is that the slope of the ground surface is gentle. In this study, we aim to clarify the mechanism of the long-distance ground flow by using seismic response analysis and simplified liquefaction analysis. Three sites with different damage conditions will be targeted. Then, a one-dimensional soil column model is created from the boring log, and the response characteristics are analyzed by seismic response analysis. Also, a geological survey of the site confirmed that there are mountains of 2000 m in height around the site, thus water easily accumulates in the alluvial fan under the mountains. It is also assumed that pressurized groundwater exists. There are some possible causes of the flow: one is the formation of an aquifer beneath the surface layer, which may have initiated the flow; the other is the liquefaction from the surface layer to the cap layer, which may have caused the cap to come off and pressurized groundwater to be ejected, leading to long-distance ground flow. In this study, we consider the liquefaction of the surface layer, and the large relative displacement between the cap layer and the layer below it created water paths, and the pressurized groundwater is supplied continuously to the upper layer through the water paths. As a result, in terms of ground seismic response analysis results, shear strain and relative displacement were larger at the damaged point, but no remarkable results were obtained. However, in the simple liquefaction analysis, it was suggested that the liquefaction layer increased when the pressure is taken into account, leading to long-distance ground flow.

1. Introduction

A moment magnitude (Mw) 7.5 earthquake occurred at 10:02 UTC on September 28, 2018 in Central Sulawesi, Republic of Indonesia, causing long-distance ground flow in five areas in and around Palu city, about 80 km south of the epicenter, as well as numerous other ground flow events. The long-distance ground flow was about 3.5 km × 1.5 km in area. As an example, Figure 1 shows pre- and post-earthquake photographs of the Petobo area and Lolu area the target area of this study. A common feature of these areas is that they have gentle slopes with an average slope of approximately 1 to 5 %, and the slope of the aforementioned flow area is reported to be 1.9 % [1]. Geological investigations at the site have confirmed that water easily collects in the surrounding area due to the presence of 2000 m-high mountains and a fan-shaped area below the mountains [2], and the existence of pressurized groundwater has been inferred from the confined aquifer layers and interviews with local residents [3]. Various methods have been used to elucidate the mechanism of long-distance ground flow, and multiple causes have been considered. For example, it is possible that the flow was initiated by the formation of a water layer during the dissipation of excess pore pressure due to liquefaction of the ground beneath the surface layer. There is also a sandy silt layer, which could serve as a cap layer, and be disturbed by liquefaction of the underlying ground, allowing groundwater to continuously flow into the surface layer from the deeper pressurized aquifer layer, leading to long-distance ground flow [2].

Kumagai et al. [4] conducted a ground seismic response analysis using an equivalent linear analysis method that takes frequency characteristics into account, and evaluated ground behavior by changing the input ground motion level. However, although relative displacement and maximum shear strain increased with seismic level, no noticeable trend was observed. In a study by Kiyota et al. [5], liquefaction strength tests of samples collected in the field and simplified liquefaction determination method considering the applied pressurized groundwater were conducted. The causes of this phenomenon have been investigated analytically in only a few studies. In this study, we investigate three sites with different damage conditions using nonlinear time domain analysis and a simplified liquefaction determination method in order to elucidate the mechanism of long-distance ground flow occurrence and to understand the trend of seismic response characteristics at each site.

(a) Damage in Petobo area (Source: Google Earth)

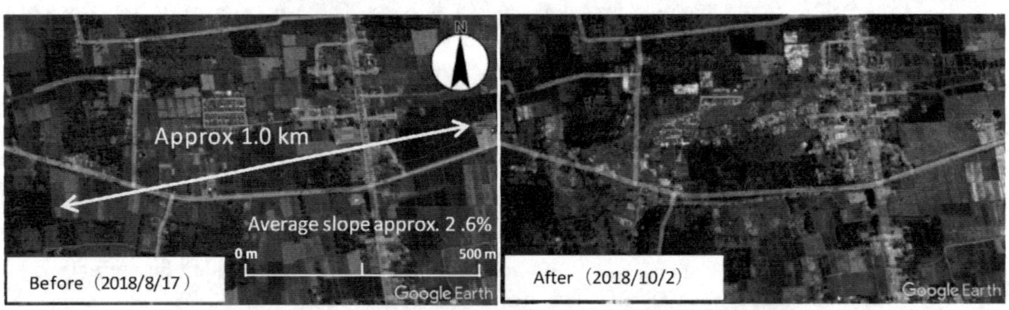

(b) Damage in Lolu area (Source: Google Earth)

Figure1: Targeted damage area

2. Methods for studying seismic behavior

In this study, three sites are considered: one in the non-flow area near the long-distance ground flow area (Balaroa area, Palu City) and two in the flow area (Petobo area and Lolu area, Palu City).

2.1. Ground seismic response analysis method and input ground motion

Unlike the study by Kumagai et al. [4], TDAP III [6] is used to perform a nonlinear time domain analysis that takes into account the nonlinear characteristics of the ground. A one-dimensional soil column model with plane strain elements is used to examine the differences in the analytical results. The details of the model are described below.

The 2018 Sulawesi earthquake was observed only at the ground surface, and there are no observed waves at the engineering seismic base layer of the earthquake, which is necessary for ground seismic response analysis. Therefore, the NS component of the seismic waveform observed at G.L. -83 m on Kobe Port Island during the 1995 Hyogo-ken Nanbu Earthquake, which was an inland crustal earthquake similar to the present earthquake and caused liquefaction, is used as the input seismic motion [7]. Figure 2 shows the acceleration time history before multiplying by an input factor. The maximum acceleration at the ground surface at the seismic observation point for this earthquake is approximately 280 Gal. To align the maximum acceleration levels at the ground surface, we use the input seismic motion with the input factor of 0.3, referring to the study by Kumagai et al [4].

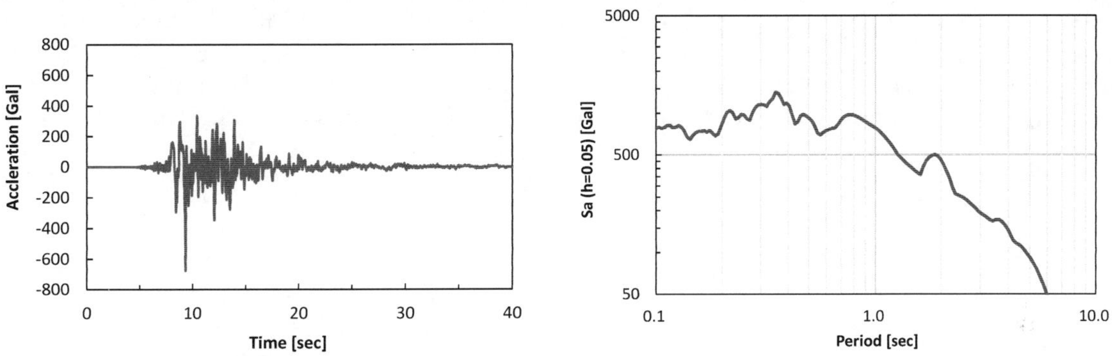

(a) Acceleration time history **(b) Acceleration response spectrum**

Figure2: The 1995 Hyogo-ken Nanbu Earthquake: Kobe Port Island G.L.-83 [m] (NS)

2.2. Creation of one-dimensional soil column model

A one-dimensional(1D) soil column model with one degree of freedom in only one horizontal direction is set up by referring to local borehole logs [8]. Figure 3 shows a simplified 1D soil column model and N value for the Petobo, Lolu, and Balaroa districts. In the Petobo and Lolu areas, the alternation of silty sand and sandy silt layers, which is the cause of low permeability, may increase the pressure, while in the Balaroa area, the gravel layer with good permeability accounts for a large proportion of the soil, and there is low pressure.

The 1D soil column model with nonlinear characteristics is created based on the above, and an elastic soil layer is added below the 1D soil column to a depth of 50 m. The soil layer is divided into 0.5 m increments from the ground surface. Shear wave velocities are calculated using PS logging data for the Lolu and Balaroa areas, and N values for the Petobo area using Equation (1a, b). The density of each soil layer is set based on the borehole log and are shown in Table 1. The shear wave velocity is set to 300 m/s for N value greater than 50 in the Petobo area. The shear modulus is then calculated from Equation (2). The shear wave velocity of the additional elastic soil layer is the same as that of the lowest soil layer for which nonlinear characteristics are considered.

$$sandy\ soil\ layer: V_s = 80N^{\frac{1}{3}} \tag{1a}$$

$$clay\ soil\ layer: V_s = 100N^{\frac{1}{3}} \tag{1b}$$

$$G_0 = \rho {V_s}^2 \tag{2}$$

where V_s: shear wave velocity [m/s], N: Standard Penetration Test N value, G_0: shear modulus [kN/m^2], ρ: density [g/cm^3]

In the nonlinear time domain analysis, there are various equations for the nonlinear relationship between shear strain and shear stress in the ground. The modified Hardin-Drnevich model (modified HD model) [9] is adopted for the skeleton curve, and equation (3) is adopted for the hysteresis loop, which is a virtual skeleton curve adjusted so that the equivalent damping ratio by hysteretic damping satisfies the h-γ relationship.

$$h = h_{max}(1 - G/G_0)^{\beta_1} \tag{3}$$

where h : damping ratio, h_{max} : maximum damping ratio, G/G_0 : shear modulus ratio, β_1: arbitrary constant (=1)

In this study, h_{max} = 21% for sandy soil and h_{max} = 17% for cohesive soil are adopted with reference to the study by Koyamada et al [10] because laboratory test of the in-situ samples has not been carried out, and there are no actual data available. Figure 4 shows the dynamic deformation characteristics to be set. For boundary conditions, the bottom of the model is fixed, and the points of contact at the same height of the model are constrained to have equal horizontal displacement. Dynamic ground seismic response analysis is then performed with a one-way horizontal input from the bottom of the model.

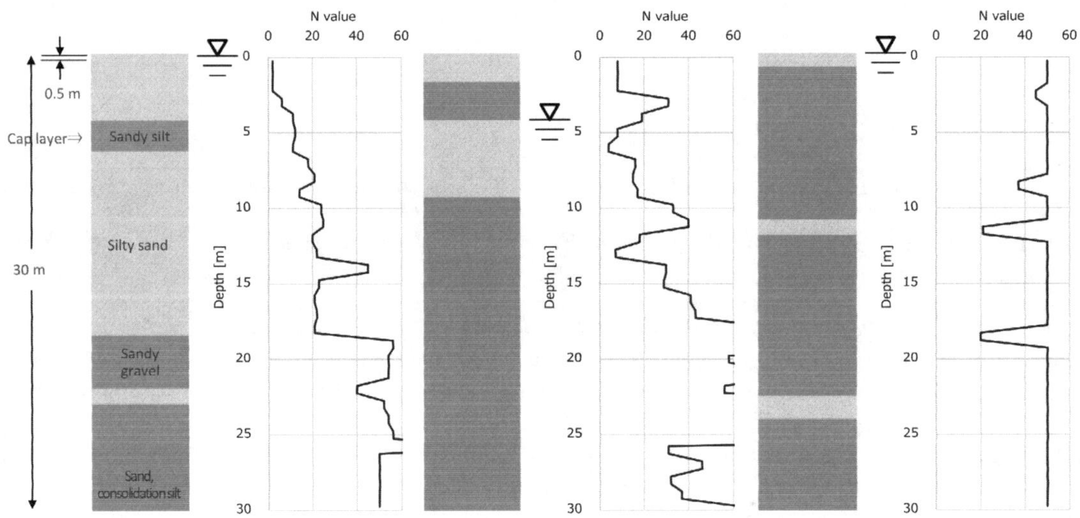

(a) Petobo **(b) Lolu** **(c) Balaroa**

Figure3: Conceptual diagram of 1D soil column model and N value

Table1: Density used for 1D soil column model

Soil	Density [g/cm^3]
Silty sand	1.84
Sandy silt	1.84
Sandy gravel	2.14
Sand/ Consolidated silt	1.84

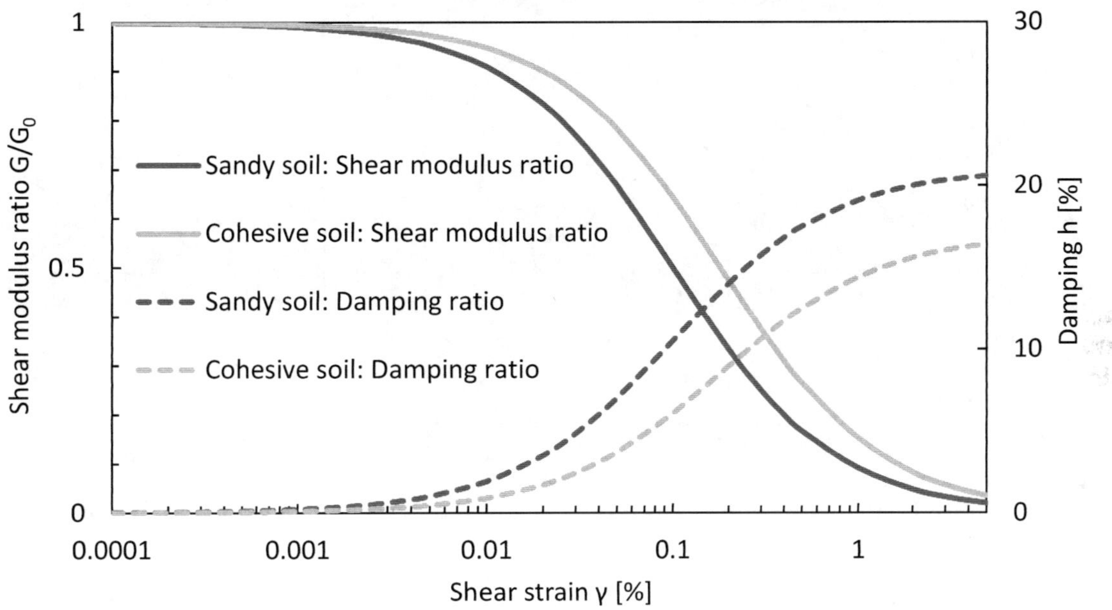

Figure4: Dynamic deformation characteristics

2.3. Simplifies liquefaction determination method

Based on the 1D soil column model used in the ground seismic response analysis, the simplified liquefaction determination (FL method) is applied to the soil layers with liquefaction potential, as shown in Equation (4).

$$F_L = R/L \tag{4}$$

where F_L : Resistance to liquefaction ($F_L < 1$: liquefaction occurs, $F_L > 1$: no liquefaction) R: dynamic shear strength ratio, L: seismic shear stress ratio

At the damaged points, the results of simple liquefaction evaluation according to the Specification for Highway Bridges [11] and the results of simplified liquefaction determination method taking the pressure into account are compared. Since the undamaged points are mainly composed of gravel layers with good permeability, it is considered that no actual pressure is generated, so only the simplified liquefaction evaluation according to the Specification for Highway Bridges is performed. The horizontal seismic coefficient used is the depth distribution of the maximum acceleration obtained from the ground seismic response analysis. The pressure is taken into account by subtracting the shear stress from the effective overburden pressure when calculating the seismic shear stress ratio. Since the value of the actual pressure that occurred before the earthquake is unknown, a judgment is made by assuming the ratio of the effective overburden pressure.

IOP Conf. Series: Earth and Environmental Science 1244 (2023) 012035 doi:10.1088/1755-1315/1244/1/012035

3. Ground seismic response analysis results

The depth profiles of maximum shear strain in the Petobo, Lolu, and Balaroa areas are shown in Figure 5 to 7, respectively. The stratigraphic classification of the results is the same as in Figure 3.

3.1. Damage point 1 (Petobo area)

The maximum shear strain tends to be smaller in the cap layer (sandy silt layer) and larger in the silty sand layers above and below. This indicates that liquefaction may have occurred in the silty sand layers.

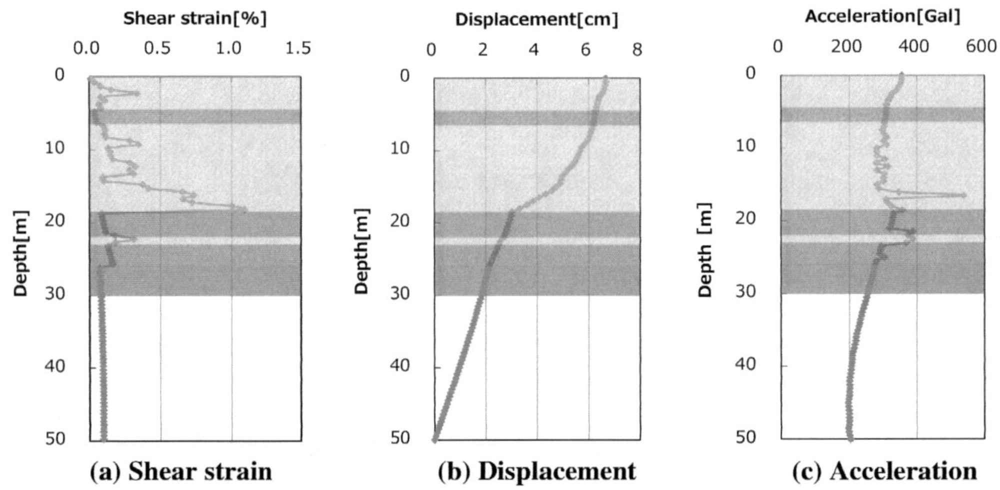

Figure5: Depth distribution of ground seismic response analysis results in Petobo area

3.2. Damage point 1 (Petobo area)

As in the damaged site 1, the maximum shear strain tends to be smaller in the sandy silt layer and larger in the layers above and below. In addition, liquefaction occurrence should be considered in the same way.

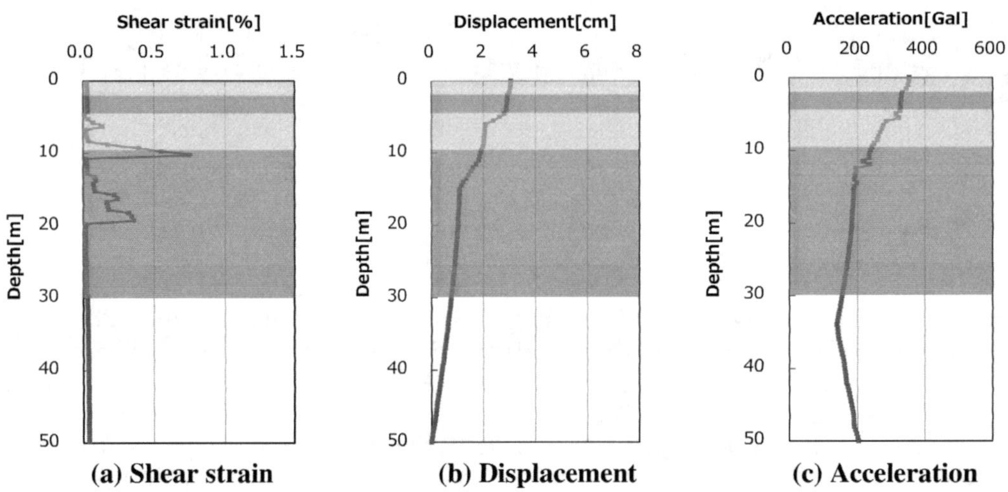

Figure6: Depth distribution of ground seismic response analysis results in Lolu area

3.3. Non damage point (Balaroa area)

The seismic response differs significantly from that of the damaged sites, and the overall maximum acceleration and maximum shear strain generated are small. This suggests that the possibility of liquefaction is small.

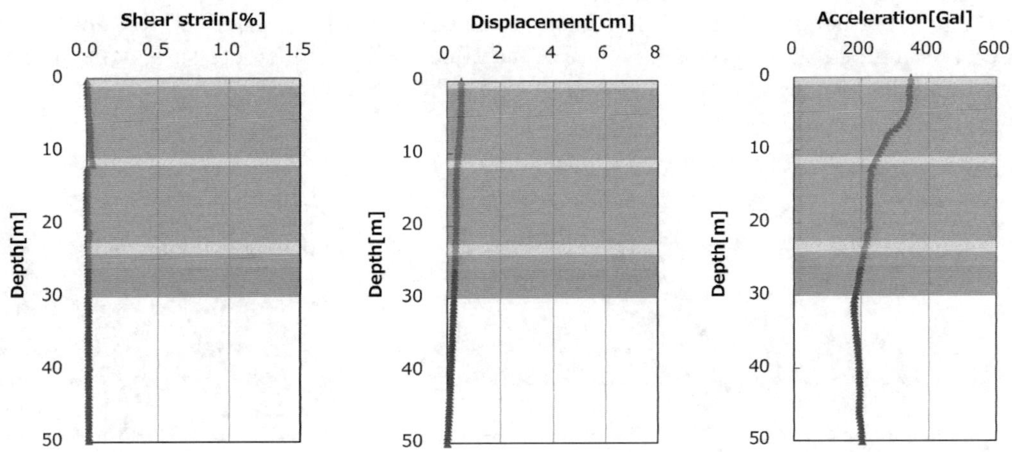

(a) Shear strain **(b) Displacement** **(c) Acceleration**

Figure7: Depth distribution of ground seismic response analysis results in Balaroa area

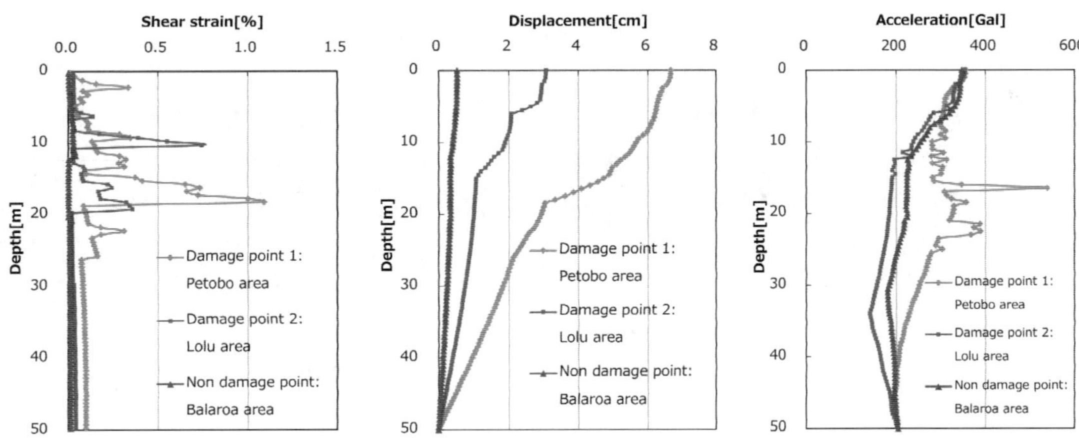

(a) Shear strain **(b) Displacement** **(c) Acceleration**

Figure8: Depth Distribution of ground seismic response analysis results

4. Simplified liquefaction determination results

Figure 9, 10 and 11 show the results of the simplified liquefaction determination for the Petobo, Lolu, and Balaroa areas, respectively. The respective assumptions for the pressure are shown in the legend of the figure. The strata classification is the same as in Figure 3.

4.1. Damage point 1 (Petobo area)

If the pressure is not considered, liquefaction occurs mainly in the surface layer above the sandy silt layer. Assuming that the pressure is 90% of the effective overburden pressure, liquefaction occurs not only in the surface layer but also in the layer below the sandy silt layer.

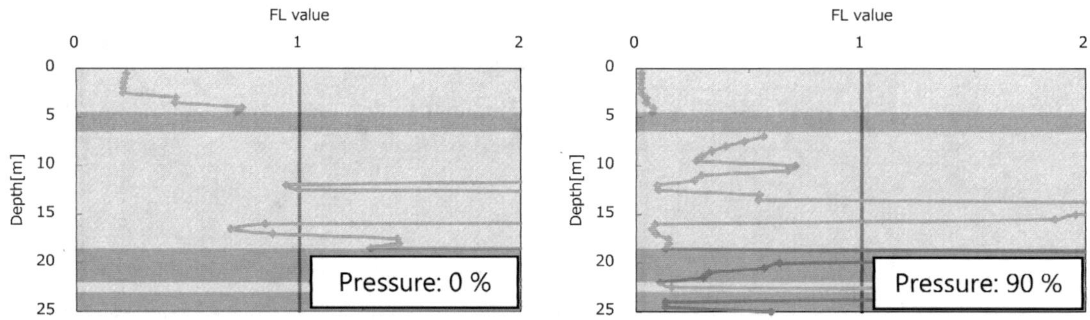

Figure9: Depth distribution of simplified liquefaction analysis results in Petobo area

4.2. Damage point 2 (Lolu area)

If the overburden pressure is not considered, a portion of the silty sand layer will liquefy. Assuming that the pressure is 70% of the effective overburden pressure, the silty sand layer liquefies along the depth.

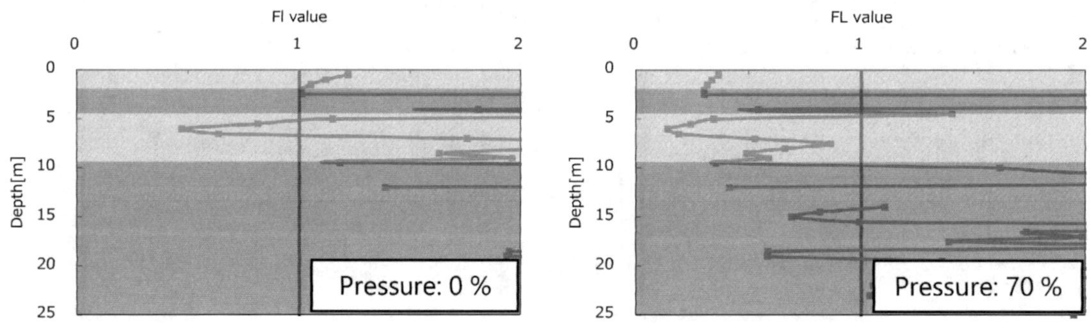

Figure10: Depth distribution of simplified liquefaction analysis results in Lolu area

4.3. Non damage point (Balaroa area)

When the pressure is not taken into account, the majority of the subject ground does not liquefy. In addition, since this site has very good permeability and no flow is actually occurring, no consideration is given to the pressure.

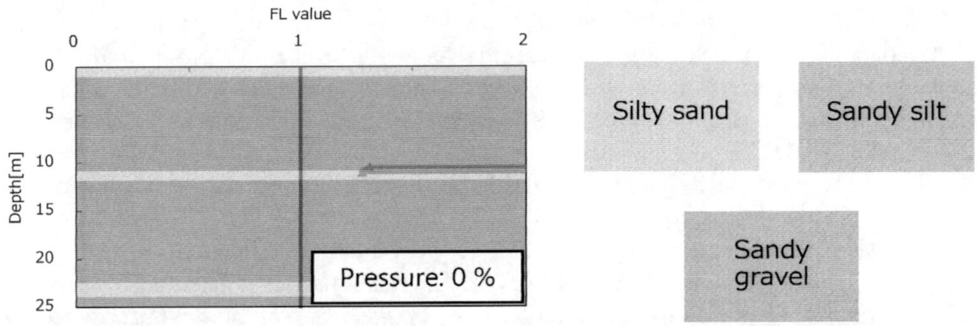

Figure11: Depth distribution of simplified liquefaction analysis results in Balaroa area

5. Summary

In this study, ground seismic response analyses and simplified liquefaction determinations were conducted at three sites with different damage conditions.

In the damage point 1, which is located in the Petobo area, the simplified liquefaction determination results show that only the surface layer above the sandy silt layer liquefies when the pressure is not taken into account10. and that this does not lead to long-distance ground flow. When considering the pressure, liquefaction potential increases in the layer below the sandy silt layer. It can be suggested that after liquefaction, pressurized water continues to be supplied to the surface layer from the aquifer layer under pressure. This may influence the occurrence of long-distance ground flow. In the damage point 2, which is located in the Lolu area, part of the silty sand layer liquefies when the pressure is not taken into account. When the pressure was assumed to some extent, the silty sand layer liquefies in its entirety. In the no damage point, which is located in the Balaroa area, all layers of the subject soil did not liquefy, which is reasonable because this area is outside of the ground flow area. The influence of ground pressure on the ground is quite high, although it needs to be confirmed by further studies using local seismic records.

As a possible development of the study, the 1D soil column model used in the ground seismic response analysis and the results of the analysis should be carefully examined, and the results of field investigations and laboratory tests should be further reflected and analyzed using local earthquake observation records. In addition, effective stress analysis may be conducted to study the behavior including liquefaction during earthquakes.

6. Acknowledgements

This research was supported by JSPS Grants-in Aid for Scientific Research JP19KK0108. We also thank the Kobe Port Island Earthquake Observatory for allowing us to use their observation records of the 1995 Hyogo-ken Nanbu Earthquake in Kobe, Japan. The authors would like to express their gratitude.

References

[1] Miyazima, M., Hendra, S., Yoshida, M., Ono, Y., Kosa, K., Ida S. O., Martini and Irdhiani. (2019). "Geotechnical damage in the 2018 Sulawesi earthquake, Indonesia", *Geoenvironment Disasters*, Japan.

[2] Yoshida, M., Hendra S., Miyazima, M., Ono, Y., Kousa, K., Serikawa, Y., Fuzikawa, H. (2020) "*Factors to liquefaction-induced ground flow during the 2018 Sulawesi earthquake, Indonesia*", JSCE paper series A1 (Structural and earthquake engineering), Vol.76, No.4, I_647-I_655.

[3] Kiyota, T., Furuichi, H., Risiqi F. H., Tada. N., and Hasbulluah, N. (2020). "*Overview of long-distance flow-slide caused by the 2018 Sulawesi earthquake, Indonesia*" Soils and Foundations, 60, 3, pp.722-735.

[4] Kumagai, T., Ikeda, T., Kiyota, T. (2020). "Numerical study on dynamic behavior of surface ground near the long-distance ground flow area caused by the 2018 Sulawesi earthquake, Indonesia" production research, Vol72, No.6, pp.415-418.

[5] Kiyota, T., Shiga, M., Katagiri, T., Furuichi, H., Hasbullah, N. "*Effect of artesian pressure on liquefaction-induced flow-slide*" *A case study of the 2018 Sulawesi Earthquake, Indonesia"* Performance-based Design in Earthquake Geotechnical Engineering (PBD-IV), accepted.

[6] ARK Information systems (2018). *TDAPIII manual version.10.*

[7] Kobe-shi Development Bureau (1995). Report on the investigation of landfill ground deformation due to the Hyogo-ken Nanbu Earthquake (Port Island, Rokko Island) (In Japanese).

[8] Japan International cooperation Agency (JICA) (2021). "Final Report: Project for Development of Regional Disaster Resilience Plan in Central Sulawesi in the Republic of Indonesia" Appendix III.

[9] Kokusho, T., Sakurai, A. (1978). "*About the Modified Hardin-Drnevich model* (In Japanese)" 33rd JSCE Annual Academic Lecture Outline III, pp.116-117.

[10] Koyamada, K., Miyamoto, Y., Miura, K. (2003). "*Non-linear properties of surface soils evaluated from collected samples (In Japanese)*" 38th The Japan National Conference on Geotechnical Engineering, pp.2077-2078.

[11] Japan Road Association (2017). Specifications for Highway Bridges part V Seismic design, pp.110-119.

IOP Conf. Series: Earth and Environmental Science 1244 (2023) 012036 doi:10.1088/1755-1315/1244/1/012036

Response Spectrum Design Based on Site Class and Site-Specific Analysis in the Transition Zone

W Dwiyantoro[1,2], T F Fathani[1] and A D Adi[1]

[1]Department of Civil and Environmental Engineering, Universitas Gadjah Mada, Yogyakarta, Indonesia
[2]Directorate General of Human Settlements, Ministry of Public Works and Housing, South Jakarta, Indonesia

E-mail: wahyu.dwiyantoro@mail.ugm.ac.id

The building structure design needs to pay attention to the seismic code and all its consequences. Building A which will be built in Sleman, Special Region of Yogyakarta is 12.20 km from the Opak Fault causing the Bantul Earthquake of Mw 6.3 in 2006, needs seismic analysis properly using SNI 1726:2019. Based on the site near fault criteria on seismic code, Building A does not belong to the site near fault category, so it used the Probabilistic Seismic Hazard Analysis method. However, considering its distance and magnitude, the value is in the range of the requirements limit. Hence, it is still relevant when compared to the Deterministic Seismic Hazard Analysis method for designing the response spectrum. Based on the site classification in the seismic code, a site that is vulnerable to liquefaction belongs to the site class F and therefore requires site-specific analysis. The site-specific surface response spectrum is taken from the equivalent-linear and non-linear site response analysis using Deepsoil v7.0 software. The response spectrum is also calculated based on site class according to the shear wave velocity (v_s) and N-SPT criteria. The site-specific spectral acceleration results are compared to 80% of the site class E spectral acceleration, as required. The site belongs to site class D using v_s criteria and belongs to site class E using the N-SPT criteria. Only the equivalent-linear site response spectrum in the modified maximum horizontal direction spectrum is eligible when compared to requirements. Most of the non-linear site response spectrum results or the geometric mean response spectrum of the equivalent-linear site response results can satisfy the requirements compared to 80% of the site class D spectral acceleration because the site response analysis uses the v_s parameter. The non-linear site response spectral acceleration is smaller than the equivalent-linear site response spectral acceleration.

1. Introduction

The building structure design needs to consider seismicity aspect and the influenced location effect to determine the response spectrum design which is appropriate with the site characteristic. The earthquake loads for building structure in Indonesia have already been regulated in SNI 1726:2019 [1] based on Indonesian Earthquake Source and Hazard Map 2017 [2]. According to [1], site class can be determined by the average of data value of N-SPT up to 30 m depth ($\overline{N}$), undrained shear strength (s_u), and shear wave velocity up to 30 m depth (v_{s30}). The criteria for site class F is determined by the characteristics of the soil layer profile, one of which is prone to and has the potential to collapse due to earthquake loads such as easily liquefied, highly sensitive clay, and weakly cemented soil. The site class F requires a site-specific response analysis (SSRA) and specific geotechnical investigation, especially for structures that have fundamental vibration period more than 0.5 s. The SSRA according to [1] and [3] is performed by site response analysis and risk-targeted maximum considered earthquake (MCE_R) ground motion hazard analysis. MCE_R ground motion hazard analysis shall be taken from probabilistic ground motion and deterministic ground motion. A case in a seismic design of building structure at the location of the Sleman, Special Region of Yogyakarta, which based on [1] belongs to the site class F with the liquefaction potential based on the initial indication of the liquefaction potential criteria [4]. The site is about 12.20 km from the Opak Fault which experienced a movement and caused an earthquake of M_w 6.3 in 2006 [5]. According to the site near the fault criteria in [1], this site does not belong to the site near the fault category. However, the site distance to the Opak Fault is between the 2 categories for the site near the fault criteria: those are 10 km and 15 km. It is also for historical magnitudes whose values are between M_w 6 and M_w 7. It is possible that this site belongs to the transition zone or plateau according to [6]. According to [2], the structures belonging to the sites near fault category require deterministic seismic hazard analysis (DSHA). Site-specific response analysis according to [1] and [3] are performed for seismic design at this site accommodating the site class F because of the potential of liquefaction and accommodates considerations of DSHA due to sites near the fault consideration. Previously, the research which related to response spectrum in Special Region of Yogyakarta was conducted by [7] and [8], but it only used site classes based on seismic codes and it did not use site-specific studies. Research which was conducted by [9] was almost similar to this study. However, the spectral parameter values did not refer to seismic code [1], but it was analyzed by the author. Meanwhile, the location was not in the transition zone. The site response analysis used Uniform Building Code 1997 [10]. The research used spectral matching techniques with time domain, and did not use the 11 ground motion records as required by [3] and [11]. Synthetic ground motion in Special Region of Yogyakarta which was analyzed with spectral matching technique using 11 pairs of ground motion recordings was done by [12]. However, ground motion was only analyzed on bedrock without site response analysis to obtain ground motion and response spectrum on the ground surface. Surface response spectrum research with reference to site-specific and site class in the transition zone in Sleman, Special Region of Yogyakarta has never been done before, so this research was conducted.

2. Method

Site-specific analysis was performed by MCE_R ground motion hazard analysis. The MCE_R spectral acceleration shall be taken as the smallest acceleration at each period from probabilistic and deterministic ground motions [1], [3]. Probabilistic (MCE_R) ground motion shall be taken as maximum horizontal direction of spectral acceleration from 5% damped acceleration response spectrum with 1% probability of exceedance in 50 year. The 5% damped acceleration response spectrum with 1% probability of exceedance in 50 year is calculated from the spectral acceleration from 5% damped acceleration response spectrum 2% probability of exceedance in 50 year (2500 years return period) multiplied by risk coefficient, C_R. The 5% damped acceleration response spectrum with 2500 year return period is calculated from [2]. The C_R value is determined from [1]. Deterministic (MCE_R) ground motion is calculated from 84th percentile 5% damped spectral acceleration in the direction of maximum horizontal response which ordinate value will be scaled if the largest spectral acceleration value is less than 1.5 amplification factor F_a [1], [3]. The 84th percentile 5% damped spectral acceleration was performed using DSHA through a four-step process [13]. Ground motion modification was conducted referring to [11] and [3] using amplitude scaling method. Spectral matching was not used because it is not suitable when applied to sites near faults [3]. Equivalent-linear and non-linear site response analysis was performed by Deepsoil v7.0 software [14] to each pair of ground motions that have been scaled to produce surface ground motion and surface response spectrum. As a comparison, the surface response spectrum was also performed by procedure according to the site class based on $\overline{N}$ criteria and v_{s30} criteria.

3. Design response spectrum based on site class criteria

Response spectrum were calculated using the parameters of MCE_R spectral acceleration mapped for short period (S_s) and 1.0 second period (S_1) taken from [1]. The S_s and S_1 values were 1.10 g and 0.50 g respectively. The $\overline{N}$ and v_{s30} value were 12.32 and 317.70 m/s respectively. According to [1], the site belonged to the site class E from the $\overline{N}$ criteria. Based on the v_{s30} value, the site belonged to the site class D. The site coefficients, F_a and F_v for the site class D at S_s = 1.10 g and S_1 = 0.50 g, were 1.06 and 1.80 respectively. The site coefficients, F_a and F_v for the site class E at S_s = 1.10 g and S_1 = 0.50 g, were 1.02 and 2.20 respectively. Spectral acceleration at each period was calculated from spectral acceleration multiplied by the associated site coefficients. Spectral response acceleration design at each period was taken 2/3 from spectral acceleration at associated period. Spectral response acceleration design at short period (S_{Ds}) and 1.0 seconds period (S_{D1}) values for the site class D were 0.78 g and 0.60 g respectively. The S_{Ds} and S_{D1} values for the site class E were 0.75 g and 0.73 g, respectively. The design response spectrum based on site class criteria is presented in Figure 1.

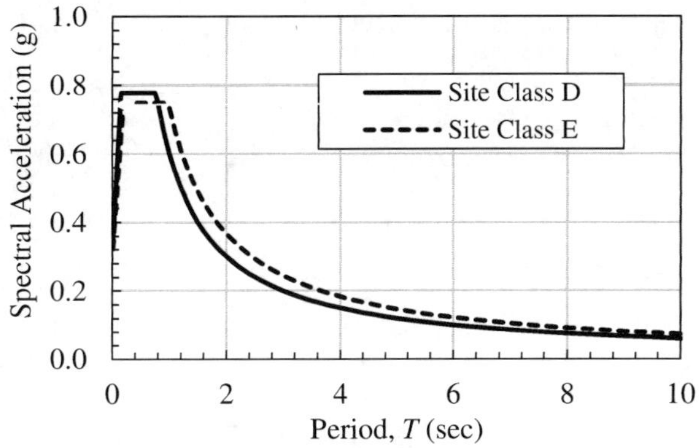

Figure 1. Design response spectrum based on site class criteria

Based on Figure 1, it is known that the peak acceleration of site class D response spectrum is greater than the peak acceleration of site class E response spectrum. The peak response spectrum for the site class D also tends to be concentrated in a smaller period than the site class E.

4. Risk-targeted maximum considered earthquake (MCE_R) ground motion analysis

The 5% damped acceleration response spectrum with 2500 year return period for probabilistic (MCE_R) ground motion was calculated from [2]. Response spectrum were calculated using the parameters of spectral response acceleration at 0.2 seconds (SA(0.2)) and 1.0 seconds (SA(0.1)) with site coefficient, F_a and F_v each equal to 1.0 because of bedrock assumption [1]. The value of each spectral acceleration is multiplied by the directivity factor to produce spectral acceleration in the maximum horizontal response direction with reference to [1]. The directivity factor 1.0 on the PGA, 1.1 on the SA(0.2), and 1.3 on the SA(1.0). The value of the directivity factor for spectral acceleration in a period of less than 0.2 seconds was linearly interpolated from 1.0 to 1.1. The directivity factor values for the spectral acceleration in the period between 0.2 seconds and 1.0 seconds were interpolated linearly from 1.1 to 1.3. Furthermore, the acceleration value of the response spectrum in the maximum horizontal response direction is multiplied by the C_R according to [1]. The C_R value in the period less than or equal to 0.2 seconds uses the C_{RS} value of 0.95. The C_R value in a period of more than or equal to 1.0 seconds uses a C_{R1} value of 0.95. The acceleration value of the response spectrum for periods of more than 0.2 seconds and less than 1.0 seconds using interpolation of the C_{RS} and C_{R1} values and the value is constant at 0.95. The final probabilistic (MCE_R) response spectrum is shown in Figure 2. The 84th percentile 5% damped acceleration response spectrum for deterministic (MCE_R) ground motion was calculated from an empirical approach using the ground motion prediction equation (GMPE) based on the relationship between magnitude and source distance. Determination of earthquake sources around the site using initial identification from the IRIS Earthquake Browser catalog [15]. The earthquake history is then plotted in the earthquake source catalog from [2] to determine the characteristics and parameters of the earthquake source (Table 1). The GMPE equation used to assess the controlling earthquake source is based on Kanno *et al.* 2006 [16]. This equation only uses the moment magnitude (M_w) and source distance (R) parameters and can be used for shallow events and deep events. The criteria used to determine the controlling earthquake source are PGA, SA(0.2) and SA(1.0) presenting in Table 1. Based on Table 1, the value of PGA and SA(0.2) for the source of the Opak fault earthquake is greater than the other two earthquake sources, so that the source of the Opak fault earthquake is considered as a controlling earthquake source at the site. Furthermore, to obtain the spectral acceleration for each period, the Chiou-Youngs 2014 GMPE equation is used [17]. This equation is one of equations used by [2]. The Chiou-Youngs 2014 GMPE equation [17] is a development of the Chiou-Youngs 2008 GMPE [18] which is used for fault and shallow crustal earthquake sources with strike slip, reverse slip, and normal mechanisms.

Table 1. The earthquake sources used and the spectral acceleration values based on the Kanno *et al.* 2006 GMPE

Source Zone	Source Name	Type	M_w	Distance, R (km)	PGA (g)	SA(0.2) (g)	SA(1.0) (g)
1	Opak Fault	line	6.60	12.20	0.85	2.16	0.69
2	Klaten Earthquake 1979	point	5.20	48.60	0.17	0.46	0.05
3	Megathrust West-Central Java	area	8.70	97.60	0.73	1.58	2.03

This equation was developed based on earthquake events that occurred in California, Japan, New Zealand, Taiwan, and Turkey with a magnitude between M_w 3.1 to M_w 7.6 [2]. The Chiou-Youngs 2008 GMPE equation uses an earthquake database with a radius of 70 km so that it is relatively more concentrated in a small radius than the other 2 GMPE equations used by [2] for fault and shallow crustal earthquake sources, namely Boore-Stewart-Seyhan-Atkinson 2014 [19] and Campbell-Bozorgnia 2014 [20]. The calculation of the spectral acceleration performed by the Pacific Earthquake Engineering Research (PEER) NGA West2 website calculator [21] with the input parameters of the Opak fault is presented in Table 2 [2].

Table 2. Input parameters for spectral acceleration calculation using Chiou-Youngs 2014 GMPE

Damping Ratio	Fault Type	M_w	Width (km)	DIP (deg)	v_{s30} (m/s)	Epsilon
0.05	Strike Slip	6.60	9.90	60	317.70	1

The epsilon value is 1 for a calculation with standard deviation of +1 which is equal to the 84th percentile. The spectral acceleration of the 84th percentile 5% damped results then multiplied by the directivity factor with the same value as applied to the probabilistic spectral acceleration to produce spectral acceleration in the maximum horizontal response direction. The maximum spectral acceleration value of the response spectrum in the maximum horizontal response direction compared to 1.5 F_a in gravity units, with the F_a value for the site class F of 1.0 [3]. The value of the maximum spectral acceleration in the maximum horizontal response direction is 1.31 g < 1.50 g, so that each value of the spectral acceleration in the maximum horizontal response direction needs to be multiplied by a factor of 1.50 g/1.31 g = 1.15 to produce the maximum spectral acceleration value equal to 1.50 g. The deterministic spectral acceleration can be seen in Figure 2. The site-specific MCE_R spectral acceleration in each period shall be taken as the lesser of spectral acceleration from probabilistic and deterministic response spectrum as shown in Figure 2. The probabilistic MCE_R response spectrum is more dominant in about 0.2 second to 3 second period.

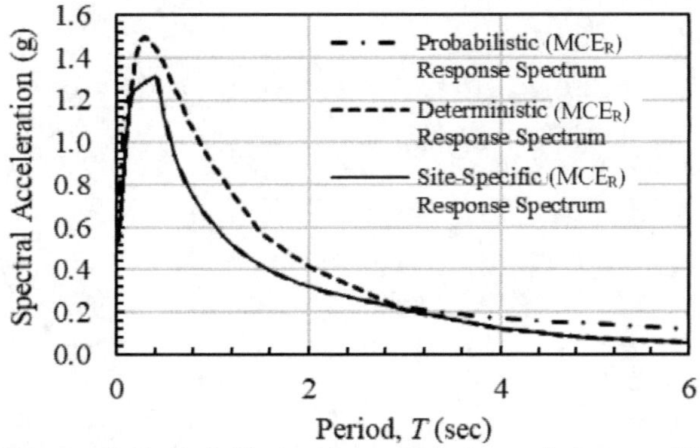

Figure 2. Probabilistic, deterministic, and site-specific (MCE$_R$) response spectrum

5. Ground motion

The ground motion selection refers to the deaggregation results in Yogyakarta. Deaggregation results are used to determine the earthquake hazard with the most controlling magnitude and distance in a site [22]. This process is needed to select the ground motion record data which then will be performed by amplitude scaling, so that synthetic ground motion and response spectrum are obtained that are suitable for the site. The deaggregation value uses secondary data in previous studies due to the limitations of the available earthquake data catalog. The deaggregation results refers to [22] and available for the spectral acceleration period of 0.2 seconds and 1.0 seconds with a probability of exceedance 2% in 50 years. The deaggregation result used for consideration of the selection of ground motion are the closest to spectral acceleration period of 1.0 seconds because they are closest to the fundamental period of the structure whose value is 1.12 s. The source of the shallow crustal earthquake in the deaggregation results is M_w 6.8 and R = 17 km. The source of the subduction earthquake on the results of the deaggregation is M_w 7.4 and R = 210 km.

Ground motion record was taken from the PEER NGA West 2 ground motion database [21], because the availability of ground motion records in Indonesia was still very limited. The number of ground motion records used was a minimum of 11 referring to [11] and [3]. In addition to the criteria of magnitude and distance used as considerations in the selection of ground motion recordings, the v_{s30} is used to indicate site characteristics. The v_{s30} value should be at least 750 m/s because in the following ground motion analysis is assumed to be in the outcropping rock. However, due to the limitation of available records, sites with v_{s30} < 750 m/s were used. The same method is used in determining the distance criteria that becomes a priority after v_{s30}. The distance parameter uses the terminology of rupture distance (R_{Rup}) and Joyner-Boore distance (R_{JB}). The ground motion recordings used can be seen in Table 3. Period range for ground motion modification was determined by [1], [11], and [3], which was between 20% of the minimum fundamental period ($T_{\min}$) and 200% of the maximum fundamental period ($T_{\max}$) for the principal horizontal directions of responses. The $T_{\min}$ and $T_{\max}$ values in this structure were 0.8 seconds and 1.12 seconds respectively. The period range for scaling was 0.16 seconds to 2.24 seconds. Ground motion modification was performed by amplitude scaling method. This method referred to [1], [3] by using rotD100 spectral data [21] on each record then scaled and averaged until the value of the spectral acceleration in the reviewed period was greater than or equal to the value of 90% of the target spectrum. The target spectrum used the site-specific MCE$_R$ spectral acceleration. The scale used for each spectral data rotD100 (Table 3) is then used to scale the acceleration time history on the same record. The scale factor applied was attempted to follow the [23], which was not to exceed 5. However, due to the limitation of ground motion recordings in accordance with the desired earthquake mechanism, several pairs of ground motion recordings used a scale factor more than 5. The Kocaeli earthquake recording uses a fairly large scale, in order to accommodate the vibration characteristics with short period (T_s) and base period (T_0) values at large values due to the limitations of ground motion recording data at a large distance.

Table 3. Used ground motion records and its scale

No.	Earthquake name	Focal mechanism	Year	Magnitude (M_w)	Distance (km) R_{JB}	Distance (km) R_{Rup}	v_{s30} (m/s)	Scale
1	San Fernando	Reverse	1971	6.61	21.50	21.50	969.07	1.00
2	Loma Prieta	Reverse oblique	1989	6.93	12.15	18.51	713.59	1.00
3	Totori	Strike slip	2000	6.61	15.58	15.59	967.27	2.20
4	Chi-chi	Reverse	1999	6.20	21.11	22.14	652.85	9.00
5	El-Mayor	Strike slip	2010	7.20	89.69	90.14	743.00	5.00
6	Darfield	Strike slip	2010	7.00	25.21	25.67	649.67	2.00
7	Northridge-01	Reverse	1994	6.69	23.10	23.64	996.43	2.80
8	Iwate	Reverse	2008	6.90	16.26	16.27	825.83	1.50
9	Cristhchurch	Reverse oblique	2011	6.20	13.91	16.13	649.67	2.00
10	Hector Mine	Strike slip	1999	7.13	10.35	11.66	726.00	1.50
11	Kocaeli	Strike slip	1999	7.51	65.53	65.53	612.78	15.00

6. Site response analysis

This analysis was performed to calculate the amplification characteristic of the soil profile at the site. The soil profile used based on the results of the borlog correlated with the shear wave velocity from the seismic downhole test results can be seen in Table 4.

Table 4. Input data of soil profile

Layer No.	Description	Layer thickness (m)	Total unit weight (kN/m^3)	Shear wave velocity (m/sec)	Depth at top layer (m)	Vertical effective stress (kPa)
1	Sandy Clay	2.0	12.16	145.46	0	0
2	Loose Sand-1	2.0	12.16	252.61	2.0	24.32
3	Loose Sand-2	1.0	12.16	270.61	4.0	48.65
4	Loose Sand-3	3.0	16.48	300.17	5.0	60.81
5	Medium Dense Sand-1	2.0	16.48	314.70	8.0	110.26
6	Medium Dense Sand-2	3.0	16.37	338.86	10.0	143.22
7	Medium Dense Sand-3	2.0	16.73	352.07	13.0	192.34
8	Dense Sand-1	3.0	16.73	336.92	15.0	225.81
9	Dense Sand-2	2.0	16.67	352.44	18.0	276.01
10	Dense Sand-3	2.0	16.67	337.16	20.0	309.36
11	Dense Sand-4	3.0	18.66	361.43	22.0	342.71
12	Dense Sand-5	3.5	18.66	449.21	25.0	398.67
13	Dense Sand -6	7.5	18.66	435.79	28.5	463.97
14	Dense Sand - Interpolated 1	7.0	20.52	550.00	36.0	603.89
15	Dense Sand - Interpolated 2	7.0	22.26	650.00	43.0	747.53
16	Bedrock	-	24.00	750.00	50.0	903.35

The bedrock depth was determined using previous study refers to [24]. The bedrock depth at the site is based on the closest data in the study of [24] is estimated to be 50 m. Seismic downhole testing was carried out to a depth of 29.5 m or layer 13. Layers 14 and 15 were obtained by interpolating linearly the shear wave velocity to bedrock. The total unit weight and layers 14 and 15 are interpolated to suit the shear wave velocity. Layer thicknesses 13, 14, and 15 are evenly distributed with reference to the interpolation of shear wave velocity values to bedrock.

Site response analysis using Deepsoil v7.0 software was performed in non-linear and equivalent linear method. The reduction of shear modulus and damping curve referred to the Darendeli 2001 approach

IOP Conf. Series: Earth and Environmental Science 1244 (2023) 012036 doi:10.1088/1755-1315/1244/1/012036

[25]. Consideration of outcropping rock in this case was applied by applying elastic bedrock with a wave speed velocity of 750 m/s, a volume weight of 24 kN/m^3, and a damping ratio of 1%. All surface ground motions from the site response analysis were then transformed into spectral accelerations. The non-linear and equivalent linear site response spectral accelerations were then geometrically averaged and multiplied by the directivity factor to produce the maximum horizontal response direction spectral acceleration. The final spectral acceleration value was multiplied by 2/3 according to [1].

The final spectral acceleration value is multiplied by 2/3 factor according to [1] and modified by rearranging using the spectral acceleration parameters in the 0.2 second and 1.0 second period. The spectral acceleration in each spectra is then compared with the requirements of [1], which is a minimum of 80% site class E spectral acceleration. As an additional comparison, the value of 80% site class D spectral acceleration is used because the site belongs to the site class D based on the v_{s30} criteria. The equivalent-linear and non-linear site response analysis results has been presented in Figure 3. Based on Figure 3, the spectral acceleration of the non-linear site response analysis is smaller than the results of the site response linear equivalent analysis. The surface spectral response acceleration compared to the site class D and site class E spectral response acceleration has been presented in Figure 4. Based on Figure 4, only the linear equivalent site response spectrum in the modified maximum horizontal response direction satisfy the requirements for 80% acceleration of the site class E response spectrum.

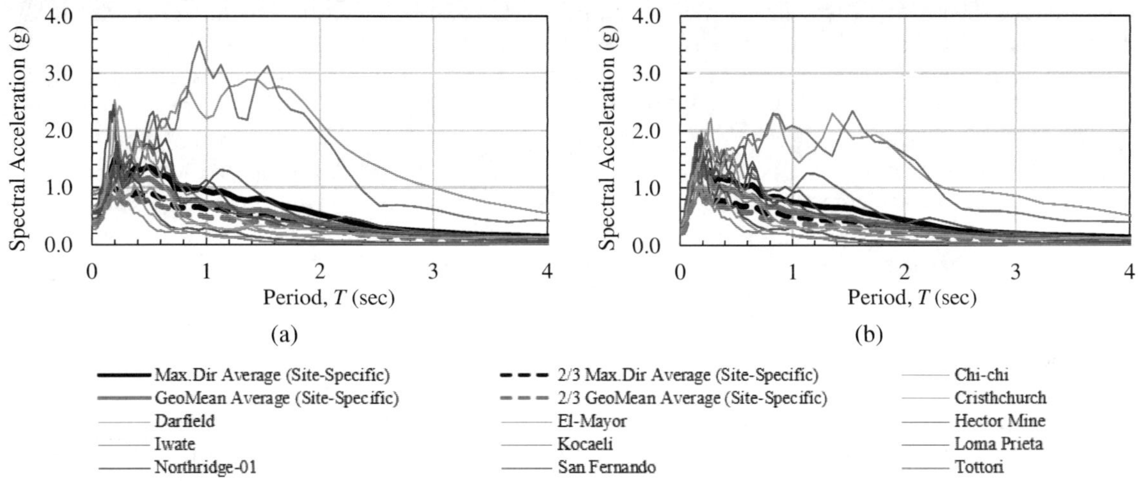

Figure 3. Surface response spectrum: (a)Equivalent-linear result, (b) Non-linear result

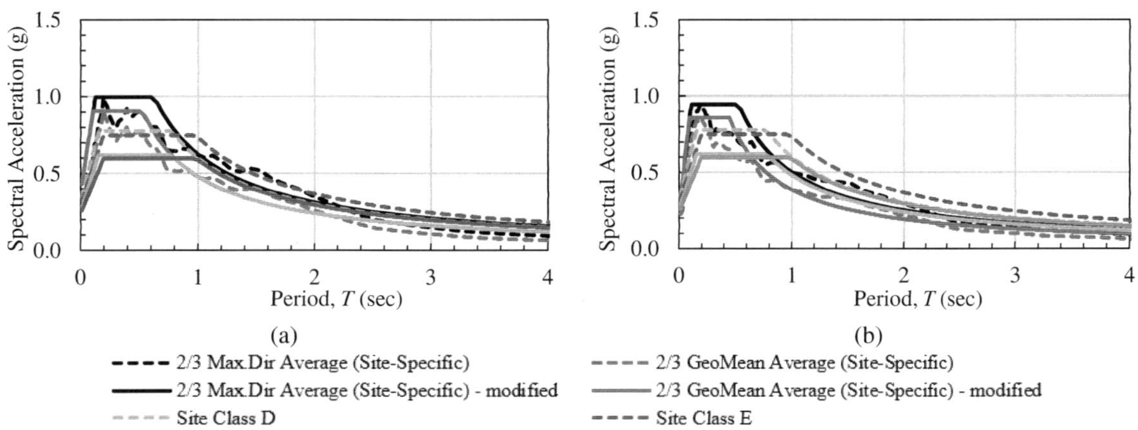

Figure 4. Comparation between surface site-specific response spectrum and surface response spectrum site class D, site class E: (a) Equivalent- linear result, (b) Non-linear result

7. Discussion

Both the determination of MCE_R and the selection of ground motion recordings affected the results of site-specific analysis. In this analysis, MCE_R was conducted by using probabilistic MCE_R from 5% damped with 2500 years return period earthquake map which was processed and produced a spectral acceleration which was lower than deterministic in a period of about 0.2 seconds to 3 seconds (Figure 2). MCE_R used the lesser spectral acceleration of probabilistic and deterministic, so that the characteristic response spectrum in the period of about 0.2 seconds to 3 seconds followed the probabilistic results.

Based on Figure 3b, were only the equivalent-linear site response spectrum in the modified maximum horizontal direction spectrum fulfill the requirements. All non-linear response spectrum results did not fulfill the requirements. The response spectrum of the equivalent-line site response results site which were geometrically averaged, both before and after modification and the response spectrum of the maximum horizontal direction spectrum equivalent-linear site response results before being modified did not fulfill the requirements. The fundamental matter causing the spectral acceleration value unqualified was the T_s value which was smaller than the requirement resulting in the peak spectral acceleration to be concentrated in a lower period than the requirement.

Furthermore, the value of T_s on the response spectrum of the site class E was influenced by the amplification factor of F_a and F_v for the site class E. This site belonged to the site class E if it was classified based on the N-SPT criteria. While the site response analysis used the v_s value from seismic downhole test, not using the empirical correlation of the N-SPT value. Site class based on classification according to v_{s30} was site class D. Hence, the amplification characteristics in the site response analysis results tended to lead to the site class D. Therefore, it would be better if the site-specific results were compared to the response spectrum for the site class D. Conversely, the site response analysis may use the value of v_s from the empirical correlation of the N-SPT, but it is not recommended, because more representative data is available from the seismic downhole test results. As a suggestion, site-specific analysis can be carried out with MCE_R earthquakes that have larger T_s and a ground motion recording database from other sources or simulation results [11] in order to get closer to the requirements of the deaggregation results.

8. Conclusion

Site specific analysis needs to pay attention to the input response spectrum of MCE_R with response spectrum which will be used as a comparison according to the site class. This case especially if the site class classified as N-SPT mean and v_{s30} are different, and the site class classified as N-SPT results in worse conditions (close to site class F) compared to the v_{s30} classification site class.

Based on the analysis result, the response spectrum of site-specific results meeting the requirements were only the equivalent-linear site response spectra in the modified maximum horizontal direction spectrum. If it needs a non-linear response spectrum from site response results or a geometric mean response spectrum from a linear equivalent site response, another analysis should be carried out referring to the discussion section.

References

[1] Badan Standardisasi Nasional 2019 *SNI* 1726:2019 *Tata cara perencanaan ketahanan gempa untuk struktur bangunan gedung dan nongedung* (Badan Standardisasi Nasional, Jakarta, Indonesia)

[2] National Center of Earthquake Studies 2017 *Earthquake Source and Hazard Maps* (Research and Development Agency of the Ministry of Housing and Public Works, Bandung, Indonesia)

[3] American Society of Civil Engineers 2017 *Minimum Design Loads and Associated Criteria for Buildings and Other Structures* (American Society of Civil Engineers, Reston, Virginia, United States of America)

[4] Rahardjo P P 2021 *Seismic Hazard Study and Site-Specific Response Analysis Universitas Nahdlatul Ulama Yogyakarta* (Geotechnical Engineering Consultant, Bandung, Indonesia)

[5] Earthquake Engineering Research Institute 2006 *The* M_w 6.3, *Java, Indonesia, Earthquake of May* 27, 2006 (EERI Earthquake Special Report-Agustus 2006, Oakland, California)
[6] Leyendecker E, Hunt R, Frankel A and Rukstales K 2000 Development of Maximum Considered Earthquake Ground Motion Maps *J. Earthq. Spectra.* **16**, 21–40
[7] Sungkono K K D 2019 Response Spektra Gempa Kota Yogyakarta, Surakarta dan Semarang Berdasarkan Peta Gempa SNI 2012 dan Peta Gempa 2017 *Jurnal Teknik Sipil* **4**(1), 39-44
[8] Patria A S N 2021 Perbandingan Parameter Spektrum Respons Desain SNI 1726:2012 dan SNI 1726:2019 pada 29 Kabupaten/Kota di Pulau Jawa *Jurnal Keilmuan Teknik Sipil* **4**(1), 126-134
[9] Makrup L 2016 The Earthquake Ground Motion and Response Spectra Design for Sleman, Yogyakarta, Indonesia with Probabilistic Seismic Hazard Analysis and Spectral Matching in Time Domain *American Journal of Civil Engineering* **4**(6), 298-305
[10] Uniform Building Code 1997 *Uniform Building Code Volume 2 Structural Engineering Design Provisions* (International Conference of Building Officials, 5360 Workman Mill Road, Whittier, California 90601-2298, Printed In USA)
[11] Badan Standardisasi Nasional 2020 *SNI* 8899:2020 *Tata cara pemilihan dan modifikasi gerak tanah permukaan untuk perencanaan gedung tahan gempa* (Badan Standardisasi Nasional, Jakarta, Indonesia)
[12] Prawirodikromo W, Teguh M, Makrup L and Suryo B 2020 Development of Synthetic Ground Motion at a Specific Site in Yogyakarta Town, Indonesia Utilizing the PSHA Method *E3S Web Conferences* **156** 02011
[13] Reiter L 1990 *Earthquake Hazard Analysis: Issues and Insights* (Columbia University Press, New York, United States of America)
[14] Hashash Y M A, Musgrove M I, Harmon J A, Ilhan O, Xing G, Numanoglu O, Groholski D, Phillips C A and Park D 2020 *DEEPSOIL V7.0, User Manual* (Board of Trustees of University of Illinois at Urbana-Champaign, Urbana, IL)
[15] Incorporated Research Institutions for Seismology 2022 IRIS Earthquake Browser, https://ds.iris.edu/ieb/ (accessed Jun. 24, 2022)
[16] Kanno T, Narita A, Morikawa N, Fujiwara H and Fukushima Y 2006 A new attenuation relation for strong ground motion in Japan based on recorded data *Bull. Seismol. Soc. Am.* **96(**3), 879–897
[17] Chiou B S and Youngs R R 2014 Update of the Chiou and Youngs NGA Model for the Average Horizontal Component of Peak Ground Motion and Response Spectra *Earthq. Spectra.* **30**(3), 1117–1153
[18] Chiou B S, Youngs R R 2008 An NGA Model for the Average Horizontal Component of Peak Ground Motion and Response Spectra *Earthquake Spectra.* **24**(1), 173–215
[19] Boore D M, Stewart J P, Seyhan E and Atkinson G M 2014 NGA-West2 Equations for Predicting PGA, PGV, and 5% Damped PSA for Shallow Crustal Earthquakes *Earthq. Spectra.* **30**(3), 1057–1085
[20] Campbell K W and Bozorgnia Y 2014 NGA-West2 Ground Motion Model for the Average Horizontal Components of PGA, PGV, and 5% Damped Linear Acceleration Response Spectra *Earthq. Spectra.* **30**(3), 1087–1115
[21] Pacific Earthquake Engineering Research Center 2022 NGA-West2 -- Shallow Crustal Earthquakes in Active Tectonic Regimes, https://ngawest2.berkeley.edu/ (accessed Apr. 16, 2022)
[22] Sunardi B 2015 Percepatan Tanah Sintetis Kota Yogyakarta Berdasarkan Deagregasi Bahaya Gempa *Jurnal Lingkungan dan Bencana Geologi.* **6**(3), 211-228
[23] Rezaeian S, Zhong P S, Hartzell and Zareian F 2015 Validation of Simulated Earthquake Ground Motions Based on Evolution of Intensity and Frequency Content *Bull. Seismol. Soc. Am.* **105**, 3036-3049
[24] Perdhana R and Nurcahya B E 2006 Seismic microzonation based on microseismic data and

damage distribution of 2006 Yogyakarta Earthquake *E3S Web Conf.* **76**, 4–7
[25] Darendeli M B 2001 *Development of a New Family of Normalized Modulus Reduction and Material Damping Curves* (The University of Texas at Austin, United States)

Acknowledgement

We would like to express our sincerely gratitude for Ministry of Public Works and Housing and Balai Prasarana Permukiman Wilayah Daerah Istimewa Yogyakarta which have supported the data and all parts of this article.

EARTHQUAKE RISK POTENTIAL CAUSED BY ACTIVE FAULTS IN SUMATRA (Case Study: Earthquakes in the Land of West Sumatra Province and its vicinity)

Murjaya, Jaya[1,3] , Rahman, Aditya Setyo[1] , Sili, Petrus Demon[1], Karnawati, Dwikorita[1,2] , Supriyanto[3], Fatimah, Siti[3], Sutiyono[1], Sugianto, Dedi[1], Ahadi, Suaidi[1]

[1]Agency for Meteorology, Climatology, and Geophysics, Jakarta, 10610, Indonesia. [2]Geological Engineering Department, Faculty of Engineering, Gadjah Mada University, Yogyakarta 55281, Indonesia

[3]Geoscience Department, Natural and Science Faculty, University of Indonesia, Depok, 16424, Indonesia

Setyo.bmkg@gmail.com, murjaya1@gmail.com

Abstract. The land of West Sumatra province is one of prone area stricken by the destructive earthquakes. Based on historical records, the land of West Sumatra province has a high seismic risk related to the earthquake events along active faults, such as the Sianok, Barumun, Sumani, Siulak and Angkola faults. Some destructive earthquakes occurred in these faults and caused fatalities in West Sumatra area and its vicinity, e.g., the earthquakes near Bukit Tinggi district in 1926, 1977, 2007 (doublet earthquakes-Tabel 1) and 2022. A previous study shows that the Sumani, Siulak, Sianok, Barumun and Angkola faults has a potential magnitude 7.1-7.7, a geodetic slip-rate of 6-14 mm/y respectively and these move generally strike slip. Using these magnitude scenarios to make the earthquakes impact modelled, we find the maximum intensity is about of VII-VIII MMI scale, the Peak Ground Acceleration (PGA) is about of 42-64 % g, and Peak Ground Velocity (PGV) is about of 30-50 cm/s for the assumption of 5-15 km hypocentral depth. Also with using the scaling law relation of earthquake parameters and magnitude scenarios the earthquake return period can be calculated by a kinematic model with the source mechanism was modelled as a strike slip faults. Estimation of the earthquake return period with magnitude (Mw) scenarios about 7.1-7.7 for these active faults is about of 114 until 612 years for strike slip model respectively. This method can be developed to make identically assessment for others active fault in Sumatra region and the result can be used as one of a consideration to support of planning of the long term landscape in the prone of earthquakes area.

Keywords: *shake map, Peak Ground Acceleration, Peak Ground Velocity, return period*

1. Introduction

The land of West Sumatra province in Indonesia is one of the destructive earthquakes prone area. Based on historical records, the land of West Sumatra province is high seismicity region and has some active faults that is often trigger earthquakes, such as the Sianok, Barumun, Sumani, Siulak and Angkola faults. In the recent years, the some destructive earthquakes occurred in Pariaman District of the West Sumatra province and its vicinity. There are destructive earthquakes occurred on 1926, 1977, 2007(doublet earthquakes-Tabel 1) and the last February 25, 2022 (that has magnitude (Mw) 6.1 and hypocentral depth 10 km [1]. That earthquake caused about of 380 people injured, 1 people died and more than 2000 (buildings damaged). The aftershocks distribution is identical direction with the straightness of the northwest edge of the Sianok Fault. That earthquake triggered also the landslide in the several part of the Pariaman District, West Sumatra.

The destructive earthquakes in February 25, 2022 (Mw 6.1) is located closed by the earthquakes epicentral has occurred on March 06, 1977 and June 28, 1926 as like show at Fig. 1. The destructive and significant earthquakes in Fig. 1 located nearby the Sianok, Angkola and Barumun Faults in West Sumatra Province.

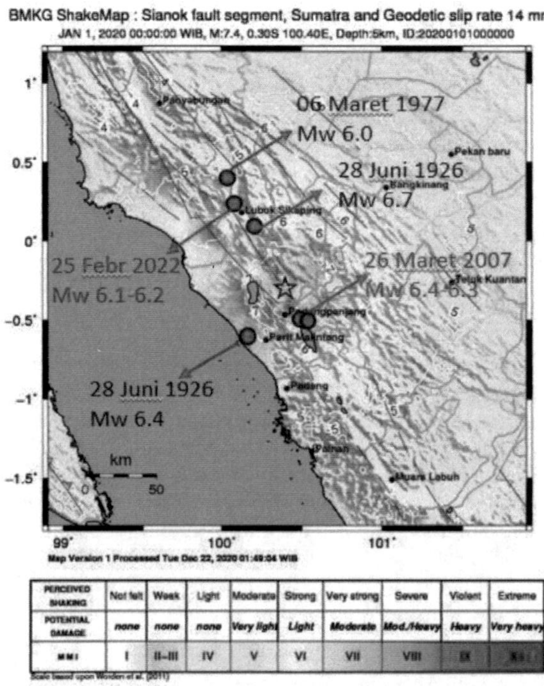

Figure 1: Significant Earthquakes distribution in West Sumatra Province since 1900-2022. Red circle is epicentre of earthquake, and colours background shows shaking level in MMI scale.

Table 1. List of the destructive and significant earthquakes in West Sumatra [2, 3, 4].

Number	Time	Magnitude	Depth (km)	Remark
1	June 28, 1926	6.4	-	USGS [4]
2	June 28, 1926	6.7	-	More than 354 people died. In Agam District 472 buildings collapse, and in Padang Panjang 2.383 buildings collapse and 247 people died. Maximum intensity about VIII-IX MMI [3].
3	March 6, 1977	6.0	-	More than 1000 buildings collapse in Sinurat and Talu village, and surface rupture is about 5 – 75 m. The maximum intensity is estimated about VIII MMI [3].
4	March 26, 2007	6.4	-	About 67 people died and 826 people injured. More than 43.719 buildings all types collapse in

				Bukit Tinggi, Payahkumbu and Solok districts. This effect included caused by second earthquake event (doublet earthquakes)[2]
5	March 26, 2007	6.3	-	
6	February 25, 2022	6.2	-	About 380 people injured and 1people died. More than 2000 buildings several types damaged.

Table 1 show that is the destructive earthquakes [2, 3, 4] caused by the active faults therefore we need of a paleoseismic study of in Sianok, Angkola, Siulak and Barumun Faults and a slip monitoring that faults respectively.

The tectonics of West Sumatra which part of Sumatra island are dominated in the western with oblique subduction of the Australian plate to the northwestward beneath the Eurasia plate and move about of 5-7 cm/year [5]. The Australian plate subduct to the northwestward direction from the Sumatra Trench to Sumatra land and it is achieving depths about of 100-300 km. In the Western part zone between Sumatra Trench and volcanic-arc of Sumatra inland located fore-arc region. The most dominant factors of the tectonic pattern to form a fore-arc region is affected by subduction movements. The affected of oblique subduction in this region caused to form of Sumatra faults in the Sumatra inland. These faults has an important tectonic aspect because it is considered as a boundary between micro Sunda plate and Indo-Australian plate [6].

This paper purposes to assess of the (paleo) earthquakes of active faults in West Sumatra Province and its vicinity. The assessment involved the macroseismic (MMI scale), Peak Ground Acceleration model (gal) and Peak Ground Velocity model (cm/s) of earthquake if there are sources occurred in the Sianok, Barumun, Sumani, Siulak and Angkola faults. The assessment to make of the return period estimate using the magnitude scenarios and slip rate data [7]. The all assessment result is used to support a land use planning which more secure from the earthquake threat.

Figure 2 shows the earthquakes distribution caused by Sumatra active faults since 1900-2022. In this paper we discussed only of the earthquake which occurred in Sumatra land especially the earthquakes with source from the active faults in West Sumatra area and its vicinity as like as is described by Book of Peta Sumber dan Bahaya Gempa Indonesia Tahun 2017 [7]. That book explained of all the active faults and slip velocity in Indonesia including in Sumatra area. This paper made an assessment of the earthquake effects if the earthquakes occurred in the Sianok, Barumun, Sumani, Siulak and Angkola faults which located in the West Sumatra area and its vicinity. We used the magnitude scenarios of each fault to analyze of the effects including the estimate of the shakemap, PGA and PGV. Then we made also the return period model (Tr) using a slip velocity rate of each fault.

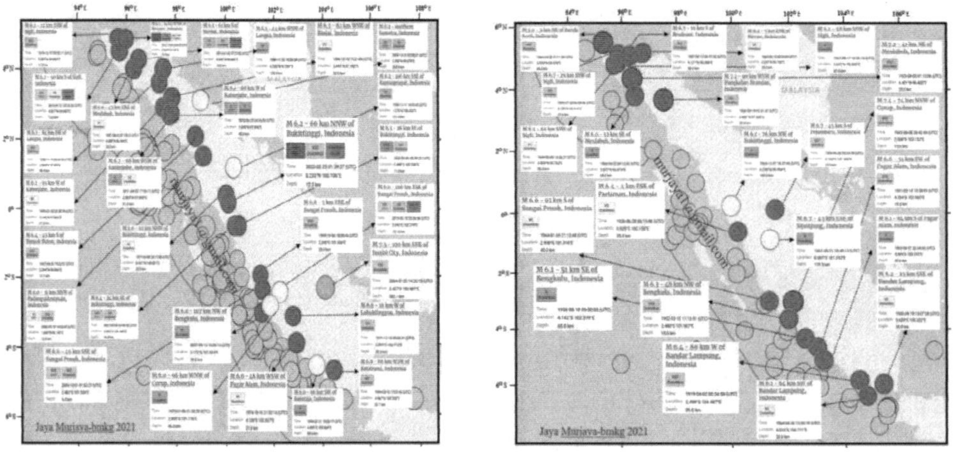

(a) (b)

Figure 2: The earthquakes historical records (a) and (b), the Sumatra land has the strong earthquakes history along Sumatra active faults, including the Sianok, Barumun, Sumani, Siulak and Angkola faults in West Sumatra province. (a) The earthquakes historical records since 1970- 2022 and (b) The earthquakes historical records since 1900-1969.

2. Data and Method

2.1. Data

In this paper is used all of the earthquakes data that occurred in and or near the active faults in West Sumatra) based on the Indonesian earthquakes catalogue data [2, 3, 4] (Table 1 and Fig. 2). Then the data of a geodetic slip rate of the active faults in West Sumatra was taken from Book of Peta Sumber dan Bahaya Gempa Indonesia Tahun 2017 [7] which reveals that the Sumani, Siulak, Sianok, Barumun and Angkola faults has a estimation magnitude about of 7.1-7.7 and a geodetic slip rate of 6-14 mm/y respectively. The magnitude data is converted to determine of seismic moment and the static stress drop and it is used to estimate of recurrence time of the earthquake in that faults respectively. Also we used the fatalities and earthquake damage data of earthquake on February 25, 2022 from BPBD (The Local Government Disaster Management Agency) of West Sumatra Province to confirm of our intensity model (shakemap model).

2.2. Scaling Law and Estimation of Earthquake Impact Model

For estimation of potential impact of active fault earthquakes like as shakemap, PGA and PGV in West Sumatra region used model and scaling law. We assumed that all of the active fault earthquakes has the strike slip nodal plane. Whereas to predict of Tr of each active fault earthquakes we used scenario magnitude and slip velocity data [7].

2.3. Static Theory of Stress Drop

During co-seismic, the tectonic energy will be released after the critical stage is passed over. This energy is stored along the inter-seismic period since the occurring a last large earthquake event in that region. Figure 3 indicates the illustration of accumulation and released stresses. This process was started from the initial stress (σ_1), then occurring of accumulation stress along the inter-seismic period until it reached the critical stress (σ_o). After the critical stress is passed over, the tectonic energy will be released as an earthquake and it will drop to initial stress stage. The difference of both stress stage is called as a static stress drop ($\Delta\sigma_s$). This process will be returned until the next earthquake in the future. Interseismic of two earthquakes event it is called as a return period of earthquake (Tr) or recurrence time.

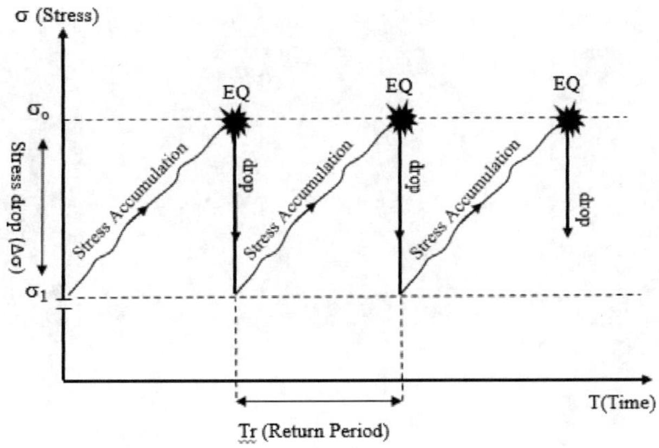

Figure 3: Concept of the initial stress (σ_1), critical stress (σ_0), stress drops and earthquake return period (Tr) relations (modified from Haskel [8]).

3. Result and Discussion

In this paper we will assess the earthquakes risk potential if the earthquakes occurred in the Sumani, Siulak, Sianok, Barumun and Angkola faults. We used the Maximum Credible Earthquake (MCE) scenarios between 7.1-7.7 and a geodetic slip rate about of 6-14 mm/y respectively.

3.1. Shaking Impact of Earthquake

Secondary effects of the earthquake is a shaking impact on buildings and others such as ground shaking, liquefaction, landslide etc. Figure 4 indicates the intensity map or shakemap (MMI scale), Peak Ground Acceleration (PGA) (% g), and Peak Ground Velocity (PGV) (cm/s) simulate models of the strike-slip earthquake scenario (MCE 7.4) for Sianok Fault with h of about 15 km. Then the hypocenter model is assumed as a point source located along the Sianok Fault. If the MCE scenario is of about 7.4 and h 15 km respectively, it will give the impact of shaking in epicenter area of about VII-VIII MMI, PGA of about 49 % g, and PGV of about 36 cm/s respectively. This result is similar with the values relation between intensity (VII-VIII MMI), PGA (0.18 - 0.65g) and PGV (16-60 cm/s) from USGS. The others result of model for Angkola, Barumun, Sumani and Siulak faults shows in Figures 5, 6,7 and 8.

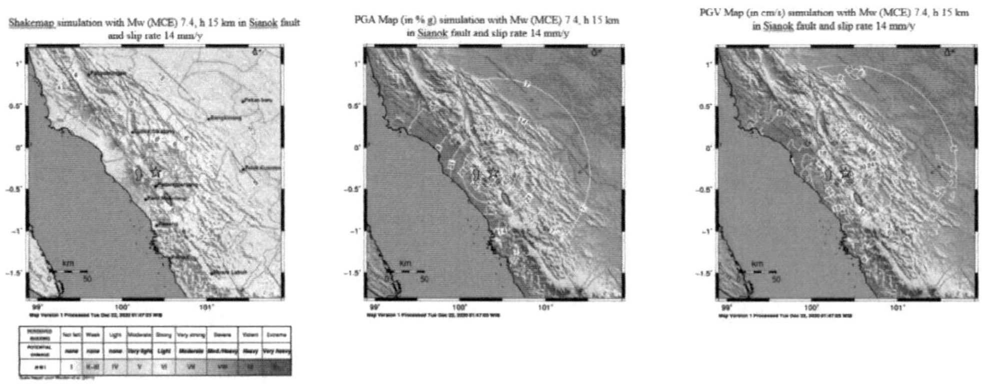

Figure 4: The result model of a strike slip earthquake (MCE 7.4 and h 15 km) impacts for the Sianok Fault. There are Intensity map (MMI), PGA (% g) map and PGV map (cm/s).

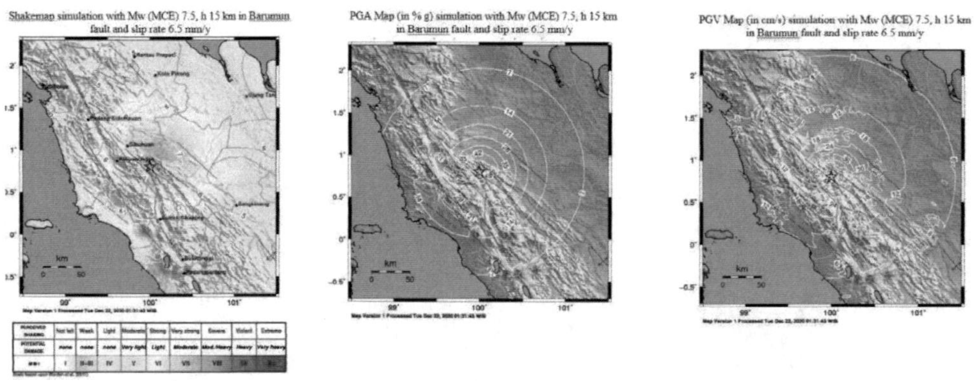

Figure 5: The result model of a strike slip earthquake (MCE 7.5 and h 15 km) impacts for the Barumun Fault. There are Intensity map (MMI), PGA (% g) map and PGV map (cm/s).

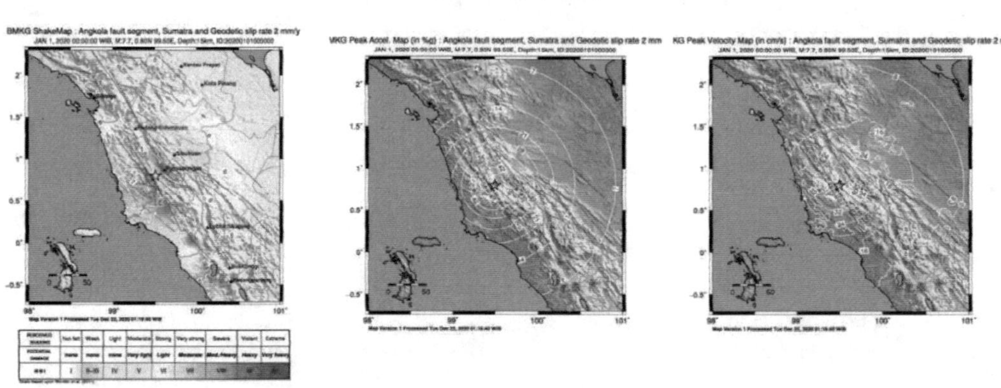

Figure 6: The result model of a strike slip earthquake (MCE 7.7 and h 15 km) impacts for the Angkola Fault. There are Intensity map (MMI), PGA (% g) map and PGV map (cm/s).

IOP Conf. Series: Earth and Environmental Science 1244 (2023) 012037 doi:10.1088/1755-1315/1244/1/012037

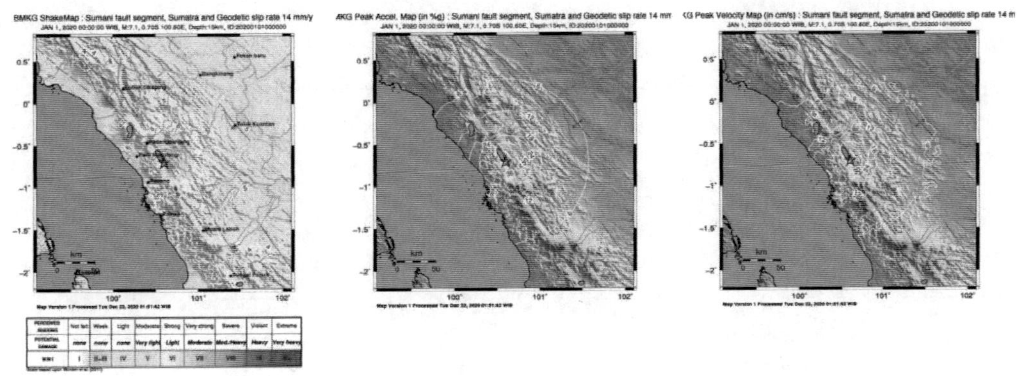

Figure 7: The result model of a strike slip earthquake (MCE 7.1 and h 15 km) impacts for the Sumani Fault. There are Intensity map (MMI), PGA (% g) map and PGV map (cm/s).

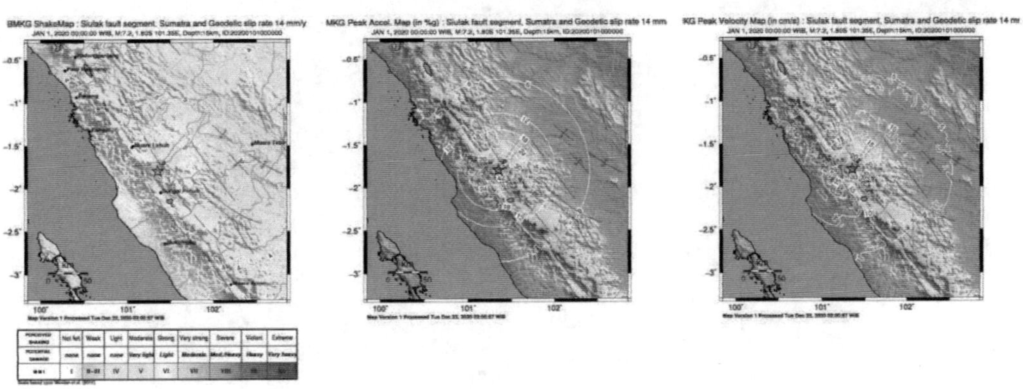

Figure 8: The result model of a strike slip earthquake (MCE 7.2 and h 15 km) impacts for the Siulak Fault. There are Intensity map (MMI), PGA (% g) map and PGV map (cm/s).

3.2. Estimation of the Earthquake Recurrence Time

To estimate of the earthquake recurrence time (Tr) in this paper it used the MCE scenarios of magnitude about 7.1 - 7.7 and a geodetic slip rate of 6-14 mm/y respectively. Davidson and Scholz [9] revealed that the earthquake recurrence time equal an earthquake moment divided by moment rate accumulation or source time function.

Table 2: Estimation of the recurrence time for several MCE scenarios uses scaling law relation of Sianok, Barumun, Angkola, Sumani and Siulak active fault earthquakes.

Model	MCE (magnitude scenarios)	Strike slip (Focal Mechanism)	Tr SF ($years_1$)	Tr BR($years_2$)	Tr AK ($years_3$)	Tr SU ($years_4$)	Tr SL ($years_5$)
	7.7	Strike Slip	-	-	612	-	-
	7.6	Strike Slip	-	-	-	-	-
	7.5	Strike Slip	-	438	474	-	-
	7.4	Strike Slip	172	-	-	-	-
	7.3	Strike Slip	-	323	350	-	-
Kinematic	7.2	Strike Slip	131	-	-	-	131

	7.1	Strike Slip	-	246	266	114	-
	7.0	Strike Slip	100	-	-	-	100
	6.9	Strike Slip	-	186	202	87	-
	6.8	Strike Slip	75	-	-	-	75
	6.7	Strike Slip	-	141	153	66	-
	6.6	Strike Slip	58	-	-	-	58
	6.5	Strike Slip	-	107	116	50	-
	6.4	Strike Slip	43	-	-	-	43

[1]SF is Sianok Fault, West Sumatra Indonesia
[2]BR is Barumun Fault, West Sumatra Indonesia
[3]AK is Angkola Fault, West Sumatra Indonesia
[4]SU is Sumani Fault, West Sumatra Indonesia
[5]SL is Siulak Fault, West Sumatra Indonesia
Tr is return period estimation

Then Figures 9a, 9b, 9c, 9d and 9e describe the recurrence time simulation result of the Sianok, Angkola, Barumun, Sumani and Siulak Faults used the slip rate geodetic data of about 14 mm/yr, 6 mm/y, 6.5 mm/y, 14 mm/y and 14 mm/y respectively [7].

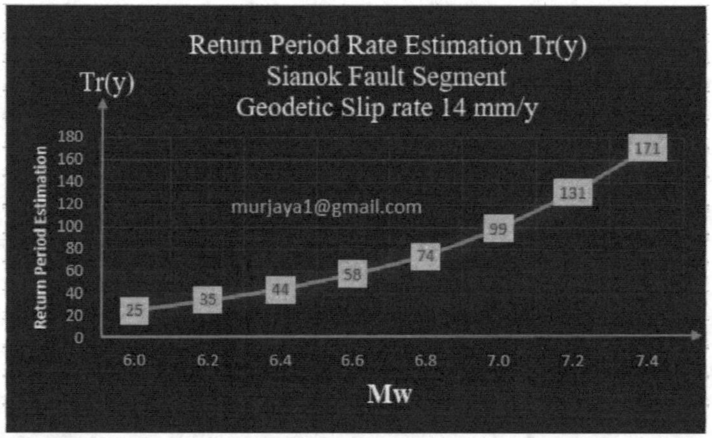

Figure 9a: Simulation of the Earthquake Return Period Estimation for a Sianok Fault in West Sumatra. The estimation was calculated by the kinematic model with magnitude scenarios of about 6.0 ≤ Mw ≤ 7.4 and a geodetic slip rate of 14 mm/y.

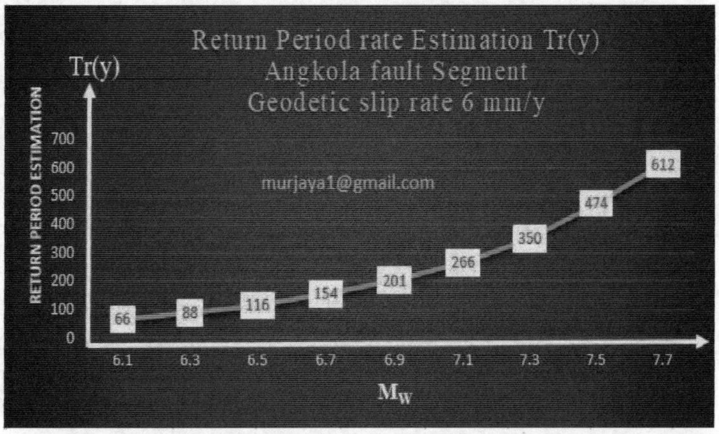

Figure 9b: Simulation of the Earthquake Return Period Estimation for a Angkola Fault in West Sumatra. The estimation was calculated by the kinematic model with magnitude scenarios of about 6.1 ≤ Mw ≤ 7.7 and a geodetic slip rate of 6 mm/y

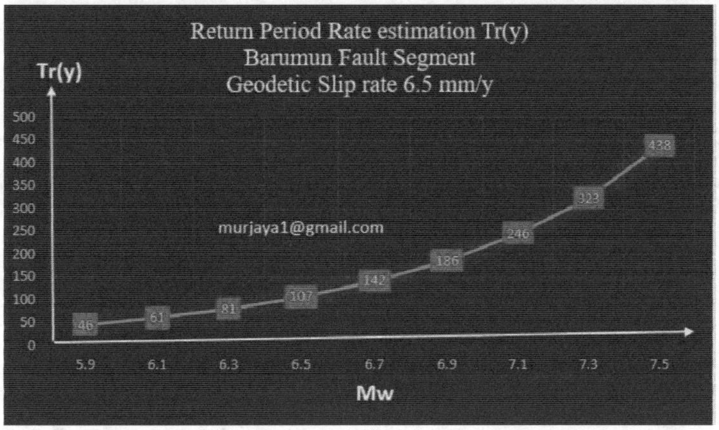

Figure 9c: Simulation of the Earthquake Return Period Estimation for a Barumun Fault in West Sumatra. The estimation was calculated by the kinematic model with magnitude scenarios of about 5.9 ≤ Mw ≤ 7.5 and a geodetic slip rate of 6.5 mm/y.

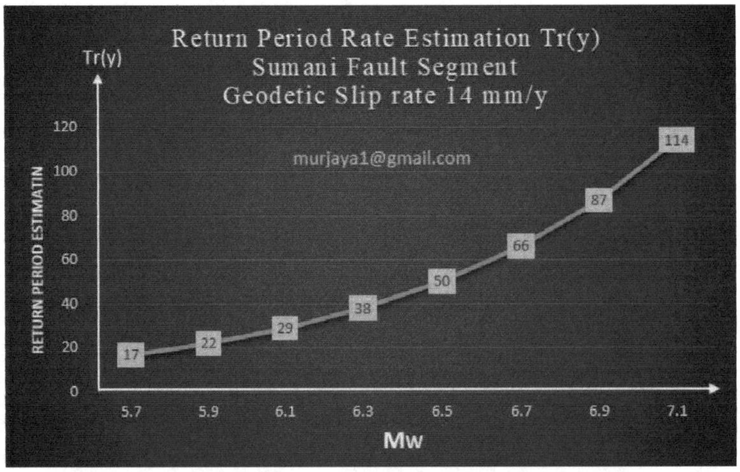

Figure 9d: Simulation of the Earthquake Return Period Estimation for a Sumani Fault in West Sumatra. The estimation was calculated by the kinematic model with magnitude scenarios of about 5.7 ≤ Mw ≤ 7.1 and a geodetic slip rate of 14 mm/y.

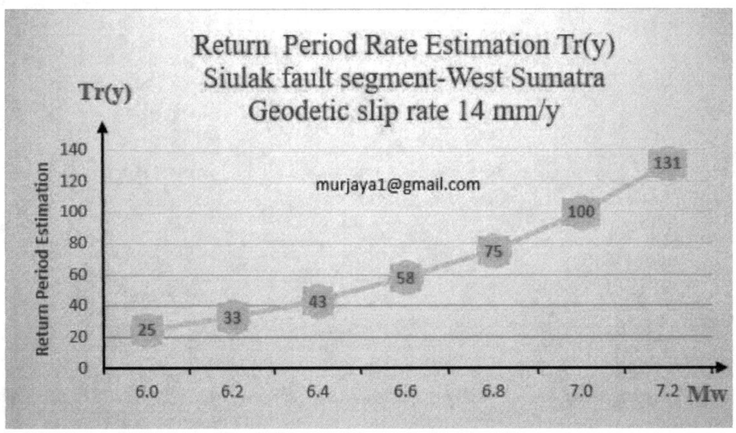

Figure 9e: Simulation of the Earthquake Return Period Estimation for a Siulak Fault in West Sumatra. The estimation was calculated by the kinematic model with magnitude scenarios of about 6.0 ≤ Mw ≤ 7.2 and a geodetic slip rate of 14 mm/y.

The Pariaman earthquake in West Sumatra (Mw 6.2 and h 10 km) on Feruary 25, 2022 has a maximum intensity of VI (until VII) MMI based on model in epicenter region. We compared of the shaking impact model to the field survey results which conducted by team after earthquake occurred. They reported the earthquake damage impact starting from almost southeast to northwest follows a Sianok Fault orientation. Moreover, the distribution of damage aligned with the result of microzonation survey by BMKG team. They revealed that the infrastructure damage generally under construction building. Whereas the PGA result from model for Mw 6.2 scenario is about of 18-21%g and PGV about 10-11.5 cm/s for hypocenter depth 10 km. This results is compared with field survey results of Yogyakarta earthquake May 27, 2006 (Mw 6.3 or mb 5.9) which it has been done by Fathani (2007) [10] and USGS [4]. He has been done several empirical formulas to determine of PGA using mb 5.9 and Mw 6.3. There are empirical formulas by Fukushima-Tanaka (1990), Campbell (1981), Matuschka (1980), Esteva (1974) and Donovan (1973). Based on several empirical analysis he found the value of PGA is about of 0.194-0.322 g. These values in accordance with the PGA modeling result such as explained above. The shaking impact (MMI scale) and g value (gal) relations has been explained by Housner (1970), Gere and Shah (1984), and Yeats et al (1997) in Fathani (2007)[10]. To refer their result for M_L 6.0, the impact of earthquake has an intensity about of VI-VII MMI scale and a PGA about of 0.18-0.21 g. Then we make model earthquake impacts with source in (and around) Sianok, Angkola, Barumun, Sumani and Siulak faults with magnitude scenario based on data from Indonesian Earthquake Study Center [7] as like showed on figure 4-8.

Figure 4 shows our simulation models of the shaking impact caused by Sianok active fault earthquake with MCE magnitude 7.4 and h 15 km. There are a shakemap (MMI scale), PGA map (in % g), and PGV map (cm/s) respectively. In this model the hypocenter is assumed as a point source which located in Sianok fault (strike slip fault). If we used the MCE scenario 7.4 (Sianok fault) and h 15 km, it will be resulted an intensity, PGA and PGV in epicenter area about VII-VIII MMI, 49 % g, and 36 cm/s respectively. By using an identical method, Figure 5 shows a result for the Barumun earthquake scenario M 7.5 and h 15 km, the shaking impact such as intensity, PGA and PGV in epicenter surrounding area about VII-VIII MMI, 49% g, and 42 cm/s. Figure 6 shows aresult for the Angkola fault earthquake scenario M 7.7 and h 15 km, the intensity, PGA and PGV in epicenter surrounding area about VII-VIII MMI, 63% g, and 50 cm/s. Figure 7 shows a result for the Sumani earthquake scenario M 7.1 and h 15 km, the intensity, PGA and PGV in epicenter surrounding area about VII MMI, 42% g, and 30 cm/s. Figure 8 shows a result for the Siulak earthquake scenario M 7.2 and h 15 km, the intensity, PGA and PGV in epicenter surrounding area about VII MMI, 42% g, and 30 cm/s.

These PGA values result and if we compared with empirical of PGA which used on Yogyakarta earthquake 2006 event [10], we conclude that all of active fault earthquakes above (Mw scenarios $\geq$ 7) will cause of significant damage to buildings. The result is able to become one of consideration to the earthquake risk mitigation and land use program in the future. The high intensity, PGA and PGV can be interpreted that if an earthquake occurred along these faults is able to cause heavy damage for infrastructure like building, bridge, electric and irrigation grid, etc. Beside that for supporting the earthquake risk mitigation program in the future, the landslide and liquefaction as impact of an active fault earthquake also need to be considered.

After calculating the some fundamental of earthquake risk for active faults in West Sumatra Province as like earthquake/shakemap, PGA, PGV, we tried to estimate of Tr for these faults. We used geodetic data of these faults, there are Sianok, Angkola, Barumun, Sumani and Siulak faults and whish its slip rate 14 mm/y, 6 mm/y, 6.5 mm/y, 14 mm/y and 14 mm/y respectively[7]. Table 2 shows the result of the return period (Tr) estimation for Sianok, Angkola, Barumun, Sumani and Siulak faults Based on table 2 showed that Tr for Sianok fault earthquake (M 7.4 and slip rate 14 mm/y) is about of 172 years and less for the smaller magnitude. For Angkola fault (M 7.7 and slip rate 6 mm/y), Barumun fault (M 7.5 and slip rate 6.5 mm/y), Sumani fault (M 7.1 and slip rate 14 mm/y) and Siulak fault (M 7.2 and slip rate 14 mm/y) has Tr is about of 612, 438, 114, and 131 years respectively and less for the smaller magnitude.

4. Conclusion

Based on Indonesian Earthquakes Catalogue or historical earthquakes, West Sumatra Province is one of earthquake prone region in Indonesia. The seven destructive earthquakes at least caused seriously damaged in West Sumatra Province, six destructive earthquakes occurred caused by active faults in land of Sumatra and one destructive earthquake with source in Mentawai island (northwest Padang) and caused large damage in Padang City and its vicinity, namely the 2009 Padang earthquake (Mw 7.6).

Then we assumed that all active faults in West Sumatra region as like as Sianok, Angkola, Barumun, Sumani and Siulak faults can cause earthquakes in the future. Based on simulation with using scenario magnitude (Mw 7.1-7.7) and slip rate 6-14 mm/y, the simulation result shows that the intensity VII-VIII MMI, PGA and PGV values is about of 42-63% and 30-50 cm/s respectively. Meanwhile the estimation of Tr for maximum magnitude scenario these faults is about 114-612 years and less for the smaller magnitude. This study is hoped to be able to support the earthquake risk mitigation program in the future in West Sumatra and its vicinity region.

Acknowledgements
We are grateful to Indonesian Agency for Meteorology, Climatology, and Geophysics (BMKG) and other institutions for data supporting in this preliminary study.

References
[1] The Indonesia Agency for Meteorology, Climatology, and Geophysics (BMKG). (2022). Pusat Gempabumi dan Tsunami-PGN. Jakarta.
[2] BMKG. (2020). Indonesian Earthquakes Catalogue. Jakarta.
[3] Pusat Vulkanologi dan Mitigasi Bencana Geologi. (2008). *Katalog Gempabumi Merusak di Indonesia tahun 1629 - 2007*. Badan Geologi. Departemen Energi dan Sumber Daya Mineral.
[4] United State Geological Survey Website (2022). http//www.usgs.gov earthquake.
[5] Newcomb, K. R. and McCann, W. R., (1987). "Seismic History and Seismotectonics of Sunda Arc." *Journal of Geophysical Research: Solid Earth, Volume 92, Issue B1, pp. 421-439.*
[6] Sukendar Asikin. (1985). *Geologi Struktur Indonesia*. Lab. Geologi Dinamis-Geologi Institut Teknologi Bandung. Bandung.
[7] PUSGEN (Indonesian Earthquake Study Center). (2017). *Peta Sumber dan Bahaya Gempa Indonesia Tahun 2017*. Consortium of some relevance of Indonesia earthquake institutions.
[8] Shibazaki.B. (2006). *Lecture Note –References Related to Scaling Relations of Earthquakes*. International Institute of Seismology and earthquake Engineering. Building Research Institute. Tsukuba, Japan.
[9] Davidson, F.C. and Scholz, C.H. (1985). "Frequency Moment Distribution of earthquake arc: A Test the Characteristic Earthquake Model." *Bulletin of the Seismological Society of America. Vol.75. No.5. pp. 1349-1361*. October 1985.
[10] Fathani, T.F. (2007). "The Determination of Peak Ground Accelaration at Opak-Oyo Fault, Bantul Regency, Yogyakarta." *Proceeding AUN/SEED-Net*. The Yogyakarta Earthquake 2006. Geological Engineering Department, Faculty of Engineering, Universitas Gadjah Mada. Indonesia.

doi:10.1088/1755-1315/1244/1/012038

Building Treatments in Numerical Modeling of Dam Break Induced Flow in Urban Area

C Sandi[1], M B Adityawan[2], A K Nurendyastuti[3], M Farid[4], D Harlan[2], A A Kuntoro[2], and Widyaningtias[2]

[1] Graduate School of Water Resources Management, Faculty of Civil and Environmental Engineering, Bandung Institute of Technology, Bandung, Indonesia
[2] Center for Water Resources Development, Bandung Institute of Technology, Bandung, Indonesia
[3] Graduate School of Civil Engineering, Faculty of Civil and Environmental Engineering, Bandung Institute of Technology, Bandung, Indonesia
[4] Center for Coastal and Marine Development, Bandung Institute of Technology, Bandung, Indonesia

calvinsandi458@gmail.com

Abstract. Dam break is one of the most dangerous disasters, which cause a large mass of water from the reservoir is suddenly released to the downstream. Although dam break events are still rare, the occurrences can cause severe damage to the downstream area, especially if the downstream is an urban area. Urban flood due to dam break has a different characteristic compared to the flood in nature. Modeling the flood is good way to learn the urban flood characteristic. Therefore, further studies are needed to develop an urban dam break model. Numerical modeling is one of the most effective methods to simulate the flood. In numerical modeling, buildings can be modeled with 3 approaches, as a wall boundary, as an area with high manning value, and as a contour. This paper discusses about two-dimensional flow of a dam break which flows through a sloping channel and single building, particularly comparing 3 approaches to modeled the building. The case is based on laboratory experiment of previous study. The model was developed with 2D Shallow Water Equations (SWE) which are based on the conservation of mass and momentum derived from Navier Stokes equation. The equations then discretized using Mac-Cormack numerical scheme. Numerical filter is used to reduce the numerical instability or oscillation in the model. The numerical model results are then compared to the experimental result from previous study. The accuracy of the numerical model was checked using Root Mean Square Error (RMSE) to determine the best methods to modeled the building. The results indicate that modeling the building as a wall boundary have the best results with lowest RMSE value and also shown the closest comparison of water level compared to the experimental results. By conducting this study, it is hoped to assist the urban dam break modeling in the future.

1. Introduction

Dam has a multipurpose function, such as irrigation, hydropower, etc., which make it a valuable infrastructure for a nation. Indonesia's government has given great attention to the availability of water resources infrastructure, especially dam. It is proven that from 2015 to 2021, 27 dams have been built in several areas in Indonesia. However, every dam has a hazard potential, namely dam break. Therefore, this factor must be considered in designing a dam. There are several factors that can caused dam break, such as geotechnical failure, construction strength, even natural disaster including earthquake [1]. Dam break will cause a large mass of water release to the downstream in a sudden, causing a heavy damage to its downstream [2], especially if the downstream is an urban area, as in the case of the Way Ela dam failure in 2013 [3]. Moreover, the lack of mitigation system became a contributing factor in causing a heavy loss in the affected area [4]. Thus, a mitigation plan for dam break must be conducted in designing a dam through risk analysis [5]. It has been proven from history that implementing mitigation plans can reduce losses significantly.

However, analyzing urban flood has its own challenge, as it has a different characteristic from flood in nature. One of the factors that influence is the roughness factor, which significantly affect the inundation patterns [6]. For instance, a study shown that the velocity of water that go through a building's layout is much slower than the one that go through an open area [7]. Numerical modeling is one of the best methods to analyze the urban flood characteristics. This method offers low cost and easy to use, yet still provides a good accuracy [8][9]. To model an urban flood, building must be defined. According to studies by [2][10], building in numerical modeling can be modeled with three approaches, namely a wall boundary approach, a high manning value approach, and a contour approach. Despite its wide used, the best building approach in term of accuracy has yet to be determined.

Therefore, the objective of this study is to compare three building approaches to obtain the most suitable approach to simulate buildings in a numerical model by comparing the simulation results with experimental results. Numerical modeling in this study was conducted using MATLAB will be carried out to study the flow due to dam break against a building. The governing equation is based on the St. Venant 2D Equation that discretized using Finite Difference Mac Cormack scheme with additional numerical filter. The Mac Cormack scheme was chosen because it has the best accuracy base on a study conducted by [5][11][12]. Hansen Numerical Filter chosen to reduce the oscillation and help stabilize the numerical model [8][13][14]. For the experiment case, a dam break through a sloped channel and single building is chosen, which was conducted and studied by [15].

2. Research Methodology

2.1. Governing Equations

The equation for numerical simulation is using 2D Saint Venant equation, which consists of continuity and momentum equation, as shown in as follows:

Continuity:

$$\frac{\partial h}{\partial t}+\frac{\partial (hu)}{\partial x}+\frac{\partial (hv)}{\partial y}=0 \tag{1}$$

Momentum x-direction:

$$\frac{\partial u}{\partial t}+u\frac{\partial u}{\partial x}+v\frac{\partial u}{\partial y}+g\frac{\partial h}{\partial x}-g(So_x-Sf_x)=0 \tag{2}$$

Momentum y-direction:

$$\frac{\partial u}{\partial t}+u\frac{\partial u}{\partial x}+v\frac{\partial u}{\partial y}+g\frac{\partial h}{\partial x}-g(So_x-Sf_x)=0 \tag{3}$$

IOP Conf. Series: Earth and Environmental Science 1244 (2023) 012038 doi:10.1088/1755-1315/1244/1/012038

2.2. *Numerical Scheme*

As mentioned previously, the Mac-Cormack scheme is used to solve the 2D Saint Venant equation. This scheme contains two steps, namely predictor and corrector steps. In the predictor step, the forward difference scheme is used, while in the corrector step, the backward difference scheme is used. The value in the next time step for each hydrodynamic parameter is the average of the predictor and corrector step results [2]. The discretization process of the equation is given below.

The predictor step is as follows:

- Continuity

$$h^*_{i,j}=h^t_{i,j}-\frac{\Delta t}{\Delta x}(u^t_{i+1,j}h^t_{i+1,j}-u^t_{i,j}h^t_{i,j})-\frac{\Delta t}{\Delta y}(v^t_{i,j+1}h^t_{i,j+1}-v^t_{i,j}h^t_{i,j}) \quad (4)$$

- Momentum x-direction

$$u^*_{i,j}=u^t_{i,j}-\frac{\Delta t}{\Delta x}u^t_{i,j}(u^t_{i+1,j}-u^t_{i,j})-\frac{\Delta t}{\Delta y}v^t_{i,j}(u^t_{i,j+1}-u^t_{i,j})-\frac{\Delta t}{\Delta x}g(h^t_{i+1,j}-h^t_{i,j})-\frac{\Delta t}{\Delta x}g(z_{i+1,j}-z_{i,j})-gsf^{\,t}_{x_{i,j}}\Delta t \quad (5)$$

- Momentum y-direction

$$v^*_{i,j}=v^t_{i,j}-\frac{\Delta t}{\Delta y}v^t_{i,j}(v^t_{i,j+1}-v^t_{i,j})-\frac{\Delta t}{\Delta x}u^t_{i,j}(v^t_{i+1,j}-v^t_{i,j})-\frac{\Delta t}{\Delta y}g(h^t_{i,j+1}-h^t_{i,j})-\frac{\Delta t}{\Delta y}g(z_{i,j+1}-z_{i,j})-gsf^{\,t}_{y_{i,j}}\Delta t \quad (6)$$

The corrector step is as follows:

- Continuity

$$h^{**}_{i,j}=h^t_{i,j}-\frac{\Delta t}{\Delta x}(u^*_{i,j}h^*_{i,j}-u^*_{i-1,j}h^*_{i-1,j})-\frac{\Delta t}{\Delta y}(v^*_{i,j}h^*_{i,j}-v^*_{i,j-1}h^*_{i,j-1}) \quad (7)$$

- Momentum x-direction

$$u^{**}_{i,j}=u^t_{i,j}-\frac{\Delta t}{\Delta x}u^*_{i,j}(u^*_{i,j}-u^*_{i-1,j})-\frac{\Delta t}{\Delta y}v^*_{i,j}(u^*_{i,j}-u^*_{i,j-1})-\frac{\Delta t}{\Delta x}g(h^*_{i,j}-h^*_{i-1,j})-\frac{\Delta t}{\Delta x}g(z_{i,j}-z_{i-1,j})-gsf^{\,*}_{x_{i,j}}\Delta t \quad (8)$$

- Momentum y-direction

$$v^{**}_{i,j}=v^t_{i,j}-\frac{\Delta t}{\Delta y}v^*_{i,j}(v^*_{i,j}-v^*_{i,j-1})-\frac{\Delta t}{\Delta x}u^*_{i,j}(v^*_{i,j}-v^*_{i-1,j})-\frac{\Delta t}{\Delta y}g(h^*_{i,j}-h^*_{i,j-1})-\frac{\Delta t}{\Delta y}g(z_{i,j}-z_{i,j-1})-gsf^{\,*}_{y_{i,j}}\Delta t \quad (9)$$

The value of h, u, and v for the next time step are as follows:

- Continuity

$$h^{t+1}_{i,j}=\frac{h^*_{i,j}+h^{**}_{i,j}}{2} \quad (10)$$

- Momentum x-direction

$$u_{i,j}^{t+1}=\frac{u_{i,j}^{*}+u_{i,j}^{**}}{2} \tag{11}$$

- Momentum y-direction

$$v_{i,j}^{t+1}=\frac{v_{i,j}^{*}+v_{i,j}^{**}}{2} \tag{12}$$

2.3. Numerical Filter

Numerical filter is applied to help stabilize the numerical model. The instability of the model can be caused by shock waves, which may occur in downstream of the dam. Unstable model will give an oscillating or undefined results. The filter itself will act as an artificial dissipation, as there is no energy dissipation in numerical modeling [16]. In this study, Hansen numerical filter was used. The filter will be applied to the water velocity (x and y direction) and water depth parameters using C coefficient by the following equation:

$$F_{i,j} = CF_{i,j} + (1 - C)\frac{(F_{i-1,j} + F_{i+1,j} + F_{i,j-1} + F_{i,j+1})}{4} \tag{13}$$

3. Result and Discussion

Before the experiment case was modeled, the reliability of Mac Cormack scheme with Hansen Filter was tested in 1D case. The case used to test the scheme is an 1D analytical dam break calculation from a study by [17]. The comparison of results from this study and analytical results is shown in Figure 2.

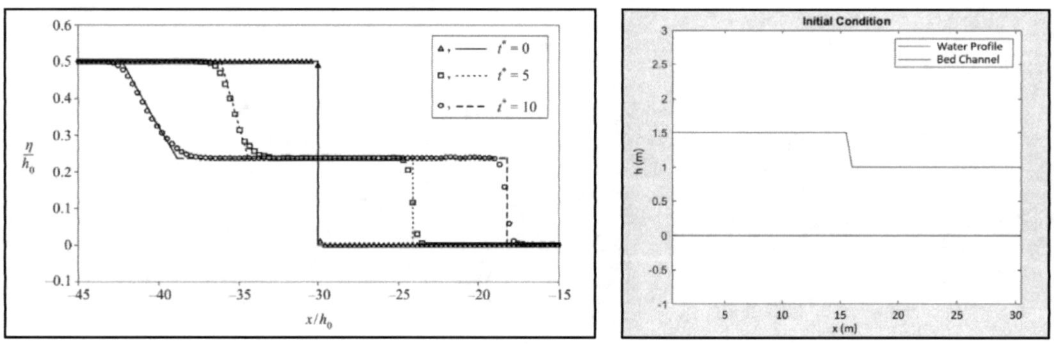

Figure 1. Analytical Results (left) and Initial Condition Set Up (right)

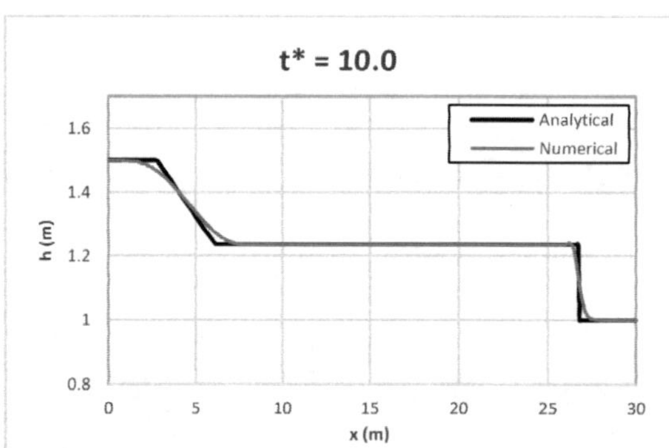

Figure 2. Analytical Results and Numerical Results at t*=10

The results indicates that the numerical results using Mac Cormack scheme have shown good results compared to the analytical results. Therefore, the Mac Cormack scheme is reliable for 1D numerical simulation. The Mac Cormack scheme with Hansen Filter is then tested to test the reliability in 2D simple case numerical simulation. The case used in the second test is a 2D circular dam break, which then compared to the analytical results [18]. The comparison of results is shown in Figure 4.

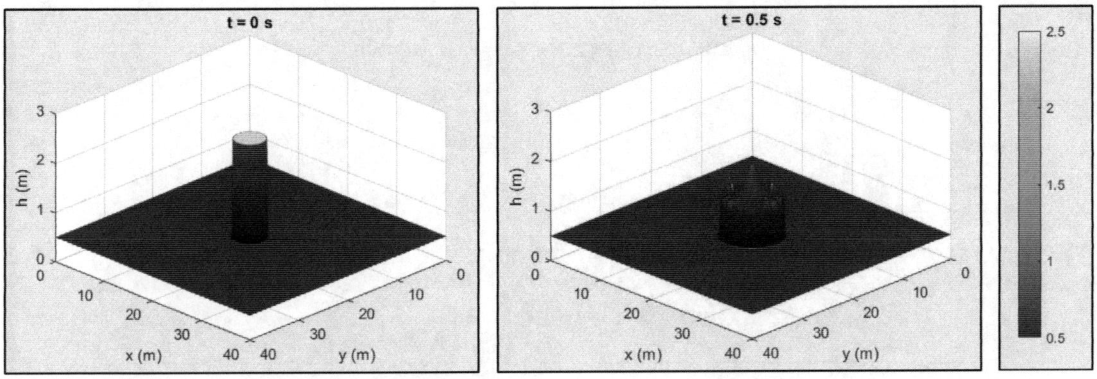

Figure 3. Computed Image of the Flow at t = 0 s (left) and t = 0.5 s (right)

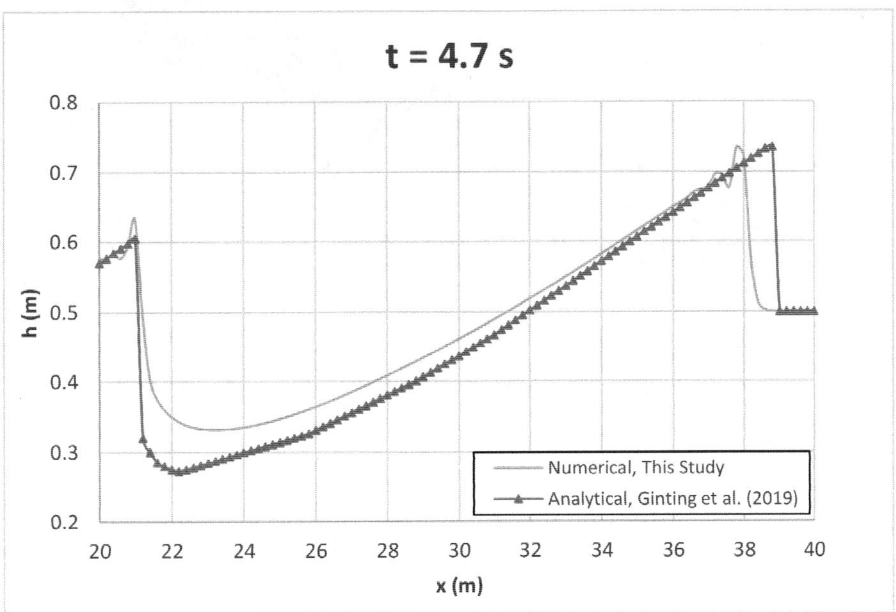

Figure 4. Analytical Results and Numerical Results at t = 4.7 s

In general, the results indicates that the numerical results using Mac Cormack scheme have shown good results compared to the analytical results [18]. Therefore, the Mac Cormack scheme is considered reliable and can be use in the main case which is a more complex model. The main model case to be analyse is an experimental case. The experiment was conducted in a flume with a size of 20 x 1.2 m. A reservoir with 0.1 m initial water depth, was placed in the upstream, followed by a 1:2 slope leading up into a flat floodplain. Hansen Filter coefficient (C) was set to 0.99. The downstream boundary was free created by a step for discharge.

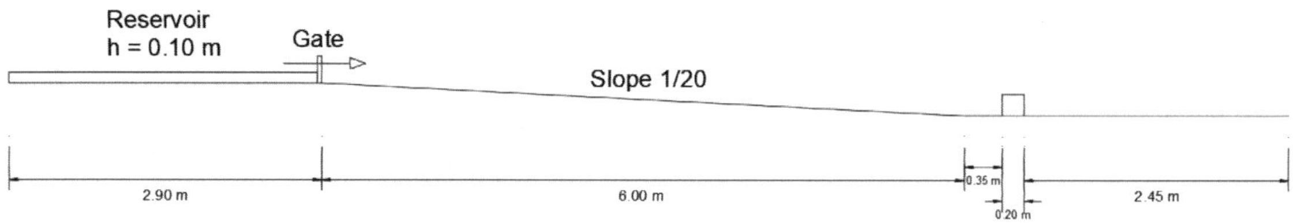

Figure 5. Experimental Case Sketch

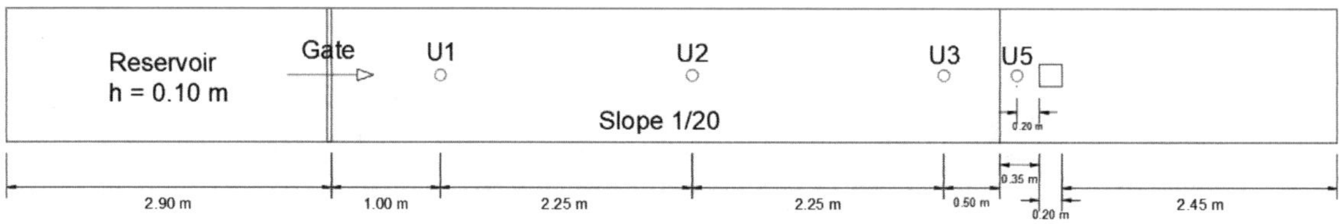

Figure 6. Control Points Position

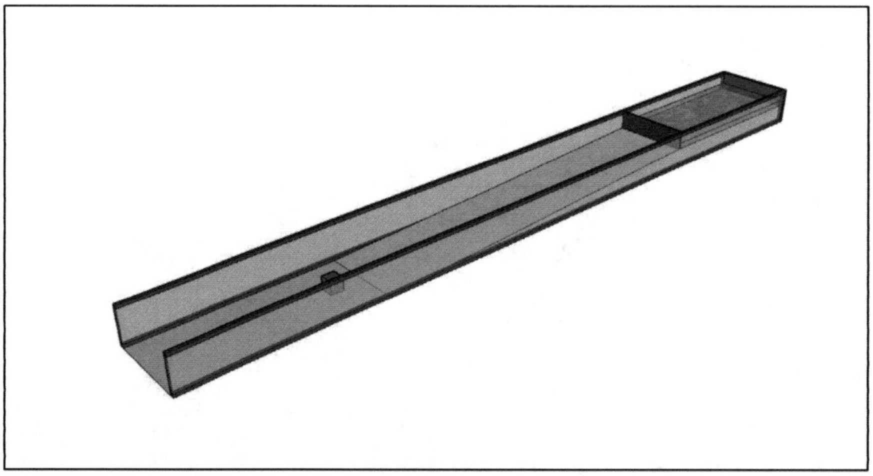

Figure 7. 3D Illustration of Experimental Case

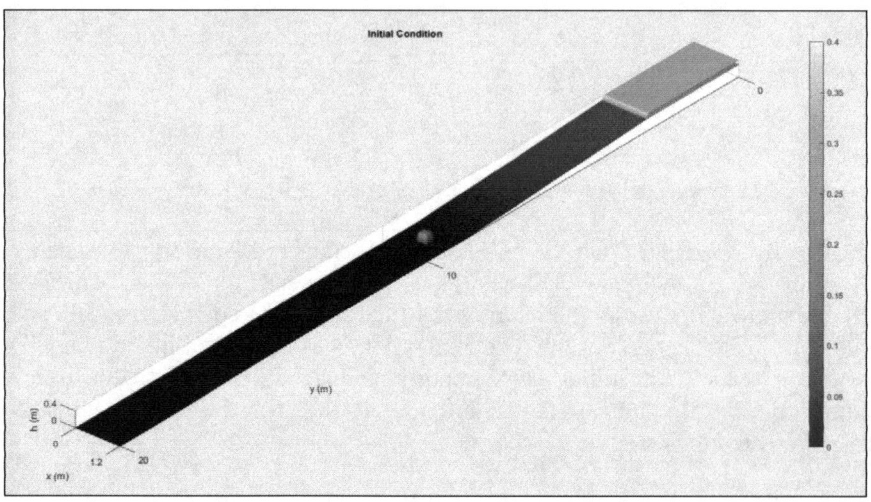

Figure 8. Set Up Initial Condition

Water depth is monitored at four control points, namely U1, U2, U3, and U5 as shown in Figure 6. The grid size of 5 cm was used to discretized the numerical model. The model will be simulated for 20 seconds. with 0.001 seconds time step used. The results of numerical modelling for 3 different approaches to define the building are described in the following sections.

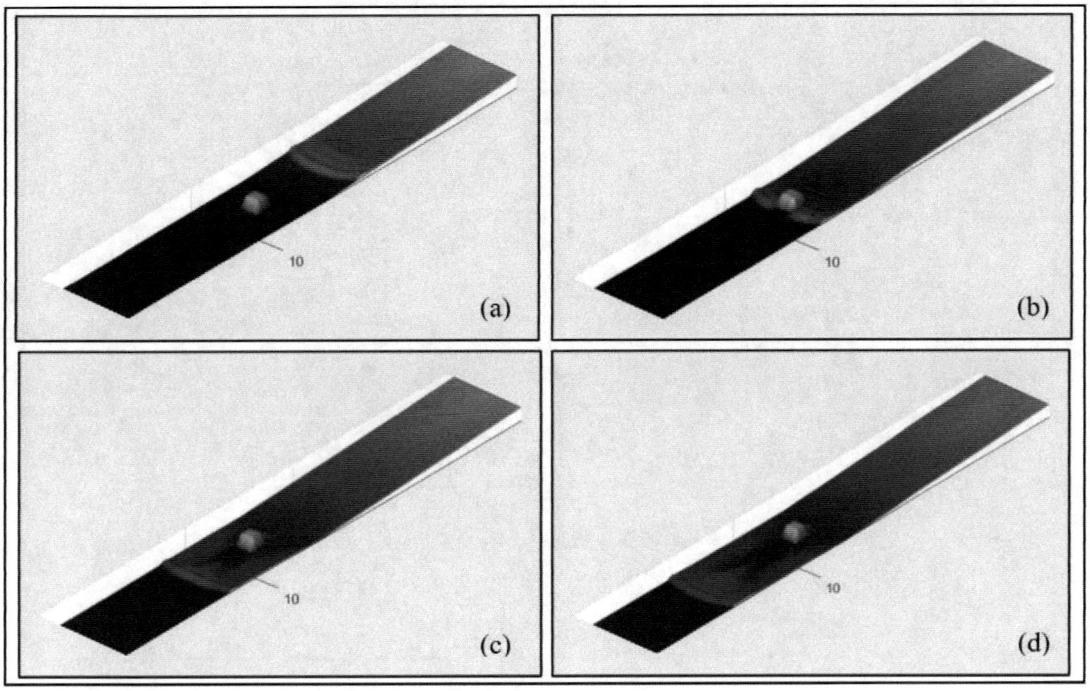

Figure 9. Computed Image of the Flow at t = 4 s (a), t = 5 s (b), t = 6 s (c), t = 7 s (d)

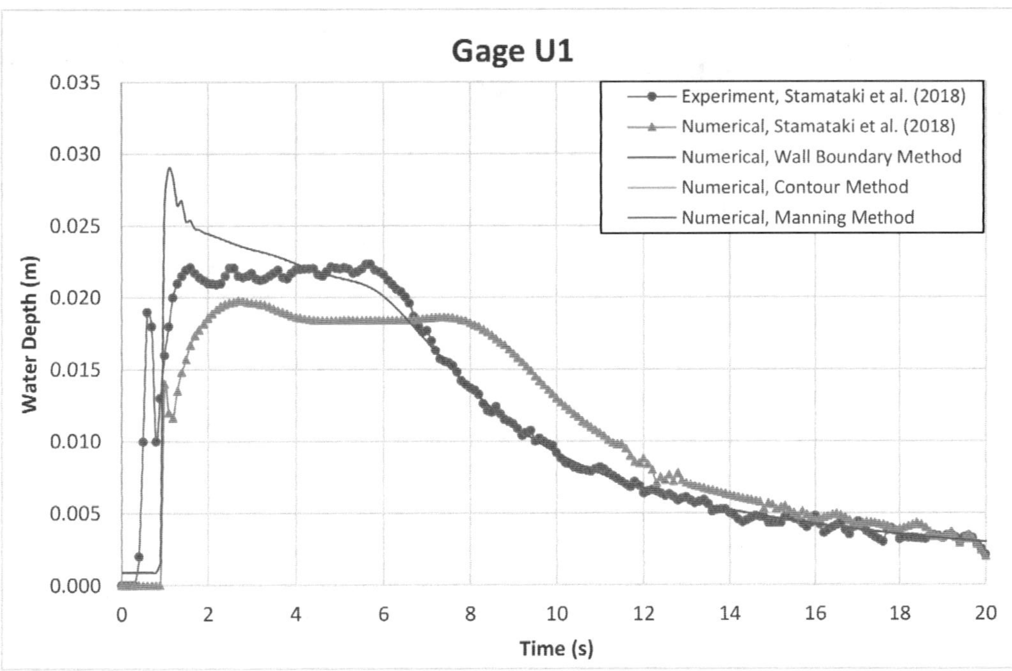

Figure 10. Results at Gage U1

IOP Conf. Series: Earth and Environmental Science 1244 (2023) 012038 doi:10.1088/1755-1315/1244/1/012038

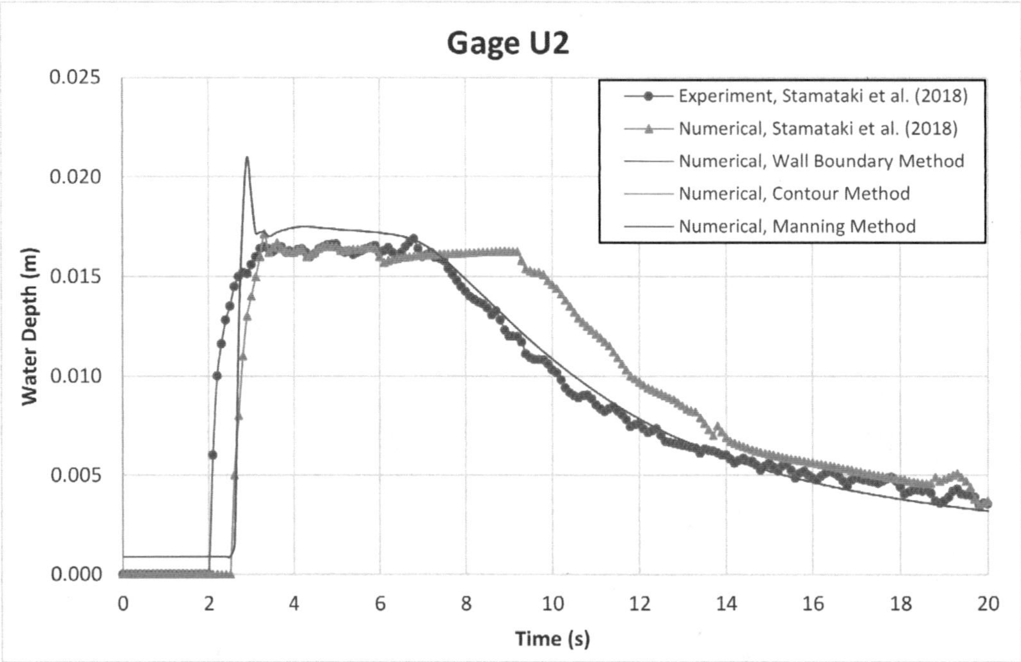

Figure 11. Results at Gage U2

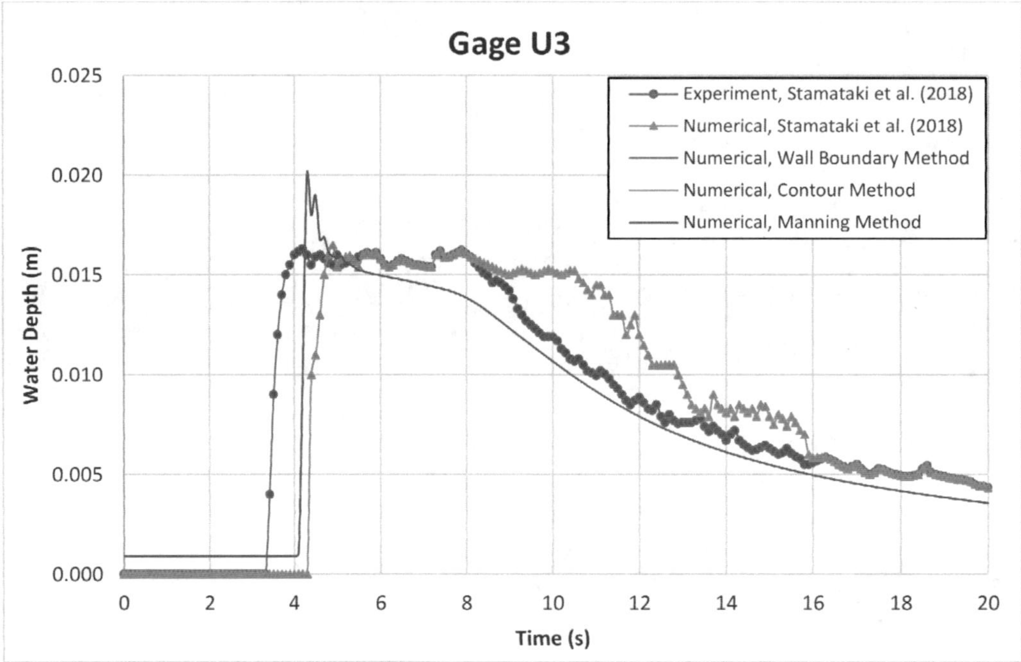

Figure 12. Results at Gage U3

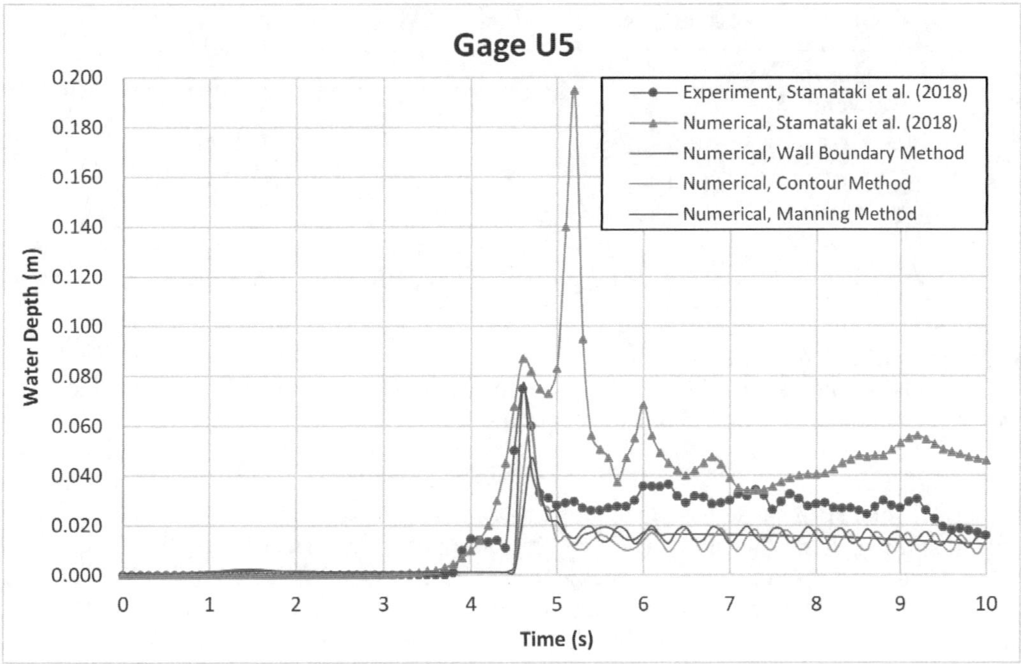

Figure 13. Results at Gage U5

The model results from every building approach have shown good results compared to the experimental results from [15]. In general, the model results in this study also shown a better result, since the water fluctuations from this study has a closer result compared to the experimental results. In addition, the numerical results from previous study have a delay in all control point and have a very high water fluctuations in control point U5. Meanwhile, this study shown a similar water fluctuation in control point U5 compared to the experiment results until 5 seconds of the simulation, which the water just hit the building. The results obtained after 5 seconds of the simulation generally have a low water level compared to the experiment results.

The difference in results that occur is due to the use of different building approach and scheme in the numerical model. In this study, finite difference was used to solve the equation, while the previous study used OpenFOAM software to simulate the case which use the finite volume method to solve the equation. The upstream water fluctuation (U1-U3) shows no difference in the results of each building approach because the building has no effect on the flow at upstream control points. The control point affected by the building is Gage U5, which has shown a different result on each building approach as shown in Figure 7.

The RMSE (Root Mean Square Error) method was used to determine the best building approach to define a building on numerical simulation. This method calculates the magnitude of the difference between the experimental results with the numerical results in the form of error. Thus, the approach that has the lowest RMSE can be stated as the best approach, since it has the highest accuracy compared to the experimental results. The RMSE equation is as follows:

$$\mathrm{RMSE} = \sqrt{\frac{\sum_{i=1}^{n}\left(y_{experimental} - y_{numerical}\right)^2}{n}} \tag{14}$$

Table 1. Recapitulation of RMSE Value

Control Point	RMSE		
	Wall Boundary	**Manning**	**Contour**
U1	0.00270	0.00270	0.00270
U2	0.00205	0.00205	0.00205
U3	0.00274	0.00274	0.00274
U5	0.01103	0.01228	0.01290
Total	**0.01852**	**0.01977**	**0.02039**

The RMSE value in every control point for every building approach is shown in Table 1. From Table 1, it is obtained that the wall boundary approach has the best accuracy, which is indicated by the the lowest RMSE value compared to other approaches.

4. Conclusion

The numerical model results compared to experimental results from [15] have shown good comparison, and in general this study has better result than the numerical model from previous study. Furthermore, from the results of the main case, it can be concluded that defining a building with the wall boundary approach give the best result with lowest RMSE value compared to Manning and Contour approach. Although the wall boundary approach gives the best result, this approach has a disadvantage since the water cannot overflow the building because of the wall boundary condition. Thus, the wall boundary approach is not recommended to be used to define a short building, it can only be used on a tall building or in a case where the water will not overflow into the building.

For further development for numerical case on urban area with more complex scenario, the building can be modeled using the wall boundary approach for a tall building. Until now, there were many dam break studies in the real dam situation, that was conducted using numerical modeling. Thus, it is hoped that this study can assist other researchers who are conducting a dam break study using numerical modeling on a real dam situation case or on a simplified experiment case.

Acknowledgments

This study was supported by Institut Teknologi Bandung Research 2022 (Riset ITB 2022), P2MI ITB 2022 and Riset Unggulan 3P 2022.

References

[1] Maitsa T R, Mardika M G I, Adityawan M B, Harlan D, Kusumastuti D and Kuntoro A A 2020 2D numerical simulation of urban dam break and its effect to building using lax scheme with numerical filter *E3S Web of Conferences* (Padang) **156** 04003

[2] Kusuma M S B, Farid M and Adityawan M B 2008 Numerical model study of two dimension flow generated by a dam break *International Conference on Earthquake and Disaster Mitigation* (Jakarta) p 415

[3] Yakti B P, Adityawan M B, Farid M, Suryadi Y, Nugroho J and Hadihardaja I K 2018 2D modeling of flood propagation due to the failure of Way Ela Natural Dam *MATEC Web of Conferences, The Third International Conference on Sustainable Infrastructure and Built Environment* (Bandung: EDP Sciences) **147** 03009

[4] Purnama M R, Adityawan M B, Farid M, Chrysanti A, Rayadi B T and Suryadi Y 2021 Development of tsunami inundation map for the coast of Palu City *IOP Conference Series: Earth and Environmental Science* (Jakarta) **737** 012049

[5] Farid M, Yakti B P, Rizaldi A and Adityawan M B 2016 Finite difference numerical scheme for simulating dam break flow *The 5th HATHI International Seminar Water Resilience in a Changing World* (Bali) p 173

[6] Adityawan M B, Rahmasari P, Chrysanti A, Farid M, Yakti B P and Purnama M R 2021 Effect

of land cover on tsunami overland flow propagation: a case study of Painan, West Sumatra, Indonesia *International Journal on Advanced Science, Engineering and Information Technology* **11** 5 p 1704

[7] Sihombing Y I, Adityawan M B, Chrysanti A, Widyaningtias, Farid M, Nugroho J, Kuntoro A A and Kusuma M A 2019 Tsunami overland flow characteristic and its effect on Palu Bay due to the Palu Tsunami 2018 *Journal of Earthquake and Tsunami* **14** p 1

[8] Putri P I D, Iskandar R F, Adityawan M B, Kardhana H and Indrawati D 2020 2D shallow water model for dam break and column interactions *Journal of the Civil Engineering Forum* **6** 3 p 237

[9] Zendrato N L H, Harlan D, Adityawan M B and Natakusumah D K 2019 1D numerical modelling of dam break using finite element method *MATEC Web of Conferences* **270** 04022

[10] Haile A T and Rientjes T H M 2005 Effects of Lidar DEM resolution in flood modeling *ISPRS WG III/3, III/4, V/3 Workshop "Laser Scanning 2005"* (Enschede) p 168

[11] Zendrato N L H, Chrysanti A, Yakti B P, Adityawan M B, Widyaningtias and Suryadi Y 2018 Application of finite difference schemes to 1D St. Venant for simulating weir overflow *MATEC Web of Conferences* **147** 03011

[12] Natakusumah D K, Kusuma M S B, Darmawan H, Adityawan M B and Farid M 2007 Pemodelan hubungan hujan dan aliran permukaan pada suatu DAS dengan metoda beda hingga *PROC. ITB Sains & Tek.* (Bandung) **39 A** 1 & 2

[13] Harlan D, Adityawan M B, Natakusumah D K and Zendrato N L H 2019 Application of numerical filter to a Taylor Galerkin finite element model for movable bed dam break flows *International Journal of GEOMATE* **16** 57 p 209

[14] Mellivera A, Zain K, Adityawan M B, Harlan D, Farid M and Yakti B P 2020 Development of FTCS artificial dissipation for dam break 2D modelling *Jurnal Teknik Sipil* **27** 1 p 1

[15] Stamataki I, Zang J, Buldakov E, Kjeldsen T and Stagonas D 2018 Study of dam break flow interaction with urban settlements over a sloping channel *E3S Web of Conferences* (Lyon-Villeurbanne) **40** 06006

[16] Adityawan M B and Tanaka H 2012 Bed stress assessment under solitary wave run-up *Earth Planets and Space* **64** 12 p 945

[17] Li Y and Raichlen F 2002 Non-breaking and breaking solitary wave run-up *Journal of Fluid Mechanics* **456** p 295

[18] Ginting B M and Mundani R P 2019 Comparison of shallow water solvers: applications for dam-break and tsunami cases with reordering strategy for efficient vectorization on modern hardware *Water 2019* (Munich) **11** 639

ISSUES AND PROBLEMS AFFECTING POST-DISASTER RECONSTRUCTION ACTIVITIES IN INDONESIA

H Prawenti[1*], F Susilowati[2], R M Jannah[3], S H Susanto[4]

[1]Department of Civil Engineering, Universitas Tidar, Central Java, Indonesia

[234]Department of Civil Engineering, Universitas Tidar, Central Java, Indonesia

*corresponding author: herlitapr@untidar.ac.id

Abstract. Efforts to mitigate the danger of disasters continue. Both long-term and short-term Disaster Risk Reduction (DRR) activities have been undertaken by governments and International Organizations. One type of long-term post-disaster activity is rehabilitation and reconstruction, which are an important part of the catastrophe recovery phaseLong-term reconstruction activities are activities that have uncertainty. The reconstruction necessitates the integration of long-term development plans that account for various catastrophic risks and the possibility of mitigating future dangers. Based on past research, uncertainty variables show a problem with the activity of reconstruction activities. Numerous countries, including Malaysia, Nepal, America, and Estonia, have conducted research on the challenges and issues associated with construction activity. This research variable was developed from several studies conducted in several countries to obtain more comprehensive variables for study in the disaster-prone. Assessment instruments were developed using questionnaires distributed to a sample of respondents who met the initial criteria. Relative importance index (RII) is used in data processing procedures, and this paper examines the concerns and difficulties experienced during the reconstruction process. The data analysis found that the most dominant aspect that influences the construction period is the delay in project completion. With the value of the Relative Importance Index (RII) at the High-Medium (H–M) level (0.6 RI 0.8) with a value of 0.769 and all research variables are included in the High-Medium (H–M), namely ($0.6 \leq RI \leq 0.8$).

Keywords: *disaster risk reduction, long-term recovery, reconstruction activities, relative importance analysis*

1. Introduction

Disasters are destructive in nature and result in losses. The biggest loss experienced from the consequences of disasters is death. Within the framework, The purpose of organizing the reconstruction is to build return in the long term permanently part or all of physical and non-physical facilities and infrastructure, along with all institutional and service systems that have been devastated by disasters, so that their condition recovers and their functions can run well and the community can be better protected from various disaster threats.

Reconstruction planning begins at the post-disaster stage to formulate systematic steps to carry out a coordinated, comprehensive, and thorough reconstruction process by incorporating elements of disaster risk reduction (prevention, mitigation, and preparedness) from the start. There are at least ten strategies in reconstruction implementation, ranging from involving as much community participation as possible, both from the disaster-affected community and society in general, to ensure the availability of information regarding all reconstruction activities for all stakeholders in order to build communication in order to ensure accountability and transparency of the reconstruction process[1]. Coordination between stakeholders is also needed to support the implementation of reconstruction through mobilising both human, social, and economic resources. The question is whether all the resources necessary to fulfil these strategies have been properly utilised to support the implementation of the reconstruction.

Priority is given to long-term reconstruction projects in order to improve their condition and function and to protect the community from various disaster threats better. The reconstruction stage is an activity carried out after a disaster. And it has the long-term function of reducing disaster risk. Compared to new construction, reconstruction of operational facilities faces higher challenges, especially when many projects involved[2]. One of the related issues regarding the recovery period is shown in Figure 1 [3]

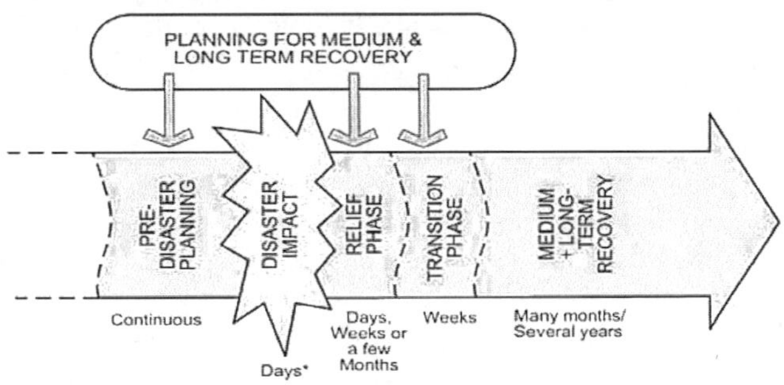

Figure 1 Recovery Plan Time Requirements

Source : Mind the Gap! Post-disaster reconstruction and the transition from humanitarian relief.p.9, RICS by the Max Lock Center, 2006

It takes years to initiate long-term recovery. Whereas long-term recovery is urgently needed for the affected communities. This study aimed to ascertain the management of reconstruction issues in several regions of Indonesia, as well as the extent to which they influenced reconstruction activities. According to the data contained in the monitoring page for Palu rehabilitation and reconstruction, the problem lies in the provision of land for permanent housing relocation, where up to 1 year after the disaster occurred, problems with land have not been fully resolved. This, of course, can disrupt the reconstruction process, resulting in affected communities needing to wait several years to get their lives back. Several related studies were carried out in several disaster areas in Indonesia; this study used research variables used in several countries to find problems with reconstruction and target respondents more largely both in terms of agencies and regions. Finding issues and problems that have a significant influence during reconstruction activities can help formulate future strategies and minimize negative impacts on the reconstruction process.

2. Indonesian disaster overview

Throughout 2020, several types of disasters, namely floods, tornadoes, landslides, earthquakes, and tidal waves or abrasions, caused damage. Such damage consists of lightly damaged, moderately damaged, and severely damaged. Data on the amount of damage caused by natural disasters in Indonesia throughout 2020 can be seen in table 1

Table 1. Data on the amount of damage caused by natural disasters in Indonesia in 2020

Type of Damage	Total damage (unit)
House	42,762
Public Facility	1,542

Source : Badan Nasional Penanggulangan Bencana (BNPB)

The damage consisted of categories of light damage, moderate damage, and heavy damage. A total of 26,196 houses were lightly damaged (LD), 10,394 were heavily damaged (HD), and 6,172 were moderately damaged (MD). In addition, some infrastructure damage also occurred due to the disaster. The number of damage from disasters throughout 2020 to public facilities infrastructure was 1,542 units, including 727 worship facilities, 672 educational facilities, 442 bridges, 143 health facilities, and 134 office facilities. Apart from the damage, there were also recorded fatalities consisting of 536 injuries, 370 people died, and 39 people lost. The disaster caused substantial damage in several locations, necessitating restoration and reconstruction

3. Post-disaster Reconstruction

Regulation of the Head of the National Disaster Management Agency Number 11 of 2008 states that Reconstruction is the formulation of policies and efforts as well as well-planned, consistent and sustainable concrete steps to rebuild all infrastructure, facilities, and institutional systems permanently, both at the government and community levels, with the main target of growing economic, social and economic activities culture, the establishment of law and order, and the rise of the role and participation of civil society in all aspects of social life in the post-disaster area. Rehabilitation and reconstruction are very important parts of the disaster recovery phase.

Rehabilitation and reconstruction activities are aimed at achieving long-term recovery. Disaster recovery is a very significant stage in the disaster management cycle, whereby the recovery of support from government and non-governmental institutions in the aftermath of a disaster usually begins to recede and the affected communities struggle on their own [4].

However, the selected reconstruction activities are prioritized to long-term reconstruction activities, not just reconstruction for the fulfilment of the time. Reconstruction activities are activities that have uncertainty. These uncertainty variables can cause issues in the reconstruction activity; therefore, this study will identify the issues and problems faced at the reconstruction stage in the Indonesian territory.

4. Research methods

The method of this study consists of 3 (three) stages. First, the previous research study, then in the second stage, the dissemination of questionnaires from research variables that have been obtained through literature studies; and the third is looking for the influence of these variables on the failure of reconstruction implementation using RII (Relative Importance Index).

4.1.Data Analysis
The variable of this research is a literature study from previously conducted studies. From research conducted by Dzulkarnaen 2014 by identifying the challenges faced in reconstruction projects in several regions, namely Aceh, Sri Lanka, Bam, Wenchuan and Haiti, 12 challenges were found that non-governmental organizations had to face. In the research conducted by Pamidimukkala 2020 identified 177 journals, 3 main challenges were obtained, namely *the general challenge which consists of 6 aspects which are divided into 16 challenge variables, the social challenge challenge which consists of 5 aspects which is divided into 13 challenge variables*, the economic challenge *challenge which consists of 4 aspects divided into 10 challenge variables, and physical challenges challenges which consist of 5 aspects divided into 12 challenge variables.* The research conducted by Shresta 2021 obtained 7 aspects divided into 13 variables, regarding the issue found in urban housing reconstruction. Bilau Research, 2015, by reviewing previous research, obtained 8 issues from the implementation of post-disaster housing reconstruction.

These studies developed into six aspects, namely: problems that occurred during the construction period, which consisted of 19 issues; problems with management and planning, which consisted of 5 issues; problems with regulation, which consisted of 4 issues; and problems with land, which consisted of 4 issues; health and safety issues for the reconstruction site, which consists of 5 issues; and financial problems during the reconstruction process.

4.2.Respondent Profile
Presentation data was obtained through the distribution of questionnaires. A questionnaire is several written questions used to obtain information from respondents. Questionnaire by providing a list of indicators of research variables, which are then selected by respondents on a scale of 1 to 5. Where a scale of 1 does not have an influence up to scale 5, it greatly affects the failure of the implementation of reconstruction. Then the data is processed using quantitative descriptive analysis.
Data kuantita
The government has established a disaster management agency, with levels ranging from national and regional to implementing units[5]. Each of these levels has authority and function in disaster management (Figure 2).

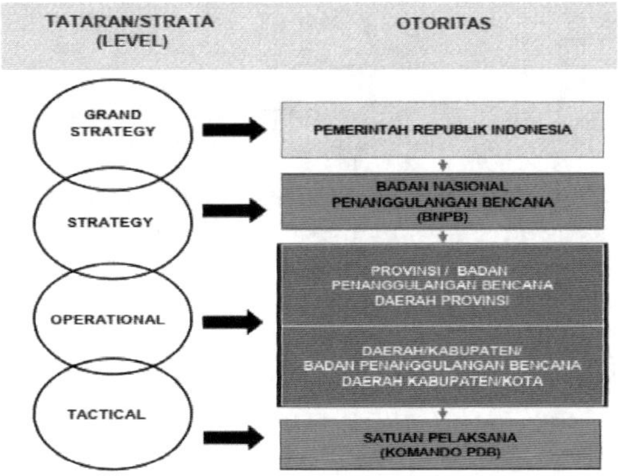

Figure 2 Level in Disaster Management Agency
Source : Konsep rencana Kontinjensi Bencana, BNPB, Herlianto 2021

The questionnaire was distributed to the DPU of central Java, the Ministry of Central PUPR, the DPUPR of several regions in Sumatera, several contractors and consultants who have been involved in the implementation of reconstruction in Indonesian as an implementing unit, BPBD of Central Java Region and BNPB as a Disaster Management Agency. The Returned and completed questionnaires totalled 26 questionnaires. The respondent's profile based on work institution can be seen in Figure 3

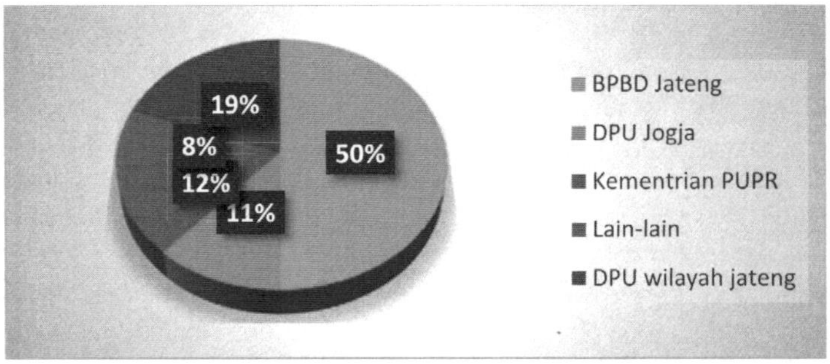

Figure 3. Respondent's profile based on work institution

The most dominant respondents in filling out this questionnaire were from the Regional Disaster Management Agency (BPBD) in several regions of Central Java, which is about half of the total number of respondents. The respondents "Miscellaneous" were respondents outside the grouping in figure 3, namely respondents who had worked at the Aceh Rehabilitation and Reconstruction Agency and the Semarang City Government. While the respondent's profile is based on the length of work experience can be seen in figure 4

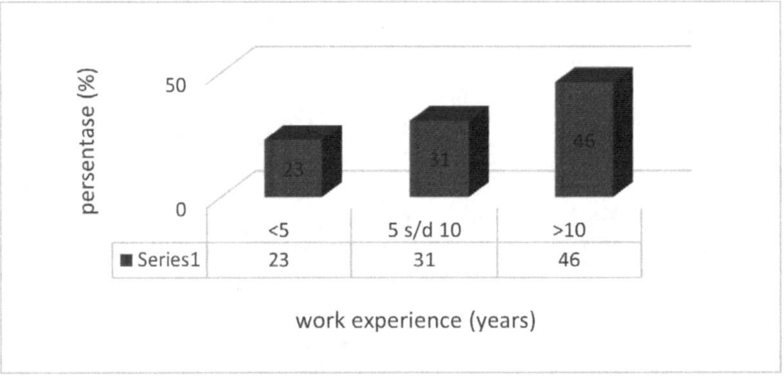

Figure 4. Respondent's profile based on the length of work experience

Almost half of the respondents have more than ten years of work experience. One-third of respondents have 5 to 10 years of work experience, and approximately one-fifth of the respondents have work experience under five years.

4.3. Validity Test

The validity test is used to test the extent of the accuracy or correctness of a questionnaire as a measuring tool for research variables.Validity test using Pearson Product Moment correlation. The specified significance level is 95%. Significance test through hypothesis testing where H0 stated that there was no relationship between these variables and problems in reconstruction (Invalid), and H1 stated that there was a relationship between these variables and problems in reconstruction (Valid). By comparing the p value with the alpha value and r count with r table, the test results are obtained. For degrees of freedom (df) = n - k, where n represents the number of research samples, and r table 0.3882. By testing, H0 is accepted if the r count value $\leq$ r table and p-value ≥ 0.05 (Invalid), while H0 is rejected if the r count value > r table and p value < 0.05 (Valid). The validity test using Minitab 2.0 software shows the p-value for all variables is 0.000 and r count > r table (Table 2). Consequently, all variables are valid.

IOP Conf. Series: Earth and Environmental Science 1244 (2023) 012039 doi:10.1088/1755-1315/1244/1/012039

4.4. Reliability Test

A reliability test aims to determine the dependability of a measuring instrument, i.e., if the instrument remains constant when remeasured. Cronbach's alpha is frequently applied to reliability tests on scales and questionnaires. This was typically regarded as 0.70 (five times) or >0.70 (three instances); however, one publication vaguely mentioned "the acceptable values of 0.7 or 0.6 [6].

The authors employed a wide variety of qualitative descriptors to interpret estimated alpha values. In some literature, alpha values were described as excellent (0.93 to 0.94), strong (0.91 to 0.93), reliable (0.84 to 0.90); robust (0.81); fairly high (0.76 to 0.95); high (0.73 to 0.95); good (0.71 to 0.91); relatively high (0.70 to 0.77); slightly low (0.68); reasonable (0.67 to 0.87); adequate (0.64 to 0.85); moderate (0.61 to 0.65); satisfactory (0.58 to 0.97); acceptable (0.45 to 0.98); sufficient (0.45 to 0.96); not satisfactory (0.4 to 0.55) and low (0.11) [7].

Reliability test using minitab20 software by producing an alpha value of 0.995. Moreover, All alpha values are greater than 0.7, indicating that all reliability coefficients are acceptable and the internal consistency of the criteria included in the scale is excellent.

4.5. Relative Importance Index (RII) Analysis

The feedback of the respondents was analyzed using the Microsoft Excel tool. In addition, this study used relative index analysis to rank the criteria according to their relative significance

The following formula is used to determine the relative index

$$Relative\ Index\ \ (RI) = \sum \frac{w}{A\,N} \tag{1}$$

where (w) represents the weighting assigned by each respondent on a scale from one to five, with one representing the least weight and five the highest. The value (A) is the highest sample weight, and (N) is the total sample size.

Based on survey results, a ranking analysis was conducted to determine the relative significance of impacts. It should be noted that the ratings on the scale show the order of importance of the criterion, not the relative weight of each rating. Relative index analysis is utilized to rank criteria based on their relative significance. Recapitulation of data analysis can be seen ini table 2.

Table 2. Data Analysis Recapitulation

Kode	Aspects	Indicators	Reference	Rvalue	Mean	StDv	Relative Indices	Rank Within Aspects
A								
A1	Construction Period	Delay in starting time project implementation	[8]	0.821	3.23	1.14	0.646	8
A2		Delay in project completion	[8]	0.726	3.85	1.05	0.769	1
A3		Limited Access to Resources	[8][9]	0.969	3.65	1.13	0.731	4
A4		Challenge in competition obtaining resources	[8][9]	0.901	3.38	1.20	0.677	7
A5		Construction Work Quality Issue	[9][10]	0.774	3.77	1.21	0.754	2
A6		Quality material Issue	[9][10]	0.871	3.62	1.30	0.723	5
A7		Absence of proper technical assistance	[9]	0.938	3.73	1.19	0.746	3

Kode	Aspects	Indicators	Reference	Rvalue	Mean	StDv	Relative Indices	Rank Within Aspects
A8		The inexperienced construction management team	[9][10]	0.966	3.77	1.34	0.754	2
A9		Cost Overruns	[8]	0.967	3.62	1.42	0.723	5
A10		Lack of Preliminary Assessment	[8]	0.916	3.38	1.33	0.677	7
A11		Absence of coordination between stakeholders	[8][10]	0.949	3.73	1.04	0.746	3
A12		Monitoring and Control Issues	[10]	0.962	3.65	1.13	0.731	4
A13		Logistics and supply chain Issues	[10]	0.941	3.50	1.33	0.700	6
A14		Risk Management Issues	[10]	0.973	3.62	1.42	0.723	5
A15		Human resources issues	[10]	0.971	3.77	1.21	0.754	2
A16		OHS Management Issue	[10]	0.949	3.73	1.04	0.746	3
A17		Inadequate project plan	[9]	0.971	3.77	1.21	0.754	2
A18		Time pressure challenge	[11]	0.973	3.50	1.33	0.700	6
A19		The complexity of building design	[9]	0.973	3.50	1.33	0.700	6
B	Management and Planning							
B1		Unclear reconstruction strategy	[9]	0.961	3.42	1.33	0.685	3
B2		Lack of Disaster Damage Assessment Database	[8]	0.945	3.58	1.27	0.715	1
B3		Inadequate operational management related to post-disaster recovery	[9]	0.721	3.38	0.94	0.677	4
B4		Less effective management	[9]	0.831	3.54	1.14	0.708	2
B5		There are no guidelines regarding Reconstruction activities	[11]	0.853	3.54	1.3	0.708	2
C	Regulation							
C1		Several policies related to the implementation of the reconstruction do not yet exist	[8]	0.987	3.69	1.29	0.738	1
C2		Institutional capacity for inadequate planning and regulation	[9]	0.945	3.58	1.14	0.715	4
C3		Ineffective policies of local government involvement	[11]	0.962	3.65	1.13	0.731	2
C4		Absence of Policy Involvement from the Government	[9]	0.945	3.58	1.27	0.715	4
D	Land							
D1		Issues with land ownership/land legality	[8][11]	0.987	3.69	1.29	0.738	1
D2		High land value	[9]	0.951	3.54	1.42	0.708	5
D3		Lack of land for reconstruction	[9]	0.931	3.58	1.27	0.715	3

Kode	Aspects	Indicators	Reference	Rvalue	Mean	StDv	Relative Indices	Rank Within Aspects
D4		Inadequate Land Size	[11]	0.878	3.12	1.31	0.623	6
D5		Poor land conditions	[9][11]	0.749	3.62	1.13	0.723	3
E	Health and Safety							
E1		Issues in health and safety management	[9][10]	0.902	3.50	1.27	0.700	3
E2		Environmental Risks	[9]	0.928	3.62	0.94	0.723	1
E3		Exposure to hazardous materials	[9]	0.826	3.27	1.28	0.654	4
E4		Geological properties of residential land issues	[10]	0.966	3.50	1.33	0.700	3
E5		Remoteness and geographical conditions	[10]	0.931	3.58	1.27	0.715	2
F	Financial							
F1		Financial Affordability Constraints	[11]	0.945	3.27	1.43	0.654	2
F2		Late disbursement/payment	[11]	0.951	3.38	1.24	0.677	1
F3		Problem with financial arrangements	[10]	0.883	3.27	1.37	0.654	2

The recapitulation data is presented in graphical form so that each indicator can be compared (Figure 5)

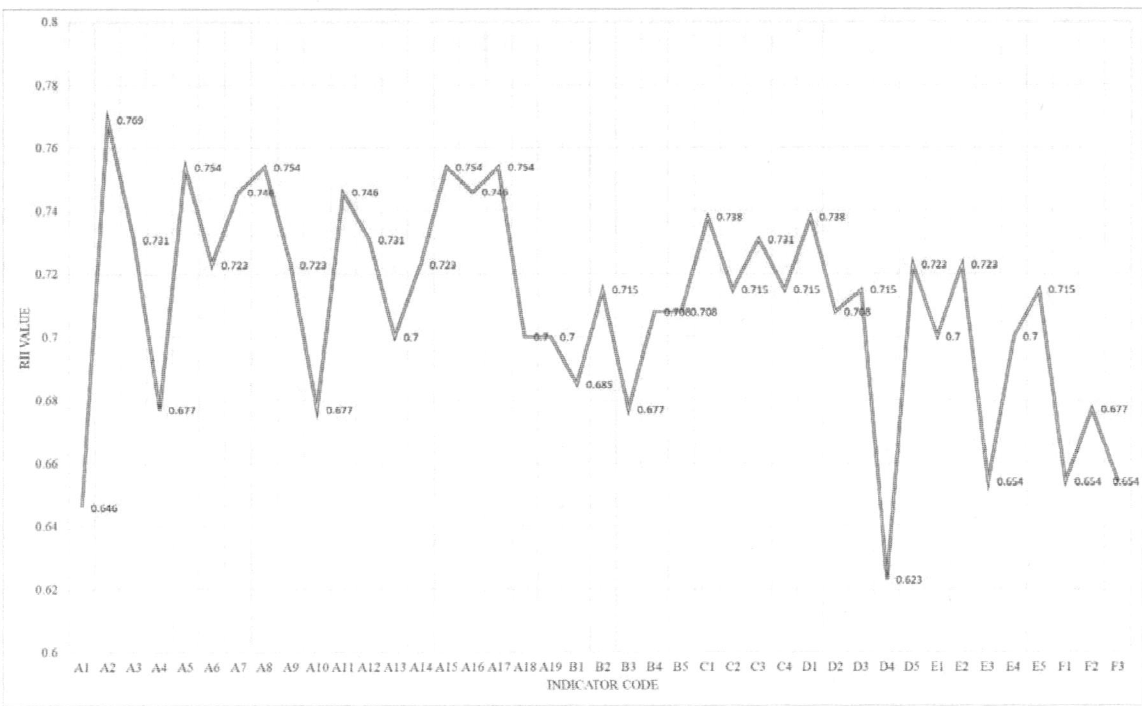

Figure 5. Recapitulation Data Chart

5. Discussion
Five significant levels are derived from the RI value: high (H) ($0.8 \leq RI \leq 1$), high-medium (H–M) ($0.6 \leq RI \leq 0.8$), medium (M) ($0.4 \leq RI \leq 0.6$), medium-low (M-L) ($0.2 \leq RI \leq 0.4$) and low (L) ($0 \leq RI \leq 0.2$) [12].

As can be seen in figure 5, all the relative indices values are in the High Medium (H–M) (0.6 RI 0.8). With the RI value for the Issues and Problem variables, the largest is in the category that occurred during the construction period. Based on the ranking (R) of relative indices (RI) for each aspect, problems in project completion delays are the most significant issues that occur in construction implementation (Code A). Then problems with quality construction work, inexperienced construction management team, problems with human resources, and inadequate project plans have the same Relative Indices (RI). In the aspect of management and planning (Code B), the lack of a disaster damage assessment database is the most significant issue. In the regulatory aspect (Code C), Several policies related to the implementation of the reconstruction do not yet exist are the most influential aspects, with a Relative indices (RI) value of 0.738. On the land aspect (Code D), problems with land ownership and land legality. From Health and Safety aspects (Code C), problems with environmental risk give the highest RI value. And in the financial aspect (Code F), the delay in disbursement/payment with a Relative indices (RI) value of 0.677, where this value still has a lower value than others.

Meanwhile, based on the ranking of relative indices (RI) of the overall from all respondent criteria (table 2), dominance occurs in aspects that occur during the construction period, where's that have a relative indices (RI) value of greater than 0.75 are all in the construction aspect with an average relative indices (RI) of 0.725. Meanwhile, the lowest average value of relative indices (RI) is in the financial aspect where the value of relative indices (RI) of all variables less than 0.7.

In similar research that has been carried out previously, challenges to resources in reconstruction in Gaza, the category of reconstruction projects ranks third out of 6 main categories. In this category, land issues are ranked second, and challenges to proper planning are ranked third[13]. Bilau 2015 suggests considerable challenges in the planning and implementation of reconstruction, which requires an appropriate institutional structure and involves affected communities. During construction, the implementation of reconstruction faced the same challenges to management as in general

6. Conclusion
This study takes research variables from various literature studies that have been carried out from several regions of cases in several countries. The author reviews these variables as variable issues and challenges in implementing the reconstruction studied for the Indonesian disaster area. A total of 41 research variables were taken and grouped into six aspects. Data collection using questionnaires that were distributed to several predetermined respondents. The Relative Importance Index (RII) method is used for data processing after the 41 variables are tested for validity and reliability. This method is used to see the most significant influence on the ranking. Based on the relative indices (RI), all research variables are included in the high-medium (H–M), namely ($0.6 \leq RI \leq 0.8$), From the processing of the data, it was found that issues regarding delays and human resources were the most influential in the implementation of reconstruction.

Acknowledgements
The author would like to thank the Regional Disaster Management Agency and the Ministry of Public Works for their willingness to share their knowledge and experience in this research. I would also express my gratitude to Tidar University for providing support and assistance in this research.

References
[1] BNPB, 'Peraturan Kepala BNPB No 11 Tahun 2008 Tentang Pedoman Rehabilitasi dan Rekonstruksi Pasca Bencana', *Bnpb*, p. 70, 2008.

[2] M. Attalla, T. Hegazy, and E. Elbeltagi, 'In-House Delivery of Multiple-Small Reconstruction Projects', *J. Manag. Eng.*, vol. 20, no. 1, pp. 25–31, 2004, doi: 10.1061/(asce)0742-597x(2004)20:1(25).

[3] RICS, 'Mind the gap! Post-disaster reconstruction and the transition from humanitarian relief', no. June, pp. 1–105, 2006.

[4] T. R. Works, 'Rehabilitation and', vol. 80, no. 12, pp. 384–396, 2000.

[5] M. H. Widyaiswara, 'Konsep rencana kontinjensi bencana', 2021.

[6] R. A. L. F. van Griethuijsen *et al.*, 'Global patterns in students' views of science and interest in science', *Res. Sci. Educ.*, vol. 45, no. 4, pp. 581–603, 2015, doi: 10.1007/s11165-014-9438-6.

[7] K. S. Taber, 'The Use of Cronbach's Alpha When Developing and Reporting Research Instruments in Science Education', *Res. Sci. Educ.*, vol. 48, no. 6, pp. 1273–1296, 2018, doi: 10.1007/s11165-016-9602-2.

[8] D. Ismail, T. A. Majid, R. Roosli, and N. A. Samah, 'Project Management Success for Post-disaster Reconstruction Projects: International NGOs Perspectives', *Procedia Econ. Financ.*, vol. 18, no. September, pp. 120–127, 2014, doi: 10.1016/s2212-5671(14)00921-6.

[9] P. Apurva, K. Sharareh, and S. Elnaz, 'Challenges in Post-Disaster Housing Reconstruction: Analysis of Urban vs. Rural Communities', pp. 103–110, 2020, doi: 10.3311/ccc2020-061.

[10] A. A. Bilau, E. Witt, and I. Lill, 'A Framework for Managing Post-disaster Housing Reconstruction', *Procedia Econ. Financ.*, vol. 21, no. 15, pp. 313–320, 2015, doi: 10.1016/s2212-5671(15)00182-3.

[11] C. B. Shrestha, B. Parajuli, and S. Malani, 'Unravelling constraints of urban housing reconstruction', *Prog. Disaster Sci.*, vol. 11, p. 100178, 2021, doi: 10.1016/j.pdisas.2021.100178.

[12] O. P. Akadiri, 'Development of a multi-criteria approach for the selection of sustainable materials for building projects', *PhD Thesis - Univ. Wolverhampt.*, pp. 1–437, 2011, [Online]. Available: http://wlv.openrepository.com/wlv/bitstream/2436/129918/1/Akadiri_PhD thesis.pdf.

[13] N. Saleh, A. Enshassi, and M. Sundermeier, 'Challenges Hindering the Resourcing for Post-Disaster Housing Reconstruction in Gaza', *J. Constr. Dev. Ctries.*, vol. 26, no. 2, pp. 183–210, 2021, doi: 10.21315/jcdc2021.26.2.9.

Design of Vertical Evacuation Building at Painan City Using Results From Tsunami Propagation Modeling

P Rahmasari[1], D Kusumastuti[2], and M B Adityawan[3]

[1]Department of Civil Engineering, Politeknik Negeri Pontianak, Indonesia,
[2]Faculty of Civil and Environmental Engineering, Institut Teknologi Bandung, Indonesia,
[3]Center for Water Resources Development, Institut Teknologi Bandung, Indonesia.

Corresponding author : 2294puspitarahmasari@gmail.com

Abstract. Fault under the Mentawai Islands has high potential to cause a large earthquake and triggering tsunami. Painan City, is located near the fault. With geographical condition on a bay surrounded by hills in the upstream area, a tsunami may cause a flow channeling effect and increase the inundation distance. An alternative of mitigation effort is to build a Vertical Evacuation Building (VEB) to help the community to reach a safe elevation within the evacuation time. Therefore, VEB should be designed to withstand tsunami impacts and to resist large earthquakes. This study conducts the design and analysis of the performance of multi-functional VEB structures at Painan City. The tsunami propagation simulation was carried out, and parameters obtained were used for structural design. Three VEB models were developed, i.e., (i) model with walls under tsunami and earthquake loads, (ii) model without walls under tsunami and earthquake loads, and (iii) model without walls under earthquake load only as the lateral load. Non-linear static pushover analyses were later conducted to evaluate the structural performances. The study revealed that debris force affects the most compared to other tsunami force components. Structural openings also affect the tsunami load received by the structure. Column moment force for models designed with tsunami load is greater than that of earthquake, while the difference is less significant for beam. The internal force differences occur along the tsunami inundated depth, thus influence the component design. From the pushover analysis, all models were found to meet the performance targets for earthquake.

1. Introduction

Lies between two tectonic plates, the Australian plate, and the Eurasian plate, make Indonesia exposed to earthquakes and tsunamis. The Mentawai fault has the potential to cause an earthquake larger than 8Mw in the near decades followed by a destructive tsunami [1]. Mentawai megathrust is close to the coast of West Sumatra, thus can pose a significant threat to communities in nearby towns. One of the cities that will be affected by the risk of a tsunami due to the earthquake source in the Mentawai Islands is Painan City [2-4]. Painan is a small town in Pesisir Selatan Regency, with an area of 108.92 km^2 and a population of 15982 people. Painan's location, which is very close to the earthquake source and near the bay, as well as its geographical condition which is surrounded by hills will exacerbate the impact of the tsunami on this city. Research showed that Painan will be severely affected by the tsunami with a maximum inundation depth of 6.37 m in the city area [7].

Mitigation efforts need to be done to minimize the impact of the tsunami that will occur. An alternative is to provide shelter as a strategy to save the community. The Vertical Evacuation Building (VEB) has the potential to save many lives, especially during a tsunami with a close source that does not provide enough time for evacuation (considering that the warning time is only less than 30 minutes from the earthquake as the first warning of an impending tsunami). This building functions as a rescue building so that the community can immediately reach a safe height to avoid being hit by the tsunami.

In addition to being able to withstand the tsunami, the VEB must also be designed in such a way that it can survive when it is hit by a large earthquake, hence the design of the VEB must also meet the requirements of earthquake-resistant buildings. The design of building structures with tsunami loading itself is very few in Indonesia. In addition, there is no locally standardized regulation on how to design a tsunami-resistant building structure. Therefore, in this study, a design procedure on VEB was conducted with earthquake and tsunami loads, including a comparative analysis of the performance of the multifunctional vertical evacuation building structure in Painan City.

2. Metodhology

The flowchart of this study is shown in Figure 1 below.

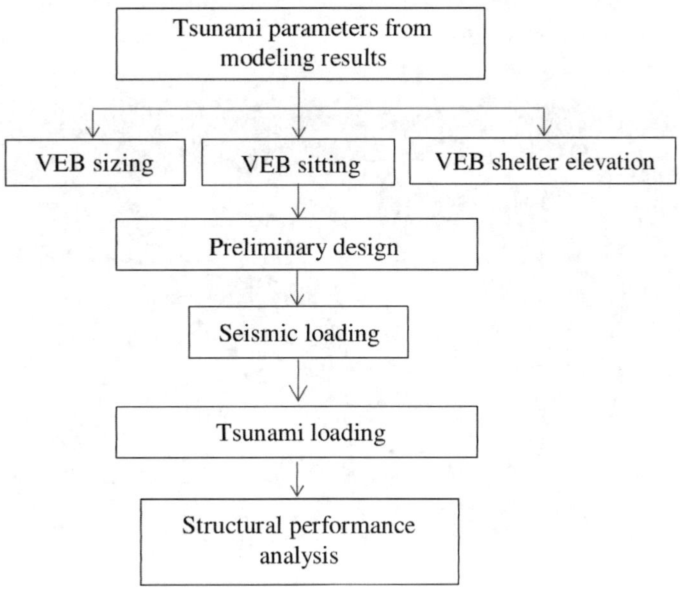

Figure 1. Research Flow Chart

3. Vertical Evacuation Building

3.1. Tsunami Hazard

In this study, the VEB were assumed to be located in Painan. The city was selected due to its geological and geographical conditions, which makes it prone to tsunami hazard. A river also runs through the city. Therefore, this study with VEB scenario may provide significant benefit for mitigation in this area.

The tsunami scenario trigerred by an 8.92 Mw at the Mentawai fault shows that the generated tsunami may reach up to 1875 m inland, the wave arrives on the coast 17 minutes after the rupture with a maximum water depth of approximately 8.3 m, the averaged inundation depth in the city area is approximately 2 - 4 m, with a flow velocity of approximately 5 m/s [7]. This tsunami scenario is used in this study.

3.2. VEB Location

Figure 2 shows that Painan has an area of 108,92 km^2 [9], which is marked in the red line. It has a population of 15982 people. It was assumed that 90% of the resident lives in the city area, which is marked in the yellow line. It was estimated that approximately 30% or 4316 people will head to VEB when there is a tsunami warning while the rest will head to the nearest highlands, the boundary area is marked in the white line. Based on these population considerations, the design of VEB then was carried out according to FEMA P-646-2012.

Four VEB structures were planned, each VEB consists of 6 floors measuring 30 x 18 m with a span of 6 m. The short side of the building was designed to be parallel to the coastline to minimize tsunami forces that will hit the VEB. These VEB were designed as multi-function facilities, so that each floor can be used for other purposes while the 5th and 6th floors were the tsunami refugee area. Hence, a total area of 4320 m^2 was obtained and the minimum refugee area requirement was achieved.

The first wave arrives on the coast approximately 17 minutes after the rupture [7]. On average, sick people or people with disabilities are assumed to be able to walk at a speed of 3.22 km/hour [10]. The furthest distance that can be achieved by this group of people and the available evacuation time were the basis for consideration in determining the location of VEB. Thus, the maximum required spacing between VEB is 2.25 km. Each VEB was designed to be located on an average of 439 m distance or equivalent to approximately 8 minutes walk from each other (Figure 3). The VEB location was assumed to experience the maximum inundation elevation, which was used to determine the minimum elevation of the refugee area (Figure 4).

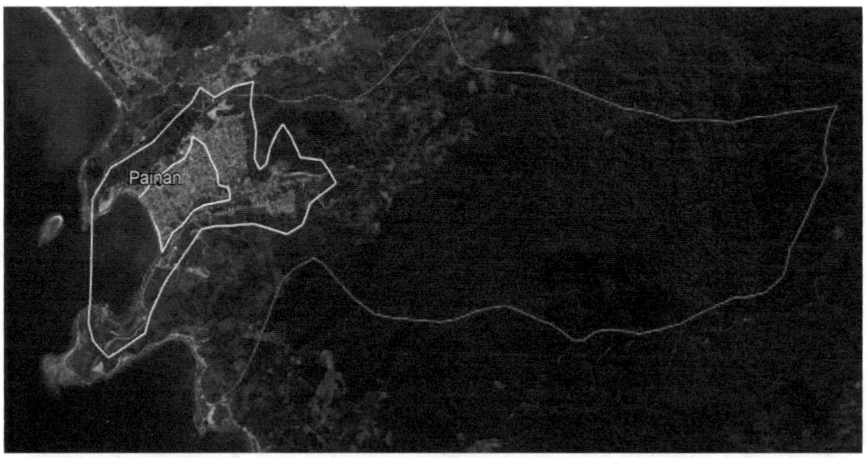

Figure 2. Boundary Areas as Consideration for VEB Size Determination

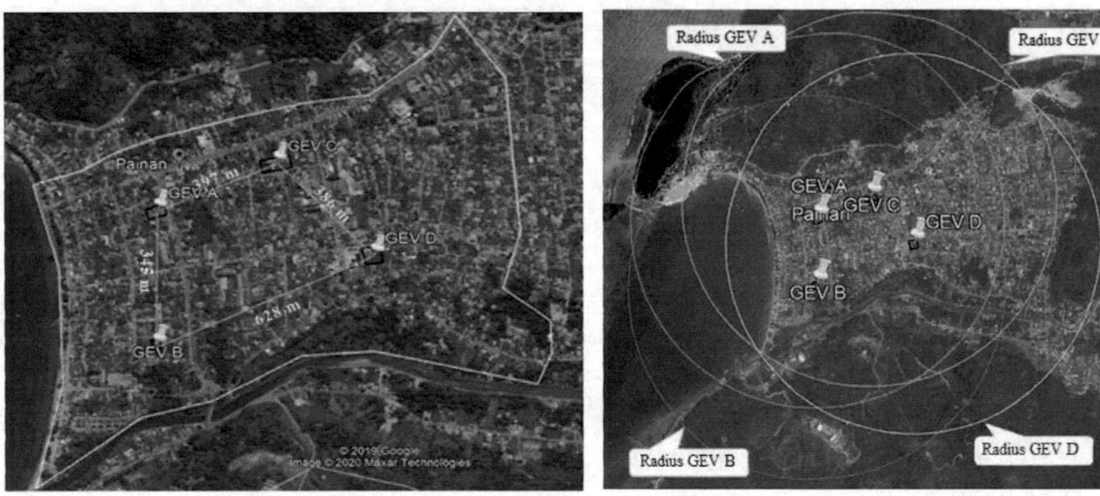

Figure 3. VEB Locations and Coverage Area

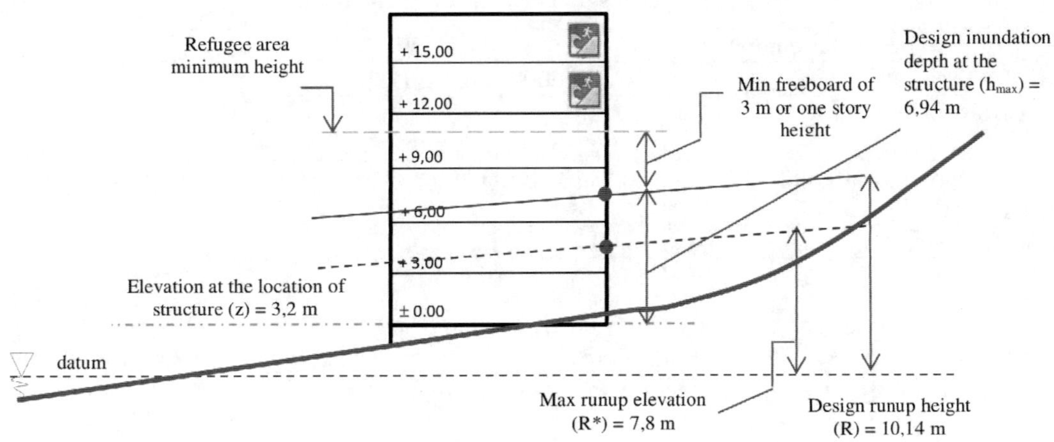

Figure 4. Tsunami Parameter on VEB.

4. Structural Modeling

The building is a six-story reinforced concrete frame structure with 6 m span. Each floor has 3 m height (Figure 5). Structural modeling was conducted using ETABS v. 18.1.10. The properties of structural elements are presented in Table 1.

Three structural models were developed for analyses:

1. M_1 : Building is assumed to be watertight, all walls are intact and any openings such as windows are ignored.
2. M_2 : Building is assumed to have minimal resistance, all walls are detached from the building during tsunami.
3. M_3 : Building is assumed to not experience tsunami

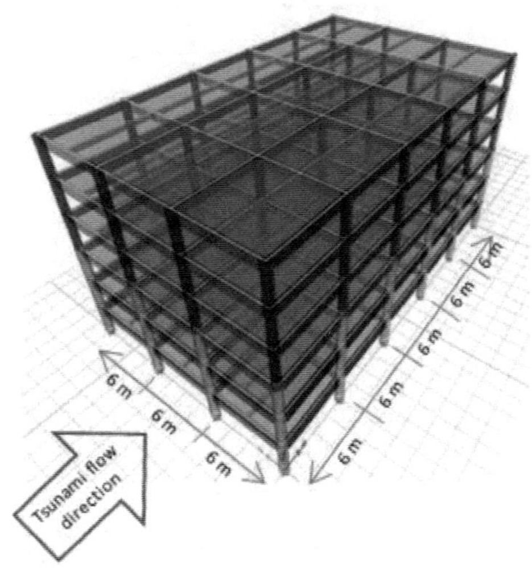

Figure 5. 3D Typical Structural Modeling

SNI 1727-2013 [11] was used to determine the design loads of the building. Considering their functions, live load of 2.5 kN/m^2 was applied for 1st – 4th floor, assuming the function as offices, and 4.79 kN/m^2 for the rest, assuming the function as shelter. Seismic load was calculated based on SNI 1726-2012 [12] using the response spectrum for Painan city obtained from http://puskim.pu.go.id/Aplikasi/desain_spektra_indonesia_2011/, with site class D.

Table 1. Properties of The Structural Elements.

Structural elements	Properties
Column size	C1 (D600 mm), C2 (D700 mm)
Beam size	B1 (400 mm x 700 mm), B2 (350 mm x 600 mm), B3 (250 mm x 400 mm)
Compressive strength of concrete (fc')	35 MPa for column, 30 MPa for beam and slab
Yield stress of steel rebars (fy)	400 MPa

The tsunami flow direction was assumed perpendicular to the coastline. Tsunami loads are calculated based on FEMA P-646 [10]. Forces generated by tsunami include hydrostatic force, buoyant force, hydrodynamic force, impulse force, debris impact force, damming debris force, uplift forces on elevated floors, and extra gravity force.

Table 2 below shows tsunami loads applied to M_1 and M_2 models. There are significant differences in the amount of buoyant forces, hydrostatic forces, and impulse forces to the structure between M_1 and M_2. Furthermore from the modeling, the debris force gives the critical lateral force to the structure. It acts as a concentrated load on the structural element. With 5 m/s velocity, if the debris is a 40ft container, the force concentrated on one element will go up to 3583.853 kN. Whereas with debris in form of a wood log, the force concentrated on one element was found to be 213.612 kN. The floating debris was assumed to be in wood log form weighing 450 kg in this study.

Not all tsunami force hits the structure at the same time altogether. Tsunami forces were applied to the models using load combinations on FEMA P-646 to obtain the tsunami loads resultant (T_s). This resultant load is then combined with gravity loads to obtain the design load combination. It should be noted that earthquake load is also not combined with tsunami load because it's very unlikely for both loads to occur simultaneously.

Table 2. Recapitulation of Tsunami Loads Applied to VEB Model 1 and Model 2.

		M_1			
Tsunami Load		Forces			Applied As - To
Hydrostatic	*Fhs*	edge column	112.334	kN/m	Applied as line load along wetted part of the exterior columns.
		middle column	224.669	kN/m	
Buoyant	*Fb*	corner column	674.006	kN	To the base floor. However, because it wasn't modeled, the load was entered as upward point load at the base floor column according to floor tributary.
		edge column	1348.012	kN	
		middle column	2696.023	kN	
Hydrodynamic	*Fd'*	edge column	29.637	kN/m	Applied as line load along the wetted part to exterior columns not applied Fs or Fdm force
		middle column	59.275	kN/m	
Impulse	*Fs'*	edge column	44.456	kN/m	Applied as line load along the wetted part = 6.94 m to the rearest row columns of the building (from the wave direction)
		middle column	88.912	kN/m	
Debris	*Fi*		213.612	kN	Applied as point load at the water surface = 6.94 m to all columns affected by Fd.
Damming Debris	*Fdm'*		39.517	kN/m	Applied as line load to wetted columns at the front row (from the wave direction) which are assumed to be hit.
Uplift	*Fu*		-		Fu = 0, No elevated floor is submerged.
Extra Gravity	*Fr*		-		Fr = 0, Because the VEB assumed to be 100% watertight. Thus, no water is trapped in the building.

			M_2		
Tsunami Load			Forces		Applied As - To
Hydrostatic	*Fhs*		-		Fhs = 0. No water height difference between the outer side and inner side of the walls
Buoyant	*Fb*		-		Fb = 0, No volume of water displaced.
Hydrodynamic	*Fd'*		4.149	kN/m	Applied as line load to all submerged columns which Fs or Fd forces are not.
Impulse	*Fs'*		6.224	kN/m	Applied as line load to all submerged columns located at the rearest row (from the wave direction).
Debris	*Fi*		213.612	kN	Applied as point load at the water surface = 6.94 m to all columns affected by Fd.
Damming Debris	*Fdm'*		39.517	kN/m	Applied as line load to wetted columns at the front row (from the wave direction) which are assumed to be hit.
Uplift	*Fu*	3rd floor	0.01303	kN/m^2	To all flooded floor system
		2nd floor	0.01135	kN/m^2	
Extra Gravity	*Fr*		-		Fr = 0, Because no outer walls assumed, that when the outside elevation drops immediately, no water is held back.

5. Structural Analysis

As a tsunami shelter, VEB should comply to earthquake-resistant building requirements to resist large earthquake prior to the tsunami. VEB is also expected to have sufficient reserve capacity to resist the subsequent tsunami loading effects. Using Performance Based Design approach, the structures responses due to earthquake were evaluated. Referring to FEMA 356 and ATC 40, there are three levels of performance,

1. Immediate Occupancy (IO) : Structure retains its original strength and stiffness, negligible damage.
2. Life Safety (LS) : significant structural damage, but not result the collapse of structural elements, Major cracks occur, Sructural deformation is visible,Building can be reused after restoration.
3. Collapse Prevention (CP) : Damage to the core of the structure, there may be a significant degradation of the stiffness and lateral strength of the structure. Large permanent lateral deformation. The building is on the verge of collapse. It may be irreparable and unsafe to use again because activities in the building can leads to collapse.

VEB as an essential facility should be at least Life Safety performance for the Maximum Considered Earthquake (MCE) [10].

With the aforementioned approach, a nonlinear static pushover analysis was conducted to all models to obtain the capacity curve. Then, the performance point, that is the intersection between demand and capacity curves, can be obtained. Based on this point, the performance level can be determined. Several important facts from the analysis are presented here.

5.1. Plastic Hinges Formation

Plastic hinge development is one key to find out which part of the building will most likely suffer damage in the first place. As the color ranges from A (black) to E (red), the hinge damage severity increases. As the design adheres to strong-column weak-beam concept. plastic hinges appearance in some columns are undesirable. Up to performance point, the study showed that all structural element in M_1, M_2, M_3 are Immediate Occupancy level, although plastic hinge appeared in some columns (Figure 4).

5.2. Capacity Curve

Figure 5 shows the comparisons of M_1, M_2 and M_3 capacity curves in X and Y directions. Although all models have similar structural dimension, M_1 shows the highest capacity. This strength difference lies in elements directly affected by the tsunami i.e. B1 and C1 which were installed with more reinforcement than M_2 and M_3. Tsunami load in M_2 becomes smaller because the flow was unobstructed by the structure, therefore the elements internal forces were dominated by earthquake loads. M_2 and M_3 have the identical capacity because they both have identical structural strength.

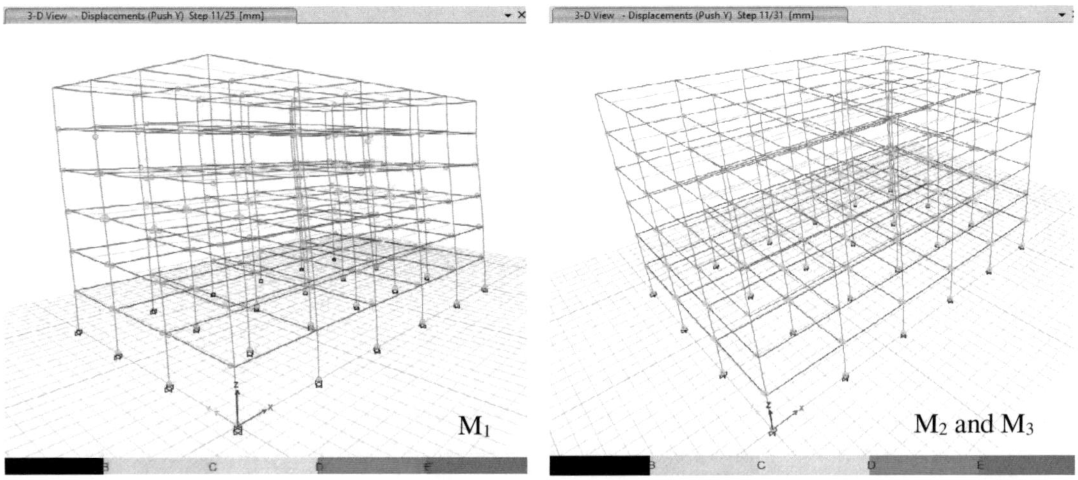

Figure 6. Development of Plastic Hinges at Peformance Point.

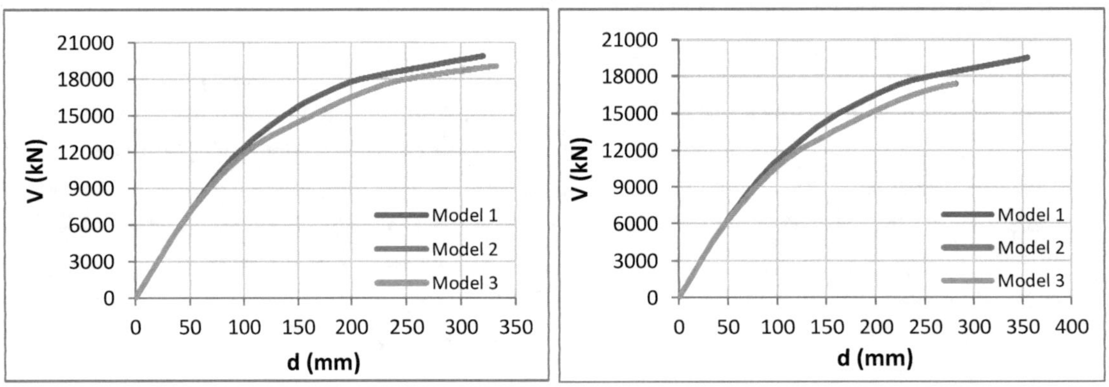

(a) X direction (b) Y direction

Figure 7. M_1, M_2 and M_3 Capacity Curves

5.3. Performance Level

Using the capacity curve and the performance point for all models, the performance level were obtained. Table 3 shows that all model is in Immediate Occupancy level in X-direction. In Y-direction (weak axis), M_1 has the best performance i.e. Immediate Occupancy level, while M_2 and M_3 are in Life safety level. Thus, these three models have met the performance targets required by codes as earthquake-resistant buildings.

Table 3. Structural Performance Levels

	Performance Level	
Model	X Direction	Y Direction
M_1	IO	IO
M_2	IO	LS
M_3	IO	LS

Considering that the structures were all performed satisfactorily under seismic load, the next check was to ensure that models M_1 and M_2 could resist the tsunami load. FEMA P646-2012 allows the building to be damaged up to the level of Collapse Prevention performance due to tsunami, where the building is on the verge of collapse with the structure experiencing severe damage but is not permitted to collapse after the tsunami with a maximum estimated return period of 2500 years. Analysis reveals that both models M_1 and M_2 were able to resist the load, hence have adequate capacities. The design procedures from FEMA P646-2012 can be considered adequate for designing VEB in this study.

6. Conclution

A study of Vertical Evacuation Building was conducted to evaluate the design process and the structural performance. Several conclusions can be obtained as follows:

1. Debris impact force is the tsunami load that affects structure the most. The floating debris acts as a concentrated load on the structural element. The magnitude of this load depends on the type of debris, the debris orientation at impact, and the flow velocity. With 5 m/s velocity and debris in form of a wood log, the force concentrated on one element was found to be 213.612 kN.
2. For structures resisting tsunami loads, with and without walls presence on M_1 and M_2 models to block the tsunami flow greatly affects the tsunami load received by the structure. Significant differences were found in buoyant force, hydrostatic force, and impulse force.
3. Compared to the model designed for earthquake only (M_3), the element internal forces on M_2 are somewhat similar despite experiencing tsunami. Both beam and column internal forces were dominated by earthquake loads. Since the tsunami flow in M_2 was unobstructed by the structure therefore tsunami load becomes smaller.
4. Pushover analysis shows that M_1 has a better performance level in Immediate Occupancy for both X and Y directions, while M_2 and M_3 in Immediate Occupancy for X direction and Life Safety for Y direction. Therefore, these VEBs have met the performance targets required by regulations as earthquake-resistant buildings.

7. References

[1] Pusat Studi Gempa Nasional (PuSGen). (2017)."*Peta Sumber dan Bahaya Gempa Tahun 2017. 1st edition : Pusat Penelitian dan Pengembangan Perumahan dan pemukiman.*" Badan Penelitian dan Pengembangan Kementerian Pekerjaan Umum dan Perumahan Rakyat, Indonesia, pp. 182,189.

[2] J C Borrero. (2006). "*Tsunami Indundation Modelling for Wastern Sumatera*". PNAS, vol. 103 no. 52. 19677.

[3] J McCloskey, A Antonioli, A Piatanesi. (2008). "*Tsunami Threat in the Indian Ocean from a Future Megathrust Earthquake West of Sumatra.*" EPSL, 265(1):61-81.

[4] A Muhari, F Imamura, D H Natawidjaja, S Diposaptono, H Latief, J Post, and F A Ismail (2010), "*Tsunami Mitigation Efforts With pTA In West Sumatra Province, Indonesia.*" J. of Earthquake and Tsunami, Vol. 04, No. 04, pp. 341-368.

[5] U.S. Geological Survey USGS, 2017

[6] R Collings, D Lange, A Rietbrock, F Tilmann, D Natawidjaja, B Suwargadi, M Miller, and J Saul. (2012). "*Structure and Seismogenic Properties of the Mentawai Segment of The Sumatra*

Subduction Zone Revealed by Local Earthquake Traveltime Tomography." Journal of Geophysical Research: Solid Earth Volume 117, Issue B1

[7] M B Adityawan, P Rahmasari, C Asrini, etc. (2021). "*Effect of Land Cover on Tsunami Overland Flow Propagation: A Case Study of Painan, West Sumatra, Indonesia*." International Journal Advanced Science Engineering Information Technology". Vol.11. No. 5. ISSN: 2088-5334

[8] Linlin Li, Z. Huang, Q. Qiu, Natawidjaja D. H., and Kerry S. (2012). "*Tsunami-Induced Coastal Change: Scenario Studies for Painan,West Sumatra, Indonesia*". Journal of Earth, Planets and Space, Vol. 64, pp. 799-816

[9] Badan Pusat Statistik Kabupaten Pesisir Selatan. (2019). "Kecamatan IV Jurai Dalam Angka". ISBN: 13020.1916

[10] Federal Emergency Management Agency. (2012). *FEMA P-646: Guidelines for Design of Structures for Vertical Evacuation from Tsunamis.* Applied Technology Council, Redwood City, California, USA.

[11] Badan Standarisasi Nasional. (2013). *SNI 1727-2013: Beban Minimum Untuk Perancangan Bangunan Gedung dan Struktur Lain.* Jakarta.

[12] Badan Standarisasi Nasional. (2012). *SNI 1726-2012: Tata Cara Perencanaan Ketahanan Gempa untuk Struktur Bangunan Gedung dan Non Gedung*. Jakarta.

[13] Federal Emergency Management Agency, (2000). *FEMA 356. Prestandard and Commentary for the Seismic Rehabilitation of Buildings*. Washington, D.C: USA.

[14] Yeh, H., Barbosa, A. R., Ko, H., & Cawley, J. G. (2014). "*Tsunami Loadings On Structures: Review And Analysis".* Coastal Engineering Proceedings, 1(34):4. DOI: 10.9753/icce.v34.currents.4

Implementation of a quick assessment application called InaRISK for damaged houses after the West Pasaman Earthquake (2022)

Fauzan[1*], N D Carol[1], F A Ismail[1], and Y Narny[2]

[1]Civil Engineering Department, Universitas Andalas, Padang, Indonesia
[2]Faculty of Humanities, Universitas Andalas, Padang, Indonesia

*Corresponding author: fauzan@eng.unand.ac.id

Abstract. West Sumatra, located at the boundary of two of Earth's shifting tectonic plates , is prone to natural disasters, especially earthquake. On 25th February 2022, an earthquake centered 18 km northeast of Simpang Empat, the capital city of West Pasaman, with a magnitude of 5.2 M, struck off at 08.35 WIB. This earthquake claimed lives and damaged buildings. After the earthquake, the need for rehabilitation and reconstruction is indispensable. Various methods are used in assessing the damaged houses, however, the assessment still use the manual method but is assessed as having many shortcomings. In the case of West Pasaman earthquake, the start of application-based to collect data is improvement to the manual method with its practicality and time-saving capability. This application, developed by National Agency for Disaster Countermeasure (BNPB) is called InaRisk. Although this application has made a breakthrough, improvement for this application is required as there are still drawbacks especially it cannot be be used without internet connection. This paper discusses the usage of a quick assessment application called InaRISK for damaged houses after the West Pasaman earthquake. A new assessment method developed by Andalas University Team called Earthquake Rapid Assessment (E-RA) is also discussed in this paper.

1. Introduction

Indonesia is an archipelago country where the three worlds plate meets, the Indo-Australian plate moving northward, the Eurasian plate roughly moving south, and the Pacific plate roughly to the west [1]. According to plate tectonic theory, tectonic activities are concentrated in plate boundaries [2] and the highest activity is found where two or more plates collide in plate meeting points [3]. Therefore, the islands of Indonesia are susceptible to natural disasters and particularly earthquakes [4]. Major earthquakes that once occured in Indonesia are Aceh Earthquake (2004), West Sumatra Earthquake (2009), Lombok and Palu Earthquake (2021), Mamuju Earthquake, and the latest is West Pasaman Earthquake (2022). These earthquakes claimed lives and damaged buildings, either light, moderate, or heavy. After the earthquake, the need for rehabilitation and reconstruction is indispensable by obtaining data on the damaged houses as soon as possible so it could be used by governments to take inventory of the public property damage. The damaged properties are categorized into three levels, namely slightly damaged, moderately damaged, and heavily damaged, as shown in figure 1.

a. Light damaged house b. Moderate damaged house

c. Heavily damaged house

Figure 1. Damaged houses category.

From the post-earthquake observation experience in Lombok and Palu (2021), Mamuju Earthquake (2021), and West Pasaman Earthquake (2022), during this time, various survey methods were carried out with the most common standard is of visual examination such FEMA 154 [5] and FEMA 310 [6] or using "Quick Visual Assessment" method from National Agency for Disaster Countermeasure (BNPB) regulation number of 2017 [7] as had been done in Lombok dan Palu (2021). However, these methods take longer time to collect the data and obtains so diverse data that it is difficult for governments to establish the valid data. This causes tardiness in providing much-needed donation to the community in need. Comprehensive methods for projecting a building at risk of damage after an earthquake should take into account long-term aspects and possible future disasters [8]. Learning from that experience, National Agency for Disaster Countermeasure (BNPB) developed an application called InaRISK with surveying feature called post-disaster assessment (JITUPASNA) that was implemented directly in West Pasaman Earthquake (2022) that caused damage to ±6,627 citizen's houses [9].

2. Method

Post-disaster assessment (JITUPASNA) survey with InaRISK app in West Pasaman Earthquake (2022) was implemented in Pasaman and West Pasaman district. This activity aimed to obtain data of people whose houses are affected by post-earthquake damage. This survey activity employed InaRISK app developed by National Agency for Disaster Countermeasure (BNPB) to support disaster countermeasure activities and disaster-related research surveys. The method of activity involved filling out the e-form in post-disaster assessment (JITUPASNA) survey with the usage of InaRisk app. E-form JITUPASNA uses forms based on post-disaster data home forms by National Agency for Disaster Countermeasure (BNPB) that can be seen in figure 2.

1. House Form

Province	:	Photo
City	:	
Sub-district	:	
Sector	: ☐Settlement ☐ Infrastructure ☐ Economy ☐ Social ☐ Cross Sector	
Competence	: ☐ Central ☐ Province ☐ City ☐ private	
Asset Name	:	
Address	:	
Coordinate	:	
Damage Data		Photo
Level of Damage	: ☐ Heavy damage ☐ Moderate Damage ☐ Light Damage	
Volume	: (unit/meter/or other unit)*	
Unit price	: Rp (local unit price per m/unit/or other)*	
Building Content	: Rp	
Officer Name	:	

Damage

Building Element	Damage Type	Component Quantity	Damage weight (%)
Foundation/Sloof	☐ Cracks is visible on the concrete surface (1 - 2mm)	1;2; ≥ 3	3;4;5
	☐ Concrete destruction with visible reinforcement	1;2; ≥ 3	5;8;11
	☐ Stagnation and/or destruction of foundation	1;2; ≥ 3	11;14;15
Column/Beam	☐ Cracks is visible on the concrete surface (0,2 - 1 mm)	1;2; ≥ 3	7;9;11
	☐ Visible cracks (1 - 2mm)	1;2; ≥ 3	7;9;11
	☐ Destruction of concrete, buckled reinforcement or deformation of column Building Element	1;2; ≥ 3	12;26;35
Roof frame/ Roof truss	☐ Roof batten is partially or fully destroyed	1;2; ≥ 3	3;5;11
	☐ Roof truss is partially or fully collapse	1;2; ≥ 3	11;14;15
Roof cover	☐ Fall partially or fully	1; ≥ 2	5;10
Wall	☐ Stucco delaminates or fall over	1;2; ≥ 3	3;4;5
	☐ Cracks is visible ,through the rocks	1;2; ≥ 3	5;8;11
	☐ The wall is partially or fully fall over	1;2; ≥ 3	11;14;15
Floor	☐ Small crack that spreads partially or fully	1; ≥ 3	2;3
	☐ Large crack/split/destroyed partially or fully	1;2; ≥ 3	8;9;10

Figure 2. Post-disaster Housing damage assessment form by BNPB.

The stages of activity to survey the damaged houses using the InaRISK app are as follows:

2.1. Preparation

The activity began with the distribution of 100 college-students into several teams for the entire site of Nagari Kajai, Nagari Aua Kuniang and Kinali in West Pasaman Regency for 7 days. The departure starts on March 02, 2022. The survey location can be seen in figure 3.

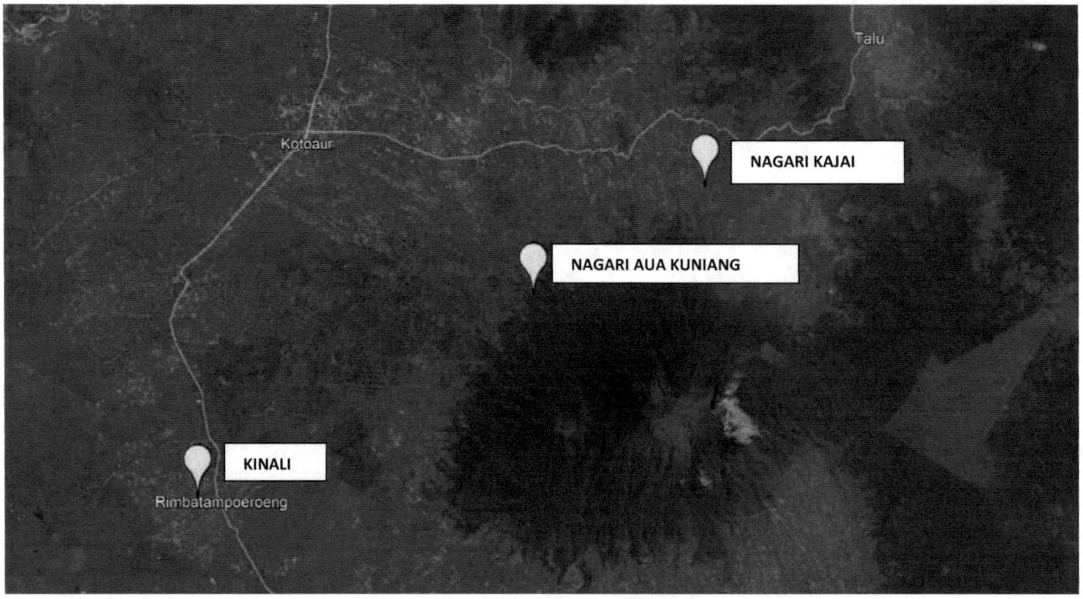

Figure 3. Survey location (Google maps 2022)

2.2. Implementation

The implementation of the activity is the filling out the e-survey post-disaster assessment (JITUPASNA) forms directly through the InaRISK app using the surveyor personal cell phone containing assessment of damaged houses. Some photos of the activities can be seen in figure 4.

Figure 4. Surveying activity.

2.3. Evaluation

The evaluation of this activity involves obtaining data on damaged houses in the aftermath of the West Pasaman Earthquake (2022) in accordance with the category of home damage that can be filed as valid data for government assistance grants. The data obtained will be issued to the public in the form of a map of the damaged houses and its category with a percentage in each area that can be seen on the InaRISK app shown in figure 5 (InaRISK 2022).

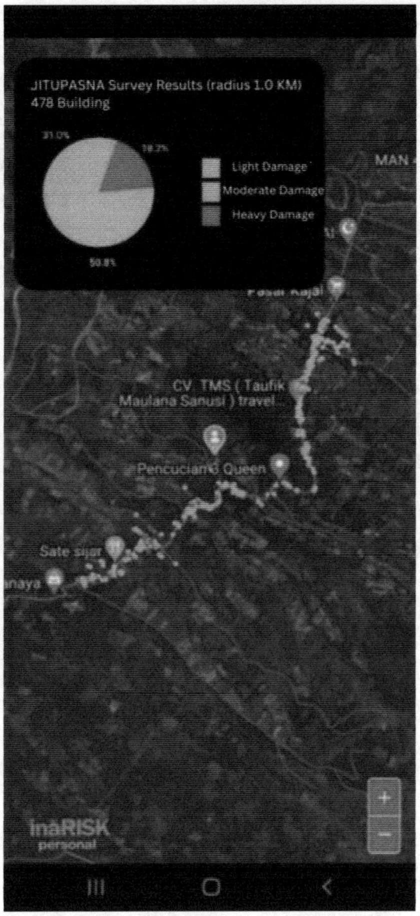

Figure 5. The map of damaged houses in percentage on the InaRISK app (InaRISK 2022).

3. Results and discussion

3.1. InaRISK

InaRISK app is an android and iOS online application containing a region's hazard level information of natural disasters and is completed with an action recommendation to do as its interactive participation [10]. InaRISK app greatly aids governments in disseminating information and educating people about the risks of harm around [11]. The InaRISK also contains additional features, particularly the ones that will be discussed this time of post-disaster assessment (JITUPASNA) survey. InaRISK uses the National Agency for Disaster Countermeasure (BNPB) regulation number of 2017 on plan drafting and rehabilitation and reconstruction of the post-disaster and also technical guidelines for earthquake-resistant homes and buildings by the directorate general of invention by the ministry of public works and people's housing of 2006 as a reference in JITUPASNA.

The Implementation of InaRISK Personal specifically the post-disaster assessment (JITUPASNA) survey consisting of an assessment of the rate of houses damaged post-earthquake in West Pasaman Earthquake has a workflow of applications, as shown in figure 6 (InaRISK 2022).

IOP Conf. Series: Earth and Environmental Science 1244 (2023) 012041 doi:10.1088/1755-1315/1244/1/012041

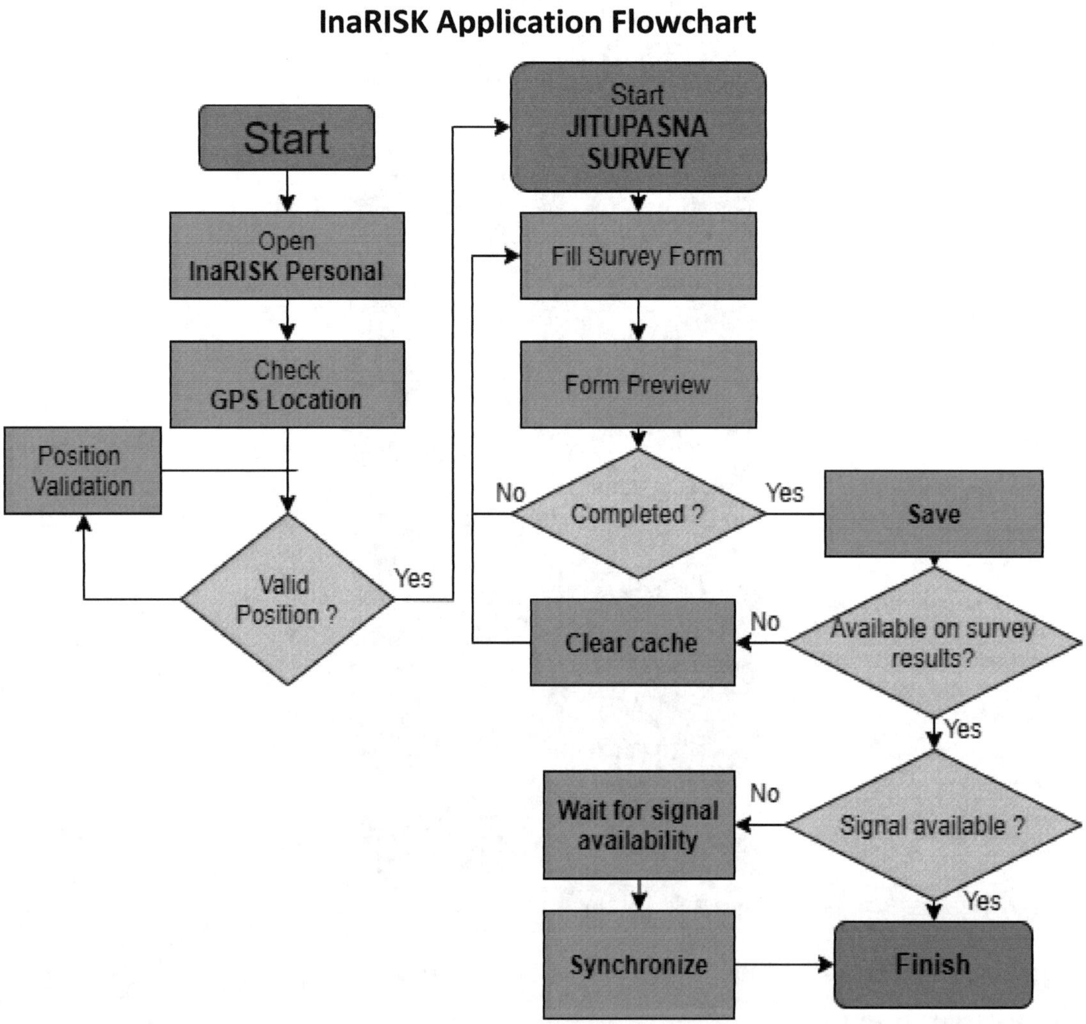

Figure 6. JITUPASNA survey workflow in InaRISK app (InaRISK 2022).

The e-form of post-disaster assessment (JITUPASNA) survey is illustrated in figure 7 (InaRISK 2022).

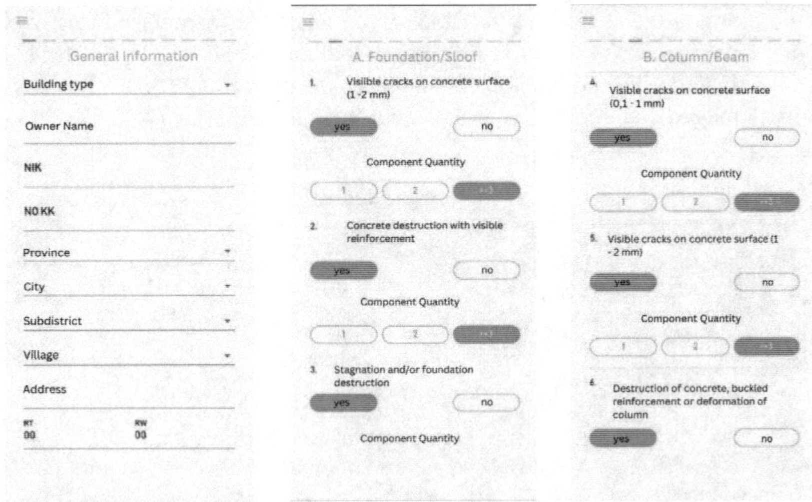

Figure 7. JITUPASNA survey e-form (InaRISK 2022)

JITUPASNA, the assessment form is divided to 10 parts that consist of general information of the home owner, foundation/sloof, column/beam, truss, roof, wall, floor, building contents, disaster effect cost and catastrophic loss of income. This form is filled by typing in the provided area or choosing the criteria of the damaged by selecting the numbers: 1 as light damage, 2 as moderate damage or 3 as heavy damage. After filling out the e-form of post-disaster assessment (JITUPASNA) survey, the app will calculate the cumulative weight of the damage in the form of answers that can be reviewed, as in figure 8 (InaRISK 2022).

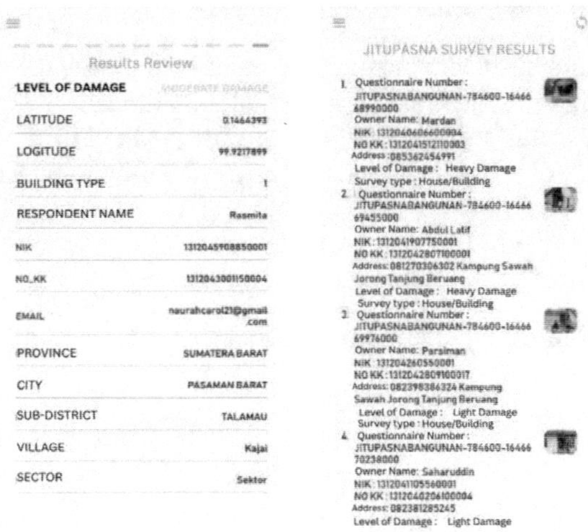

Figure 8. JITUPASNA survey form preview and results (InaRISK 2022)

As example in looking at the use of the InaRISK app, one of the resident houses in Nagari Kajai on behalf of Mrs. Nisma was recorded with the conditions that can be seen in and after filling the e-form of JITUPASNA survey. According to the app calculations, the house of Mrs. Nisma was included in the category of heavily damaged category, as seen in figure 9 (InaRISK 2022).

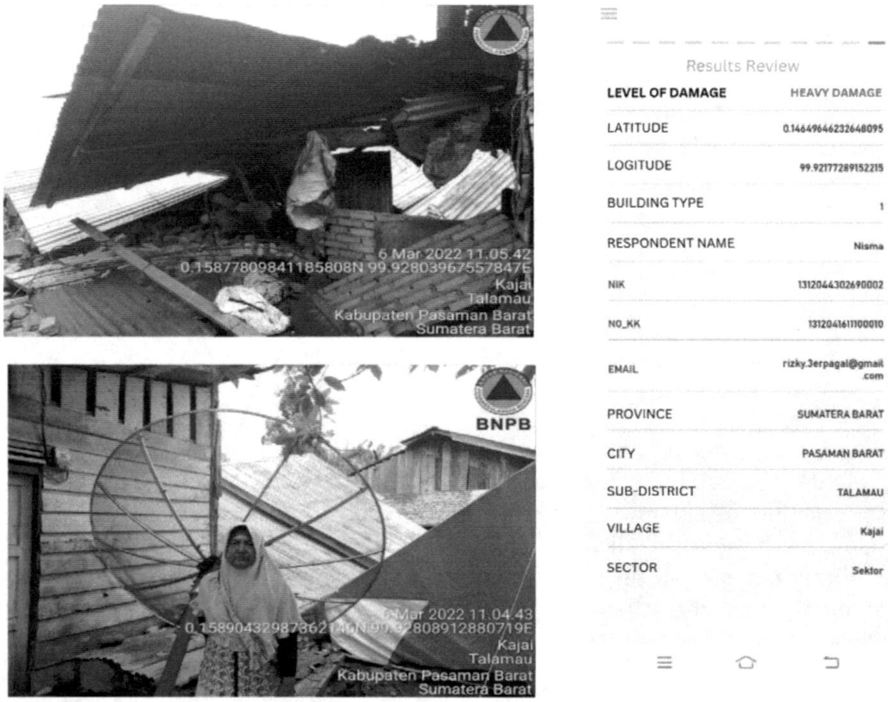

Figure 9. The condition of Mrs.Nisma, citizen of Nagari Kajai Post-West Pasaman Earthquake 2022 and the InaRISK results (InaRISK 2022)

The data on damaged houses in post-West Pasaman Earthquake during the field surveys in the area of Nagari Kajai and Kinali in West Pasaman and Nagari Aua Kuniang in Pasaman District for 7 days is shown in table 1.

Table 1. Houses damage category data.

Nagari Name	Damage Category	Amount
Kajai	Light	746
	Moderate	499
	Heavy	354
Kinali	Light	111
	Moderate	21
	Heavy	1
Aua Kuniang	Light	763
	Moderate	259
	Heavy	97

From the data, it can be seen that the total data of the damaged houses was 2,851 houses, namely 1,620 of lightly damaged, 779 of moderately damaged and 452 of heavily damaged. This become evidence as one of the purposes of InaRISK app with its post-disaster assessment (JITUPASNA) feature as a support to the post-disaster recovery activity because this app was able to obtain the amount of data on houses damaged after the West Pasaman Earthquake immediately (2022).

3.2. Usage evaluation of the inarisk app

The usage of InaRISK application during the field surveys reveals many shortcomings, especially the feature post-disaster assessment (JITUPASNA) in the form and post-assessment such as:

- Surveyor or assessor cannot edit data from the forms that have been synchronized with the central data. Therefore, if a form has been filled incorrectly during the survey, another surveyor must enter the same data and contact the call center to remove the wrong data so as not to disturb the data validation.
- The unavailability of section to record the building owner's telephone number in the general information section for validating the data after survey.
- Due to connection or internet problem in the sites, the usage of the application must be either online or offline. The InaRISK application can be used online, but while in offline condition, the form does not issue a questionnaire's number or record the data as a null which has not been synchronized with the central data, as shown in figure 10 (InaRISK 2022).

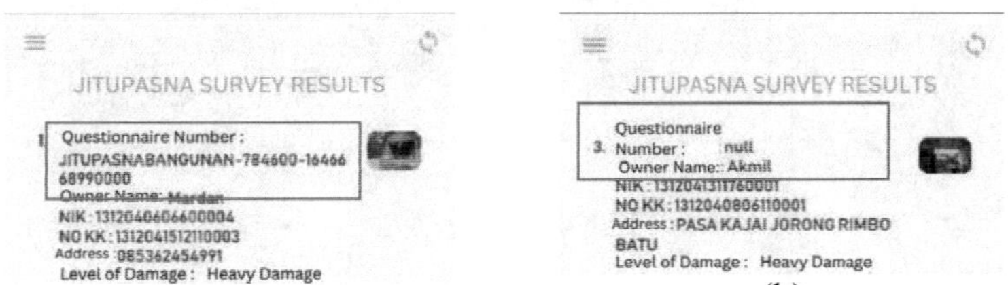

(a) (b)

Figure 10. (a) The assessment result in online condition and (b) the assessment result in offline condition (InaRISK 2022)

- If there is disturbance of internet connection while filling out a survey form, the results will be lost and the application display it as "no data such" as can be seen in figure 11(InaRISK 2022).

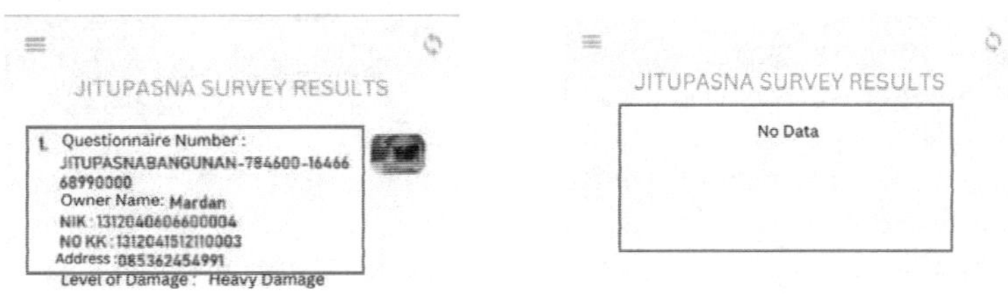

(a) (b)

Figure 11. (a) Data Comparison in online surveys and (b) offline survey (InaRISK 2022)

Some of these shortcomings significantly affect the data as it will interfere the validation of data collection that are ready to be used for the government to administer relief funds. This shortfall leads to an improvement in the next admissible applications in anticipation of errors during data rectification.

3.3. Usage evaluation of the InaRisk app

As an improvement to the InaRISK, the development of develop a new application named E-RA (Earthquake Rapid Assessment) is planned. It is an application that focused on assessment activities (post-earthquake examination), thus facilitating widespread use. With a field consideration that has emerged from the activity of the use of the InaRISK application, the E-RA is designed with a system that makes it easier for users not only in online condition but also in offline condition. The E-RA app focuses solely on a post-earthquake assessment and can produce high levels of data in order to support the rate of success in rehabilitation and construction of the future earthquake earthquake disaster. Initial view design of the E-RA application can be seen in figure 12.

Figure 12. E-RA application display

The methods to develop the E-RA are summarized into several stages, that are as follows:

- Experiment Stages
  In this stage, library review and literature studies are gathered as assessment points in calculations in the quick assessment method and design the initial display design, forms and employment line on the E-RA (earthquake rapid assessment) application.
- Production Stages
  An application stage based on information collected or established reference. For this E-RA application, it is applied to the rapid application development (RAD).
- Testing and Analysis of Experiment Stages
  The stages consist of a series of systematic activities that are designed to test and evaluate the application's usefulness according to established purposes.

4. Conclusion

Based on the above discussion, it is concluded that the performance of JITUPASNA InaRISK BNPB to accelerate the data collection of damaged houses after an earthquake disaster in support of the success of the rehabilitation and reconstruction program is satisfactory. However, there are drawbacks of the InaRISK application as it cannot be used in offline (without internet connection) condition, resulting in the data form cannot be edited after it is synchronized with the central data and the unavailability of section to record the building owner's telephone number. Improvement is needed in some areas of this application to sustain the validation of the obtained data.

National Agency for Disaster Countermeasure (BNPB) is expected as well to improve the aspects of the InaRISK app, so that in the event of a further earthquake, such problems will no longer interfere with the performance of data collection of the application and in future the data generated from it can help governments to deliver help more quickly to communities that need help.

In spite of some of the problem that arose from InaRISK app, there has been an application development to focus on the assessment (examination) activity named E-RA by Andalas University Team, which is an abbreviation for earthquake rapid assessment. It is hoped that in the future this application will be the beginning of a development in assessing the damaged houses using the survey app that can facilitate rehabilitation activities and reconstruction of especially after an earthquake disaster.

Acknowledgements

The authors gratefully acknowledge the financial support from Civil Engineering Department, Engineering Faculty, Andalas University, which has funded this research. The authors also acknowledge

National Agency for Disaster Countermeasure (BNPB) which has given an opportunity to do this research.

References

[1] Hayes G P 2013 *Seismicity of the Earth 1900-2012* Sumatera and Vinicity, USGS.
[2] Banyunegoro V H, Alatas, Z A, Jihad A, Eridawati, and Umar, M 2019 Probabilistic Seismic Hazard Analysis for Aceh Region *IOP Conf. Ser.: Earth Env. Sci.* **273** 012015
[3] Banyunegoro V H, Umar, M, and Idris Y 2020 Seismic microtremor experiment to determine seismic vulnerability of North Aceh *IOP Conf. Ser.: Mat. Sci. Eng.* **846** 12053
[4] Banyunegoro V H, Firdausi R, Umar M, and Idris Y 2021 Shallow One-Dimensional Shear Wave Velocity Structure in Lhokseumawe Aceh-Indonesia *J. Phys.: Conf. Ser.* **1811** 012078
[5] FEMA 2002 *FEMA 154: Rapid Visual Screening of Buildings for Potential Seismic Hazards – A Handbook* (United States: Federal Emergency Management Agency)
[6] FEMA 1998 *Handbook for the Seismic Evaluation of Buildings A Prestandard* (United States: Federal Emergency Management Agency)
[7] BNPB 2017 *National Agency for Disaster Countermeasure Regulation Number 5 2017 about Preparation of Post-Disaster Rehabilitation and Reconstruction Plans* (Jakarta: National Agency for Disaster Countermeasure)
[8] Kijko A, Retief S J P and Graham G 2003 Seismic Hazard Risk Assessment Tulbagh, South Africa: Part II – Assessment of Seismic Risk *Natural Hazard* **30** 25
[9] Julius A M, Pribadi S, Saputra A A, Prayitno B S, Ahadi S, Hermanto D, Arifin H, Satria L A, & Zevanya C S S 2022 An On-Site Post-Event Survey of the 2022 Mw 6.1 West Pasaman Sumatera Destructive Earthquake NTU *J. of Renewable Energy* **2(1)** 39–49
[10] BNPB 2022 User Guide Survey Jitupasna (Jakarta: National Agency for Disaster Countermeasure)
[11] Chicie, Safuan, & Musa A A 2022 InaRisk Personal Application Development ULIL ALBAB: Multidisciplinary Sci. J. **1(7)** 2277–2287

AUTHOR INDEX